Introduction to Classical Electrodynamics 3

Dedicated to the memory of my parents
Dedicated to the memory of Daniel Arnaudon
Dedicated to Prof. Amitabha Chakrabarti

Series Editor
Frédérique de Fornel

Introduction to Classical Electrodynamics 3

Boucif Abdesselam

ISTE

WILEY

First published 2026 in Great Britain and the United States by ISTE Ltd and John Wiley & Sons, Inc.

ISTE Ltd
27-37 St George's Road
London SW19 4EU
UK

www.iste.co.uk

John Wiley & Sons, Inc.
111 River Street
Hoboken, NJ 07030
USA

www.wiley.com

Library of Congress Control Number: 2025945758

British Library Cataloguing-in-Publication Data
A CIP record for this book is available from the British Library
ISBN 978-1-83669-002-3

The manufacturer's authorized representative according to the EU General Product Safety Regulation is Wiley-VCH GmbH, Boschstr. 12, 69469 Weinheim, Germany, e-mail: Product_Safety@wiley.com.

Table of Contents

Preface ix

Chapter 1. Waveguides 1

1.1. The different types of waveguides 1
1.2. Description of a guided wave 2
1.2.1. Propagation equation and field expressions 2
1.2.2. Classification of propagation modes 6
1.3. TE and TM modes in a hollow conducting waveguide 8
1.3.1. Rectangular waveguide 9
1.3.2. Attenuation coefficient in a lossy waveguide 24
1.3.3. Propagation between two parallel planes 29
1.3.4. Cylindrical waveguide 33
1.4. Dielectric waveguides 37
1.4.1. Propagation between two parallel dielectric planes 37
1.4.2. Step index optical fiber 44
1.4.3. Modes with azimuthal symmetry 45
1.4.4. Modes with azimuthal dependence 48
1.4.5. Propagation modes in weakly guiding fibers 54
1.5. Resonant cavities 59
1.5.1. Rectangular metallic resonant cavity 60
1.5.2. Cylindrical resonant cavity 65

Chapter 2. Transmission Lines 71

2.1. Parallel-plate transmission line 71
2.2. General formulation 73
2.3. Modeling of lossless transmission lines 78
2.4. Modeling of a transmission line with losses 79
2.5. Wave equations and sinusoidal solutions 82

2.6. Effects of losses in transmission lines 86
2.6.1. Low-loss limit 87
2.6.2. Distortionless lines 87
2.6.3. Solutions with arbitrary impedance at the line output 88
2.7. Resonance effect in a transmission line 94
2.8. Energy aspect in a transmission line 95

Chapter 3. Electromagnetic Radiation 99

3.1. Retarded potentials 99
3.1.1. Inhomogeneous wave equations 99
3.1.2. Solutions of the inhomogeneous wave equations 101
3.1.3. Solutions by Fourier analysis 102
3.2. Radiation from a localized oscillating source 106
3.2.1. Radiation from an electric dipole 113
3.2.2. Radiations from the magnetic moment and the electric quadrupole 119
3.2.3. Multipole radiation of order $l \geq 2$ 127
3.2.4. Energy and angular momentum of multipolar fields 132
3.3. Radiation from an extended source 135

Chapter 4. Radiation from Point Charges 145

4.1. Lińard–Wiechert potentials 145
4.2. Fields of a moving point charge 147
4.3. Radiation from a charge in uniform motion 152
4.4. Radiation from a slowly moving charge 153
4.5. Radiation from a relativistic charge 154
4.5.1. Velocity and acceleration parallel 155
4.5.2. Acceleration perpendicular to velocity 158
4.5.3. General case 166
4.6. Frequency distribution of the radiated energy 170
4.7. Charge in instantaneous circular motion 177
4.8. Particle in circular motion 183
4.9. Thomson scattering of radiation 188
4.10. Cherenkov radiation 196

Chapter 5. Kirchhoff Diffraction 201

5.1. Initial value problem: Kirchhoff's integral representation 201
5.1.1. Green's function for the time-dependent wave equation 201
5.1.2. Cauchy problem 204
5.1.3. Poisson solution 205
5.1.4. Kirchhoff representation of the field 206

5.2. Kirchhoff integral for diffraction 207
5.3. Vector form of the Kirchhoff integral 212
5.4. Babinet's principle of complementary screens 217
5.5. Diffraction by a circular aperture 220
5.6. Scattering by a conducting sphere in the short-wavelength limit 231

Chapter 6. Theory of Special Relativity 241

6.1. Postulates of special relativity and Lorentz transformation 241
6.2. FitzGerald–Lorentz length contraction, time dilation and proper time 248
6.3. Law of addition of velocities 250
6.4. Four-dimensional Formalism 255
6.4.1. Brehme–Lorentz diagram 255
6.4.2. Transformations and Lorentz group 259
6.4.3. Contravariant coordinates 262
6.4.4. Covariant coordinates 263
6.4.5. Change of basis 264
6.4.6. Four-velocity, four-momentum and four-acceleration 266
6.4.7. Tensors 269
6.4.8. Differentiation, vector analysis and integration 272
6.5. Relativistic dynamics 274
6.5.1. Relativistic Lagrangian for a free particle 274
6.5.2. Four-momentum 275
6.5.3. Application to collisions 278
6.5.4. Minkowski four-force 280

Chapter 7. Covariant Formulation 285

7.1. A brief remark on field theory 285
7.1.1. Transition from a discrete system to a continuous one 285
7.1.2. Continuous system 286
7.1.3. Euler–Lagrange equations 287
7.1.4. Hamiltonian formalism 288
7.1.5. Gauge invariance 290
7.1.6. Infinitesimal symmetry 291
7.1.7. Energy–momentum tensor of a field 292
7.1.8. Angular momentum tensor 293
7.1.9. Electromagnetic interaction 293
7.2. Free particle in a field 294
7.2.1. Four-potential 294
7.2.2. Equation of relativistic dynamics 296
7.2.3. Electromagnetic field tensor 298
7.2.4. Lorentz force 300
7.2.5. Change of reference frames for the field 302

7.2.6. Invariants of the electromagnetic field 305
7.2.7. First group of Maxwell equations 306
7.3. Sources of the field: second group of Maxwell equations 307
7.3.1. Interaction between a field and a current 308
7.3.2. Free field action (field self-interaction) 309
7.3.3. Magnetization and polarization 311
7.4. Energy–momentum tensor 312
7.4.1. Conservation law 314
7.4.2. Time component: conservation of energy 315
7.4.3. Spatial components: conservation of momentum 316
7.5. Electromagnetic mass 318
7.6. Radiation reaction force from energy balance 321
7.7. Abraham–Lorentz model 325
7.8. Abraham–Lorentz equation of motion 332
7.9. Dirac radiation equation 334
7.10. And the story continues... 335

References 337

Index 343

Preface

There is no place in this new kind of physics both for the field and matter, for the field is the only reality.

Albert Einstein

An *electric charge* is an abstract concept, comparable to the concept of *mass*, introduced to explain certain behaviors in physics. However, unlike mass, an electric charge can take two experimental forms that are arbitrarily called *positive* and *negative*. Two charges of the same nature, both positive or both negative, for instance, repel each other, while two opposite charges attract each other. This phenomenon is called *electrostatic interaction*.

The history of electric charges goes back to the ancient Greeks who observed that rubbing certain objects against fur produced an imbalance in the total electric charge. These "electrified" objects would also attract other light objects, like hair. What is more, if the objects are rubbed against each other for long enough, it is even possible to create a spark.

Electric charges can be experimentally measured. Particles observed in nature have charges that are integer multiples of an *elementary charge*, which is a *fundamental physical constant*. An elementary charge is the electric charge on a proton or, equivalently, the opposite of the electric charge on an electron. It is expressed in *Coulombs* (C), or in $A \times s$ in the international system of units. The value of an electric charge, considered to be invisible, and its *discrete nature* were first discovered in 1909 by Robert Andrews Millikan.

The elementary charge, denoted by e, has an approximate value $e = 1.602176634 \times 10^{-19}$C. The exact value of the elementary charge is currently defined by: $e = \sqrt{\frac{2h\alpha}{\mu_0 c}}$, where h is *Planck's constant*, with $h = 6.626070040 \times 10^{-34}$Js; α is the *fine structure constant*, with $\alpha = 7.2973525664 \times 10^{-3}$; μ_0 is the *vacuum magnetic permeability*, with $\mu_0 = 4\pi \times 10^{-7}\text{Hm}^{-1}$; and c is the speed of light in a vacuum, with $c = 299{,}792{,}458$ m/s. The existence of quarks was postulated in the beginning of 1960 and they are believed to possess a fractional electric charge $\left(\pm\frac{1}{3}e\right)$ or $\left(\pm\frac{2}{3}e\right)$, but they are confined within hadrons, whose charge and may be zero, and may be equal to the elementary charge, or may be an integer multiple of the elementary charge. Quarks have not yet been separately detected, but they are assumed to have existed in a free state in the very first instants of the Universe (the quark epoch). Charge is an invariant in the theory of special relativity. A particle of charge q, whatever its velocity, will always retain the same charge q.

There are two branches of study:

– *Electrostatics* is the branch of physics that studies the phenomena created by static electric charges for the observer. In physics, an electrically charged particle creates a vector field that is called an *electric field*. In the presence of a charged particle, the local properties of a space are therefore modified. This is therefore an accurate translation of the concept of a field. If there is another charge in this same field, it will be subject to the action of the electric force (one of the four forces of nature) exerted at a distance by the particle: the electric field is in a way the mediator of this distant action. If there is a movement of electric charges, generally electrons, within a conductive material under the effect of a potential difference at the ends of the material, an electric current is produced. The concept of electric current was first introduced by Benjamin Franklin. He was the first to imagine electricity as a type of invisible fluid present in all matter. He postulated that friction of an insulating surface causes the fluid to move and that a flow of this fluid constitutes an electric current. A material is made up of a large number of electric charges, but these charges balance out among themselves, that is, there is the same number of electrons (negative) as protons (positive). At normal temperatures, the material is electrically neutral. When a static electricity effect is produced, this means that there has been a displacement of charges from material A to material B. This is the *electrification phenomenon*. Excess or missing charges, that is, *uncompensated charges*, are responsible for electrical effects on a body (the example of the rubbed stick). There are two kinds of materials:

a) A material is said to be a *perfect conductor* if, when electrified, the uncompensated charges move freely through the material.

b) If the uncompensated charges do not move freely and remain fixed in the place where they were deposited, we say that it is a *perfect insulator (or dielectric)*.

In reality, a real material is obviously lies between these two limit states:

– *Magnetism* represents all physical phenomena in which objects exert attractive or repulsive forces on other materials. The electric currents and magnetic moments of the fundamental elementary particles are at the origin of the magnetic field generated by these forces. The magnetic field is therefore a vector quantity (characterized by an intensity and a direction) defined for each point in space. It is determined by the position and orientation of magnets, electromagnets and the displacement of electric charges. The presence of this field is seen in the existence of a force acting upon the moving electric charges (called the *Lorentz force*), and various effects that affect certain materials (paramagnetism, diamagnetism, or ferromagnetism, depending on the case). Magnetic susceptibility is the quantity that determines how much a magnetic field interacts with a material.

Electricity and magnetism are intimately related. In 1820, Ørsted discovered the relationship between electricity and magnetism: *a wire carrying an electric current is capable of deflecting the magnetized needle of a compass*. Subsequently, Ampère, based on Ørsted's work, discovered and formulated a number of laws governing the relationship between magnetism and electrodynamics. In 1831, Faraday discovered that if an electric current produces a magnetic field, the reverse is also true. An electric current can be produced by setting a magnetic field in motion, according to *Lenz's law*.

This marked the birth of electromagnetism. It is therefore the field of physics that studies the interactions between electrically charged particles, whether at rest or in motion, using the concept of the electromagnetic field. Electromagnetism can be defined as the branch that studies the electromagnetic field and its interactions with charged particles. The electric field and the magnetic field are the two components of the electromagnetic field described by electromagnetism. The equations that describe the evolution of the electromagnetic field are called *Maxwell equations*. Electric and magnetic fields waves can propagate freely in space and in most materials. These are called *electromagnetic waves* and correspond to all manifestations of light across wavelengths (radio waves, microwaves, infrared, visible domain, ultraviolet rays, X-rays and gamma rays). Maxwell wrote: *The agreement of the results seems to show that light and magnetism are affections of the same substance, and that light is an electromagnetic disturbance propagated through the field according to electromagnetic laws*. From this point of view, optics can be seen as an application of electromagnetism. The electromagnetic interaction is also one of the four fundamental interactions. Together with quantum mechanics, it enables us to understand the existence, cohesion and stability of atoms and molecules.

From the point of view of fundamental physics, the theoretical development of classical electromagnetism led to the theory of special relativity at the beginning of the 20th century. The need to reconcile electromagnetic theory and quantum mechanics led to the construction of quantum electrodynamics, in which electromagnetic interaction is interpreted as an exchange of particles called *photons*. In particle physics, *electromagnetic interaction* and *weak interaction* are unified within the *electroweak theory*.

Note to readers

Electromagnetic field theory is often poorly or insufficiently taught in university physics programs. The heavy dependence on mathematics, especially vector calculus, integral calculus, complex functions and special functions, etc., add to the difficulty in teaching this subject and may obscure certain phenomena such that students get mired in mathematical difficulties and lose sight of practical applications and vice versa. This modest contribution aims to lay out the essential ideas of the theory in the most logical order. Special importance is given to the mathematical developments of this theory. The experimental side of the theory is only lightly approached. We will not insist on units of measurement. These three volumes will contain many examples (the most classic). I have tried to give as much detail as possible in the calculations. Across the three volumes, the discussion is essentially divided into three main topics: (1) charges and/or dipoles, as the source of the electrostatic field; (2) currents and/or magnetization as sources of the magnetic field; (3) electrodynamics, where the electric field and magnetic field are of equal importance. This book discusses many applications of electromagnetic theory. For instance, there are the solutions of Laplace and Poisson equations in various coordinate systems, dielectric medias, magnetized medias, waveguides, transmission lines, electromagnetic radiation, radiation by point-like charges, Kirchhoff diffraction, special relativity, etc. For the reader's convenience, we have listed all of the mathematical tools used in the book into six appendices, which will be available to download online: vector analysis, tensors, special/hypergeometric functions, etc. References are presented at the end of each volume. Formulas are given here in the international system of units. We have tried to adopt the notations most commonly used in the literature. Graphs and 3D representations have been produced using Maple.

This three-volume set is meant for physics students, engineering students, as well as mathematics students who are interested in problems in electromagnetic theory. It can also be used by anyone interested in physics in general, or who simply wants to learn more out of intellectual curiosity. There are many books on the same subject and therefore it seems reasonable to repeat some classical developments that we find excellent, rather than striving for originality at any cost.

Volume 1 will cover the following topics:

– electrostatic field in vacuum (Gauss' theorem, etc.);

– conductor at equilibrium;

– electrokinetics;

– magnetic field in vacuum (Laplace force, Biot–Savart law, the Ampère theorem, etc.);

– electric and magnetic moments;

– boundary problems for static potentials in vacuum.

Volume 2 will study the following points:

– static potential with a source – Poisson equation (image method, Green function, etc.);

– electromagnetic induction phenomena;

– electromagnetic waves in vacuum;

– electromagnetic energies;

– polarization, polarized media (dielectric);

– magnetization, magnetized media;

– electromagnetic fields in matter.

Volume 3 will analyze the following concepts:

– waveguides;

– transmission lines;

– electromagnetic radiation;

– radiation from point-like charges;

– Kirchhoff diffraction;

– theory of special relativity;

– covariant formulation.

The online appendices will be available at the following URL: www.iste.co.uk/abdesselam/electrodynamics1.zip.

The appendices will cover:

– some vector analysis results (Appendix 1);

– tensors (Appendix 2);

– the Dirac function (Appendix 3);

– orthogonal functions (Appendix 4);

– beta and gamma functions (Appendix 5);

– special functions (Appendix 6).

For online appendices, please see: www.iste.co.uk/abdesselam/electrodynamics1.zip.

Acknowledgments

This work would never have seen the light of day without the encouragement of many people. I am particularly grateful to my students, colleagues and many others. I thank Frédérique de Fornel for her patience and advice. Finally, I dedicate this book to my wife, Nadia, and my children, Abir, Ahmed Nour, Mohammed Sami and Samah. I would like to express my gratitude to them.

Boucif Abdesselam
October 2025

Chapter 1

Waveguides

1.1. The different types of waveguides

Guided propagation involves channeling an electromagnetic wave within a space bounded by metallic or dielectric interfaces from a transmitter to a receiver. The main advantage of guided propagation is the transmission of the wave shielded from external disturbances (atmospheric and others); and, sometimes, with low attenuation. Wave guiding is achieved using devices called *waveguides*, whose structure varies depending on the frequency range of the waves:

– hollow waveguides with a rectangular cross-section for millimeter and centimeter waves;

– optical fibers for microwaves;

– coaxial cables for high frequencies.

Waveguides are generally composed of a vacuum or dielectric medium, bounded by metallic or dielectric walls. In such a medium, the waves must satisfy the Maxwell equations and the boundary conditions on the metallic or dielectric walls.

The most widely used waveguides are the following:

– *Metallic waveguides*: these are metal tubes with a circular or rectangular cross-section.

– *Dielectric waveguides and optical fibers*: a dielectric waveguide is a device that is made up of the juxtaposition of several dielectric materials with different refractive indices. The material with the highest refractive index is generally placed in the center of the guide. An electromagnetic wave that enters the guide essentially stays within this high-index material.

– *Coaxial cables*: a coaxial cable takes the form of an *inner conductor*, surrounded by an *insulation* and then a *braided and wrapped conductive shield*. All of this is wrapped within a *protective cover*. Coaxial cables will be studied in Chapter 2.

1.2. Description of a guided wave

1.2.1. *Propagation equation and field expressions*

We will see if it is possible to propagate a monochromatic wave with angular frequency ω, within the guide and along the Oz axis. The medium inside the guide, which will support the propagation of the electromagnetic wave, must have the lowest possible absorbance – it generally consists of air or a dielectric material. To simplify, let us assume that this propagation medium is an LIH (linear, isotropic, and homogeneous) dielectric, with electric permittivity ε and magnetic permeability μ (in the case of a vacuum, we have $\varepsilon = \varepsilon_0$ and $\mu = \mu_0$). The behavior of an electromagnetic wave within the waveguide is studied by solving the Maxwell equations (in the propagation medium), taking into account the boundary conditions imposed on the wave (on the fields) on the internal faces delimiting the propagation medium.

The propagation medium is a perfect LIH dielectric material (i.e. without free charges and no free current). At any point in this medium, the Maxwell equations are written as:

$$\begin{aligned}
&\vec{\nabla}\cdot\vec{\mathbf{D}} = 0\\
&\vec{\nabla}\cdot\vec{\mathbf{B}} = 0\\
&\vec{\nabla}\times\vec{\mathbf{H}} = \frac{\partial\vec{\mathbf{D}}}{\partial t}\\
&\vec{\nabla}\times\vec{\mathbf{E}} = -\mu\frac{\partial\vec{\mathbf{H}}}{\partial t}
\end{aligned} \qquad [1.1]$$

where $\vec{\mathbf{D}} = \varepsilon\vec{\mathbf{E}}$ and $\vec{\mathbf{B}} = \mu\vec{\mathbf{H}}$. In the case of a conducting medium, the free current term $\vec{\mathbf{J}}_{\text{free}}$ (which appears in $\vec{\nabla}\times\vec{\mathbf{H}} = \vec{\mathbf{J}}_{\text{free}} + \frac{\partial\vec{\mathbf{D}}}{\partial t}$) can be integrated into the displacement field, which becomes a generalized displacement field $\vec{\mathrm{D}}$, related to the electric field through the complex dielectric permittivity $\hat{\varepsilon}$. To obtain the propagation equation for the electromagnetic field, we apply the double curl to the electric and magnetic fields: $\vec{\nabla}\times\left(\vec{\nabla}\times\vec{\mathbf{E}}\right) = \vec{\nabla}\left(\vec{\nabla}\cdot\vec{\mathbf{E}}\right) - \Delta\vec{\mathbf{E}} = -\Delta\vec{\mathbf{E}}$, because $\vec{\nabla}\cdot\vec{\mathbf{E}} = 0$. The last equation in [1.1] also yields $\vec{\nabla}\times\left(\vec{\nabla}\times\vec{\mathbf{E}}\right) = -\mu\frac{\partial}{\partial t}\left(\vec{\nabla}\times\vec{\mathbf{H}}\right) = -\mu\varepsilon\frac{\partial^2\vec{\mathbf{E}}}{\partial t^2}$. Thus, from this, we can derive the equation for the propagation of the electric field. The same procedure leads to the propagation equation for the magnetic excitation field in the medium. The propagation equations for the fields are written as:

$$\begin{aligned}
&\Delta\vec{\mathbf{E}} - \mu\varepsilon\frac{\partial^2\vec{\mathbf{E}}}{\partial t^2} = 0\\
&\Delta\vec{\mathbf{H}} - \mu\varepsilon\frac{\partial^2\vec{\mathbf{H}}}{\partial t^2} = 0
\end{aligned} \qquad [1.2]$$

To study propagation in a waveguide, we first need to choose an analytical expression for the electromagnetic field that satisfies both the propagation equations [4.2] and the boundary conditions imposed on the medium. In the case of a cylindrical guide, cylindrical coordinates should be chosen, while in the case of a rectangular guide, a Cartesian system is preferable. In both cases, the choice of plane-wave propagation is not possible (with the exception of certain special situations) because a plane wave cannot satisfy the multiple boundary conditions imposed (the plane-wave amplitude is assumed to be constant). To do this, the fields must be decomposed into longitudinal fields $\left(\vec{\mathbf{E}}_z, \vec{\mathbf{H}}_z\right)$ (parallel to the direction of the guide, chosen here as the Oz axis) and in transverse fields $\left(\vec{\mathbf{E}}_\perp, \vec{\mathbf{H}}_\perp\right)$ (perpendicular to Oz). In the expressions for $\vec{\mathbf{E}}$ and $\vec{\mathbf{H}}$, the dependence in z (assumed to be sinusoidal) is separated from the dependence in (x, y).

We therefore consider an electromagnetic wave propagating along the Oz axis as a plane wave:

$$\begin{aligned}\vec{\mathbf{E}}\left(\vec{\mathbf{r}}, t\right) &= \Re\left[\vec{\mathbf{E}}\left(x, y\right) e^{-\mathrm{i}(\omega t - \mathrm{k}_z z)}\right]\\ \vec{\mathbf{H}}\left(\vec{\mathbf{r}}, t\right) &= \Re\left[\vec{\mathbf{H}}\left(x, y\right) e^{-\mathrm{i}(\omega t - \mathrm{k}_z z)}\right]\end{aligned} \qquad [1.3]$$

but with variable amplitude. Here, k_z is the *wavenumber* or the *propagation constant* in the Oz direction. The problem is therefore reduced to solving the Maxwell equations for the amplitudes $\vec{\mathbf{E}}\left(x, y\right)$ and $\vec{\mathbf{H}}\left(x, y\right)$, which depend on the transverse coordinates: the problem thus becomes a 2D problem. Let us start by decomposing the fields into transverse and longitudinal components:

$$\begin{aligned}\vec{\mathbf{E}} &= \vec{\mathbf{E}}_\perp + E_z \vec{\mathbf{k}}\\ \vec{\mathbf{H}} &= \vec{\mathbf{H}}_\perp + H_z \vec{\mathbf{k}}\end{aligned} \qquad [1.4]$$

where $\vec{\mathbf{E}}_\perp$ and $\vec{\mathbf{H}}_\perp$ are perpendicular to Oz. We also define the transverse gradient $\vec{\nabla}_\perp = \vec{\mathbf{i}}\frac{\partial}{\partial x} + \vec{\mathbf{j}}\frac{\partial}{\partial y} = \vec{\nabla} - \vec{\mathbf{k}}\frac{\partial}{\partial z}$. We want to show here that the transverse field $\vec{\mathbf{E}}_\perp$ (and, consequently, $\vec{\mathbf{E}}$) is fully determined if we know the E_z components of the electric field and H_z of the magnetic field.

This result is also true for the transverse field $\vec{\mathbf{H}}_\perp$. First, let us observe that:

$$\begin{aligned}\vec{\nabla} \times \vec{\mathbf{E}} &= \left(\vec{\nabla}_\perp + \vec{\mathbf{k}}\frac{\partial}{\partial z}\right) \times \left(\vec{\mathbf{E}}_\perp + E_z \vec{\mathbf{k}}\right)\\ &= \vec{\nabla}_\perp \times \vec{\mathbf{E}}_\perp + \vec{\mathbf{k}} \times \frac{\partial \vec{\mathbf{E}}_\perp}{\partial z} - \vec{\mathbf{k}} \times \vec{\nabla}_\perp E_z\end{aligned} \qquad [1.5]$$

The first term is oriented along $\overrightarrow{\mathbf{k}}$, while the other two are transverse. Next, we write the Maxwell equations for the fields [1.3], separating the transverse and longitudinal components. For the Maxwell–Faraday equation, we obtain:

$$\begin{aligned}\overrightarrow{\nabla} \times \overrightarrow{\mathbf{E}} + \mu \frac{\partial \overrightarrow{\mathbf{H}}}{\partial t} &= \overrightarrow{\nabla}_{\perp} \times \overrightarrow{\mathbf{E}}_{\perp} + \mathrm{i}k_z \overrightarrow{\mathbf{k}} \times \overrightarrow{\mathbf{E}}_{\perp} - \overrightarrow{\mathbf{k}} \times \overrightarrow{\nabla}_{\perp} E_z \\ &- \mathrm{i}\mu\omega \overrightarrow{\mathbf{H}}_{\perp} - \mathrm{i}\mu\omega H_z \overrightarrow{\mathbf{k}} = 0\end{aligned} \quad [1.6]$$

which, after separating the components, gives:

$$\overrightarrow{\nabla}_{\perp} \times \overrightarrow{\mathbf{E}}_{\perp} - \mathrm{i}\mu\omega H_z \overrightarrow{\mathbf{k}} = 0 \quad [1.7]$$

$$\overrightarrow{\mathbf{k}} \times \left(\mathrm{i}k_z \overrightarrow{\mathbf{E}}_{\perp} - \overrightarrow{\nabla}_{\perp} E_z\right) - \mathrm{i}\mu\omega \overrightarrow{\mathbf{H}}_{\perp} = 0 \quad [1.8]$$

Taking the cross product of equation [1.8] with $\overrightarrow{\mathbf{k}}$ yields:

$$\mathrm{i}k_z \overrightarrow{\mathbf{E}}_{\perp} + \mathrm{i}\mu\omega \overrightarrow{\mathbf{k}} \times \overrightarrow{\mathbf{H}}_{\perp} = \overrightarrow{\nabla}_{\perp} E_z \quad [1.9]$$

Similarly, by taking the dot product of equation [1.7] with $\overrightarrow{\mathbf{k}}$, we find:

$$\overrightarrow{\mathbf{k}} \cdot \left(\overrightarrow{\nabla}_{\perp} \times \overrightarrow{\mathbf{E}}_{\perp}\right) = \mathrm{i}\mu\omega H_z \quad [1.10]$$

Let us repeat the same procedure for the Maxwell–Ampère equation:

$$\begin{aligned}\overrightarrow{\nabla} \times \overrightarrow{\mathbf{H}} - \varepsilon \frac{\partial \overrightarrow{\mathbf{E}}}{\partial t} &= \overrightarrow{\nabla}_{\perp} \times \overrightarrow{\mathbf{H}}_{\perp} + \mathrm{i}k_z \overrightarrow{\mathbf{k}} \times \overrightarrow{\mathbf{H}}_{\perp} - \overrightarrow{\mathbf{k}} \times \overrightarrow{\nabla}_{\perp} H_z \\ &+ \mathrm{i}\varepsilon\omega \overrightarrow{\mathbf{E}}_{\perp} + \mathrm{i}\varepsilon\omega E_z \overrightarrow{\mathbf{k}} = 0\end{aligned} \quad [1.11]$$

which leads to:

$$\overrightarrow{\nabla}_{\perp} \times \overrightarrow{\mathbf{H}}_{\perp} + \mathrm{i}\varepsilon\omega E_z \overrightarrow{\mathbf{k}} = 0 \quad [1.12]$$

$$\overrightarrow{\mathbf{k}} \times \left(\mathrm{i}k_z \overrightarrow{\mathbf{H}}_{\perp} - \overrightarrow{\nabla}_{\perp} H_z\right) + \mathrm{i}\varepsilon\omega \overrightarrow{\mathbf{E}}_{\perp} = 0 \quad [1.13]$$

and consequently:

$$\mathrm{i}k_z \overrightarrow{\mathbf{H}}_{\perp} - \mathrm{i}\varepsilon\omega \overrightarrow{\mathbf{k}} \times \overrightarrow{\mathbf{E}}_{\perp} = \overrightarrow{\nabla}_{\perp} H_z \quad [1.14]$$

$$\overrightarrow{\mathbf{k}} \cdot \left(\overrightarrow{\nabla}_{\perp} \times \overrightarrow{\mathbf{H}}_{\perp}\right) = -\mathrm{i}\varepsilon\omega E_z \quad [1.15]$$

The other two Maxwell equations can be reduced to:

$$\vec{\nabla}_\perp \cdot \vec{\mathbf{E}}_\perp = -\mathrm{i}\mathrm{k}_z E_z \qquad [1.16]$$

$$\vec{\nabla}_\perp \cdot \vec{\mathbf{H}}_\perp = -\mathrm{i}\mathrm{k}_z H_z \qquad [1.17]$$

Equations [1.9–1.10] and [1.14–1.17] make it possible to express $\vec{\mathbf{E}}_\perp$ and $\vec{\mathbf{H}}_\perp$ in terms of E_z and H_z. To see this, take the cross product of $\vec{\mathbf{k}}$ with equation [1.9] and then substitute equation [1.14] into the result:

$$\frac{\mathrm{k}_z}{\omega\varepsilon}\left(\mathrm{i}\mathrm{k}_z\vec{\mathbf{H}}_\perp - \vec{\nabla}_\perp H_z\right) - \mathrm{i}\omega\mu\vec{\mathbf{H}}_\perp = \vec{\mathbf{k}} \times \vec{\nabla}_\perp E_z \qquad [1.18]$$

which gives:

$$\vec{\mathbf{H}}_\perp = \frac{1}{\omega^2 - \mathrm{k}_z^2 v^2}\left[\frac{\mathrm{i}\omega}{\mu}\vec{\mathbf{k}} \times \vec{\nabla}_\perp E_z + \mathrm{i}\mathrm{k}_z v^2 \vec{\nabla}_\perp H_z\right] \qquad [1.19]$$

where $v^2 = \frac{1}{\varepsilon\mu}$. Repeating the same procedure for [1.14] and taking into account [1.9], we get:

$$\frac{\mathrm{k}_z}{\omega\mu}\left(-\mathrm{i}\mathrm{k}_z\vec{\mathbf{E}}_\perp + \vec{\nabla}_\perp E_z\right) + \mathrm{i}\omega\varepsilon\vec{\mathbf{E}}_\perp = \vec{\mathbf{k}} \times \vec{\nabla}_\perp H_z \qquad [1.20]$$

which allows us to derive $\vec{\mathbf{E}}_\perp$:

$$\vec{\mathbf{E}}_\perp = \frac{1}{\omega^2 - \mathrm{k}_z^2 v^2}\left[-\frac{\mathrm{i}\omega}{\varepsilon}\vec{\mathbf{k}} \times \vec{\nabla}_\perp H_z + \mathrm{i}\mathrm{k}_z v^2 \vec{\nabla}_\perp E_z\right] \qquad [1.21]$$

Equations [1.19] and [1.21] therefore allow us to determine all transverse components $\vec{\mathbf{E}}_\perp$ and $\vec{\mathbf{H}}_\perp$, as soon as the two longitudinal components E_z and H_z are known. We can thus limit ourselves to solving the propagation equation only for E_z and H_z, and then deduce $\vec{\mathbf{E}}_\perp$ and $\vec{\mathbf{H}}_\perp$ using equations [1.19] and [1.21]. In particular, when $\mathrm{k}_z^2 \neq \frac{\omega^2}{v^2}$, the electromagnetic field is zero if the components E_z and H_z are simultaneously zero.

The components E_z and H_z satisfy the Helmholtz equation:

$$\left(\vec{\nabla}^2 + \varepsilon\mu\omega^2\right)\begin{pmatrix} E_z \\ H_z \end{pmatrix} = 0 \qquad [1.22]$$

Substituting expressions [1.3] into [1.22], we find:

$$\left(\vec{\nabla}_{\perp}^{2} + \gamma^{2}\right)\begin{pmatrix} E_z \\ H_z \end{pmatrix} = 0 \qquad [1.23]$$

where:

$$\gamma^2 = \varepsilon\mu\omega^2 - \mathsf{k}_z^2 \qquad [1.24]$$

We must solve this equation, taking into account the boundary conditions imposed in the Oxy plane due to the presence of the waveguide walls (conductive walls or interfaces between two dielectric media).

1.2.2. *Classification of propagation modes*

Unlike waves propagating in vacuum, electromagnetic waves are not always transverse, that is, the electric and magnetic fields are not necessarily perpendicular to the direction of propagation (the Oz axis). Different propagation modes can be distinguished:

1) *TM (transverse magnetic) mode*: the field $\vec{\mathbf{H}}$ is perpendicular to the direction of propagation, but $E_z \neq 0$. We must therefore solve equation [1.23] for E_z. This mode is also called an **E** type wave (since E_z is non-zero).

2) *TE (transverse electric) wave*: the electric field $\vec{\mathbf{E}}$ is perpendicular to the direction of propagation, but $H_z \neq 0$. We must therefore solve equation [1.23] for H_z. This mode is also called an **H** type wave (since H_z is non-zero).

3) *TEM (transverse electric and magnetic) mode*: the fields $\vec{\mathbf{E}}$ and $\vec{\mathbf{H}}$ are perpendicular to the direction of propagation, as if the wave were propagating in a vacuum (i.e. $E_z = H_z = 0$).

According to [1.19] and [1.21], for a TEM wave to exist, it is required that $\vec{\mathbf{E}}_{\perp}\gamma^2 = 0$ and $\vec{\mathbf{H}}_{\perp}\gamma^2 = 0$. For $\vec{\mathbf{E}}_{\perp} \neq 0$ and $\vec{\mathbf{H}}_{\perp} \neq 0$, we must require that $\gamma^2 = 0$. The dispersion relation for TEM waves is therefore:

$$\mathsf{k}_z^2 = \varepsilon\mu\omega^2 \qquad [1.25]$$

(the propagation constant is fixed for TEM waves). We also see from [1.7] and [1.16] that the transverse electric field $\vec{\mathbf{E}}_{\perp}$ satisfies the following relations:

$$\begin{aligned} \vec{\nabla}_{\perp} \times \vec{\mathbf{E}}_{\perp} &= 0 \\ \vec{\nabla}_{\perp} \cdot \vec{\mathbf{E}}_{\perp} &= 0 \end{aligned} \qquad [1.26]$$

These are the two-dimensional electrostatic equations in the absence of charges. We know that this implies that $\vec{\mathbf{E}}_\perp$ derives from a scalar potential, that is, there exists a function $V(x,y)$ such that $\vec{\mathbf{E}}_\perp = -\vec{\nabla}_\perp V$. Since $\vec{\nabla}_\perp \cdot \vec{\mathbf{E}}_\perp = 0$, this potential satisfies the 2D Laplace equation:

$$\Delta_\perp V = 0 \qquad [1.27]$$

Relations similar to [1.26] are satisfied by the transverse magnetic excitation field $\vec{\mathbf{H}}_\perp$. In fact, equations [1.12] and [1.17] yield (for TEM waves):

$$\begin{aligned} &\vec{\nabla}_\perp \times \vec{\mathbf{H}}_\perp = 0 \\ &\vec{\nabla}_\perp \cdot \vec{\mathbf{H}}_\perp = 0 \end{aligned} \qquad [1.28]$$

which also indicates that the magnetic excitation field $\vec{\mathbf{H}}_\perp$ derives from a potential $\Psi(x,y)$, satisfying the $2D$ Laplace equation:

$$\begin{aligned} &\vec{\mathbf{H}}_\perp = \vec{\nabla}_\perp \Psi \\ &\Delta_\perp \Psi = 0 \end{aligned} \qquad [1.29]$$

For TEM waves, the fields $\vec{\mathbf{E}}_\perp$ and $\vec{\mathbf{H}}_\perp$ are related (see [1.13–1.14]) in this way:

$$\begin{aligned} &\vec{\mathbf{H}}_\perp = \frac{\mathsf{k}_z}{\mu\omega}\vec{\mathbf{k}} \times \vec{\mathbf{E}}_\perp = Z^{-1}\vec{\mathbf{k}} \times \vec{\mathbf{E}}_\perp \\ &\vec{\mathbf{E}}_\perp = -Z\vec{\mathbf{k}} \times \vec{\mathbf{H}}_\perp \end{aligned} \qquad [1.30]$$

with $Z = \sqrt{\frac{\mu}{\varepsilon}}$. In other words, the $\vec{\mathbf{H}}$ field lines are perpendicular to the $\vec{\mathbf{E}}$ field lines. A consequence of equations [1.30] is that a TEM wave cannot propagate inside a hollow waveguide surrounded by a conductor (as will be demonstrated explicitly later for rectangular waveguides) because the electrostatic field is always zero inside a conductor. This result is quite general and can be extended to any hollow waveguide with conductive walls and a cross-section of any shape. To see this, let us consider two closed contours C_1 and C_2 of the Oxy plane, surrounding the guide at positions z_1 and z_2, respectively. Let us choose the following surfaces: (1) S_1 the surface bounded by C_1 in the Oxy plane; (2) S_2 the surface bounded by C_2 in the Oxy plane; (3) S_{env} is the surface that envelopes the guide, bounded on one side by C_1 and on the other by C_2. The integrals $\int_{S_1} \Psi\vec{\mathbf{H}}_\perp \cdot d\vec{\mathbf{S}}_1$, $\int_{S_2} \Psi\vec{\mathbf{H}}_\perp \cdot d\vec{\mathbf{S}}_2$ and $\int_{S_{\text{env}}} \Psi\vec{\mathbf{H}}_\perp \cdot d\vec{\mathbf{S}}_{\text{env}}$ are zero, because $\vec{\mathbf{H}}_\perp$ is in the Oxy plane for the first two and zero on the $\vec{\mathbf{S}}_{\text{env}}$ surface for the third. This means that:

$$\int_{S_2} \Psi\vec{\mathbf{H}}_\perp \cdot d\vec{\mathbf{S}}_2 - \int_{S_1} \Psi\vec{\mathbf{H}}_\perp \cdot d\vec{\mathbf{S}}_1 + \int_{S_{\text{env}}} \Psi\vec{\mathbf{H}}_\perp \cdot d\vec{\mathbf{S}}_{\text{env}} = \oint_S \Psi\vec{\mathbf{H}}_\perp \cdot d\vec{\mathbf{S}} = 0 \quad [1.31]$$

This last integral can be transformed into a volume integral using the divergence theorem:

$$0 = \oint_S \Psi \vec{\mathbf{H}}_\perp \cdot d\vec{\mathbf{S}} = \int_\tau \vec{\nabla}_\perp \cdot \left(\Psi \vec{\mathbf{H}}_\perp\right) d\tau = \int_\tau \left(\vec{\mathbf{H}}_\perp\right)^2 d\tau \qquad [1.32]$$

which gives $\vec{\mathbf{H}}_\perp = 0$ and then $\vec{\mathbf{E}}_\perp = 0$ in τ. Thus, a waveguide made up of a hollow conductor with no other conductor on its inside cannot conduct a TEM electromagnetic wave. These modes can exist in the presence of two or more conductors.

1.3. TE and TM modes in a hollow conducting waveguide

Consider a hollow waveguide surrounded by a conducting wall. In this case, TEM modes do not exist and only TE and TM waves are possible. To minimize energy losses inside these guides during propagation, it is assumed that the interior is vacuum (or air). Therefore, we take $\varepsilon = \varepsilon_0$ and $\mu = \mu_0$ in what follows. Let $\vec{\mathbf{n}}$ be the normal to the guide's conductive wall (directed toward the interior). Across this wall, the tangential component of $\vec{\mathbf{E}}$ must be continuous. The same applies to the normal component of $\vec{\mathbf{H}}$. Since the wall is a perfect conductor, the fields must vanish inside the conductor. Hence, by continuity, $\left(\vec{\mathbf{E}} \times \vec{\mathbf{n}}\right)_S$ and $\left(\vec{\mathbf{H}} \cdot \vec{\mathbf{n}}\right)_S$ must be zero on the inner side of the wall. Taking into account the decomposition [1.4], we deduce that:

$$\left(\vec{\mathbf{E}}_\perp \times \vec{\mathbf{n}}\right)_S = 0 \qquad [1.33]$$

$$(E_z)_S = 0 \qquad [1.34]$$

$$\left(\vec{\mathbf{H}}_\perp \cdot \vec{\mathbf{n}}\right)_S = 0 \qquad [1.35]$$

$\left(\vec{\mathbf{E}}_\perp \times \vec{\mathbf{n}}\right)_S$ and $(E_z)_S\, \vec{\mathbf{k}} \times \vec{\mathbf{n}}$ represent the transverse and longitudinal components, respectively, of $\left(\vec{\mathbf{E}} \times \vec{\mathbf{n}}\right)_S$ and they must both vanish independently. If we now calculate the scalar product of equation [1.14] with $\vec{\mathbf{n}}$, we obtain:

$$\mathrm{i}k_z \left(\vec{\mathbf{H}}_\perp \cdot \vec{\mathbf{n}}\right)_S - \mathrm{i}\omega\varepsilon \left[\vec{\mathbf{n}} \cdot \left(\vec{\mathbf{k}} \times \vec{\mathbf{E}}_\perp\right)\right]_S = \left(\vec{\mathbf{n}} \cdot \vec{\nabla}_\perp H_z\right)_S \qquad [1.36]$$

According to [1.35], the first term is zero on the wall. The mixed product of the second term is written as $\left[\vec{\mathbf{k}} \cdot \left(\vec{\mathbf{E}}_\perp \times \vec{\mathbf{n}}\right)\right]_S$; it is zero according to [1.33].

Therefore, the right-hand side of equation [1.36], which is the derivative of H_z in the $\vec{n}$ direction, must also be zero:

$$\left(\frac{\partial H_z}{\partial n}\right)_S = 0 \qquad [1.37]$$

In the same way, let us take the vector product of equation [1.9] and $\vec{n}$:

$$\mathrm{i}k_z\left(\vec{\mathbf{E}}_\perp \times \vec{n}\right)_S + \mathrm{i}\omega\mu\left[\left(\vec{\mathbf{k}} \times \vec{\mathbf{H}}_\perp\right) \times \vec{n}\right]_S = \left(\vec{\nabla}_\perp E_z \times \vec{n}\right)_S \qquad [1.38]$$

The two terms on the left in [1.38] are obviously zero, since $\left[\left(\vec{\mathbf{k}} \times \vec{\mathbf{H}}_\perp\right) \times \vec{n}\right]_S = -\vec{\mathbf{k}}\left(\vec{n} \cdot \vec{\mathbf{H}}_\perp\right)_S = 0$, which gives:

$$\left(\vec{n} \times \vec{\nabla}_\perp E_z\right)_S = 0 \qquad [1.39]$$

The problem therefore boils down to solving the Helmholtz equation [1.23], taking into account the boundary conditions [1.34] and [1.37]. As the boundary conditions are different for E_z and H_z, the solutions and eigenvalues must be different between the TM and TE modes.

1.3.1. *Rectangular waveguide*

Consider a waveguide with a rectangular cross-section (with sides a and b). The inner region of the waveguide is defined by $0 \leq x \leq a$ and $0 \leq y \leq b$.

The length of the waveguide along Oz is assumed to be infinite, as shown in Figure 1.1.

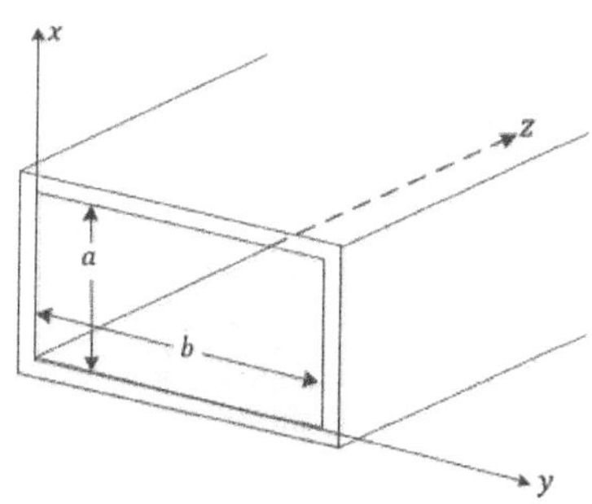

Figure 1.1. *Rectangular waveguide of height a and width b*

1.3.1.1. *TM waves ($H_z = 0$, $E_z \neq 0$)*

Let us first consider the TM modes. The Helmholtz equation is written as follows:

$$\left(\frac{\partial^2}{\partial x^2} + \frac{\partial^2}{\partial y^2} + \gamma^2\right) E_z = 0 \qquad [1.40]$$

where $\gamma^2 = \frac{\omega^2}{c^2} - \mathsf{k}_z^2$. The boundary conditions are:

$$\begin{aligned} E_z\left(x=0,y\right) &= E_z\left(x=a,y\right) = 0 \\ E_z\left(x,y=0\right) &= E_z\left(x,y=b\right) = 0 \end{aligned} \qquad [1.41]$$

The solution to [1.40] is obtained by separation of variables. We assume that the unknown function can be written in the form $E_z\left(x,y\right) = X\left(x\right)Y\left(y\right)$, where $X\left(x\right)$ (respectively, $Y\left(y\right)$) depends only on x (respectively, y). Substituting into the Helmholtz equation and dividing by XY, we obtain $\frac{X''}{X} + \frac{Y''}{Y} + \gamma^2 = 0$. The first term depends only on x, the second only on y and the third is constant. This equality can only hold if the first two terms are also constants, that is, $\frac{X''}{X} = -\mathsf{k}_x^2$, and $\frac{Y''}{Y} = -\mathsf{k}_y^2$, where the separation constants must satisfy the relation:

$$\mathsf{k}_x^2 + \mathsf{k}_y^2 = \gamma^2 \qquad \Longleftrightarrow \qquad \mathsf{k}_x^2 + \mathsf{k}_y^2 + \mathsf{k}_z^2 = \frac{\omega^2}{c^2} \qquad [1.42]$$

In particular, when $\omega = 0$, we obtain stationary solutions. If at least one of the wavenumbers is purely imaginary, the solutions are decreasing (exponential). If all three wavenumbers are real, there is pure propagation (sinusoidal solutions). The values of these wavenumbers are determined by the boundary conditions imposed on the walls in the Ox and Oy directions. The general solutions are given by:

$$\begin{aligned} X\left(x\right) &= A_1 \sin\left(\mathsf{k}_x x\right) + A_2 \cos\left(\mathsf{k}_x x\right) \\ Y\left(y\right) &= B_1 \sin\left(\mathsf{k}_y y\right) + B_2 \cos\left(\mathsf{k}_y y\right) \end{aligned} \qquad [1.43]$$

and therefore:

$$E_z\left(x,y\right) = \left[A_1 \sin\left(\mathsf{k}_x x\right) + A_2 \cos\left(\mathsf{k}_x x\right)\right]\left[B_1 \sin\left(\mathsf{k}_y y\right) + B_2 \cos\left(\mathsf{k}_y y\right)\right] \quad [1.44]$$

where A_1, A_2, B_1 and B_2 are constants. Since the rectangular waveguide is composed of perfectly conducting walls, the longitudinal component of the electric field on these walls is zero (as already mentioned in [1.41]). In order to satisfy the boundary conditions at $x = 0$ and $y = 0$, it is essential that $A_2 = B_2 = 0$. Consequently, it follows that:

$$E_z\left(x,y\right) = E_0 \sin\left(\mathsf{k}_x x\right) \sin\left(\mathsf{k}_y y\right) \qquad [1.45]$$

where E_0 is the amplitude of the field. In addition, to satisfy the other two boundary conditions in $x = a$ and $y = b$, the wavenumbers k_x and k_y must be quantified as follows:

$$\begin{aligned}\mathsf{k}_x &= \frac{m\pi}{a} \quad m = 1, 2, 3, \ldots \\ \mathsf{k}_y &= \frac{n\pi}{b} \quad n = 1, 2, 3, \ldots\end{aligned} \tag{1.46}$$

Note that if m or n is zero, the electric field E_z is zero. Therefore, the desired solution is:

$$E_z - \Re\left[E_{mn} \sin\left(\frac{m\pi x}{a}\right) \sin\left(\frac{n\pi y}{b}\right) e^{-\mathrm{i}(\omega t - \mathsf{k}_z z)}\right] \tag{1.47}$$

This represents what is called the *waveguide eigenmode*, denoted by TM_{mn}. Thus, the most general solution of the Helmholtz equation [1.22] is obtained as a linear combination of all possible eigenmodes. Since:

$$\mathsf{k}_z^2 = \frac{\omega^2}{c^2} - \mathsf{k}_x^2 - \mathsf{k}_y^2 = \frac{\omega^2}{c^2} - \pi^2 \left(\frac{m^2}{a^2} + \frac{n^2}{b^2}\right) \tag{1.48}$$

we observe the following:

1) k_z can only take discrete values.

2) k_z is real only if:

$$\omega \geq \pi c \sqrt{\frac{m^2}{a^2} + \frac{n^2}{b^2}} \tag{1.49}$$

In other words, for a given value of ω, the (m, n) eigenmode can only propagate in the waveguide if condition [14.49] is satisfied. Otherwise, k_z becomes purely imaginary, that is, $\mathsf{k}_z = \mathrm{i}\delta$ with $\delta > 0$; this introduces the factor $e^{-\delta z}$ in the expression of E_z, preventing any propagation (the wave is therefore attenuated). Therefore, there is a lower limit for ω, called the *cutoff frequency* and denoted $\omega_{m,n}$, below which no propagation is possible.

For the TM_{mn} mode, we obtain:

$$\omega_{m,n} = \pi c \sqrt{\frac{m^2}{a^2} + \frac{n^2}{b^2}} \tag{1.50}$$

Modes that are excited at a frequency lower than their cutoff frequency therefore do not propagate, but are still present over a certain distance near the excitation point of the waveguide.

Using the notation [1.50], equation [1.48] can also be written as:

$$\omega^2 = \omega_{m,n}^2 + c^2 \mathsf{k}_z^2 \qquad [1.51]$$

This is the *dispersion relation* of the waveguide; this formula resembles that of a plasma, as illustrated in Figure 1.2.

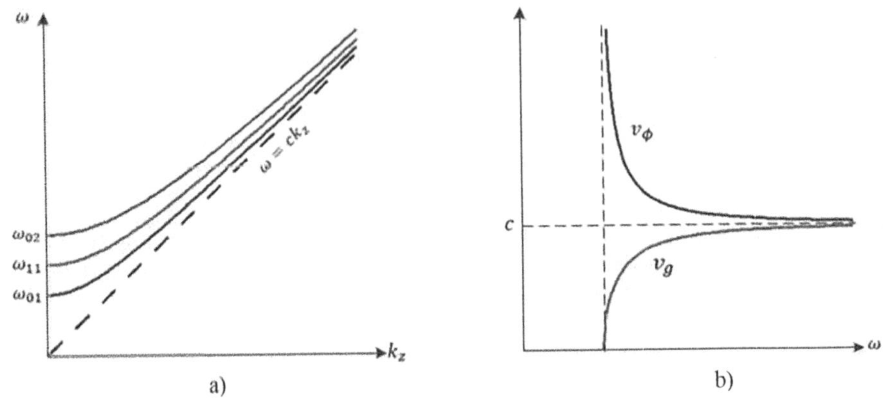

Figure 1.2. *(a) Dispersion relations for some modes of the rectangular waveguide. (b) Phase velocity and group velocity for the (m, n) mode. For a color version of this figure, see www.iste.co.uk/abdesselam/electrodynamics3.zip*

3) The phase and group velocities are given, respectively, by:

$$v_\phi = \frac{\omega}{\mathsf{k}_z} = c\frac{\omega}{\sqrt{\omega^2 - \omega_{m,n}^2}} > c$$

$$v_g = \frac{d\omega}{d\mathsf{k}_z} = c^2\frac{\mathsf{k}_z}{\omega} = \frac{c^2}{v_\phi} < c \qquad [1.52]$$

As in a plasma, the phase velocity is greater than c, while the group velocity (the physical velocity) is less than c.

The other components of the fields $\vec{\mathbf{E}}_\perp$ and $\vec{\mathbf{H}}_\perp$ can be derived from the relations [1.19] and [1.21]:

$$\vec{\mathbf{E}}_\perp = \frac{i\mathsf{k}_z c^2}{\omega^2 - \mathsf{k}_z^2 c^2}\vec{\nabla}_\perp E_z$$
$$\vec{\mathbf{H}}_\perp = \frac{i\omega}{\mu\left(\omega^2 - \mathsf{k}_z^2 c^2\right)}\vec{\mathbf{k}} \times \vec{\nabla}_\perp E_z \qquad [1.53]$$

Taking into account the expression [1.48] of the field E_z, we find:

$$\begin{cases} E_x = \Re\left[\dfrac{\mathsf{i}\mathsf{k}_x\mathsf{k}_z}{\mathsf{k}_x^2+\mathsf{k}_y^2}E_{mn}\cos\left(\mathsf{k}_x x\right)\sin\left(\mathsf{k}_y y\right)e^{-\mathsf{i}(\omega t-\mathsf{k}_z z)}\right] \\ E_y = \Re\left[\dfrac{\mathsf{i}\mathsf{k}_y\mathsf{k}_z}{\mathsf{k}_x^2+\mathsf{k}_y^2}E_{mn}\sin\left(\mathsf{k}_x x\right)\cos\left(\mathsf{k}_y y\right)e^{-\mathsf{i}(\omega t-\mathsf{k}_z z)}\right] \\ E_z = \Re\left[E_{mn}\sin\left(\mathsf{k}_x x\right)\sin\left(\mathsf{k}_y y\right)e^{-\mathsf{i}(\omega t-\mathsf{k}_z z)}\right] \end{cases} \quad [1.54]$$

$$\begin{cases} H_x = \Re\left[-\dfrac{\mathsf{i}\omega\mathsf{k}_y}{\mu c^2\left(\mathsf{k}_x^2+\mathsf{k}_y^2\right)}E_{mn}\sin\left(\mathsf{k}_x x\right)\cos\left(\mathsf{k}_y y\right)e^{-\mathsf{i}(\omega t-\mathsf{k}_z z)}\right] \\ H_y = \Re\left[\dfrac{\mathsf{i}\omega\mathsf{k}_x}{\mu c^2\left(\mathsf{k}_x^2+\mathsf{k}_y^2\right)}E_{mn}\cos\left(\mathsf{k}_x x\right)\sin\left(\mathsf{k}_y y\right)e^{-\mathsf{i}(\omega t-\mathsf{k}_z z)}\right] \\ H_z = 0 \end{cases} \quad [1.55]$$

We clearly see that the boundary conditions:

$$\begin{aligned} &E_x\left(y=0\right)=E_x\left(y=b\right)=0 \\ &E_y\left(x=0\right)=E_y\left(x=a\right)=0 \\ &H_x\left(x=0\right)=H_x\left(x=a\right)=0 \\ &H_y\left(y=0\right)=H_y\left(y=b\right)=0 \end{aligned} \quad [1.56]$$

are satisfied. The tangential components of the electric field $\overrightarrow{\mathbf{E}}$ and the normal components of the magnetic field $\overrightarrow{\mathbf{H}}$ are automatically zero on the walls. The solutions found in [1.54] and [1.55] can be interpreted as the result of the fields of four waves resulting from reflection on the waveguide walls. Indeed, let us rewrite the sine functions using De Moivre's formulas:

$$\begin{aligned} E_z = \Re\Big[&-\frac{1}{4}E_{mn}e^{-\mathsf{i}(\omega t-\mathsf{k}_x x-\mathsf{k}_y y-\mathsf{k}_z z)}+\frac{1}{4}E_{mn}e^{-\mathsf{i}(\omega t+\mathsf{k}_x x-\mathsf{k}_y y-\mathsf{k}_z z)}+ \\ &\frac{1}{4}E_{mn}e^{-\mathsf{i}(\omega t-\mathsf{k}_x x+\mathsf{k}_y y-\mathsf{k}_z z)}-\frac{1}{4}E_{mn}e^{-\mathsf{i}(\omega t+\mathsf{k}_x x+\mathsf{k}_y y-\mathsf{k}_z z)}\Big] \end{aligned} \quad [1.57]$$

where $\mathsf{k}_x = \frac{m\pi}{a}$ and $\mathsf{k}_y = \frac{n\pi}{b}$. This expansion can be interpreted as follows: the first term corresponds to an incident wave propagating in the direction of the wave vector $\overrightarrow{\mathsf{k}} = (\mathsf{k}_x, \mathsf{k}_y, \mathsf{k}_z)$. The second term is the wave resulting from the reflection of this incident wave on the face parallel to the Oyz plane at $x = a$ (reflection coefficient equal to -1 and $\mathsf{k}_x \longrightarrow -\mathsf{k}_x$). The third term is the wave obtained after reflection of the incident wave on the face located at $y = b$. Finally, the last term comes from a double successive reflection on the faces at $x = a$ and $y = b$. This decomposition shows that the angles of incidence are quantized on the walls; only these values make it possible to achieve destructive interference on the boundaries. The propagation of the electromagnetic wave in the waveguide is accompanied by the propagation of currents and charges on the inner walls. The surface charge densities on the walls of the waveguide are obtained from the discontinuity equation of the normal component of $\overrightarrow{\mathbf{D}}$:

$$\begin{aligned}
\sigma(x=0) &= \varepsilon_0 E_x(x=0) = \Re\left[\frac{\mathsf{i}\mathsf{k}_x\mathsf{k}_z\varepsilon_0}{\mathsf{k}_x^2+\mathsf{k}_y^2} E_{mn}\sin(\mathsf{k}_y y)\, e^{-\mathsf{i}(\omega t-\mathsf{k}_z z)}\right]\\
\sigma(x=a) &= -\varepsilon_0 E_x(x=a) = -\Re\left[\frac{\mathsf{i}\mathsf{k}_x\mathsf{k}_z\varepsilon_0(-1)^m}{\mathsf{k}_x^2+\mathsf{k}_y^2} E_{mn}\sin(\mathsf{k}_y y)\, e^{-\mathsf{i}(\omega t-\mathsf{k}_z z)}\right]\\
\sigma(y=0) &= \varepsilon_0 E_y(y=0) = \Re\left[\frac{\mathsf{i}\mathsf{k}_y\mathsf{k}_z\varepsilon_0}{\mathsf{k}_x^2+\mathsf{k}_y^2} E_{mn}\sin(\mathsf{k}_x x)\, e^{-\mathsf{i}(\omega t-\mathsf{k}_z z)}\right]\\
\sigma(y=b) &= -\varepsilon_0 E_y(y=b) = -\Re\left[\frac{\mathsf{i}\mathsf{k}_y\mathsf{k}_z\varepsilon_0(-1)^n}{\mathsf{k}_x^2+\mathsf{k}_y^2} E_{mn}\sin(\mathsf{k}_x x)\, e^{-\mathsf{i}(\omega t-\mathsf{k}_z z)}\right]
\end{aligned} \qquad [1.58]$$

Similarly, the surface currents are found using the discontinuity equation for the tangential component of the field $\overrightarrow{\mathbf{H}}$:

$$\begin{aligned}
K_z(x=0) &= H_y(x=0) = \Re\left[\frac{\mathsf{i}\omega\mathsf{k}_x\varepsilon_0}{\mathsf{k}_x^2+\mathsf{k}_y^2} E_{mn}\sin(\mathsf{k}_y y)\, e^{-\mathsf{i}(\omega t-\mathsf{k}_z z)}\right]\\
K_z(x=a) &= -H_y(x=a) = -\Re\left[\frac{\mathsf{i}\omega\mathsf{k}_x\varepsilon_0(-1)^m}{\mathsf{k}_x^2+\mathsf{k}_y^2} E_{mn}\sin(\mathsf{k}_y y)\, e^{-\mathsf{i}(\omega t-\mathsf{k}_z z)}\right]\\
K_z(y=0) &= -H_x(y=0) = \Re\left[\frac{\mathsf{i}\omega\mathsf{k}_y\varepsilon_0}{\mathsf{k}_x^2+\mathsf{k}_y^2} E_{mn}\sin(\mathsf{k}_x x)\, e^{-\mathsf{i}(\omega t-\mathsf{k}_z z)}\right]\\
K_z(y=b) &= H_x(y=b) = -\Re\left[\frac{\mathsf{i}\omega\mathsf{k}_y\varepsilon_0(-1)^n}{\mathsf{k}_x^2+\mathsf{k}_y^2} E_{mn}\sin(\mathsf{k}_x x)\, e^{-\mathsf{i}(\omega t-\mathsf{k}_z z)}\right]
\end{aligned} \qquad [1.59]$$

The following are some important remarks:

1) The surface currents are purely directed along the Oz direction.

2) The charges $\sigma(x=0)$, $\sigma(x=a)$, $\sigma(y=0)$ and $\sigma(y=b)$ are, respectively, related to the currents $K_z(x=0)$, $K_z(x=a)$, $K_z(y=0)$ and $K_z(y=b)$ by charge conservation laws. For example, we can verify that:

$$\frac{\partial K_z(x=0)}{\partial z} + \frac{\partial \sigma(x=0)}{\partial t} = 0 \iff K_z(x=0) = \frac{\omega}{\mathsf{k}_z}\sigma(x=0) \qquad [1.60]$$

etc.

3) If m is even (respectively, odd), the charges and currents on the opposite walls at $x=0$ and $x=a$ have opposite signs (respectively, the same sign).

4) Similarly, if n is even (respectively, odd), the opposite walls at $y = 0$ and $y = b$ carry charges of opposite signs (respectively, the same sign) and are traversed by currents flowing in opposite directions (respectively, in the same direction).

5) The field lines of $\overrightarrow{\mathbf{E}}$ in the Oxy plane are obtained from the following equation:

$$\frac{dy}{dx} = \frac{E_y}{E_x} = \frac{\mathsf{k}_y \tan(\mathsf{k}_x x)}{\mathsf{k}_x \tan(\mathsf{k}_y y)} \Longrightarrow \frac{[\cos(\mathsf{k}_x x)]^{(\mathsf{k}_y/\mathsf{k}_x)^2}}{\cos(\mathsf{k}_y y)} = \text{const.} \qquad [1.61]$$

When the surface charge is positive, the electric field points outward from the wall, whereas it points toward the wall when the surface charge is negative.

6) For the field lines of $\overrightarrow{\mathbf{H}}$ in the Oxy plane, we must solve the equation:

$$\frac{dy}{dx} = \frac{H_y}{H_x} = -\frac{\mathsf{k}_x \tan(\mathsf{k}_y y)}{\mathsf{k}_y \tan(\mathsf{k}_x x)} \Longrightarrow \sin(\mathsf{k}_x x)\sin(\mathsf{k}_y y) = \text{const.} \qquad [1.62]$$

The field configurations for the TM_{11}, TM_{12}, TM_{21} and TM_{22} modes are illustrated in Figure 1.3.

EXAMPLE 1.1.– The total charge $Q_{m,n}(z,t)$ per unit length of the waveguide carried by all walls at time t and position z for TM_{mn} mode is calculated as follows:

$$\begin{aligned} Q_{m,n}(z,t) &= \int_0^a [\sigma(y=0) + \sigma(y=b)]\,dx + \int_0^b [\sigma(x=0) + \sigma(x=a)]\,dy \\ &= \Re\left[\frac{i\mathsf{k}_z\varepsilon_0(1-(-1)^m)(1-(-1)^n)}{mn}\frac{ab}{\pi^2}E_{mn}e^{-i(\omega t - \mathsf{k}_z z)}\right] \end{aligned} \qquad [1.63]$$

The total current intensity $I(z,t)$ on all walls at time t and position z is determined using a formula analogous to [1.63]:

$$\begin{aligned} I_{m,n}(z,t) &= \int_0^a [K_z(y=0) + K_z(y=b)]dx + \int_0^b [K_z(x=0) + K_z(x=a)]dy \\ &= \Re\left[\frac{i\omega\varepsilon_0(1-(-1)^m)(1-(-1)^n)}{mn}\frac{ab}{\pi^2}E_{mn}e^{-i(\omega t - \mathsf{k}_z z)}\right] \end{aligned} \qquad [1.64]$$

It is easy to verify that the quantities $Q_{m,n}(z,t)$ and $I_{m,n}(z,t)$ are related by the continuity equation $\frac{\partial I_{m,n}}{\partial z} + \frac{\partial Q_{m,n}}{\partial t} = 0$.

We see that if m or n is even, then $Q_{m,n}(z,t)$ and $I_{m,n}(z,t)$ are zero. ■

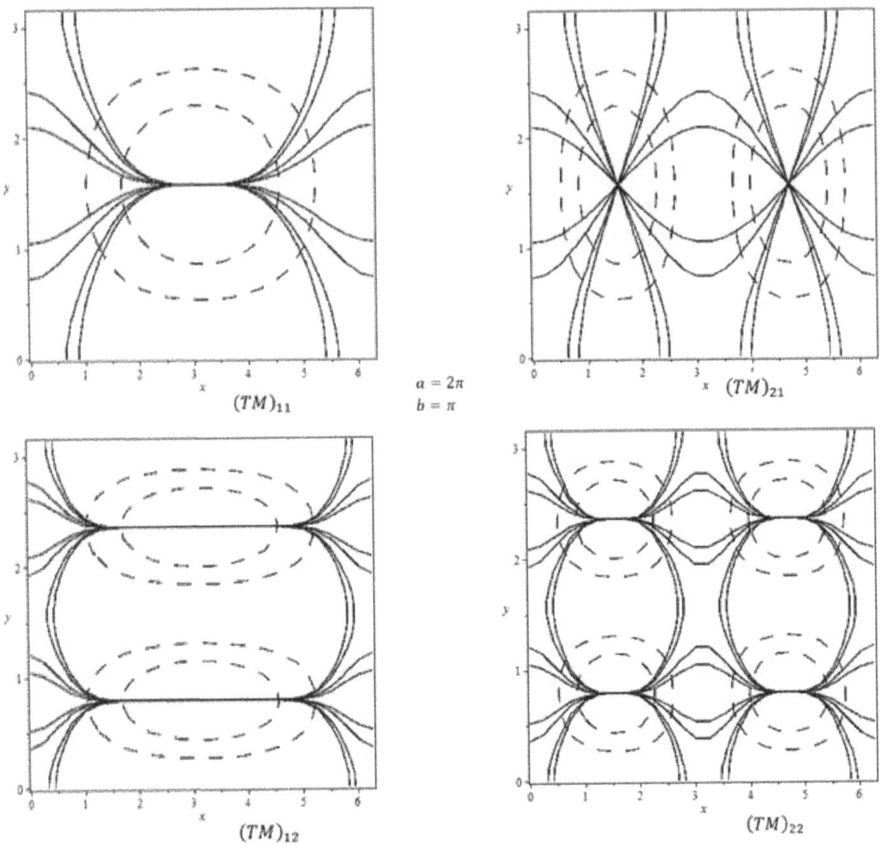

Figure 1.3. *Electric field lines (in blue) and magnetic field lines (dashed) for the TM$_{11}$, TM$_{12}$, TM$_{21}$ and TM$_{22}$= modes in the Oxy plane. For a color version of this figure, see www.iste.co.uk/abdesselam/electrodynamics3.zip*

Substituting previously found solutions [1.54] and [1.55] for fields $\overrightarrow{\mathbf{E}}$ and $\overrightarrow{\mathbf{H}}$ into the expression of the (time-averaged) Poynting vector gives:

$$\begin{aligned}\langle\overrightarrow{\mathbf{\Pi}}\rangle &= \frac{1}{2}\Re\left[\left(E_x\overrightarrow{\mathbf{i}} + E_y\overrightarrow{\mathbf{j}} + E_z\overrightarrow{\mathbf{k}}\right)e^{\mathrm{i}\mathrm{k}_z z} \times \left(H_x^\star\overrightarrow{\mathbf{i}} + H_y^\star\overrightarrow{\mathbf{j}}\right)e^{-\mathrm{i}\mathrm{k}_z^\star z}\right] \\ &= \frac{1}{2}\Re\left[\left(\left(E_x H_y^\star - E_y H_x^\star\right)\overrightarrow{\mathbf{k}} + E_z\left(H_x^\star\overrightarrow{\mathbf{j}} - H_y^\star\overrightarrow{\mathbf{i}}\right)\right)e^{\mathrm{i}(\mathrm{k}_z - \mathrm{k}_z^\star)z}\right]\end{aligned} \qquad [1.65]$$

If ω is below the cutoff frequency of a particular mode, k_z becomes purely imaginary, and the attenuation factor $e^{-|\mathrm{k}_z|z}$ then appears in the expression of the Poynting vector. For propagating modes where k_z is real, there is no dependence on z in [1.65]. In both cases, we see from [1.54] and [1.55] that the products of E_z, respectively, with $H_x^\star$ and $H_y^\star$ are purely imaginary, which means that the real parts of $E_z H_x^\star$ and $E_z H_y^\star$ are zero. Only the component of $\overrightarrow{\mathbf{\Pi}}$ directed along Oz can have a time average. The other two components (in the directions Ox and Oy) of the Poynting vector have zero time averages. Thus, we obtain:

$$\langle\overrightarrow{\mathbf{\Pi}}\rangle = \frac{\omega\varepsilon_0 \left|E_{mn}\right|^2}{2\left(\mathsf{k}_x^2+\mathsf{k}_y^2\right)^2}\Re\left[\mathsf{k}_z e^{\mathrm{i}(\mathsf{k}_z-\mathsf{k}_z^\star)z}\left(\mathsf{k}_x^2\cos^2\left(\mathsf{k}_x x\right)\sin^2\left(\mathsf{k}_y y\right)\right.\right. \\ \left.\left.+\mathsf{k}_y^2\sin^2\left(\mathsf{k}_x x\right)\cos^2\left(\mathsf{k}_y y\right)\right)\right]\overrightarrow{\mathbf{k}} \qquad [1.66]$$

If k_z is purely imaginary, we obtain $\langle\Pi_z\rangle = 0$. On the other hand, if k_z is real, the average power flux over time is not equal to zero. The total power flux directed along the Oz direction is found by integrating [1.66] over the cross-sectional area of the waveguide (here, k_z is assumed to be real):

$$\langle P\rangle = \int_0^a\int_0^b\langle\Pi_z\rangle dxdy = \frac{\omega\varepsilon_0\mathsf{k}_z ab\left|E_{mn}\right|^2}{8\left(\mathsf{k}_x^2+\mathsf{k}_y^2\right)} \qquad [1.67]$$

since m and n are different from zero. To compute [1.67], the following identities were used:

$$\int_0^a\left\{\begin{array}{l}\sin^2\left(\frac{m\pi x}{a}\right)\\ \cos^2\left(\frac{m\pi x}{a}\right)\end{array}\right\}dx = \frac{a}{m\pi}\left(\frac{m\pi x}{2a}\mp\frac{1}{4}\sin\left(\frac{2m\pi x}{a}\right)\right)\Bigg|_0^a = \frac{a}{2} \\ \int_0^b\left\{\begin{array}{l}\sin^2\left(\frac{n\pi y}{b}\right)\\ \cos^2\left(\frac{n\pi y}{b}\right)\end{array}\right\}dy = \frac{b}{n\pi}\left(\frac{n\pi y}{2b}\mp\frac{1}{4}\sin\left(\frac{2n\pi y}{b}\right)\right)\Bigg|_0^b = \frac{b}{2} \qquad [1.68]$$

EXAMPLE 1.2. SCALAR AND VECTOR POTENTIALS OF A TM WAVE – Using the relation between the magnetic field and its vector potential ($\overrightarrow{\mathbf{B}} = \overrightarrow{\nabla}\times\overrightarrow{\mathbf{A}}$), for a TM wave ($H_z = 0$), we obtain:

$$\frac{\partial A_y}{\partial x} = \frac{\partial A_x}{\partial y} \qquad [1.69]$$

which means that there exists a scalar function $F(x,y,z,t)$ defined such that $A_x = \frac{\partial F}{\partial x}$ and $A_y = \frac{\partial F}{\partial y}$. Consequently, the components E_x and E_y of the electric field are written as:

$$E_x = -\frac{\partial V}{\partial x}-\frac{\partial A_x}{\partial t} = -\frac{\partial V}{\partial x}+\mathrm{i}\omega A_x = -\frac{\partial}{\partial x}\left(V-\mathrm{i}\omega F\right) \\ E_y = -\frac{\partial V}{\partial y}-\frac{\partial A_y}{\partial t} = -\frac{\partial V}{\partial y}+\mathrm{i}\omega A_y = -\frac{\partial}{\partial y}\left(V-\mathrm{i}\omega F\right) \qquad [1.70]$$

Let us redefine the electric potential:

$$V' = V-\mathrm{i}\omega F \qquad [1.71]$$

To preserve gauge invariance, the vector potential must also be redefined. Since:

$$E_z = -\mathrm{ik}_z V + \mathrm{i}\omega A_z = -\mathrm{ik}_z V' + \mathrm{i}\omega \left(A_z - \mathrm{ik}_z F\right) \qquad [1.72]$$

we must set $A'_z = A_z - \mathrm{ik}_z F$, which means that $E_z = -\mathrm{ik}_z V' + \mathrm{i}\omega A'_z$. Let us take a function F of the form $F\left(x, y, z, t\right) = F\left(x, y\right) e^{-\mathrm{i}(\omega t - \mathrm{k}_z z)}$:

$$\begin{aligned} B_x &= \frac{\partial A_z}{\partial y} - \frac{\partial A_y}{\partial z} = \frac{\partial A_z}{\partial y} - \frac{\partial^2 F}{\partial y \partial z} = \frac{\partial}{\partial y}\left(A_z - \mathrm{ik}_z F\right) = \frac{\partial A'_z}{\partial y} \\ B_y &= \frac{\partial A_z}{\partial x} - \frac{\partial A_x}{\partial z} = \frac{\partial A_z}{\partial x} - \frac{\partial^2 F}{\partial x \partial z} = \frac{\partial}{\partial x}\left(A_z - \mathrm{ik}_z F\right) = \frac{\partial A'_z}{\partial x} \end{aligned} \qquad [1.73]$$

We thus see that the new vector potential has only one component along Oz. Hence, we obtain:

$$\begin{aligned} E_x &= -\frac{\partial V'}{\partial x}, \quad E_y = -\frac{\partial V'}{\partial y}, \quad E_z = -\mathrm{ik}_z V' + \mathrm{i}\omega A'_z \\ B_x &= \frac{\partial A'_z}{\partial x}, \quad B_y = \frac{\partial A'_z}{\partial y} \end{aligned} \qquad [1.74]$$

We can furthermore impose the Lorenz gauge on the new potentials, $\vec{\nabla} \cdot \vec{\mathbf{A}}' + \frac{1}{c^2}\frac{\partial V'}{\partial t} = \mathrm{ik}_z A'_z - \frac{\mathrm{i}\omega}{c^2} V' = 0$. This yields:

$$A'_z = \frac{\omega}{\mathrm{k}_z c^2} V' \qquad [1.75]$$

and consequently:

$$\begin{aligned} V' &= \frac{-\mathrm{ik}_z}{\frac{\omega^2}{c^2} - \mathrm{k}_z^2} E_z \\ A'_z &= \frac{-\mathrm{i}\omega}{\omega^2 - c^2 \mathrm{k}_z^2} E_z \end{aligned} \qquad [1.76]$$

It is easy to verify that these potentials also satisfy the Helmholtz wave equation.

1.3.1.2. *TE waves ($E_z = 0$, $H_z \neq 0$)*

When the electric field is entirely in the Oxy plane (transverse field), we must first solve [1.22] for H_z. As in TM modes, we propose a solution of the form $H_z = \Re\left[H_z\left(x, y\right) e^{-\mathrm{i}(\omega t - \mathrm{k}_z z)}\right]$. This yields the following:

$$\left(\frac{\partial^2}{\partial x^2} + \frac{\partial^2}{\partial x^2} + \gamma^2\right) H_z = 0 \qquad [1.77]$$

In the same way, this equation is solved by separation of variables to obtain a solution of the same form as [1.44]:

$$H_z(x,y) = \left[A_1 \sin(\mathsf{k}_x x) + A_2 \cos(\mathsf{k}_x x)\right]\left[B_1 \sin(\mathsf{k}_y y) + B_2 \cos(\mathsf{k}_y y)\right] \qquad [1.78]$$

Since $\vec{\mathbf{H}}_\perp$ must be zero in the vicinity of the walls, we must require that, in accordance with [1.37]:

$$\begin{aligned} \left.\frac{\partial H_z}{\partial x}\right|_{x=0} &= \left.\frac{\partial H_z}{\partial x}\right|_{x=a} = 0 \\ \left.\frac{\partial H_z}{\partial y}\right|_{y=0} &= \left.\frac{\partial H_z}{\partial y}\right|_{y=b} = 0 \end{aligned} \qquad [1.79]$$

which implies that:

$$\begin{cases} \left[A_1 \mathsf{k}_x \cos(\mathsf{k}_x x) - A_2 \mathsf{k}_x \sin(\mathsf{k}_x x)\right]\big|_{x=0} = 0 \\ \left[B_1 \mathsf{k}_y \cos(\mathsf{k}_y y) - B_2 \mathsf{k}_y \sin(\mathsf{k}_y y)\right]\big|_{y=0} = 0 \end{cases} \Longrightarrow \begin{cases} A_1 = 0 \\ B_1 = 0 \end{cases} \qquad [1.80]$$

and (to obtain a non-trivial solution):

$$\begin{cases} -A_2 \mathsf{k}_x \sin(\mathsf{k}_x a) = 0 \\ -B_2 \mathsf{k}_y \sin(\mathsf{k}_y b) = 0 \end{cases} \Longrightarrow \begin{cases} \mathsf{k}_x = \dfrac{m\pi}{a} \\ \mathsf{k}_y = \dfrac{n\pi}{b} \end{cases} \qquad [1.81]$$

Thus, the desired solution (the *eigenmode of the waveguide denoted TE$_{mn}$*) is:

$$H_z = \Re\left[H_{mn} \cos\left(\frac{m\pi x}{a}\right) \cos\left(\frac{n\pi y}{b}\right) e^{-\mathrm{i}(\omega t - \mathsf{k}_z z)}\right] \qquad [1.82]$$

We observe the following:

– As with the TM mode, k_x and k_y are quantized.

– k_z is given as a function of ω, k_x, and k_y by $\mathsf{k}_z = \sqrt{\frac{\omega^2}{c^2} - \mathsf{k}_x^2 - \mathsf{k}_y^2} = \sqrt{\frac{\omega^2}{c^2} - \left(\frac{m\pi}{a}\right)^2 - \left(\frac{n\pi}{b}\right)^2}$ (dispersion relation).

– k_z can be either real or purely imaginary. If k_z is real, the wave propagates purely. If k_z is purely imaginary, the wave is rapidly attenuated. The transition between pure propagation and attenuation occurs when $|\mathsf{k}_z| = 0$, at the *cutoff frequency* $\omega_c = c\sqrt{\left(\frac{m\pi}{a}\right)^2 + \left(\frac{n\pi}{b}\right)^2}$. This frequency varies for each mode (depending on m and n).

– For TE modes, either m or n (but not both) can be zero and still yield a non-trivial solution. When n equals zero, the fields do not vary with respect to y, and the electric field is purely directed along the Oy axis, while the magnetic excitation field is directed along Ox. If $a > b$, the lowest frequency occurs for the TE$_{10}$ mode, which is called the *dominant or fundamental mode*. No mode can propagate below this lowest critical frequency $\omega_{c0} = \frac{\pi c}{a}$.

– To calculate the other components of the fields $\vec{\mathbf{E}}$ and $\vec{\mathbf{H}}$, we use relations [1.19] and [1.21] with the restriction $E_z = 0$:

$$\vec{\mathbf{E}}_\perp = \frac{-\mathrm{i}\omega\mu_0}{\mathsf{k}_x^2+\mathsf{k}_y^2}\vec{\mathbf{k}} \times \vec{\nabla}_\perp H_z \qquad \vec{\mathbf{H}}_\perp = \frac{\mathrm{i}\mathsf{k}_z}{\mathsf{k}_x^2+\mathsf{k}_y^2}\vec{\nabla}_\perp H_z \qquad [1.83]$$

which gives:

$$E_x = \Re\left[-\frac{\mathrm{i}\omega\mu_0\mathsf{k}_y}{\mathsf{k}_x^2+\mathsf{k}_y^2}H_{mn}\cos(\mathsf{k}_x x)\sin(\mathsf{k}_y y)\,e^{-\mathrm{i}(\omega t-\mathsf{k}_z z)}\right]$$
$$E_y = \Re\left[\frac{\mathrm{i}\omega\mu_0\mathsf{k}_x}{\mathsf{k}_x^2+\mathsf{k}_y^2}H_{mn}\sin(\mathsf{k}_x x)\cos(\mathsf{k}_y y)\,e^{-\mathrm{i}(\omega t-\mathsf{k}_z z)}\right] \qquad [1.84]$$
$$E_z = 0$$

and:

$$H_x = \Re\left[-\frac{\mathrm{i}\mathsf{k}_z\mathsf{k}_x}{\mathsf{k}_x^2+\mathsf{k}_y^2}H_{mn}\sin(\mathsf{k}_x x)\cos(\mathsf{k}_y y)\,e^{-\mathrm{i}(\omega t-\mathsf{k}_z z)}\right]$$
$$H_y = \Re\left[-\frac{\mathrm{i}\mathsf{k}_z\mathsf{k}_y}{\mathsf{k}_x^2+\mathsf{k}_y^2}H_{mn}\cos(\mathsf{k}_x x)\sin(\mathsf{k}_y y)\,e^{-\mathrm{i}(\omega t-\mathsf{k}_z z)}\right] \qquad [1.85]$$
$$H_z = \Re\left[H_{mn}\cos(\mathsf{k}_x x)\cos(\mathsf{k}_y y)\,e^{-\mathrm{i}(\omega t-\mathsf{k}_z z)}\right]$$

As expected, we see that the tangential components of the electric field vanish at the walls.

– The electric field lines of $\vec{\mathbf{E}}$ in the Oxy plane:

$$\frac{dy}{dx} = \frac{E_y}{E_x} = -\frac{\mathsf{k}_x}{\mathsf{k}_y}\frac{\tan(\mathsf{k}_x x)}{\tan(\mathsf{k}_y y)} \implies \cos(\mathsf{k}_x x)\cos(\mathsf{k}_y y) = \text{const.} \qquad [1.86]$$

– The electric field lines of $\vec{\mathbf{H}}$ in the Oxy plane:

$$\frac{dy}{dx} = \frac{H_y}{H_x} = \frac{\mathsf{k}_y}{\mathsf{k}_x}\frac{\tan(\mathsf{k}_y y)}{\tan(\mathsf{k}_x x)} \implies \frac{[\sin(\mathsf{k}_x x)]^{(\mathsf{k}_y/\mathsf{k}_x)^2}}{\sin(\mathsf{k}_y y)} = \text{const.} \qquad [1.87]$$

– From these expressions, the surface charge and current on the walls are easily determined:

- face $x = 0, 0 \leq y \leq b$:

$$\begin{aligned}
\sigma_1 &= \varepsilon_0 E_x(x=0) = \Re\left[-\frac{\mathsf{i}\omega \mathsf{k}_y}{c^2\left(\mathsf{k}_x^2+\mathsf{k}_y^2\right)} H_{mn} \sin\left(\mathsf{k}_y y\right) e^{-\mathsf{i}(\omega t - \mathsf{k}_z z)}\right] \\
K_{1,z} &= H_y(x=0) = \Re\left[-\frac{\mathsf{i}\mathsf{k}_y \mathsf{k}_z}{\mathsf{k}_x^2+\mathsf{k}_y^2} H_{mn} \sin\left(\mathsf{k}_y y\right) e^{-\mathsf{i}(\omega t - \mathsf{k}_z z)}\right] \\
K_{1,y} &= -H_z(x=0) = \Re\left[-H_{mn} \cos\left(\mathsf{k}_y y\right) e^{-\mathsf{i}(\omega t - \mathsf{k}_z z)}\right] \\
&\frac{\partial K_{1,y}}{\partial y} + \frac{\partial K_{1,z}}{\partial z} + \frac{\partial \sigma_1}{\partial t} = 0
\end{aligned} \tag{1.88}$$

- face $x = a, 0 \leq y \leq b$:

$$\begin{aligned}
\sigma_2 &= -\varepsilon_0 E_x(x=a) = \Re\left[\frac{\mathsf{i}\omega \mathsf{k}_y (-1)^m}{c^2\left(\mathsf{k}_x^2+\mathsf{k}_y^2\right)} H_{mn} \sin\left(\mathsf{k}_y y\right) e^{-\mathsf{i}(\omega t - \mathsf{k}_z z)}\right] \\
K_{2,z} &= -H_y(x=a) = \Re\left[\frac{\mathsf{i}\mathsf{k}_y \mathsf{k}_z (-1)^m}{\mathsf{k}_x^2+\mathsf{k}_y^2} H_{mn} \sin\left(\mathsf{k}_y y\right) e^{-\mathsf{i}(\omega t - \mathsf{k}_z z)}\right] \\
K_{2,y} &= H_z(x=a) = \Re\left[(-1)^m H_{mn} \cos\left(\mathsf{k}_y y\right) e^{-\mathsf{i}(\omega t - \mathsf{k}_z z)}\right] \\
&\frac{\partial K_{2,y}}{\partial y} + \frac{\partial K_{2,z}}{\partial z} + \frac{\partial \sigma_2}{\partial t} = 0
\end{aligned} \tag{1.89}$$

- face $y = 0, 0 \leq x \leq a$:

$$\begin{aligned}
\sigma_3 &= \varepsilon_0 E_y(y=0) = \Re\left[\frac{\mathsf{i}\omega \mathsf{k}_x}{c^2\left(\mathsf{k}_x^2+\mathsf{k}_y^2\right)} H_{mn} \sin\left(\mathsf{k}_x x\right) e^{-\mathsf{i}(\omega t - \mathsf{k}_z z)}\right] \\
K_{3,z} &= -H_x(y=0) = \Re\left[\frac{\mathsf{i}\mathsf{k}_x \mathsf{k}_z}{\mathsf{k}_x^2+\mathsf{k}_y^2} H_{mn} \sin\left(\mathsf{k}_x x\right) e^{-\mathsf{i}(\omega t - \mathsf{k}_z z)}\right] \\
K_{3,x} &= H_z(y=0) = \Re\left[H_{mn} \cos\left(\mathsf{k}_x x\right) e^{-\mathsf{i}(\omega t - \mathsf{k}_z z)}\right] \\
&\frac{\partial K_{3,x}}{\partial x} + \frac{\partial K_{3,z}}{\partial z} + \frac{\partial \sigma_3}{\partial t} = 0
\end{aligned} \tag{1.90}$$

- face $y = b, 0 \leq x \leq a$:

$$\begin{aligned}
\sigma_4 &= -\varepsilon_0 E_y(y=b) = \Re\left[-\frac{\mathsf{i}\omega \mathsf{k}_x (-1)^n}{c^2\left(\mathsf{k}_x^2+\mathsf{k}_y^2\right)} H_{mn} \sin\left(\mathsf{k}_x x\right) e^{-\mathsf{i}(\omega t - \mathsf{k}_z z)}\right] \\
K_{4,z} &= H_x(y=b) = \Re\left[-\frac{\mathsf{i}\mathsf{k}_x \mathsf{k}_z (-1)^n}{\mathsf{k}_x^2+\mathsf{k}_y^2} H_{mn} \sin\left(\mathsf{k}_x x\right) e^{-\mathsf{i}(\omega t - \mathsf{k}_z z)}\right] \\
K_{4,x} &= -H_z(y=b) = \Re\left[-(-1)^n H_{mn} \cos\left(\mathsf{k}_x x\right) e^{-\mathsf{i}(\omega t - \mathsf{k}_z z)}\right] \\
&\frac{\partial K_{4,x}}{\partial x} + \frac{\partial K_{4,z}}{\partial z} + \frac{\partial \sigma_4}{\partial t} = 0
\end{aligned} \tag{1.91}$$

– The charge $Q_{12}(z,t)$ per unit length of the waveguide carried by the walls located at $x=0$ and $x=a$, at time t and position z, is equal to:

$$\begin{aligned} Q_{12}(z,t) &= \int_0^b [\sigma_1+\sigma_2]\,dy \\ &= \frac{\omega H_{mn}\left(1-(-1)^m\right)\left(1-(-1)^n\right)}{c^2\left(\mathsf{k}_x^2+\mathsf{k}_y^2\right)}\sin\left(\omega t-\mathsf{k}_z z\right) \end{aligned} \qquad [1.92]$$

while the charge $Q_{34}(z,t)$, carried by the walls located at $y=0$ and $y=b$, is the opposite. Indeed, we have:

$$\begin{aligned} Q_{34}(z,t) &= \int_0^a [\sigma_3+\sigma_4]\,dx \\ &= -\frac{\omega H_{mn}\left(1-(-1)^m\right)\left(1-(-1)^n\right)}{c^2\left(\mathsf{k}_x^2+\mathsf{k}_y^2\right)}\sin\left(\omega t-\mathsf{k}_z z\right) \end{aligned} \qquad [1.93]$$

– The total current $I_{12}(z,t)$ (in the direction of propagation) on the faces $x=0$ and $x=a$ is given by:

$$\begin{aligned} I_{12}(z,t) &= \int_0^b [K_{1,z}+K_{2,z}]\,dy \\ &= \frac{\mathsf{k}_z H_{mn}\left(1-(-1)^m\right)\left(1-(-1)^n\right)}{\mathsf{k}_x^2+\mathsf{k}_y^2}\sin\left(\omega t-\mathsf{k}_z z\right) \\ &= \frac{\mathsf{k}_z c^2}{\omega}Q_{12}(z,t) \end{aligned} \qquad [1.94]$$

and the total current $I_{34}(z,t)=\frac{\mathsf{k}_z c^2}{\omega}Q_{34}(z,t)$ on the faces $y=0$ and $y=b$ is the opposite, $I_{34}(z,t)=-I_{12}(z,t)$. The total current on the faces $x=0$ and $x=a$ (respectively, $y=0$ and $y=b$) along Oy (respectively, Ox) is zero.

– The time average over a period of the Poynting vector for the TE wave is given by:

$$\begin{aligned} \langle\vec{\mathbf{\Pi}}\rangle &= \frac{1}{2}\Re\left[\left(E_x\vec{\mathbf{i}}+E_y\vec{\mathbf{j}}\right)e^{i\mathsf{k}_z z}\times\left(H_x^\star\vec{\mathbf{i}}+H_y^\star\vec{\mathbf{j}}+H_z^\star\vec{\mathbf{k}}\right)e^{-i\mathsf{k}_z^\star z}\right] \\ &= \frac{1}{2}\Re\left[\left(\left(E_xH_y^\star-E_yH_x^\star\right)\vec{\mathbf{k}}+H_z^\star\left(E_y\vec{\mathbf{i}}-E_x\vec{\mathbf{j}}\right)\right)e^{i(\mathsf{k}_z-\mathsf{k}_z^\star)z}\right] \end{aligned} \qquad [1.95]$$

The products $H_z^\star E_y$ and $H_z^\star E_x$ are purely imaginary; this implies that the time-averaged components of the Poynting vector in the Ox and Oy directions are zero. It therefore follows that:

$$\langle \Pi_z \rangle = \frac{\omega\mu_0 |H_{mn}|^2}{2\left(\mathsf{k}_x^2+\mathsf{k}_y^2\right)^2}\Re\left[\mathsf{k}_z e^{\mathrm{i}(\mathsf{k}_z-\mathsf{k}_z^\star)z}\left(\mathsf{k}_x^2\sin^2(\mathsf{k}_x x)\cos^2(\mathsf{k}_y y)\right.\right.$$
$$\left.\left.+\mathsf{k}_y^2\cos^2(\mathsf{k}_x x)\sin^2(\mathsf{k}_y y)\right)\right] \qquad [1.96]$$

It is clear that the average $\langle \Pi_z \rangle$ is zero when k_z is purely imaginary (attenuation mode). For the propagation mode, we have $\langle \vec{\mathbf{\Pi}} \rangle \neq 0$. The power flux through the cross-sectional area of the waveguide is:

$$\langle P \rangle = \int_0^a \int_0^v \langle \Pi_z \rangle dxdy = \begin{cases} \dfrac{\omega\mu_0\mathsf{k}_z |H_{mn}|^2 ab}{8\left(\mathsf{k}_x^2+\mathsf{k}_y^2\right)} & m, n \neq 0 \\ \dfrac{\omega\mu_0\mathsf{k}_z |H_{mn}|^2 ab}{4\left(\mathsf{k}_x^2+\mathsf{k}_y^2\right)} & m \text{ or } n = 0 \end{cases} \qquad [1.97]$$

where the identities [1.68] were used once again. Note the factor of 2 in [1.97] for the TE_{m0} and TE_{0n} modes. The two indices m and n cannot both be zero at the same time because the TE_{00} mode reduces to a trivial magnetic field directed along Oz that is constant in space.

EXAMPLE 1.3. TEM WAVE INSIDE A RECTANGULAR WAVEGUIDE WITH PERFECTLY CONDUCTING WALLS.– We will see below that a TEM wave cannot propagate inside a hollow waveguide. We have already seen that for a TEM wave to exist, it is necessary that $\gamma^2 = 0$, that is, $\mathsf{k}_z^2 = \varepsilon_0\mu_0\omega^2 = \frac{\omega^2}{c^2}$, and that the transverse fields $\vec{\mathbf{E}}_\perp$ and $\vec{\mathbf{H}}_\perp$ satisfy equations [1.26], [1.28] and [1.30]. From equation $\vec{\nabla}_\perp \times \vec{\mathbf{E}}_\perp = 0$, we can assert that there must exist a function $V(x, y)$ (actually, the electric potential) such that $\vec{\mathbf{E}}_\perp = -\vec{\nabla}_\perp V$. Furthermore, this potential must satisfy the 2D Laplace equation: $\Delta_\perp V = \frac{\partial^2 V}{\partial x^2} + \frac{\partial^2 V}{\partial y^2} = 0$, since $\vec{\nabla}_\perp \cdot \vec{\mathbf{E}}_\perp = 0$. In the same way, this equation can be solved using the method of separation of variables, that is, taking $V(x, y) = X(x)Y(y)$, which gives $\frac{X''}{X} = -\frac{Y''}{Y} = -\alpha^2$, where α can, in principle, be complex. We immediately deduce the form of $V(x, y)$:

$$V(x, y) = \begin{cases} (A_1\sin\alpha x + A_2\cos\alpha x)\left(B_1 e^{\alpha y} + B_2 e^{-\alpha y}\right) & \alpha \neq 0 \\ (A_1' x + A_2')(B_1 y + B_2) & \alpha = 0 \end{cases} \qquad [1.98]$$

The constants A_1, A_2, B_1, B_2, A_1', A_2', B_1' and B_2' must be determined from the boundary conditions (satisfied by the field):

$$\left.\frac{\partial V}{\partial x}\right|_{x,y=0} = \left.\frac{\partial V}{\partial x}\right|_{x,y=b} = 0$$
$$\left.\frac{\partial V}{\partial y}\right|_{x=0,y} = \left.\frac{\partial V}{\partial y}\right|_{x=a,y} = 0 \qquad [1.99]$$

which leads to $B_1 = -B_2$, $B_1\left(e^{\alpha b} - e^{-\alpha b}\right) = 0$, $A_2 = 0$, $A_1 \sin \alpha a = 0$, $A_1' = 0$, $A_2' = 0$, $B_1' = 0$ and $B_2' = 0$. We observe that for $\alpha = 0$, the solution is identically zero. To obtain a non-trivial solution when $\alpha \neq 0$, the parameter α must satisfy the constraints $e^{2\alpha b} = 1$ and $\sin \alpha a = 0$. These two constraints cannot be verified simultaneously, which means that the constants A_1, A_2, B_1 and B_2 must be equal to zero. In both cases, the result is a zero field: a TEM wave cannot propagate inside this waveguide.

1.3.2. *Attenuation coefficient in a lossy waveguide*

In real conductors with high but finite conductivity, the fields no longer become zero instantaneously upon entering the conducting walls but instead decay over a distance δ (called *skin depth*).

To approximately estimate the fields inside the walls, we proceed as follows (the subscript c will be used to denote the fields inside the walls): (1) inside the waveguide, the fields are assumed to be described by the mode expressions obtained for perfectly conducting walls; (2) on the boundaries (the walls), the tangential component of the magnetic field $\vec{\mathbf{H}}_{\parallel}$ is continuous (due to the absence of surface currents); (3) inside the walls, the magnetic field $\vec{\mathbf{H}}_c$ decays exponentially in the direction perpendicular to the boundary and attenuates completely over a distance on the order of $\delta = \sqrt{\frac{2}{\gamma_c \mu_0 \omega}}$. Thus, the field $\vec{\mathbf{H}}_c$ is given by:

$$\vec{\mathbf{H}}_c = \left(\vec{\mathbf{H}}_{\parallel}\right)_S e^{-|\vec{\mathbf{n}}\cdot(\vec{\mathbf{r}}-\vec{\mathbf{r}}_S)|/\delta} \qquad [1.100]$$

where $\vec{\mathbf{n}}$ is the normal to the boundary surface S; (4) inside the walls, the electric field $\vec{\mathbf{E}}_c$ is calculated using a formula analogous to equation [7.134] (Volume 2):

$$\vec{\mathbf{E}}_c = (1-\mathrm{i})\sqrt{\frac{\omega\mu_0}{2\gamma_c}}\,\vec{\mathbf{n}} \times \vec{\mathbf{H}}_c = (1-\mathrm{i})\sqrt{\frac{\omega\mu_0}{2\gamma_c}}\left(\vec{\mathbf{n}} \times \vec{\mathbf{H}}_{\parallel}\right)_S e^{-|\vec{\mathbf{n}}\cdot(\vec{\mathbf{r}}-\vec{\mathbf{r}}_S)|/\delta} \qquad [1.101]$$

where γ_c is the conductivity. Near the surface, the electric field $\vec{\mathbf{E}}_c$ is related to the current density $\vec{\mathbf{J}}$ by $\vec{\mathbf{J}} = \gamma_c \vec{\mathbf{E}}_c$. This leads to a power loss (averaged over one period) in the volume τ:

$$\langle P_d \rangle = \frac{1}{2}\int_\tau \vec{\mathbf{E}}_c \cdot \vec{\mathbf{J}}^\star d\tau = \frac{1}{2}\int_\tau \gamma_c \left|\vec{\mathbf{E}}_c\right|^2 d\tau \qquad [1.102]$$

Since $\gamma_c \left|\vec{\mathbf{E}}_c\right|^2 = \mu_0 \omega \left|\vec{\mathbf{H}}_c\right|^2$, the power loss integral [1.102] can also be rewritten in the form:

$$\langle P_d \rangle = \frac{\mu_0 \omega}{2} \int_{\tau} \left|\vec{\mathbf{H}}_c\right|^2 d\tau \qquad [1.103]$$

From the real part of the complex Poynting theorem derived in section 7.2.4 of Volume 2, the divergence of the time-averaged electromagnetic power flux can be related to the time-averaged dissipated power as follows:

$$\vec{\nabla} \cdot \langle \vec{\mathbf{\Pi}} \rangle = -\langle P_d \rangle \qquad [1.104]$$

Integrating over the portion with thickness Δz of the waveguide, as shown in Figure 1.4(a), and then using the divergence theorem, we obtain:

$$\int_{\tau} \vec{\nabla} \cdot \langle \vec{\mathbf{\Pi}} \rangle d\tau = \oint_S \langle \vec{\mathbf{\Pi}} \rangle \cdot d\vec{\mathbf{S}} = \int_{z+\Delta z} \langle \Pi_z (z + \Delta z) \rangle dS - \int_z \langle \Pi_z (z) \rangle dS \qquad [1.105]$$

Since $\langle P(z) \rangle = \int_z \langle \Pi_z (z) \rangle dS$ (the total power passing through a cross-section of the guide located at position z), equation [1.105] becomes:

$$\langle P (z + \Delta z) \rangle - \langle P(z) \rangle = -\int_{\tau} \langle p_d \rangle dx dy dz \qquad [1.106]$$

Dividing by Δz and then taking the limit as $\Delta z \longrightarrow 0$, we obtain:

$$\lim_{\Delta z \longrightarrow 0} \frac{\langle P (z + \Delta z) \rangle - \langle P(z) \rangle}{\Delta z} = \frac{d\langle P \rangle}{dz} = -\int_{\tau} \langle p_d \rangle dx dy = -\langle P_{dL} \rangle \qquad [1.107]$$

where $\langle P_{dL} \rangle$ is the power dissipated per unit length. Since the fields vary as $e^{-\alpha z}$ (where α is called the *attenuation coefficient* along the waveguide), the power flux, which is proportional to the square of the fields, must vary as $e^{-2\alpha z}$; this implies that:

$$\frac{d\langle P \rangle}{dz} = -2\alpha \langle P \rangle = -\langle P_{dL} \rangle \qquad [1.108]$$

We see that α is proportional to the ratio between the power dissipated per unit length and the total electromagnetic power propagating inside the waveguide:

$$\alpha = -\frac{1}{2\langle P \rangle} \frac{d\langle P \rangle}{dz} = \frac{1}{2} \frac{\langle P_{dL} \rangle}{\langle P \rangle} \qquad [1.109]$$

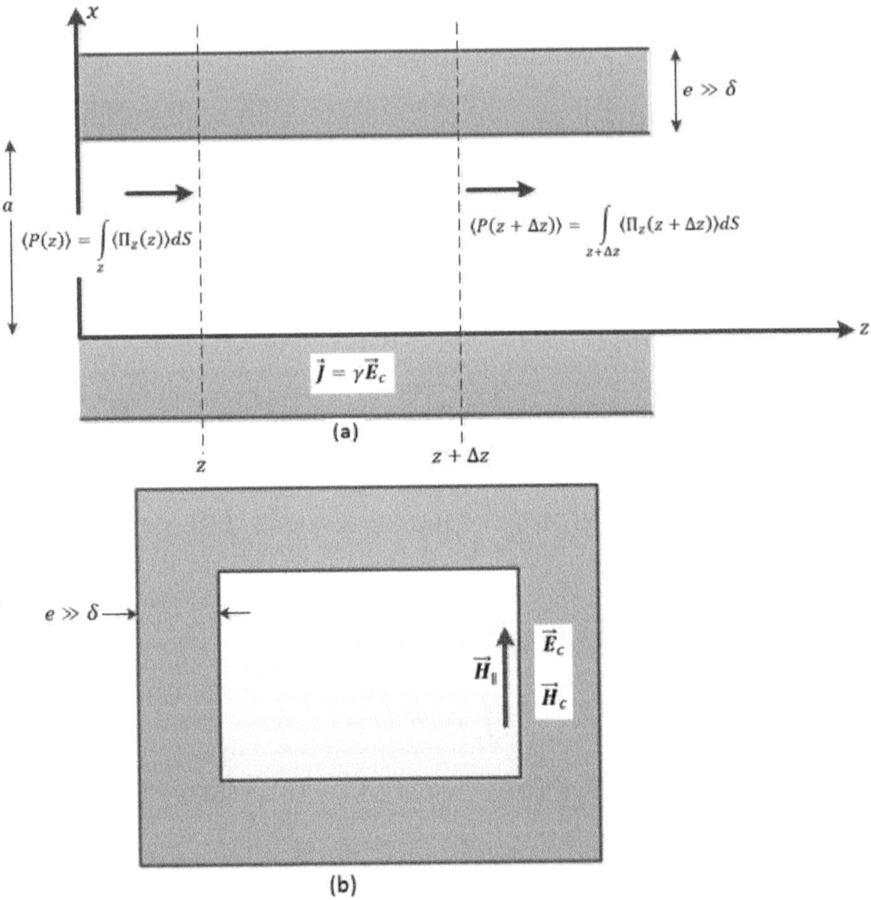

Figure 1.4. *Attenuation of a guided wave in a rectangular lossy waveguide*

Calculate $\langle P_{dL} \rangle$ for a rectangular waveguide with wall thickness e, as illustrated in Figure 1.4(b). We assume that $e \gg \delta$. The surface integral $\int_z \langle P_{dL} \rangle dxdy$ can be divided into four distinct integrals, one for each side.

Since the skin depth is much smaller than the dimensions of the waveguide, the contribution from the corners is neglected. Examine the contribution to the dissipated power of the wall at $x = a$. Inside this wall, the magnetic excitation field $\overrightarrow{\mathbf{H}}_c$ takes the form $\overrightarrow{\mathbf{H}}_c = \overrightarrow{\mathbf{H}}_{\|}(x = a)\, e^{-(x-a)/\delta}$. Consequently, the power dissipated at this wall $(x = a)$, assuming $e \gg \delta$, is given by:

$$\langle P_{dL}\rangle|_{x=a} = \frac{\mu_0\omega}{2}\int_{y=0}^{b}\int_{x=a}^{a+e}\left|\overrightarrow{\mathbf{H}}_{\parallel}(x=a)\right|^2 e^{-2(x-a)/\delta}dxdy \quad [1.110]$$

$$\simeq \frac{\mu_0\omega}{2}\int_{y=0}^{b}\int_{x=a}^{\infty}\left|\overrightarrow{\mathbf{H}}_{\parallel}(x=a)\right|^2 e^{-2(x-a)/\delta}dxdy = \frac{\mu_0\omega\delta}{4}\int_{y=0}^{b}\left|\overrightarrow{\mathbf{H}}_{\parallel}(x=a)\right|^2 dy$$

Each of the four sides contributes similarly to [1.110]. Consequently, the total (time-averaged) power dissipated per unit length is:

$$\langle P_{dL}\rangle = \frac{\mu_0\omega\delta}{4}\oint\left|\overrightarrow{\mathbf{H}}_{\parallel}\right|^2 dl \quad [1.111]$$

where the integral is taken around the perimeter of the waveguide. The field $\overrightarrow{\mathbf{H}}_{\parallel}$ can be replaced in the previous equation by the surface current $\overrightarrow{\mathbf{K}}$ of the perfect conductor using the relation $\left|\overrightarrow{\mathbf{K}}\right|^2 = \left|\overrightarrow{\mathbf{n}}\times\overrightarrow{\mathbf{H}}_{\parallel}\right|^2 = \left|\overrightarrow{\mathbf{H}}_{\parallel}\right|^2$. By making this substitution, we obtain:

$$\langle P_{dL}\rangle = \frac{\mu_0\omega\delta}{4}\oint\left|\overrightarrow{\mathbf{K}}\right|^2 dl \quad [1.112]$$

This result has a simple interpretation: assuming that the surface current $\overrightarrow{\mathbf{K}}$ is uniformly distributed over a strip of thickness δ (and zero elsewhere), we write $\overrightarrow{\mathbf{J}} = \frac{\overrightarrow{\mathbf{K}}}{\delta}$. Thus, the total power dissipated per unit length can then be calculated using a formula such as the following:

$$\langle P_{dL}\rangle = \frac{1}{2}\oint\overrightarrow{\mathbf{E}}\cdot\overrightarrow{\mathbf{J}}^{\star}\delta dl = \frac{1}{2\gamma_c}\oint\left|\overrightarrow{\mathbf{J}}\right|^2\delta dl = \frac{1}{2\gamma_c\delta}\oint\left|\overrightarrow{\mathbf{K}}\right|^2 dl \quad [1.113]$$

which is consistent with the result [1.112]. The above results also apply to circular waveguides or waveguides of any cross-section. $R_S = \frac{1}{\gamma_c\delta}$ is called *surface resistance*. Here are the main results:

– *waves TM$_{mn}$*: by substituting results [1.59] into [1.113], we obtain:

$$\begin{aligned}\langle P_{dL}\rangle &= \frac{1}{\gamma_c\delta}\int_0^b\frac{\omega^2\mathsf{k}_x^2\varepsilon_0^2\left|E_{mn}\right|^2}{\left(\mathsf{k}_x^2+\mathsf{k}_y^2\right)^2}\sin^2(\mathsf{k}_y y)\,dy \\ &\quad+\frac{1}{\gamma_c\delta}\int_0^a\frac{\omega^2\mathsf{k}_y^2\varepsilon_0^2\left|E_{mn}\right|^2}{\left(\mathsf{k}_x^2+\mathsf{k}_y^2\right)^2}\sin^2(\mathsf{k}_x x)\,dx \\ &= \frac{\left|E_{mn}\right|^2}{2\gamma_c\delta}\frac{\omega^2\varepsilon_0^2\left(\mathsf{k}_x^2 b+\mathsf{k}_y^2 a\right)}{\left(\mathsf{k}_x^2+\mathsf{k}_y^2\right)^2}\end{aligned} \quad [1.114]$$

where we have taken into account the identities [1.68]. Consequently, the attenuation coefficient of the TM_{mn} wave is equal to:

$$\alpha = \frac{\langle P_{dL}\rangle}{2\langle P\rangle} = \frac{2\varepsilon_0\omega}{\gamma_c\delta\mathsf{k}_z ab}\left(\frac{m^2b^3 + n^2a^3}{m^2b^2 + n^2a^2}\right) \quad [1.115]$$

– *waves* TE_{mn} $(m, n \neq 0)$: taking into account the results [1.88]–[1.91], we find:

$$\begin{aligned}\langle P_{dL}\rangle &= \frac{1}{\gamma_c\delta}\int_0^b \left\{\frac{\mathsf{k}_y^2\mathsf{k}_z^2\left|H_{mn}\right|^2}{\left(\mathsf{k}_x^2+\mathsf{k}_y^2\right)^2}\sin^2\left(\mathsf{k}_y y\right) + \left|H_{mn}\right|^2\cos^2\left(\mathsf{k}_y y\right)\right\} dy + \\ &\quad \frac{1}{\gamma_c\delta}\int_0^a \left\{\frac{\mathsf{k}_x^2\mathsf{k}_z^2\left|H_{mn}\right|^2}{\left(\mathsf{k}_x^2+\mathsf{k}_y^2\right)^2}\sin^2\left(\mathsf{k}_x x\right) + \left|H_{mn}\right|^2\cos^2\left(\mathsf{k}_x x\right)\right\} dx \\ &= \frac{\left|H_{mn}\right|^2}{2\gamma_c\delta}\left[\frac{\mathsf{k}_z^2\left(\mathsf{k}_x^2 a + \mathsf{k}_y^2 b\right)}{\left(\mathsf{k}_x^2+\mathsf{k}_y^2\right)^2} + a + b\right]\end{aligned} \quad [1.116]$$

and consequently:

$$\alpha = \frac{\langle P_{dL}\rangle}{2\langle P\rangle} = \frac{2}{\gamma_c\delta\mu_0\omega\mathsf{k}_z ab}\left[\frac{\mathsf{k}_z^2\left(m^2b + n^2a\right)ab}{m^2b^2 + n^2a^2} + \pi^2\frac{(a+b)\left(m^2b^2+n^2a^2\right)}{a^2b^2}\right] \quad [1.117]$$

– *waves* TE_{m0} $(m \neq 0)$:

$$\begin{aligned}\langle P_{dL}\rangle &= \frac{1}{\gamma_c\delta}\left[\int_0^b \left|H_{m0}\right|^2 dy + \int_0^a \left\{\frac{\mathsf{k}_z^2\left|H_{m0}\right|^2}{\mathsf{k}_x^2}\sin^2\left(\mathsf{k}_x x\right)\right.\right. \\ &\quad \left.\left. + \left|H_{m0}\right|^2\cos^2\left(\mathsf{k}_x x\right)\right\} dx\right] = \frac{\left|H_{m0}\right|^2}{\gamma_c\delta}\left[b + \left(\frac{\omega}{c\mathsf{k}_x}\right)^2\frac{a}{2}\right]\end{aligned} \quad [1.118]$$

$$\alpha = \frac{\langle P_{dL}\rangle}{2\langle P\rangle} = \frac{2\left(\frac{m\pi}{a}\right)^2\left[b + \left(\frac{\omega a}{m\pi c}\right)^2\frac{a}{2}\right]}{\gamma_c\delta\mu_0\omega\mathsf{k}_z ab} \quad [1.119]$$

– *waves* TE_{0n} $(n \neq 0)$:

$$\langle P_{dL}\rangle = \frac{\left|H_{0n}\right|^2}{\gamma_c\delta}\left[a + \left(\frac{\omega}{c\mathsf{k}_y}\right)^2\frac{b}{2}\right] \quad [1.120]$$

$$\alpha = \frac{\langle P_{dL}\rangle}{2\langle P\rangle} = \frac{2\left(\frac{n\pi}{b}\right)^2\left[a + \left(\frac{\omega b}{n\pi c}\right)^2\frac{b}{2}\right]}{\gamma_c\delta\mu_0\omega\mathsf{k}_z ab} \quad [1.121]$$

1.3.3. *Propagation between two parallel planes*

Consider two parallel plate electrodes (perfect conductors) separated by a small distance b and enclosing an HLI medium with permittivity ε_0 and permeability μ_0 (vacuum for simplicity), as illustrated in Figure 1.5(a). Since this spacing is much smaller than the width l and length L of the electrodes, the edge effects are neglected. In other words, the two conducting planes can be considered infinite. To obtain solutions for the guided wave propagation solutions of TM and TE waves between these two planes (taken parallel to the Oxz plane), it suffices to take $a \longrightarrow \infty$ from the previous results.

Since the system here is invariant under translation along the Ox direction, it follows that the components E_z (for the TM wave) or H_z (for the TE wave) must be independent of x. Consequently, all other components determined from [1.19] and [1.21] must also be independent of x. As this system contains two conductors, it can also support a TEM mode. This topic will be addressed in the following sections.

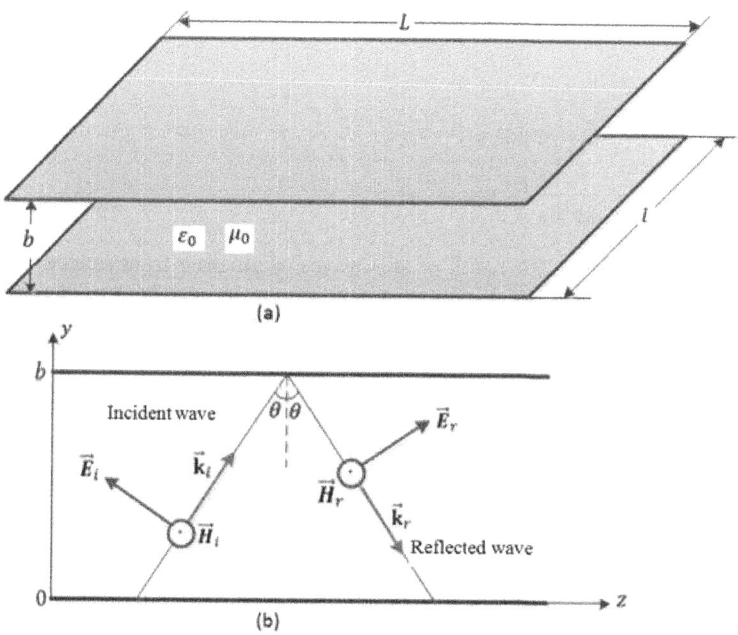

Figure 1.5. *Propagation of an EM wave between two infinite planes. For a color version of this figure, see www.iste.co.uk/abdesselam/electrodynamics3.zip*

1.3.3.1. *waves TM ($H_z = 0$, $E_z \neq$)*

To obtain a non-trivial solution for E_z when taking the limit $a \longrightarrow \infty$ in expression [1.47], we simply perform the substitution $E_{mn} \sin\left(\frac{m\pi x}{a}\right) \longrightarrow E_n$, which gives:

$$E_z = \Re\left[E_n \sin\left(\mathsf{k}_y y\right) e^{-\mathsf{i}(\omega t - \mathsf{k}_z z)}\right] \quad [1.122]$$

where n is an integer, $sfk_y = \frac{n\pi}{b}$ and $sfk_z = \sqrt{\frac{\omega^2}{c^2} - \frac{n^2\pi^2}{b^2}}$. In fact, [1.122] is the solution of the Helmholtz equation $\left(\frac{\partial^2}{\partial y^2} + \gamma^2\right) E_z = 0$, with boundary conditions $E_z\left(y = 0\right) = E_z\left(y = b\right) = 0$. We can see that $\omega_{c,n} = \frac{cn\pi}{b}$ appears here as the cut-off frequency for the n mode. If $\omega > \omega_{c,n}$, we have an unattenuated propagation. On the other hand, if $\omega < \omega_{c,n}$, we have an evanescent wave. For a given ω, all modes with a lower frequency cutoff will pass, but not the others. This is a high-pass filter.

The other components of the fields $\vec{\mathbf{E}}_\perp$ and $\vec{\mathbf{H}}_\perp$ are obtained from [1.19] and [1.21]:

$$\begin{aligned}
E_x &= H_y = 0 \\
E_y &= \Re\left[\frac{\mathsf{ik}_z}{\mathsf{k}_y} E_n \cos\left(\mathsf{k}_y y\right) e^{-\mathsf{i}(\omega t - \mathsf{k}_z z)}\right] \\
H_x &= \Re\left[-\frac{\mathsf{i}\omega}{\mu_0 c^2 \mathsf{k}_y} E_n \cos\left(\mathsf{k}_y y\right) e^{-\mathsf{i}(\omega t - \mathsf{k}_z z)}\right]
\end{aligned} \quad [1.123]$$

This wave appears as the result of two waves originating from reflections on the planes $y = 0$ and $y = b$, as illustrated in Figure 1.5(b). In fact, let us define $\mathsf{k}_y = \mathsf{k}\cos\theta = \frac{\omega}{c}\cos\theta$, $\mathsf{k}_z = \mathsf{k}\cos\theta = \frac{\omega}{c}\sin\theta$, $\vec{\mathsf{k}}_i = \mathsf{k}_y \vec{\mathbf{j}} + \mathsf{k}_z \vec{\mathbf{k}}$, $\vec{\mathsf{k}}_r = -\mathsf{k}_y \vec{\mathbf{j}} + \mathsf{k}_z \vec{\mathbf{k}}$, $E_n = \frac{2}{i}\cos\theta E_i$, $E_r = E_i$:

$$\begin{aligned}
E_x &= H_y = H_z = 0 \\
E_y &= \Re\left[E_i \sin\theta e^{-\mathsf{i}\left(\omega t - \vec{\mathsf{k}}_i \cdot \vec{\mathbf{r}}\right)}\right] + \Re\left[E_r \sin\theta e^{-\mathsf{i}\left(\omega t - \vec{\mathsf{k}}_r \cdot \vec{\mathbf{r}}\right)}\right] \\
E_z &= \Re\left[-E_i \cos\theta e^{-\mathsf{i}\left(\omega t - \vec{\mathsf{k}}_i \cdot \vec{\mathbf{r}}\right)}\right] + \Re\left[E_r \cos\theta e^{-\mathsf{i}\left(\omega t - \vec{\mathsf{k}}_r \cdot \vec{\mathbf{r}}\right)}\right] \\
H_x &= \Re\left[-\sqrt{\frac{\varepsilon_0}{\mu_0}} E_i e^{-\mathsf{i}\left(\omega t - \vec{\mathsf{k}}_i \cdot \vec{\mathbf{r}}\right)}\right] + \Re\left[-\sqrt{\frac{\varepsilon_0}{\mu_0}} E_r e^{-\mathsf{i}\left(\omega t - \vec{\mathsf{k}}_r \cdot \vec{\mathbf{r}}\right)}\right]
\end{aligned} \quad [1.124]$$

The first terms on the right-hand side of equations [1.124] represent the field components of the incident wave, while the second terms correspond to those of the reflected wave. The boundary conditions are satisfied at the planes $y = 0$ and $y = b$.

$$\begin{cases}
E_z\left(y = 0\right) = 0 \Longrightarrow E_i = E_r \\
E_z\left(y = b\right) = 0 \Longrightarrow E_i e^{\mathsf{ik}_y b} = E_r e^{-\mathsf{ik}_y b} \Longrightarrow e^{2\mathsf{ik}_y b} = 1 \Longrightarrow \mathsf{k}_y = \dfrac{n\pi}{b}
\end{cases} \quad [1.125]$$

The results are consistent with those previously obtained. The angle of incidence is therefore indeed quantized.

$$\cos\theta = \frac{cn\pi}{\omega b} \qquad [1.126]$$

We therefore conclude that only discrete values of the angle of incidence on the walls enable destructive interference on the boundaries. The following are the main results:

1) The surface charges and currents observed, respectively, on the planes $y = 0$ and $y = b$:

$$\begin{aligned}
&\sigma(y=0) = \varepsilon_0 E_y(y=0) = \Re\left[\frac{\mathrm{i}\mathsf{k}_z\varepsilon_0}{\mathsf{k}_y} E_n e^{-\mathrm{i}(\omega t - \mathsf{k}_z z)}\right] \\
&K_z(y=0) = -H_x(y=0) = \Re\left[\frac{\mathrm{i}\omega\varepsilon_0}{\mathsf{k}_y} E_n e^{-\mathrm{i}(\omega t - \mathsf{k}_z z)}\right] = \frac{\omega}{\mathsf{k}_z}\sigma(y=0) \\
&\sigma(y=b) = -\varepsilon_0 E_y(y=b) = \Re\left[-\frac{\mathrm{i}\mathsf{k}_z\varepsilon_0(-1)^n}{\mathsf{k}_y} E_n e^{-\mathrm{i}(\omega t - \mathsf{k}_z z)}\right] \\
&K_z(y=b) = H_x(y=b) = \Re\left[-\frac{\mathrm{i}\omega\varepsilon_0(-1)^n}{\mathsf{k}_y} E_n e^{-\mathrm{i}(\omega t - \mathsf{k}_z z)}\right] = \frac{\omega}{\mathsf{k}_z}\sigma(y=b)
\end{aligned} \qquad [1.127]$$

2) The time-averaged complex Poynting vector:

$$\langle\vec{\mathbf{\Pi}}\rangle = \frac{1}{2}\Re\left[E_z H_x^\star\right]\vec{\mathbf{j}} - \frac{1}{2}\Re\left[E_y H_x^\star\right]\vec{\mathbf{k}} = \frac{\mathsf{k}_z\varepsilon_0\omega\left|E_n\right|^2}{2\mathsf{k}_y^2}\cos^2(\mathsf{k}_y y)\,\vec{\mathbf{k}} \qquad [1.128]$$

3) The average power transmitted between the two planes:

$$\langle P\rangle = \int_{x=0}^{l}\int_{y=0}^{b}\langle\Pi_z\rangle dx dy = \frac{\mathsf{k}_z\varepsilon_0\omega\left|E_n\right|^2 lb}{4\mathsf{k}_y^2} \qquad [1.129]$$

4) The average power dissipated per unit length for real conductors with high conductivity:

$$\langle P_{dL}\rangle = \frac{1}{2\gamma_c\delta}\oint\left|\vec{\mathbf{K}}\right|^2 dl = \frac{\varepsilon_0^2\omega^2\left|E_n\right|^2 l}{\gamma_c\delta\mathsf{k}_y^2} \qquad [1.130]$$

5) The attenuation coefficient of the TM modes:

$$\alpha = \frac{\langle P_{dL}\rangle}{2\langle P\rangle} = \frac{2\varepsilon_0\omega}{\gamma_c\delta\mathsf{k}_z b} \qquad [1.131]$$

1.3.3.2. *Waves TE ($E_z = 0$, $H_z \neq 0$)*

For the TE modes, we must make the substitution $H_{mn} \cos(\mathsf{k}_x x) \longrightarrow H_n$ in [1.82], then calculate the other components using equations [1.19] and [1.21]:

$$\begin{aligned} H_y &= \Re\left[-\frac{\mathsf{i}\mathsf{k}_z}{\mathsf{k}_y} H_n \sin(\mathsf{k}_y y)\, e^{-\mathsf{i}(\omega t - \mathsf{k}_z z)}\right] \\ H_z &= \Re\left[H_n \cos(\mathsf{k}_y y)\, e^{-\mathsf{i}(\omega t - \mathsf{k}_z z)}\right] \\ H_x &= E_y = E_z = 0 \\ E_x &= \Re\left[-\frac{\mathsf{i}\omega\mu_0}{\mathsf{k}_y} H_n \sin(\mathsf{k}_y y)\, e^{-\mathsf{i}(\omega t - \mathsf{k}_z z)}\right] \end{aligned} \qquad [1.132]$$

The following are the essential results:

1) The surface charges and surface currents experienced, respectively, on the planes $y = 0$ and $y = b$:

$$\begin{aligned} &\sigma(y=0) = \sigma(y=b) = 0 \\ &K_z(y=0) = K_z(y=b) = 0 \\ &K_x(y=0) = H_z(y=0) = \Re\left[H_n e^{-\mathsf{i}(\omega t - \mathsf{k}_z z)}\right] \\ &K_x(y=b) = H_z(y=b) = \Re\left[(-1)^n H_n e^{-\mathsf{i}(\omega t - \mathsf{k}_z z)}\right] \end{aligned} \qquad [1.133]$$

2) The time-average of the complex Poynting vector:

$$\langle\overrightarrow{\Pi}\rangle = -\frac{1}{2}\Re\left[E_x H_z^\star\right]\overrightarrow{\mathbf{j}} + \frac{1}{2}\Re\left[E_x H_y^\star\right]\overrightarrow{\mathbf{k}} = \frac{\mathsf{k}_z\mu_0\omega\left|H_n\right|^2}{2\mathsf{k}_y^2}\sin^2(\mathsf{k}_y y)\,\overrightarrow{\mathbf{k}} \qquad [1.134]$$

3) The average power flowing between the two planes:

$$\langle P\rangle = \int_{x=0}^{l}\int_{y=0}^{b}\langle\Pi_z\rangle dx dy = \frac{\mathsf{k}_z\mu_0\omega\left|H_n\right|^2 lb}{4\mathsf{k}_y^2} \qquad [1.135]$$

4) The average power dissipated per unit length for real conductors with high conductivity:

$$\langle P_{dL}\rangle = \frac{1}{2\gamma_c\delta}\oint\left|\overrightarrow{\mathbf{K}}\right|^2 dl = \frac{\left|H_n\right|^2 l}{\gamma\delta} \qquad [1.136]$$

5) The attenuation coefficient of the TM modes:

$$\alpha = \frac{\langle P_{dL}\rangle}{2\langle P\rangle} = \frac{2\mathsf{k}_y^2}{\gamma_c\delta\mathsf{k}_z\mu_0\omega b} \qquad [1.137]$$

1.3.4. *Cylindrical waveguide*

Consider a hollow, perfectly conductive waveguide with a circular cross-section of radius of a. It is assumed that inside the waveguide $\varepsilon = \varepsilon_0$ and $\mu = \mu_0$. The position of a point inside the waveguide is identified using cylindrical coordinates (r, φ, z).

1.3.4.1. *TM modes ($H_z = 0$, $E_z \neq 0$)*

Let us first study the TM modes. The Helmholtz equation is written as:

$$\frac{1}{r}\frac{\partial}{\partial r}\left(r\frac{\partial E_z}{\partial r}\right) + \frac{1}{r^2}\frac{\partial^2 E_z}{\partial \varphi^2} + \gamma^2 E_z = 0 \qquad [1.138]$$

where $\gamma^2 = \frac{\omega^2}{c^2} - \mathrm{k}_z^2$. The boundary condition is given as:

$$E_z\left(r = a, \varphi\right) = 0 \qquad [1.139]$$

The solutions of [1.138] are obtained using the method of separation of variables and involve Bessel functions. By setting $E_z\left(r, \varphi\right) = R\left(r\right)\Phi\left(\varphi\right)$, equation [1.138] becomes $\Phi\left(\frac{d^2R}{dr^2} + \frac{1}{r}\frac{dR}{dr}\right) + \frac{R}{r^2}\frac{d^2\Phi}{d\varphi^2} + \gamma^2 R\Phi = 0$, or, equivalently, $\frac{r^2}{R}\left(\frac{d^2R}{dr^2} + \frac{1}{r}\frac{dR}{dr}\right) + \gamma^2 r^2 = -\frac{1}{\Phi}\frac{d^2\Phi}{d\varphi^2}$. The first term depends on r, while the second depends on φ. This equality can only hold if both sides are constant. Thus, we find:

$$\frac{d^2\Phi}{d\varphi^2} + \nu^2\Phi = 0 \qquad [1.140]$$

$$\frac{r^2}{R}\left(\frac{d^2R}{dr^2} + \frac{1}{r}\frac{dR}{dr}\right) + \gamma^2 r^2 = \nu^2 \qquad [1.141]$$

The solution of [1.140] is written as $\Phi\left(\varphi\right) = C_0 e^{\mathrm{i}\nu\varphi}$, where C_0 is a constant. Φ must also be periodic, that is, $\Phi\left(\varphi + 2\pi\right) = \Phi\left(\varphi\right)$, which gives $e^{2\pi \mathrm{i}\nu} = 1$. We therefore conclude that ν must be an integer. We set $\nu = m$ with $m = 0, \pm 1, \pm 2, \ldots$. Substituting $\nu = m$ into equation [1.141], we obtain $\frac{d^2R}{dr^2} + \frac{1}{r}\frac{dR}{dr} + \left(\gamma^2 - \frac{m^2}{r^2}\right)R = 0$. The general solutions are of the form $R_m\left(r\right) = A_m J_m\left(\gamma r\right) + B_m N_m\left(\gamma r\right)$, where J_m and N_m are Bessel functions of the first and second kind, respectively. Since N_m is singular near $r = 0$, we must take $B_m = 0$. Therefore, the physically acceptable solution is $R_m\left(r\right) = A_m J_m\left(\gamma r\right)$. Hence, the Helmholtz equation [1.138] provides the following solution:

$$\hat{E}_z\left(r, \varphi\right) = \Re\left[A_m J_m\left(\gamma r\right)e^{\pm im\varphi}\right] \qquad [1.142]$$

The boundary condition [1.139] then amounts to $J_m\left(\gamma a\right) = 0$, which gives $\gamma = \frac{\rho_{mn}}{a}$, where ρ_{mn} are the roots of the Bessel function J_m, that is, $J_m\left(\rho_{mn}\right) = 0$.

We therefore conclude that each root ρ_{mn} of the Bessel functions corresponds to a propagation mode. The dispersion relation for this mode is written as:

$$\omega^2 = \omega_{mn}^2 + c^2 \mathsf{k}_z^2 = \frac{c^2 \rho_{mn}^2}{a^2} + c^2 \mathsf{k}_z^2 \, (m \geq 0, n \geq 1) \tag{1.143}$$

It therefore follows that:

$$E_z(r, \varphi, z) = \Re \left[E_{mn} J_m \left(\frac{\rho_{mn} r}{a} \right) e^{\pm i m \varphi} e^{-\mathrm{i}(\omega t - \mathsf{k}_z z)} \right] \tag{1.144}$$

where E_{mn} is a complex amplitude. The other components of $\vec{\mathbf{E}}$ and $\vec{\mathbf{H}}$ are calculated from relations [1.19] and [1.21]:

$$\begin{aligned}
E_r &= \Re \left[\frac{\mathrm{i}\mathsf{k}_z a^2}{\rho_{mn}^2} \frac{\partial E_z}{\partial r} \right] = \Re \left[\frac{\mathrm{i}\mathsf{k}_z a E_{mn}}{\rho_{mn}} J_m' \left(\frac{\rho_{mn} r}{a} \right) e^{\pm i m \varphi} e^{-\mathrm{i}(\omega t - \mathsf{k}_z z)} \right] \\
E_\varphi &= \Re \left[\frac{\mathrm{i}\mathsf{k}_z a^2}{\rho_{mn}^2} \frac{1}{r} \frac{\partial E_z}{\partial \varphi} \right] = \Re \left[\mp \frac{m \mathsf{k}_z a^2 E_{mn}}{\rho_{mn}^2 r} J_m \left(\frac{\rho_{mn} r}{a} \right) e^{\pm i m \varphi} e^{-\mathrm{i}(\omega t - \mathsf{k}_z z)} \right] \\
H_r &= \Re \left[-\frac{\mathrm{i}\omega \varepsilon_0 a^2}{\rho_{mn}^2} \frac{1}{r} \frac{\partial E_z}{\partial \varphi} \right] = \Re \left[\pm \frac{m \omega a^2 \varepsilon_0 E_{mn}}{\rho_{mn}^2 r} J_m \left(\frac{\rho_{mn} r}{a} \right) e^{\pm i m \varphi} e^{-\mathrm{i}(\omega t - \mathsf{k}_z z)} \right] \\
H_\varphi &= \Re \left[\frac{\mathrm{i}\omega \varepsilon_0 a^2}{\rho_{mn}^2} \frac{\partial E_z}{\partial r} \right] = \Re \left[\frac{\mathrm{i}\omega a \varepsilon_0 E_{mn}}{\rho_{mn}} J_m' \left(\frac{\rho_{mn} r}{a} \right) e^{\pm i m \varphi} e^{-\mathrm{i}(\omega t - \mathsf{k}_z z)} \right]
\end{aligned} \tag{1.145}$$

We observe the following:

1) When ω becomes less than $\frac{c\rho_{mn}}{a}$, k_z becomes imaginary and the wave no longer propagates.

2) A double degeneracy for each mode when $m > 0$. We can consider modes in $\cos m\varphi$ or modes in $\sin m\varphi$, depending on the nature of the amplitude E_{mn}.

3) The charge accumulated on the surface of the waveguide is:

$$\begin{aligned}
\sigma(r = a) &= -\varepsilon_0 E_r (r = a) \\
&= \Re \left[-\frac{\mathrm{i}\mathsf{k}_z \varepsilon_0 a E_{mn}}{\rho_{mn}} J_m' (\rho_{mn}) e^{\pm i m \varphi} e^{-\mathrm{i}(\omega t - \mathsf{k}_z z)} \right]
\end{aligned} \tag{1.146}$$

4) The current flowing on the surface of the waveguide is:

$$\begin{aligned}
K_z(r = a) &= -H_\varphi (r = a) = \Re \left[-\frac{\mathrm{i}\omega a \varepsilon_0 E_{mn}}{\rho_{mn}} J_m' (\rho_{mn}) e^{\pm i m \varphi} e^{-\mathrm{i}(\omega t - \mathsf{k}_z z)} \right] \\
&= \frac{\omega}{\mathsf{k}_z} \sigma (r = a)
\end{aligned} \tag{1.147}$$

5) The time-averaged Poynting vector is given by:

$$\begin{aligned}\langle\overrightarrow{\mathbf{\Pi}}\rangle &= \frac{1}{2}\Re\left[\overrightarrow{\mathbf{E}}\times\overrightarrow{\mathbf{H}}^{\star}\right]\\ &= \frac{1}{2}\Re\left[-E_zH_\varphi^{\star}\overrightarrow{\mathbf{e}}_r + E_zH_r^{\star}\overrightarrow{\mathbf{e}}_\varphi + \left(E_rH_\varphi^{\star} - E_\varphi H_r^{\star}\right)\overrightarrow{\mathbf{k}}\right]\end{aligned} \qquad [1.148]$$

The term $\Re\left[E_zH_\varphi^{\star}\right]$ is zero: no energy propagates in the transverse $\overrightarrow{\mathbf{e}}_r$ direction. In the $\overrightarrow{\mathbf{e}}_\varphi$ and $\overrightarrow{\mathbf{k}}$ directions, we have:

$$\begin{aligned}\langle\Pi_\varphi\rangle &= \pm\frac{m\varepsilon_0\omega a^2\left|E_{mn}\right|^2 J_m^2\left(\frac{\rho_{mn}r}{a}\right)}{2\rho_{mn}^2 r}\\ \langle\Pi_z\rangle &= \frac{\mathrm{k}_z\varepsilon_0\omega a^2\left|E_{mn}\right|^2}{2\rho_{mn}^2}\left[J_m'^2\left(\frac{\rho_{mn}r}{a}\right) + \frac{m^2a^2}{\rho_{mn}^2r^2}J_m^2\left(\frac{\rho_{mn}r}{a}\right)\right]\end{aligned} \qquad [1.149]$$

In a circular waveguide, energy propagates in the longitudinal direction and in the azimuthal direction (in a loop).

6) For the TM_{01} mode, we have $\rho_{01} \simeq 2.405$. Using the identity $J_0' = -J_1$, the field components are written as:

$$\begin{aligned}E_r &= \Re\left[-\frac{\mathrm{i}\mathrm{k}_z aE_{01}}{\rho_{01}}J_1\left(\frac{\rho_{01}r}{a}\right)e^{-\mathrm{i}(\omega t-\mathrm{k}_z z)}\right]\\ E_z &= \Re\left[E_{01}J_0\left(\frac{\rho_{01}r}{a}\right)e^{-\mathrm{i}(\omega t-\mathrm{k}_z z)}\right]\\ E_\varphi &= H_r = H_z = 0\\ H_\varphi &= \Re\left[-\frac{\mathrm{i}\omega a\varepsilon_0E_{01}}{\rho_{01}}J_1\left(\frac{\rho_{01}r}{a}\right)e^{-\mathrm{i}(\omega t-\mathrm{k}_z z)}\right]\end{aligned} \qquad [1.150]$$

and consequently:

$$\begin{aligned}\langle\Pi_\varphi\rangle &= 0\\ \langle\Pi_z\rangle &= \frac{\mathrm{k}_z\varepsilon_0\omega a^2\left|E_{01}\right|^2 J_0'^2\left(\frac{\rho_{01}r}{a}\right)}{2\rho_{01}^2}\end{aligned} \qquad [1.151]$$

For this mode, $\omega_c = \frac{c\rho_{01}}{a}$. The cutoff wavelength is equal to $\lambda_c = \frac{2\pi c}{\omega_c} = \frac{2\pi a}{\rho_{01}} \simeq 2.61a$.

1.3.4.2. *TE modes ($E_z = 0$, $H_z \neq 0$)*

For TE modes, the Helmholtz equation for H_z is written in the same way as before. In this case, we have $E_z = 0$ and $H_z = \Re\left[H_0J_m(\gamma r)e^{\pm\mathrm{i}m\varphi}e^{-\mathrm{i}(\omega t-\mathrm{k}_z z)}\right]$ with the boundary condition $-\frac{\partial H_z}{\partial r}_{r=a} = 0$, which implies that $J_m'(\gamma a) = 0$. Therefore, γa

must be a root of J'_m, which we denote here as ρ'_{mn}. The component H_z is then written as:

$$H_z = \Re\left[H_{mn}J_m\left(\frac{\rho'_{mn}r}{a}\right)e^{\pm im\varphi}e^{-\mathrm{i}(\omega t-\mathrm{k}_z z)}\right] \quad [1.152]$$

which leads to the transverse fields:

$$\begin{aligned}
H_r &= \Re\left[\frac{\mathrm{i}\mathrm{k}_z a^2}{\rho'^2_{mn}}\frac{\partial H_z}{\partial r}\right] = \Re\left[\frac{\mathrm{i}\mathrm{k}_z a H_{mn}}{\rho'_{mn}}J'_m\left(\frac{\rho'_{mn}r}{a}\right)e^{\pm im\varphi}e^{-\mathrm{i}(\omega t-\mathrm{k}_z z)}\right]\\
H_\varphi &= \Re\left[\frac{\mathrm{i}\mathrm{k}_z a^2}{\rho'^2_{mn}}\frac{1}{r}\frac{\partial H_z}{\partial\varphi}\right] = \Re\left[\mp\frac{m\mathrm{k}_z a^2 H_{mn}}{\rho'^2_{mn}r}J_m\left(\frac{\rho'_{mn}r}{a}\right)e^{\pm im\varphi}e^{-\mathrm{i}(\omega t-\mathrm{k}_z z)}\right]\\
E_r &= \Re\left[\frac{\mathrm{i}\omega\mu_0 a^2}{\rho'^2_{mn}}\frac{1}{r}\frac{\partial H_z}{\partial\varphi}\right] = \Re\left[\mp\frac{m\omega a^2\mu_0 H_{mn}}{\rho'^2_{mn}r}J_m\left(\frac{\rho'_{mn}r}{a}\right)e^{\pm im\varphi}e^{-\mathrm{i}(\omega t-\mathrm{k}_z z)}\right]\\
E_\varphi &= \Re\left[-\frac{\mathrm{i}\omega\mu_0 a^2}{\rho'^2_{mn}}\frac{\partial H_z}{\partial r}\right] = \Re\left[-\frac{\mathrm{i}\omega a\mu_0 H_{mn}}{\rho'_{mn}}J'_m\left(\frac{\rho'_{mn}r}{a}\right)e^{\pm im\varphi}e^{-\mathrm{i}(\omega t-\mathrm{k}_z z)}\right]
\end{aligned} \quad [1.153]$$

The following are some essential remarks:

1) The average energy flux through the waveguide is given as:

$$\begin{aligned}
\langle\vec{\mathbf{\Pi}}\rangle &= \frac{1}{2}\Re\left[E_\varphi H_z^\star\vec{\mathbf{e}}_r - E_r H_z^\star\vec{\mathbf{e}}_\varphi + \left(E_r H_\varphi^\star - E_\varphi H_r^\star\right)\vec{\mathbf{k}}\right]\\
&= \pm\frac{m\omega a^2\mu_0\left|H_{mn}\right|^2 J_m^2\left(\frac{\rho'_{mn}r}{a}\right)}{2\rho'^2_{mn}r}\vec{\mathbf{e}}_\varphi +\\
&\quad \frac{\mathrm{k}_z\omega a^2\mu_0\left|H_{mn}\right|^2}{2\rho'^2_{mn}}\left[J'^2_m\left(\frac{\rho'_{mn}r}{a}\right) + \frac{m^2a^2}{\rho'^2_{mn}r^2}J_m^2\left(\frac{\rho'_{mn}r}{a}\right)\right]\vec{\mathbf{k}}
\end{aligned} \quad [1.154]$$

2) The charge and current experienced at the surface of the waveguide are given by:

$$\begin{aligned}
\sigma\left(r=a\right) &= -\varepsilon_0 E_r\left(r=a\right) = \Re\left[\pm\frac{m\omega a H_{mn}}{c^2\rho'^2_{mn}}J_m\left(\rho'_{mn}\right)e^{\pm im\varphi}e^{-\mathrm{i}(\omega t-\mathrm{k}_z z)}\right]\\
K_z\left(r=a\right) &= -H_\varphi\left(r=a\right) = \Re\left[\pm\frac{m\mathrm{k}_z a H_{mn}}{\rho'^2_{mn}}J_m\left(\rho'_{mn}\right)e^{\pm im\varphi}e^{-\mathrm{i}(\omega t-\mathrm{k}_z z)}\right]\\
K_\varphi\left(r=a\right) &= H_z\left(r=a\right) = \Re\left[H_{mn}J_m\left(\rho'_{mn}\right)e^{\pm im\varphi}e^{-\mathrm{i}(\omega t-\mathrm{k}_z z)}\right]
\end{aligned} \quad [1.155]$$

3) The smallest root among all Bessel functions J_m and their derivatives J'_m is the first root of J'_1; it is approximately $\rho'_{11} \simeq 1.84118$. Therefore, the mode with the lowest cutoff frequency is the TE_{11} mode (dominant mode). For this mode, we have:

$$\begin{aligned}
E_r &= \Re\left[\mp\frac{m\omega a^2\mu_0 H_{11}}{\rho_{11}'^2 r}J_1\left(\frac{\rho_{11}' r}{a}\right)e^{\pm i\varphi}e^{-i(\omega t-k_z z)}\right]\\
E_\varphi &= \Re\left[-\frac{i\omega a\mu_0 H_{11}}{\rho_{11}'}J_1'\left(\frac{\rho_{11}' r}{a}\right)e^{\pm i\varphi}e^{-i(\omega t-k_z z)}\right]\\
E_z &= 0\\
H_r &= \Re\left[\frac{ik_z a H_{11}}{\rho_{11}'}J_1'\left(\frac{\rho_{11}' r}{a}\right)e^{\pm i\varphi}e^{-i(\omega t-k_z z)}\right]\\
H_\varphi &= \Re\left[\mp\frac{k_z a^2 H_{11}}{\rho_{11}'^2 r}J_1\left(\frac{\rho_{11}' r}{a}\right)e^{\pm i\varphi}e^{-i(\omega t-k_z z)}\right]\\
H_z &= \Re\left[H_{11}J_1\left(\frac{\rho_{11}' r}{a}\right)e^{\pm i\varphi}e^{-i(\omega t-k_z z)}\right]
\end{aligned} \quad [1.156]$$

For this mode, $\omega_c = \frac{c\rho_{11}'}{a}$. The cutoff wavelength is $\lambda_c = \frac{2\pi c}{\omega_c} = \frac{2\pi a}{\rho_{11}'} \simeq 3.41a$.

4) Since $J_0' = -J_1$, the zeros of J_0' coincide with those of J_1; this implies a degeneracy between the TM_{1n} and TE_{0n} modes.

1.4. Dielectric waveguides

1.4.1. *Propagation between two parallel dielectric planes*

Let us briefly study the propagation modes TE and TM between two identical parallel planes (each with a thickness e, separated by a distance b) composed of a dielectric LIH material with permittivity ε_2 and permeability μ_0. Between the two plates, there is another dielectric medium with permittivity $\varepsilon_1 > \varepsilon_2$. We seek solutions where the fields are essentially confined within medium 1, as illustrated in Figure 1.6.

In other words, the solutions must be exponential inside the plates to prevent energy loss from the guide by radiation. The solutions are oscillatory only in medium 1 to ensure guided propagation.

1.4.1.1. *TM solutions*

The solutions are independent of x. For TM waves propagating in the Oz direction, the component E_z must be of the form:

$$E_z = \begin{cases} \Re\left[Ae^{-\alpha(y-b)}e^{-i(\omega t-k_z z)}\right] & y > b\\ \Re\left[(B\sin(k_y y) + C\cos(k_y y))\,e^{-i(\omega t-k_z z)}\right] & 0 < y < b\\ \Re\left[De^{\alpha y}e^{-i(\omega t-k_z z)}\right] & y < 0\end{cases} \quad [1.157]$$

The solutions decrease exponentially to remain mainly localized in the plates. sfk_y, sfk_z and the attenuation coefficient α obey the following restrictions (obtained from the Helmholtz equation):

$$\begin{aligned} \mathsf{k}_y^2 + \mathsf{k}_z^2 &= \frac{\omega^2 \mathsf{n}_1^2}{c^2} \\ -\alpha^2 + \mathsf{k}_z^2 &= \frac{\omega^2 \mathsf{n}_2^2}{c^2} \end{aligned} \quad [1.158]$$

where $\varepsilon_{1,2} = \varepsilon_0 \mathsf{n}_{1,2}^2$. The component k_z must be the same everywhere in order to satisfy the boundary conditions at each interface. It must also satisfy the following constraint:

$$\frac{\omega^2 \mathsf{n}_2^2}{c^2} \leq \mathsf{k}_z^2 \leq \frac{\omega^2 \mathsf{n}_1^2}{c^2} \quad [1.159]$$

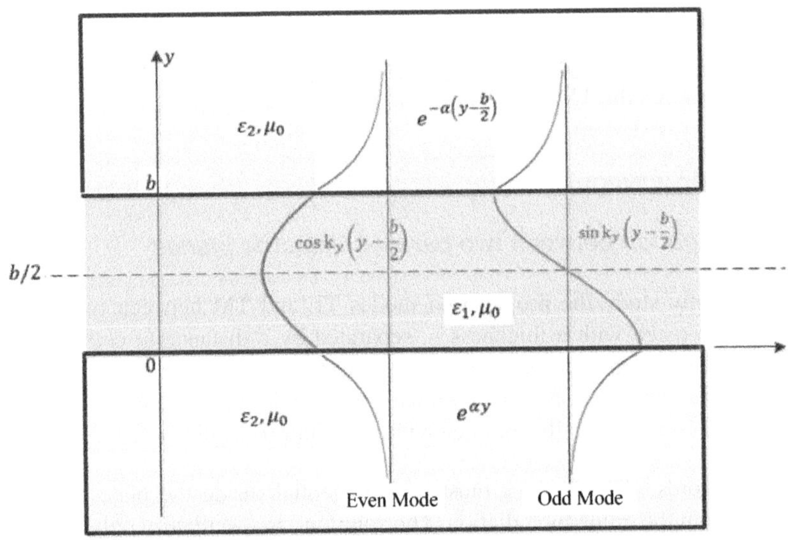

Figure 1.6. *Propagation of TM and TE modes in a dielectric medium with permittivity ε_1 placed between two parallel dielectric plates with permittivity $\varepsilon_2 < \varepsilon_1$*

The other components of the fields $\vec{\mathbf{E}}$ and $\vec{\mathbf{H}}$ can be found from [1.19] and [1.21]:

$$E_x = H_y = 0$$

$$E_y = \begin{cases} \Re\left[\dfrac{\mathsf{ik}_z}{\alpha} A e^{-\alpha(y-b)} e^{-\mathsf{i}(\omega t - \mathsf{k}_z z)}\right] & y > b \\ \Re\left[\dfrac{\mathsf{ik}_z}{\mathsf{k}_y} \left(B \cos\left(\mathsf{k}_y y\right) - C \sin\left(\mathsf{k}_y y\right)\right) e^{-\mathsf{i}(\omega t - \mathsf{k}_z z)}\right] & 0 < y < b \\ \Re\left[-\dfrac{\mathsf{ik}_z}{\alpha} D e^{\alpha y} e^{-\mathsf{i}(\omega t - \mathsf{k}_z z)}\right] & y < 0 \end{cases}$$

$$H_x = \begin{cases} \Re\left[-\dfrac{\mathrm{i}\omega\varepsilon_2}{\alpha} A e^{-\alpha(y-b)} e^{-\mathrm{i}(\omega t - \mathrm{k}_z z)}\right] & y > b \\ \Re\left[-\dfrac{\mathrm{i}\omega\varepsilon_1}{\mathrm{k}_y} \left(B\cos\left(\mathrm{k}_y y\right) - C\sin\left(\mathrm{k}_y y\right)\right) e^{-\mathrm{i}(\omega t - \mathrm{k}_z z)}\right] & 0 < y < b \\ \Re\left[\dfrac{\mathrm{i}\omega\varepsilon_2}{\alpha} D e^{\alpha y} e^{-\mathrm{i}(\omega t - \mathrm{k}_z z)}\right] & y < 0 \end{cases} \quad [1.160]$$

At the interfaces $y = 0$ and $y = b$, the tangential components of the fields $\vec{\mathbf{E}}$ and $\vec{\mathbf{H}}$ are continuous:

$$\begin{aligned} &E_z\left(y = 0^-\right) = E_z\left(y = 0^+\right) \Longrightarrow C - D = 0 \\ &E_z\left(y = b^-\right) = E_z\left(y = b^+\right) \Longrightarrow A - B\sin\left(\mathrm{k}_y b\right) \\ &\qquad\qquad\qquad\qquad\qquad - C\cos\left(\mathrm{k}_y b\right) = 0 \\ &H_x\left(y = b^-\right) = H_x\left(y = b^+\right) \Longrightarrow \frac{\alpha\varepsilon_1}{\mathrm{k}_y\varepsilon_2} B + D = 0 \\ &E_z\left(y = b^-\right) = E_z\left(y = b^+\right) \Longrightarrow A - \frac{\alpha\varepsilon_1}{\mathrm{k}_y\varepsilon_2} B\cos\left(\mathrm{k}_y b\right) \\ &\qquad\qquad\qquad\qquad\qquad + \frac{\alpha\varepsilon_1}{\mathrm{k}_y\varepsilon_2} C\sin\left(\mathrm{k}_y b\right) = 0 \end{aligned} \quad [1.161]$$

The system of equations [1.161] admits a non-trivial solution if:

$$\det\begin{vmatrix} 0 & 0 & 1 & -1 \\ 1 & -\sin\left(\mathrm{k}_y b\right) & -\cos\left(\mathrm{k}_y b\right) & 0 \\ 0 & \frac{\alpha\varepsilon_1}{\mathrm{k}_y\varepsilon_2} & 0 & 1 \\ 1 & -\frac{\alpha\varepsilon_1}{\mathrm{k}_y\varepsilon_2}\cos\left(\mathrm{k}_y b\right) & \frac{\alpha\varepsilon_1}{\mathrm{k}_y\varepsilon_2}\sin\left(\mathrm{k}_y b\right) & 0 \end{vmatrix} = 0$$

which gives:

$$\alpha = \begin{cases} \dfrac{\mathrm{k}_y\varepsilon_2}{\varepsilon_1}\tan\dfrac{\mathrm{k}_y h}{2} \\ -\dfrac{\mathrm{k}_y\varepsilon_2}{\varepsilon_1}\dfrac{1}{\tan\frac{\mathrm{k}_y b}{2}} \end{cases} \quad [1.162]$$

To obtain the allowed values of α and k_z, we must solve equations [1.158] and [1.162] simultaneously using graphical methods, as illustrated in Figure 1.7.

The solutions correspond to the intersection points of the curves (the black points). The following can be seen graphically:

1) For a guided wave to occur, it is necessary that $n\pi \leq \frac{\mathrm{k}_y b}{2} < \left(n + \frac{1}{2}\right)\pi$ for the first case, and $\left(n + \frac{1}{2}\right)\pi \leq \frac{\mathrm{k}_y b}{2} < (n+1)\pi$ for the second case.

2) First case: each time that k_y exceeds the critical value $\frac{2n\pi}{b}$, a new intersection point appears corresponding to a new propagation mode. In other words, each time the frequency ω exceeds the critical value.

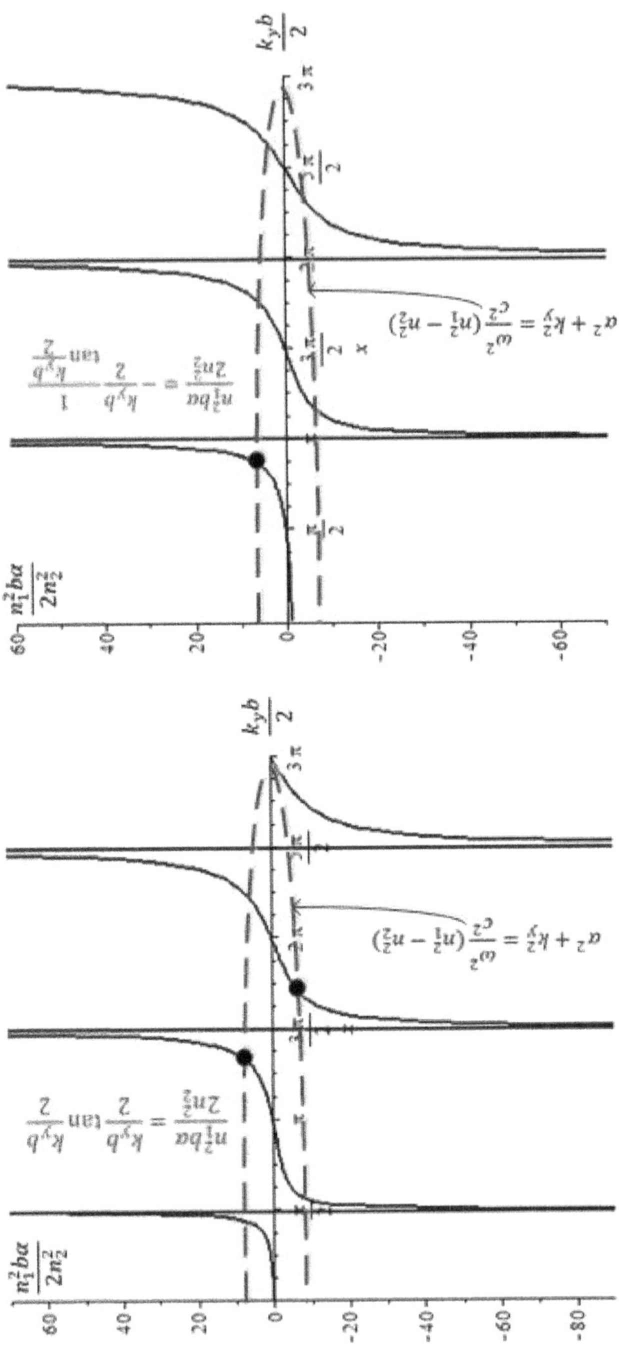

Figure 1.7. *Graphical solution of equations [1.158] and [1.162]. For a color version of this figure, see www.iste.co.uk/abdesselam/electrodynamics3.zip*

$$\omega_n = \frac{2n\pi c}{b\sqrt{\mathsf{n}_1^2 - \mathsf{n}_2^2}} \qquad [1.163]$$

a new propagation mode will appear; ω_n is the *cutoff frequency of this mode*.

3) Second case: each time the frequency ω exceeds the critical value:

$$\omega_n = \frac{(2n+1)\,\pi c}{b\sqrt{\mathsf{n}_1^2 - \mathsf{n}_2^2}} \qquad [1.164]$$

a new propagation mode is added to the others.

4) Odd solution ($\alpha = \frac{\mathsf{k}_y \varepsilon_2}{\varepsilon_1} \tan \frac{\mathsf{k}_y b}{2}$, $B = \frac{A}{\tan \frac{\mathsf{k}_y b}{2}}$, $C = -A$ and $D = -A$):

$$E_x = H_y = H_z = 0$$

$$E_y = \begin{cases} \Re\left[A\dfrac{\mathsf{ik}_z}{\alpha} e^{-\alpha(y-b)} e^{-\mathsf{i}(\omega t - \mathsf{k}_z z)}\right] & y > b \\ \Re\left[A\dfrac{\mathsf{ik}_z}{\mathsf{k}_y} \dfrac{\cos \mathsf{k}_y \left(y - \frac{b}{2}\right)}{\sin \frac{\mathsf{k}_y b}{2}} e^{-\mathsf{i}(\omega t - \mathsf{k}_z z)}\right] & 0 < y < b \\ \Re\left[A\dfrac{\mathsf{ik}_z}{\alpha} e^{\alpha y} e^{-\mathsf{i}(\omega t - \mathsf{k}_z z)}\right] & y < 0 \end{cases}$$

$$E_z = \begin{cases} \Re\left[A e^{-\alpha(y-b)} e^{-\mathsf{i}(\omega t - \mathsf{k}_z z)}\right] & y > b \\ \Re\left[A\dfrac{\sin \mathsf{k}_y \left(y - \frac{b}{2}\right)}{\sin \frac{\mathsf{k}_y b}{2}} e^{-\mathsf{i}(\omega t - \mathsf{k}_z z)}\right] & 0 < y < b \\ \Re\left[-A e^{\alpha y} e^{-\mathsf{i}(\omega t - \mathsf{k}_z z)}\right] & y < 0 \end{cases} \qquad [1.165]$$

$$H_x = \begin{cases} \Re\left[-A\dfrac{\mathsf{i}\omega\varepsilon_2}{\alpha} e^{-\alpha(y-b)} e^{-\mathsf{i}(\omega t - \mathsf{k}_z z)}\right] & y > b \\ \Re\left[-A\dfrac{\mathsf{i}\omega\varepsilon_1}{\mathsf{k}_y} \dfrac{\cos \mathsf{k}_y \left(y - \frac{b}{2}\right)}{\sin \frac{\mathsf{k}_y b}{2}} e^{-\mathsf{i}(\omega t - \mathsf{k}_z z)}\right] & 0 < y < b \\ \Re\left[-A\dfrac{\mathsf{i}\omega\varepsilon_2}{\alpha} e^{\alpha y} e^{-\mathsf{i}(\omega t - \mathsf{k}_z z)}\right] & y < 0 \end{cases}$$

This solution is called *odd* because E_z is directed oppositely in the two half-planes above and below the line $y = \frac{b}{2}$.

5) Even solution ($\alpha = -\frac{\mathsf{k}_y \varepsilon_2}{\varepsilon_1} \frac{1}{\tan \frac{\mathsf{k}_y b}{2}}$, $B = A \tan \frac{\mathsf{k}_y b}{2}$, $C = A$ and $D = A$):

$$E_x = H_y = H_z = 0$$

$$E_y = \begin{cases} \Re\left[A\dfrac{\mathrm{i}\mathrm{k}_z}{\alpha}e^{-\alpha(y-b)}e^{-\mathrm{i}(\omega t-\mathrm{k}_z z)}\right] & y>b \\ \Re\left[-A\dfrac{\mathrm{i}\mathrm{k}_z}{\mathrm{k}_y}\dfrac{\sin \mathrm{k}_y\left(y-\frac{b}{2}\right)}{\cos\frac{\mathrm{k}_y b}{2}}e^{-\mathrm{i}(\omega t-\mathrm{k}_z z)}\right] & 0<y<b \\ \Re\left[-A\dfrac{\mathrm{i}\mathrm{k}_z}{\alpha}e^{\alpha y}e^{-\mathrm{i}(\omega t-\mathrm{k}_z z)}\right] & y<0 \end{cases}$$

$$E_z = \begin{cases} \Re\left[Ae^{-\alpha(y-b)}e^{-\mathrm{i}(\omega t-\mathrm{k}_z z)}\right] & y>b \\ \Re\left[A\dfrac{\cos \mathrm{k}_y\left(y-\frac{b}{2}\right)}{\cos\frac{\mathrm{k}_y b}{2}}e^{-\mathrm{i}(\omega t-\mathrm{k}_z z)}\right] & 0<y<b \\ \Re\left[Ae^{\alpha y}e^{-\mathrm{i}(\omega t-\mathrm{k}_z z)}\right] & y<0 \end{cases} \qquad [1.166]$$

$$H_x = \begin{cases} \Re\left[-A\dfrac{\mathrm{i}\omega\varepsilon_2}{\alpha}e^{-\alpha(y-b)}e^{-\mathrm{i}(\omega t-\mathrm{k}_z z)}\right] & y>b \\ \Re\left[A\dfrac{\mathrm{i}\omega\varepsilon_1}{\mathrm{k}_y}\dfrac{\sin \mathrm{k}_y\left(y-\frac{b}{2}\right)}{\cos\frac{\mathrm{k}_y b}{2}}e^{-\mathrm{i}(\omega t-\mathrm{k}_z z)}\right] & 0<y<b \\ \Re\left[A\dfrac{\mathrm{i}\omega\varepsilon_2}{\alpha}e^{\alpha y}e^{-\mathrm{i}(\omega t-\mathrm{k}_z z)}\right] & y<0 \end{cases}$$

This solution is called *even* because E_z is directed in the same direction in the two half-planes above and below the line $y = \frac{b}{2}$.

1.4.1.2. *TE solutions*

We follow the same procedure for the TE solutions. Let us start with:

$$E_z = E_y = H_x = 0$$

$$H_z = \begin{cases} \Re\left[A'e^{-\alpha(y-b)}e^{-\mathrm{i}(\omega t-\mathrm{k}_z z)}\right] & y>b \\ \Re\left[(B'\sin(\mathrm{k}_y y)+C'\cos(\mathrm{k}_y y))\,e^{-\mathrm{i}(\omega t-\mathrm{k}_z z)}\right] & 0<y<b \\ \Re\left[D'e^{\alpha y}e^{-\mathrm{i}(\omega t-\mathrm{k}_z z)}\right] & y<0 \end{cases}$$

$$E_x = \begin{cases} \Re\left[A'\dfrac{\mathrm{i}\omega}{\alpha}e^{-\alpha(y-b)}e^{-\mathrm{i}(\omega t-\mathrm{k}_z z)}\right] & y>b \\ \Re\left[\dfrac{\mathrm{i}\omega}{\mathrm{k}_y}(B'\cos(\mathrm{k}_y y)-C'\sin(\mathrm{k}_y y))\,e^{-\mathrm{i}(\omega t-\mathrm{k}_z z)}\right] & 0<y<b \\ \Re\left[-\dfrac{\mathrm{i}\omega}{\alpha}D'e^{\alpha y}e^{-\mathrm{i}(\omega t-\mathrm{k}_z z)}\right] & y<0 \end{cases} \qquad [1.167]$$

$$H_y = \begin{cases} \Re\left[A'\dfrac{\mathrm{i}\mathrm{k}_z}{\alpha}e^{-\alpha(y-b)}e^{-\mathrm{i}(\omega t-\mathrm{k}_z z)}\right] & y>b \\ \Re\left[\dfrac{\mathrm{i}\mathrm{k}_z}{\mathrm{k}_y}(B'\cos(\mathrm{k}_y y)-C'\sin(\mathrm{k}_y y))\,e^{-\mathrm{i}(\omega t-\mathrm{k}_z z)}\right] & 0<y<b \\ \Re\left[-\dfrac{\mathrm{i}\mathrm{k}_z}{\alpha}D'e^{\alpha y}e^{-\mathrm{i}(\omega t-\mathrm{k}_z z)}\right] & y<0 \end{cases}$$

where E_x E_y, H_x and H_y are obtained using [1.19] and [1.21]. The components H_z and E_x are continuous at the interfaces $y = 0$ and $y = b$:

$$\begin{cases} H_z\left(y=0^-\right) = H_z\left(y=0^+\right) \Longrightarrow C' - D' = 0 \\ H_z\left(y=b^-\right) = H_z\left(y=b^+\right) \Longrightarrow A' - B'\sin\left(\mathsf{k}_y b\right) - C'\cos\left(\mathsf{k}_y b\right) = 0 \\ E_x\left(y=0^-\right) = E_x\left(y=0^+\right) \Longrightarrow D' + \dfrac{\alpha}{\mathsf{k}_y}B' = 0 \\ E_x\left(y=b^-\right) = E_x\left(y=b^+\right) \Longrightarrow A' - \dfrac{\alpha}{\mathsf{k}_y}B'\cos\left(\mathsf{k}_y b\right) \\ \qquad + \dfrac{\alpha}{\mathsf{k}_y}C'\sin\left(\mathsf{k}_y b\right) = 0 \end{cases} \quad [1.168]$$

H_y imposes the same constraints. Equations [1.168] have a solution unless $\alpha = \mathsf{k}_y \tan\frac{\mathsf{k}_y b}{2}$ or $\alpha = -\mathsf{k}_y \frac{1}{\tan\frac{\mathsf{k}_y b}{2}}$.

1) Odd solution $\left(\alpha = \mathsf{k}_y \tan\frac{\mathsf{k}_y b}{2}\right)$ $\left(B' = \frac{A'}{\tan\frac{\mathsf{k}_y b}{2}}, C' = -A', D' = -A'\right)$:

$$E_z = E_y = H_x = 0$$

$$H_z = \begin{cases} \Re\left[A' e^{-\alpha(y-b)} e^{-\mathrm{i}(\omega t - \mathsf{k}_z z)}\right] & y > b \\ \Re\left[A' \dfrac{\sin \mathsf{k}_y\left(y-\frac{b}{2}\right)}{\sin\frac{\mathsf{k}_y b}{2}} e^{-\mathrm{i}(\omega t - \mathsf{k}_z z)}\right] & 0 < y < b \\ \Re\left[-A' e^{\alpha y} e^{-\mathrm{i}(\omega t - \mathsf{k}_z z)}\right] & y < 0 \end{cases}$$

$$E_x = \begin{cases} \Re\left[A' \dfrac{\mathrm{i}\omega}{\alpha} e^{-\alpha(y-b)} e^{-\mathrm{i}(\omega t - \mathsf{k}_z z)}\right] & y > b \\ \Re\left[A' \dfrac{\mathrm{i}\omega}{\mathsf{k}_y} \dfrac{\cos \mathsf{k}_y\left(y-\frac{b}{2}\right)}{\sin\frac{\mathsf{k}_y b}{2}} e^{-\mathrm{i}(\omega t - \mathsf{k}_z z)}\right] & 0 < y < b \\ \Re\left[A' \dfrac{\mathrm{i}\omega}{\alpha} e^{\alpha y} e^{-\mathrm{i}(\omega t - \mathsf{k}_z z)}\right] & y < 0 \end{cases} \quad [1.169]$$

$$H_y = \begin{cases} \Re\left[A' \dfrac{\mathrm{i}\mathsf{k}_z}{\alpha} e^{-\alpha(y-b)} e^{-\mathrm{i}(\omega t - \mathsf{k}_z z)}\right] & y > b \\ \Re\left[A' \dfrac{\mathrm{i}\mathsf{k}_z}{\mathsf{k}_y} \dfrac{\cos \mathsf{k}_y\left(y-\frac{b}{2}\right)}{\sin\frac{\mathsf{k}_y b}{2}} e^{-\mathrm{i}(\omega t - \mathsf{k}_z z)}\right] & 0 < y < b \\ \Re\left[A' \dfrac{\mathrm{i}\mathsf{k}_z}{\alpha} e^{\alpha y} e^{-\mathrm{i}(\omega t - \mathsf{k}_z z)}\right] & y < 0 \end{cases}$$

2) Even solution $\left(\alpha = -\frac{\mathsf{k}_y}{\tan\frac{\mathsf{k}_y b}{2}}\right)$ $\left(B' = A' \tan\frac{\mathsf{k}_y b}{2}, C' = A', D' = A'\right)$:

$$E_z = E_y = H_x = 0$$

$$H_z = \begin{cases} \Re\left[A' e^{-\alpha(y-b)} e^{-\mathrm{i}(\omega t - \mathsf{k}_z z)}\right] & y > b \\ \Re\left[A' \dfrac{\cos \mathsf{k}_y \left(y - \frac{b}{2}\right)}{\cos\frac{\mathsf{k}_y b}{2}} e^{-\mathrm{i}(\omega t - \mathsf{k}_z z)}\right] & 0 < y < b \\ \Re\left[A' e^{\alpha y} e^{-\mathrm{i}(\omega t - \mathsf{k}_z z)}\right] & y < 0 \end{cases}$$

$$E_x = \begin{cases} \Re\left[A' \dfrac{\mathrm{i}\omega}{\alpha} e^{-\alpha(y-b)} e^{-\mathrm{i}(\omega t - \mathsf{k}_z z)}\right] & y > b \\ \Re\left[-A' \dfrac{\mathrm{i}\omega}{\mathsf{k}_y} \dfrac{\sin \mathsf{k}_y \left(y - \frac{b}{2}\right)}{\cos\frac{\mathsf{k}_y b}{2}} e^{-\mathrm{i}(\omega t - \mathsf{k}_z z)}\right] & 0 < y < b \\ \Re\left[-A' \dfrac{\mathrm{i}\omega}{\alpha} e^{\alpha y} e^{-\mathrm{i}(\omega t - \mathsf{k}_z z)}\right] & y < 0 \end{cases} \qquad [1.170]$$

$$H_y = \begin{cases} \Re\left[A' \dfrac{\mathrm{i}\mathsf{k}_z}{\alpha} e^{-\alpha(y-b)} e^{-\mathrm{i}(\omega t - \mathsf{k}_z z)}\right] & y > b \\ \Re\left[-A' \dfrac{\mathrm{i}\mathsf{k}_z}{\mathsf{k}_y} \dfrac{\sin \mathsf{k}_y \left(y - \frac{b}{2}\right)}{\cos\frac{\mathsf{k}_y b}{2}} e^{-\mathrm{i}(\omega t - \mathsf{k}_z z)}\right] & 0 < y < b \\ \Re\left[-A' \dfrac{\mathrm{i}\mathsf{k}_z}{\alpha} e^{\alpha y} e^{-\mathrm{i}(\omega t - \mathsf{k}_z z)}\right] & y < 0 \end{cases}$$

1.4.2. *Step index optical fiber*

In this section, we examine propagation modes in a circular waveguide composed entirely of dielectric materials. Consider an infinite cylinder of radius a made of a dielectric material of dielectric constant ε_1 (of index n_1) and surrounded by another dielectric medium with permittivity $\varepsilon_2 < \varepsilon_1$ (i.e. $\mathsf{n}_2 < \mathsf{n}_1$) that extends over the region $a < r < b$, as illustrated in Figure 1.8. These media are supposed to be non-magnetic $\mu = \mu_0$. This system is the simplest model of an *step index optical fiber*. The first medium (ε_1) is the *core* of the fiber, and the second medium (ε_2) is the *cladding*. Since the fields decrease exponentially in the cladding as a function of r over a short distance, we can assume that the cladding is infinite.

Unlike wave guides that are wrapped in a conductor, we consider the fields both within the core of the waveguide and in the cladding and then apply continuity conditions to the core-cladding interface of: (1) the tangential components of $\overrightarrow{\mathbf{H}}$; (2) the tangential components of $\overrightarrow{\mathbf{E}}$ (since $\overrightarrow{\mathbf{J}}_{\text{free}} = 0$); (3) the normal component of $\overrightarrow{\mathbf{H}}$ ($\overrightarrow{\mathbf{B}}_{\text{normal}} = \mu_0 H_r \overrightarrow{\mathbf{e}}_r$); and (4) the normal component of $\overrightarrow{\mathbf{D}}$ (since $\sigma_{\text{free}} = 0$). These continuity conditions are more complex than for a conducting wall and, therefore, the TM and TE modes generally do not exist separately, but are coupled in what are called *HE and EH modes* (in the first case, H_z is dominant, while in the second case, E_z is dominant). Only modes that are not dependent on φ (modes with azimuthal symmetry) can be separated into TE and TM modes. In each of these media, E_z and H_z satisfy the Helmholtz equation:

$$\begin{aligned}&\left(\vec{\nabla}_\perp^2+\left[\frac{\mathrm{n}_1^2\omega^2}{c^2}-\mathrm{k}_z^2\right]\right)\begin{pmatrix}E_z\\H_z\end{pmatrix}=0,\ r<a\\&\left(\vec{\nabla}_\perp^2+\left[\frac{\mathrm{n}_2^2\omega^2}{c^2}-\mathrm{k}_z^2\right]\right)\begin{pmatrix}E_z\\H_z\end{pmatrix}=0,\ r>a\end{aligned} \qquad [1.171]$$

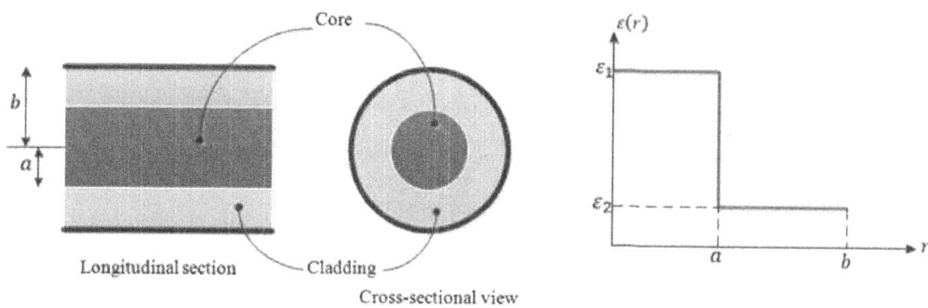

Figure 1.8. *Structure of a step index optical fiber structure*

We define $\gamma^2=\frac{\mathrm{n}_1^2\omega^2}{c^2}-\mathrm{k}_z^2$ and, to avoid an oscillatory solution (and therefore an energy loss from the fiber due to radiation) in the cladding, we take $\beta^2=-\frac{\mathrm{n}_2^2\omega^2}{c^2}+\mathrm{k}_z^2$. From these expressions, we see the necessity of having $\varepsilon_1>\varepsilon_2$ ($\mathrm{n}_1>\mathrm{n}_2$). Moreover, the propagation constant k_z must satisfy the condition $\frac{\mathrm{n}_2^2\omega^2}{c^2}<\mathrm{k}_z^2<\frac{\mathrm{n}_1^2\omega^2}{c^2}$. In order for the fields to remain finite as $r\longrightarrow 0$ and $r\longrightarrow\infty$, the solutions to equations [1.171] must be of the form:

$$\begin{aligned}E_z&=\begin{cases}A_mJ_m(\gamma r)\,e^{\pm im\varphi} & r<a\\ B_mK_m(\beta r)\,e^{\pm im\varphi} & r>a\end{cases}\\H_z&=\begin{cases}C_mJ_m(\gamma r)\,e^{\pm im\varphi} & r<a\\ D_mK_m(\beta r)\,e^{\pm im\varphi} & r>a\end{cases}\end{aligned} \qquad [1.172]$$

The function K_m does not diverge at infinity. This choice ensures that no energy propagates outward from the fiber, as the fields decay exponentially with r in the cladding.

1.4.3. *Modes with azimuthal symmetry*

Let us first consider the propagation modes without dependence on φ ($m=0$). The solutions of the Helmholtz equation (E_z and H_z) and the other field components obtained from the relations [1.19] and [1.21] are equal to (we have used the identities $J_0' = -J_1$ and $K_0' = -K_1$):

$$r<a:\begin{cases} E_z = A_0 J_0(\gamma r) \\ E_r = -\dfrac{\mathrm{i}k_z}{\gamma} A_0 J_1(\gamma r) \\ E_\varphi = \dfrac{\mathrm{i}\omega\mu_0}{\gamma} C_0 J_1(\gamma r) \end{cases} \qquad \begin{cases} H_z = C_0 J_0(\gamma r) \\ H_r = -\dfrac{\mathrm{i}k_z}{\gamma} C_0 J_1(\gamma r) \\ H_\varphi = -\dfrac{\mathrm{i}\omega\varepsilon_1}{\gamma} A_0 J_1(\gamma r) \end{cases}$$

$$r>a:\begin{cases} E_z = B_0 K_0(\beta r) \\ E_r = \dfrac{\mathrm{i}k_z}{\beta} B_0 K_1(\beta r) \\ E_\varphi = -\dfrac{\mathrm{i}\omega\mu_0}{\beta} D_0 K_1(\beta r) \end{cases} \qquad \begin{cases} H_z = D_0 K_0(\beta r) \\ H_r = \dfrac{\mathrm{i}k_z}{\beta} D_0 K_1(\beta r) \\ H_\varphi = \dfrac{\mathrm{i}\omega\varepsilon_2}{\beta} B_0 K_1(\beta r) \end{cases} \quad [1.173]$$

It is clear that the two modes TE and TM are distinct in this case.

To that end, we treat the two modes separately:

– *TE modes*($E_z = 0$): for the TE modes, the previous equations reduce to the following (fixing $A_0 = B_0 = 0$):

$$r<a:\begin{cases} E_r = E_z = H_\varphi = 0 \\ E_\varphi = \dfrac{\mathrm{i}\omega\mu_0}{\gamma} C_0 J_1(\gamma r) \\ H_r = -\dfrac{\mathrm{i}k_z}{\gamma} C_0 J_1(\gamma r) \\ H_z = C_0 J_0(\gamma r) \end{cases} \qquad r>a:\begin{cases} E_r = E_z = H_\varphi = 0 \\ E_\varphi = -\dfrac{\mathrm{i}\omega\mu_0}{\beta} D_0 K_1(\beta r) \\ H_r = \dfrac{\mathrm{i}k_z}{\beta} D_0 K_1(\beta r) \\ H_z = D_0 K_0(\beta r) \end{cases} \quad [1.174]$$

The continuity equations for H_r, H_z and E_φ, at the interface $r = a$, lead, respectively, to the following coupled equations:

$$\begin{aligned} &-\frac{\mathrm{i}k_z}{\gamma} C_0 J_1(\gamma a) = \frac{\mathrm{i}k_z}{\beta} D_0 K_1(\beta a) \\ &C_0 J_0(\gamma a) = D_0 J K_0(\beta a) \\ &\frac{\mathrm{i}\omega\mu_0}{\gamma} C_0 J_1(\gamma a) = -\frac{\mathrm{i}\omega\mu_0}{\beta} D_0 K_1(\beta a) \end{aligned} \quad [1.175]$$

It can be seen that H_r and E_φ yield the same constraints. Eliminating the ratio $\frac{C_0}{D_0}$ between the first two equations, we obtain:

$$-\frac{J_1(\gamma a)}{\gamma a J_0(\gamma a)} = \frac{K_1(\beta a)}{\beta a K_0(\beta a)} \quad [1.176]$$

– *TM modes* ($H_z = 0$): for the TM modes (choosing $C_0 = D_0 = 0$), we get:

$$r < a : \begin{cases} E_\varphi = H_r = H_z = 0 \\ E_r = -\dfrac{\mathsf{i}k_z}{\gamma} A_0 J_1\left(\gamma r\right) \\ E_z = A_0 J_0\left(\gamma r\right) \\ H_\varphi = -\dfrac{\mathsf{i}\omega\varepsilon_1}{\gamma} A_0 J_1\left(\gamma r\right) \end{cases} \qquad r > a : \begin{cases} E_\varphi = H_r = H_r = 0 \\ E_r = \dfrac{\mathsf{i}k_z}{\beta} B_0 K_1\left(\beta r\right) \\ E_z = B_0 K_0\left(\beta r\right) \\ H_\varphi = \dfrac{\mathsf{i}\omega\varepsilon_2}{\beta} B_0 K_1\left(\beta r\right) \end{cases} \qquad [1.177]$$

The continuity of D_r, E_z and H_φ, at the interface $r = a$, enforces the following (the components of D_r and H_φ yield the same constraints):

$$\begin{aligned} &-\frac{\mathsf{i}k_z\varepsilon_1}{\gamma} A_0 J_1\left(\gamma r\right) = \frac{\mathsf{i}k_z\varepsilon_2}{\beta} B_0 K_1\left(\beta r\right) \\ &A_0 J_0\left(\gamma r\right) = B_0 K_0\left(\beta r\right) \\ &-\frac{\mathsf{i}\omega\varepsilon_1}{\gamma} A_0 J_1\left(\gamma r\right) = \frac{\mathsf{i}\omega\varepsilon_2}{\beta} B_0 K_1\left(\beta r\right) \end{aligned} \qquad [1.178]$$

and consequently, by eliminating the ratio $\frac{A_0}{B_0}$ between these equations:

$$-\frac{\varepsilon_1 J_1\left(\gamma a\right)}{\gamma a J_0\left(\gamma a\right)} = \frac{\varepsilon_2 K_1\left(\beta a\right)}{\beta a K_0\left(\beta a\right)} \qquad [1.179]$$

Equations [1.176] and [1.179] can be solved graphically. Noting that:

$$\gamma^2 + \beta^2 = \left(\varepsilon_1 - \varepsilon_2\right)\mu_0\omega^2 = \left(\mathsf{n}_1^2 - \mathsf{n}_2^2\right)\frac{\omega^2}{c^2} \qquad [1.180]$$

we observe that $|\gamma|$ and $|\beta|$ are each less than $\frac{\omega}{c}\sqrt{\mathsf{n}_1^2 - \mathsf{n}_2^2}$. Let us first plot the functions $-\frac{J_1(\gamma a)}{\gamma a J_0(\gamma a)}$ and $\frac{K_1(\beta a)}{\beta a K_0(\beta a)}$ as functions of their respective arguments γa and βb, as illustrated in Figures 1.9(a) and (b). From these graphs, we observe that $\frac{J_1}{J_0}$ diverges and changes sign at the level of the roots of J_0. Moreover, since $J_0\left(x\right) \underset{x\to 0}{\longrightarrow} 1$ and $J_1\left(x\right) \underset{x\to 0}{\longrightarrow} \frac{x}{2}$, the ratio $\frac{J_1(x)}{xJ_0(x)} \underset{x\to 0}{\longrightarrow} \frac{1}{2}$. Similarly, given that $K_{0,1}\left(x\right) \underset{x\to\infty}{\longrightarrow} \sqrt{\frac{\pi}{2x}}e^{-x}$, $K_0\left(x\right) \underset{x\to 0}{\longrightarrow} -\ln\frac{x}{2}$ and $K_1\left(x\right) \underset{x\to 0}{\longrightarrow} \frac{1}{x}$, we deduce $\frac{K_1(x)}{xK_0(x)} \underset{x\to\infty}{\longrightarrow} \frac{1}{x} \longrightarrow 0$ and $\frac{K_1(x)}{xK_0(x)} \underset{x\to 0}{\longrightarrow} -\frac{\ln\frac{x}{2}}{x^2} \longrightarrow +\infty$. The graphical solution proceeds as follows in the case of TE modes (the same procedure applies for the TM modes): in a coordinate system, we plot the function $\frac{J_1\left(\sqrt{x}\right)}{\sqrt{x}J_0\left(\sqrt{x}\right)}$ as a function of x. On the same graph, we plot the function $\frac{K_1\left(\sqrt{y}\right)}{\sqrt{y}K_0\left(\sqrt{y}\right)}$ as a function of $y = \frac{\omega a}{c}\sqrt{\mathsf{n}_1^2 - \mathsf{n}_2^2} - x$. The intersections (the black points) of the two curves correspond to the solutions of equation [1.176], as shown in Figure 1.9(c).

If $\frac{\omega a}{c}\sqrt{\mathsf{n}_1^2-\mathsf{n}_2^2} < x_{01} = 2.405$, no intersection is possible. The first intersection occurs when $\frac{\omega a}{c}\sqrt{\mathsf{n}_1^2-\mathsf{n}_2^2} = x_{01}$, which gives $\omega_{01} = \frac{c x_{01}}{a\sqrt{\mathsf{n}_1^2-\mathsf{n}_2^2}}$. At this frequency, we have $\beta = 0$, and consequently $\mathsf{k}_z = \frac{\mathsf{n}_2\omega_{01}}{c}$: the wave propagates in the cladding as if there were no guiding. Below this frequency, β becomes imaginary, which turns K_0 and K_1 into Bessel functions that oscillate radially and decay as $\frac{1}{\sqrt{r}}$. The fiber no longer acts as a waveguide but rather as an antenna. When ω increases beyond ω_{01}, β grows more rapidly than γ if $\mathsf{n}_1 \gg \mathsf{n}_2$. This means that the wave decays rapidly with r in the cladding. ω_{01} thus appears as the first cutoff frequency. Each time x crosses a root of J_0, a new mode appears, each with its own cutoff frequency:

$$\omega_{0n} = \frac{c x_{0n}}{a\sqrt{\mathsf{n}_1^2-\mathsf{n}_2^2}} \qquad [1.181]$$

1.4.4. *Modes with azimuthal dependence*

The most general modes (with azimuthal dependence) propagating along dielectrics are called *HE modes* and *EH modes*. They have both $E_z \neq 0$ and $H_z \neq 0$. In particular, it can be shown that the HE_{11} mode is dominant and has no cutoff frequency. Let us now examine all of this in detail. In general, the components of the fields $\vec{\mathbf{E}}$ and $\vec{\mathbf{H}}$ take the form:

$$r < a: \begin{cases} E_z = A_m J_m(\gamma r)\, e^{\pm i m\varphi} \\ E_r = \left[\mp\frac{m\omega\mu_0}{\gamma^2 r} C_m J_m(\gamma r) + \frac{\mathsf{i}\mathsf{k}_z}{\gamma} A_m J_m'(\gamma r)\right] e^{\pm i m\varphi} \\ E_\varphi = \left[\mp\frac{m\mathsf{k}_z}{\gamma^2 r} A_m J_m(\gamma r) - \frac{\mathsf{i}\omega\mu_0}{\gamma} C_m J_m'(\gamma r)\right] e^{\pm i m\varphi} \\ H_z = C_m J_m(\gamma r)\, e^{\pm i m\varphi} \\ H_r = \left[\pm\frac{m\omega\varepsilon_1}{\gamma^2 r} A_m J_m(\gamma r) + \frac{\mathsf{i}\mathsf{k}_z}{\gamma} C_m J_m'(\gamma r)\right] e^{\pm i m\varphi} \\ H_\varphi = \left[\mp\frac{m\mathsf{k}_z}{\gamma^2 r} C_m J_m(\gamma r) + \frac{\mathsf{i}\omega\varepsilon_1}{\gamma} A_m J_m'(\gamma r)\right] e^{\pm i m\varphi} \end{cases} \qquad [1.182]$$

and:

$$r > a: \begin{cases} E_z = B_m K_m(\beta r)\, e^{\pm i m\varphi} \\ E_r = \left[\pm\frac{m\omega\mu_0}{\beta^2 r} D_m K_m(\beta r) - \frac{\mathsf{i}\mathsf{k}_z}{\beta} B_m K_m'(\beta r)\right] e^{\pm i m\varphi} \\ E_\varphi = \left[\pm\frac{m\mathsf{k}_z}{\beta^2 r} B_m K_m(\beta r) + \frac{\mathsf{i}\omega\mu_0}{\beta} D_m K_m'(\beta r)\right] e^{\pm i m\varphi} \\ H_z = D_m K_m(\beta r)\, e^{\pm i m\varphi} \\ H_r = \left[\mp\frac{m\omega\varepsilon_2}{\beta^2 r} B_m K_m(\beta r) - \frac{\mathsf{i}\mathsf{k}_z}{\beta} D_m K_m'(\beta r)\right] e^{\pm i m\varphi} \\ H_\varphi = \left[\pm\frac{m\mathsf{k}_z}{\beta^2 r} D_m K_m(\beta r) - \frac{\mathsf{i}\omega\varepsilon_2}{\beta} B_m K_m'(\beta r)\right] e^{\pm i m\varphi} \end{cases} \qquad [1.183]$$

The continuity of the components E_z, H_z, E_φ, H_φ, D_r and B_r, at the interface $r = a$, imposes the following equalities, respectively:

$$\begin{aligned}
&A_m J_m = B_m K_m && \text{(i)}\\
&C_m J_m = D_m K_m && \text{(ii)}\\
&\mp\frac{m\mathsf{k}_z}{\gamma^2 a}A_m J_m - \frac{\mathsf{i}\omega\mu_0}{\gamma}C_m J_m' = \pm\frac{m\mathsf{k}_z}{\beta^2 a}B_m K_m + \frac{\mathsf{i}\omega\mu_0}{\beta}D_m K_m' && \text{(iii)}\\
&\mp\frac{m\mathsf{k}_z}{\gamma^2 a}C_m J_m + \frac{\mathsf{i}\omega\varepsilon_1}{\gamma}A_m J_m' = \pm\frac{m\mathsf{k}_z}{\beta^2 a}D_m K_m - \frac{\mathsf{i}\omega\varepsilon_2}{\beta}B_m K_m' && \text{(iv)}\\
&\varepsilon_1\left(\mp\frac{m\omega\mu_0}{\gamma^2 a}C_m J_m + \frac{\mathsf{i}\mathsf{k}_z}{\gamma}A_m J_m'\right) = \varepsilon_2\left(\pm\frac{m\omega\mu_0}{\beta^2 a}D_m K_m - \frac{\mathsf{i}\mathsf{k}_z}{\beta}B_m K_m'\right) && \text{(v)}\\
&\pm\frac{m\omega\varepsilon_1}{\gamma^2 a}A_m J_m + \frac{\mathsf{i}\mathsf{k}_z}{\gamma}C_m J_m' = \mp\frac{m\omega\varepsilon_2}{\beta^2 a}B_m K_m - \frac{\mathsf{i}\mathsf{k}_z}{\beta}D_m K_m' && \text{(vi)}
\end{aligned}$$

[1.184]

where we have used the notation $J_m = J_m(\gamma a)$, $J_m' = J_m'(\gamma a)$, $K_m = K_m(\gamma a)$ and $K_m' = K_m'(\gamma a)$. Let us rewrite equation [1.184iii] in the following form:

$$\mp\frac{m\mathsf{k}_z}{a}\left(\frac{A_m J_m}{\gamma^2} + \frac{B_m K_m}{\beta^2}\right) = \mathsf{i}\omega\mu_0\left(\frac{C_m J_m'}{\gamma} + \frac{D_m K_m'}{\beta}\right) \qquad [1.185]$$

Substituting into [1.185] the coefficients B_m and D_m obtained from equations [1.184i] and [1.184ii], we obtain:

$$\mp\frac{m\mathsf{k}_z A_m}{a}\left(\frac{1}{\gamma^2} + \frac{1}{\beta^2}\right) = \mathsf{i}\omega\mu_0 C_m\left(\frac{J_m'}{\gamma J_m} + \frac{K_m'}{\beta K_m}\right) \qquad [1.186]$$

Similarly, the same procedure can be applied to equations [1.184iv], [1.184v] and [1.184vi]:

$$\begin{aligned}
&\pm\frac{m\mathsf{k}_z}{a}\left(\frac{C_m J_m}{\gamma^2} + \frac{D_m K_m}{\beta^2}\right) = \mathsf{i}\omega\left(\frac{\varepsilon_1 A_m J_m'}{\gamma} + \frac{\varepsilon_2 B_m K_m'}{\beta}\right) && \text{(i)}\\
&\pm\frac{m\omega\mu_0}{a}\left(\frac{\varepsilon_1 C_m J_m}{\gamma^2} + \frac{\varepsilon_2 D_m K_m}{\beta^2}\right) = \mathsf{i}\mathsf{k}_z\left(\frac{\varepsilon_1 A_m J_m'}{\gamma} + \frac{\varepsilon_2 B_m K_m'}{\beta}\right) && \text{(ii)} \quad [1.187]\\
&\mp\frac{m\omega}{a}\left(\frac{\varepsilon_1 A_m J_m}{\gamma^2} + \frac{\varepsilon_2 B_m K_m}{\beta^2}\right) = \mathsf{i}\mathsf{k}_z\left(\frac{C_m J_m'}{\gamma} + \frac{D_m K_m'}{\beta}\right) && \text{(iii)}
\end{aligned}$$

which yields:

$$\begin{aligned}
&\pm\frac{m\mathsf{k}_z C_m}{a}\left(\frac{1}{\gamma^2} + \frac{1}{\beta^2}\right) = \mathsf{i}\omega A_m\left(\frac{\varepsilon_1 J_m'}{\gamma J_m} + \frac{\varepsilon_2 K_m'}{\beta K_m}\right) && \text{(i)}\\
&\pm\frac{m\omega\mu_0 C_m}{a}\left(\frac{\varepsilon_1}{\gamma^2} + \frac{\varepsilon_2}{\beta^2}\right) = \mathsf{i}\mathsf{k}_z A_m\left(\frac{\varepsilon_1 J_m'}{\gamma J_m} + \frac{\varepsilon_2 K_m'}{\beta K_m}\right) && \text{(ii)} \quad [1.188]\\
&\mp\frac{m\omega A_m}{a}\left(\frac{\varepsilon_1}{\gamma^2} + \frac{\varepsilon_2}{\beta^2}\right) = \mathsf{i}\mathsf{k}_z C_m\left(\frac{J_m'}{\gamma J_m} + \frac{K_m'}{\beta K_m}\right) && \text{(iii)}
\end{aligned}$$

The constants A_m and C_m can now be eliminated. Multiplying [1.186] by [1.188i] and then simplifying by $A_m C_m$, we obtain:

$$\left(\frac{J'_m}{\gamma J_m}+\frac{K'_m}{\beta K_m}\right)\left(\frac{\varepsilon_1 J'_m}{\gamma J_m}+\frac{\varepsilon_2 K'_m}{\beta K_m}\right)=\frac{m^2\mathsf{k}_z^2}{\mu_0 a^2\omega^2}\left(\frac{1}{\gamma^2}+\frac{1}{\beta^2}\right)^2$$
$$=\frac{m^2\mathsf{k}_z^2\omega^2}{\mu_0 a^2 c^4\gamma^2\beta^2}\left(\mathsf{n}_1^2-\mathsf{n}_2^2\right)^2 \qquad [1.189]$$

Similarly, proceeding in the same way with formulas [1.188ii] and [1.188iii], we find:

$$\left(\frac{J'_m}{\gamma J_m}+\frac{K'_m}{\beta K_m}\right)\left(\frac{\varepsilon_1 J'_m}{\gamma J_m}+\frac{\varepsilon_2 K'_m}{\beta K_m}\right)=\frac{m^2\omega^2\mu_0}{a^2\mathsf{k}_z^2}\left(\frac{\varepsilon_1}{\gamma^2}+\frac{\varepsilon_2}{\beta^2}\right)^2 \qquad [1.190]$$

The expressions [1.189] and [1.190] are identical. In fact, dividing [1.186] by [1.188iii] (respectively, [1.188i] by [1.188iii]), we deduce that:

$$\left(\frac{\varepsilon_1}{\gamma^2}+\frac{\varepsilon_2}{\beta^2}\right)^2=\frac{\mathsf{k}_z^2}{\mu_0\omega^2}\left(\frac{1}{\gamma^2}+\frac{1}{\beta^2}\right)^2 \qquad [1.191]$$

which show that [1.189] and [1.190] are indeed the same. Equation [1.189] provides the relationship between the constant k_z and the frequency ω. For a given frequency ω and a given value of m, it allows us to determine the propagation constant k_z. In other words, the solutions of [1.189] yield the dispersion relations of the possible modes. The relation [1.189] is called the *dispersion relation* of the step index optical fiber or the *characteristic equation* of the propagation modes. In fact, for each value of m, this equation admits a finite number of solutions that satisfy the guiding condition $\frac{\mathsf{n}_2^2\omega^2}{c^2} < \mathsf{k}_z^2 < \frac{\mathsf{n}_1^2\omega^2}{c^2}$. This multiplicity of solutions results from the oscillatory nature of the Bessel functions J_m. Thus, each solution of [1.189] can be characterized by two indices (m, n), where m characterizes the azimuthal dependence of the electromagnetic field and n corresponds to the *n*th solution of [1.189] associated with the given value of m. It should be noted that the total number of modes that can propagate depends on the frequency of the incident wave in the fiber. In general, the number of modes increases with frequency. Once the solutions of [1.189] are obtained, the constants B_m, C_m and D_m can be expressed in terms of A_m, and the explicit expressions of the field components for each propagation mode can then be determined. From [1.184i], [1.184ii] and [1.186], we deduce that:

$$\begin{aligned}B_m &= \frac{J_m(\gamma a)}{K_m(\beta a)}A_m\\ C_m &= \pm\frac{\mathsf{ik}_z}{\mu_0\omega}PA_m\\ D_m &= \pm\frac{\mathsf{ik}_z}{\mu_0\omega}\frac{J_m(\gamma a)}{K_m(\beta a)}PA_m\end{aligned} \qquad [1.192]$$

where:

$$P = \frac{m\left(\frac{1}{\gamma^2 a^2} + \frac{1}{\beta^2 a^2}\right)}{\frac{J'_m(\gamma a)}{\gamma a J_m(\gamma a)} + \frac{K'_m(\beta a)}{\beta a K_m(\beta a)}} \qquad [1.193]$$

When the azimuthal number m is equal to zero, the constraints [1.176] and [1.179] are satisfied, since the right-hand side of [1.189] vanishes. This implies that one or the other of the two factors on the left must cancel. If the term $\frac{J'_m}{\gamma J_m} + \frac{K'_m}{\beta K_m}$ is zero, this corresponds to the TE modes. On the other hand, if the term $\frac{\varepsilon_1 J'_m}{\gamma J_m} + \frac{\varepsilon_2 K'_m}{\beta K_m}$ is zero, this corresponds to the modes TM. We conclude that when $m = 0$, the TE and TM modes are not coupled. The right-hand side of equation [1.189] therefore acts to couple the TM and TE solutions, resulting in HE and EH mode types. When the two refractive indices n_1 and n_2 are very close, the right-hand side of [1.189] can be neglected as a first approximation. Consequently, we obtain equations similar to [1.176] and [1.179], namely, $-\frac{J'_m(\gamma a)}{\gamma a J_m(\gamma a)} = \frac{K'_m(\beta a)}{\beta a K_m(\beta a)}$ or $-\frac{\varepsilon_1 J'_m(\gamma a)}{\gamma a J_m(\gamma a)} = \frac{\varepsilon_2 K'_m(\beta a)}{\beta a K_m(\beta a)}$. The graphical solutions of these equations show no cutoff frequency for $m \geq 1$.

Taking into account equations [1.192], the longitudinal fields take the following form:

$$\begin{aligned} E_z &= \begin{cases} A_m J_m(\gamma r)\, e^{\pm im\varphi} e^{-\mathrm{i}(\omega t - \mathrm{k}_z z)} & r < a \\ A_m \frac{J_m(\gamma a)}{K_m(\beta a)} K_m(\beta r)\, e^{\pm im\varphi} e^{-\mathrm{i}(\omega t - \mathrm{k}_z z)} & r > a \end{cases} \\ H_z &= \begin{cases} \pm \frac{\mathrm{ik}_z P}{\mu_0 \omega} A_m J_m(\gamma r)\, e^{\pm im\varphi} e^{-\mathrm{i}(\omega t - \mathrm{k}_z z)} & r < a \\ \pm \frac{\mathrm{ik}_z P}{\mu_0 \omega} A_m \frac{J_m(\gamma a)}{K_m(\beta a)} K_m(\beta r)\, e^{\pm im\varphi} e^{-\mathrm{i}(\omega t - \mathrm{k}_z z)} & r > a \end{cases} \end{aligned} \qquad [1.194]$$

and, consequently:

$$\left|\frac{H_z}{E_z}\right| = \frac{\mathrm{k}_z P}{\mu_0 \omega} \qquad [1.195]$$

If $\left|\frac{H_z}{E_z}\right| > 1$, the field is denoted EH, while for $\left|\frac{H_z}{E_z}\right| < 1$, the field is denoted HE. The transverse components of the electromagnetic field, given by equations [1.182] and [1.183], involve derivatives of Bessel functions. These derivatives can be eliminated using the recurrence relations: $J'_m(x) = \frac{m}{x} J_m(x) - J_{m+1}(x) = -\frac{m}{x} J_m(x) + J_{m-1}(x)$ and $K'_m(x) = \frac{m}{x} K_m(x) - K_{m+1}(x) = -\frac{m}{x} K_m(x) - K_{m-1}(x)$. Here are the results:

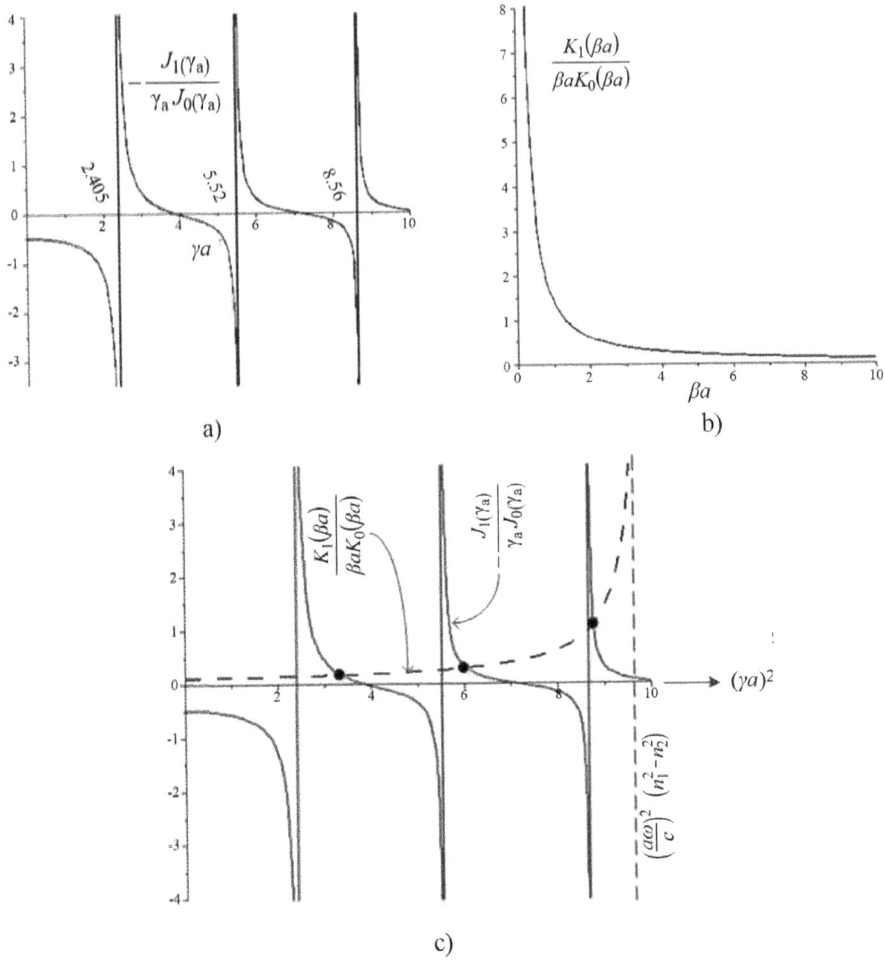

Figure 1.9. *(a and b) Behavior of the functions $\frac{J_1(\gamma a)}{\gamma a J_0(\gamma a)}$ and $\frac{K_1(\beta a)}{\beta a K_0(\beta a)}$ as functions of γa and βa, respectively. (c) Graphical solutions of equation [1.176]. The function $\frac{J_1(\sqrt{x})}{\sqrt{x} J_1(\sqrt{x})}$ appears in red, and the function $\frac{K_1(\sqrt{y})}{\sqrt{y} K_1(\sqrt{y})}$ in blue (the graph of $\frac{K_1(\sqrt{y})}{\sqrt{y} K_1(\sqrt{y})}$ is oriented to the left and plotted over the interval $\left[0, \frac{\omega a}{c}\sqrt{n_1^2 - n_2^2}\right]$). The solutions (intersection points) are marked with a black dot. For a color version of this figure, see www.iste.co.uk/abdesselam/electrodynamics3.zip*

– *Modes* EH_{mn}: knowing that $J'_m(\gamma r) = \frac{m}{\gamma r} J_m(\gamma r) - J_{m+1}(\gamma r)$ and $K'_m(\beta r) = \frac{m}{\beta r} K_m(\beta r) - K_{m+1}(\beta r)$, the fields for the EH_{mn} modes can be expressed as follows:

- In the core ($r < a$), we have (where $\mathcal{F}_{\pm m} = e^{\pm \mathrm{i} m\varphi} e^{-\mathrm{i}(\omega t - \mathrm{k}_z z)}$ is the phase term):

$$\begin{cases} E_r = \frac{\mathrm{i}\mathrm{k}_z}{\gamma} A_m \left[\frac{m}{\gamma r} (1-P) J_m(\gamma r) - J_{m+1}(\gamma r) \right] \mathcal{F}_{\pm m} \\ E_\varphi = \mp \frac{\mathrm{k}_z}{\gamma} A_m \left[\frac{m}{\gamma r} (1-P) J_m(\gamma r) + P J_{m+1}(\gamma r) \right] \mathcal{F}_{\pm m} \\ H_r = \pm \frac{\omega \varepsilon_1}{\gamma} A_m \left[\frac{m}{\gamma r} \left(1 - \frac{\mathrm{k}_z^2 c^2}{\omega^2 n_1^2} P \right) J_m(\gamma r) + \frac{\mathrm{k}_z^2 c^2}{\omega^2 n_1^2} P J_{m+1}(\gamma r) \right] \mathcal{F}_{\pm m} \\ H_\varphi = \frac{\mathrm{i}\omega \varepsilon_1}{\gamma} A_m \left[\frac{m}{\gamma r} \left(1 - \frac{\mathrm{k}_z^2 c^2}{\omega^2 n_1^2} P \right) J_m(\gamma r) - J_{m+1}(\gamma r) \right] \mathcal{F}_{\pm m} \end{cases}$$

[1.196]

- In the cladding ($r > a$), we have:

$$\begin{cases} E_r = -\frac{\mathrm{i}\mathrm{k}_z}{\beta} \frac{J_m(\gamma a)}{K_m(\beta a)} A_m \left[\frac{m}{\beta r} (1-P) K_m(\beta r) - K_{m+1}(\beta r) \right] \mathcal{F}_{\pm m} \\ E_\varphi = \pm \frac{\mathrm{k}_z}{\beta} \frac{J_m(\gamma a)}{K_m(\beta a)} A_m \left[\frac{m}{\beta r} (1-P) K_m(\beta r) + P K_{m+1}(\beta r) \right] \mathcal{F}_{\pm m} \\ H_r = \mp \frac{\omega \varepsilon_2}{\beta} \frac{J_m(\gamma a)}{K_m(\beta a)} A_m \left[\frac{m}{\beta r} \left(1 - \frac{\mathrm{k}_z^2 c^2}{\omega^2 n_2^2} P \right) K_m(\beta r) \right. \\ \left. + \frac{\mathrm{k}_z^2 c^2}{\omega^2 n_2^2} P K_{m+1}(\beta r) \right] \mathcal{F}_{\pm m} \\ H_\varphi = -\frac{\mathrm{i}\omega \varepsilon_2}{\beta} \frac{J_m(\gamma a)}{K_m(\beta a)} A_m \left[\frac{m}{\beta r} \left(1 - \frac{\mathrm{k}_z^2 c^2}{\omega^2 n_2^2} P \right) K_m(\beta r) - K_{m+1}(\beta r) \right] \mathcal{F}_{\pm m} \end{cases}$$

[1.197]

– *Modes HE$_{mn}$*: since $J_m'(\gamma r) = \frac{m}{\gamma r} J_m(\gamma r) + J_{m-1}(\gamma r)$ and $K_m'(\beta r) = -\frac{m}{\beta r} K_m(\beta r) - K_{m-1}(\beta r)$, the fields for the modes HE$_{mn}$ can be expressed as follows:

- In the core ($r < a$), we have:

$$\begin{cases} E_r = \frac{\mathrm{i}\mathrm{k}_z}{\gamma} A_m \left[-\frac{m}{\gamma r} (1+P) J_m(\gamma r) + J_{m-1}(\gamma r) \right] \mathcal{F}_{\pm m} \\ E_\varphi = \mp \frac{\mathrm{k}_z}{\gamma} A_m \left[\frac{m}{\gamma r} (1+P) J_m(\gamma r) - P J_{m-1}(\gamma r) \right] \mathcal{F}_{\pm m} \\ H_r = \pm \frac{\omega \varepsilon_1}{\gamma} A_m \left[\frac{m}{\gamma r} \left(1 + \frac{\mathrm{k}_z^2 c^2}{\omega^2 n_1^2} P \right) J_m(\gamma r) - \frac{\mathrm{k}_z^2 c^2}{\omega^2 n_1^2} P J_{m-1}(\gamma r) \right] \mathcal{F}_{\pm m} \\ H_\varphi = \frac{\mathrm{i}\omega \varepsilon_1}{\gamma} A_m \left[-\frac{m}{\gamma r} \left(1 + \frac{\mathrm{k}_z^2 c^2}{\omega^2 n_1^2} P \right) J_m(\gamma r) + J_{m-1}(\gamma r) \right] \mathcal{F}_{\pm m} \end{cases}$$

[1.198]

- In the cladding ($r > a$), we have:

$$\begin{cases} E_r = \frac{\mathsf{i}\mathsf{k}_z}{\beta}\frac{J_m(\gamma a)}{K_m(\beta a)}A_m\left[\frac{m}{\beta r}(1+P)K_m(\beta r)+K_{m-1}(\beta r)\right]\mathcal{F}_{\pm m} \\ E_\varphi = \pm\frac{\mathsf{k}_z}{\beta}\frac{J_m(\gamma a)}{K_m(\beta a)}A_m\left[\frac{m}{\beta r}(1+P)K_m(\beta r)+PK_{m-1}(\beta r)\right]\mathcal{F}_{\pm m} \\ H_r = \mp\frac{\omega\varepsilon_2}{\beta}\frac{J_m(\gamma a)}{K_m(\beta a)}A_m\left[\frac{m}{\beta r}\left(1+\frac{\mathsf{k}_z^2c^2}{\omega^2n_2^2}P\right)K_m(\beta r)\right. \\ \left.+\frac{\mathsf{k}_z^2c^2}{\omega^2n_2^2}PK_{m-1}(\beta r)\right]\mathcal{F}_{\pm m} \\ H_\varphi = \frac{\mathsf{i}\omega\varepsilon_2}{\beta}\frac{J_m(\gamma a)}{K_m(\beta a)}A_m\left[\frac{m}{\beta r}\left(1+\frac{\mathsf{k}_z^2c^2}{\omega^2n_2^2}P\right)K_m(\beta r)\right. \\ \left.+K_{m-1}(\beta r)\right]\mathcal{F}_{\pm m} \end{cases} \quad [1.199]$$

1.4.5. *Propagation modes in weakly guiding fibers*

1.4.5.1. *Dispersion relation*

When the difference between the indices n_1 and n_2 is very small, we can write in the first approximation that $\mathsf{n}_1 \simeq \mathsf{n}_2$ and consequently we must have $\mathsf{k}_z \simeq \frac{\mathsf{n}_1\omega}{c} \simeq \frac{\mathsf{n}_2\omega}{c}$, given the guiding condition $\frac{\mathsf{n}_2\omega}{c} < \mathsf{k}_z < \frac{\mathsf{n}_1\omega}{c}$. Under this approximation, the dispersion relation [1.189] becomes:

$$\left(\frac{J_m'(\gamma a)}{\gamma a J_m(\gamma a)}+\frac{K_m'(\beta a)}{\beta a K_m(\beta a)}\right)^2 \simeq m^2\left(\frac{1}{(\gamma a)^2}+\frac{1}{(\beta a)^2}\right)^2 \quad [1.200]$$

The latter is divided into two relations:

$$\frac{J_m'(\gamma a)}{\gamma a J_m(\gamma a)}+\frac{K_m'(\beta a)}{\beta a K_m(\beta a)} = m\left(\frac{1}{\gamma^2a^2}+\frac{1}{\beta^2a^2}\right) \quad [1.201]$$

or:

$$\frac{J_m'(\gamma a)}{\gamma a J_m(\gamma a)}+\frac{K_m'(\beta a)}{\beta a K_m(\beta a)} = -m\left(\frac{1}{\gamma^2a^2}+\frac{1}{\beta^2a^2}\right) \quad [1.202]$$

These two dispersion relations correspond, respectively, to the EH and HE modes. In this weak-guidance approximation, the expression for P defined by [1.193] reduces to $P = \pm 1$. By eliminating the derivative functions J_m' and K_m' from the relation [1.201] using the identities $J_m'(x) = \frac{m}{x}J_m(x) - J_{m+1}(x)$ and $K_m'(x) = \frac{m}{x}K_m(x) - K_{m+1}(x)$, we obtain:

$$\frac{\gamma a J_m(\gamma a)}{J_{m+1}(\gamma a)}+\frac{\beta a K_m(\beta a)}{K_{m+1}(\beta a)} = 0 \quad [1.203]$$

The expression [1.203] is the characteristic equation of the EH_{mn} modes. The index n designates the *n*th root of equation [1.203]. This equation corresponds to $P = 1$. Similarly, we can eliminate the derivative functions J'_m and K'_m from [1.202] using the formulas $J'_m(x) = -\frac{m}{x}J_m(x) + J_{m-1}(x)$ and $K'_m(x) = -\frac{m}{x}K_m(x) - K_{m-1}(x)$. The relation [1.202] then becomes:

$$\frac{\gamma a J_m(\gamma a)}{J_{m-1}(\gamma a)} - \frac{\beta a K_m(\beta a)}{K_{m-1}(\beta a)} = 0 \qquad [1.204]$$

This is the characteristic equation of the modes HE_{mn} and is associated with $P = -1$.

1.4.5.2. *Degeneracy of the modes*

The dispersion relation of $\text{HE}_{m+1,n}$, which follows directly from [1.204], is written as:

$$\frac{\gamma a J_{m+1}(\gamma a)}{J_m(\gamma a)} - \frac{\beta a K_{m+1}(\beta a)}{K_m(\beta a)} = 0 \qquad [1.205]$$

By substituting into this relation the expressions $J_{m+1}(x) = \frac{2m}{x}J_m(x) - J_{m-1}(x)$ and $K_{m+1}(x) = \frac{2m}{x}K_m(x) + K_{m-1}(x)$, we obtain:

$$\frac{\gamma a J_{m-1}(\gamma a)}{J_m(\gamma a)} + \frac{\beta a K_{m-1}(\beta a)}{K_m(\beta a)} = 0 \qquad [1.206]$$

This last relation is exactly the same as [1.203] for the EH modes, but with the index $m-1$. In other words, the dispersion relation of the $\text{HE}_{m+1,n}$ modes is identical to that of the $\text{EH}_{m-1,n}$ modes. This means that these two modes have exactly the same value of k_z. This is called the degeneracy of the $\text{HE}_{m+1,n}$ and $\text{EH}_{m-1,n}$ modes.

1.4.5.3. *Expressions of the field components*

Let us now give the expressions for the field components in the weak-guidance approximation in cylindrical coordinates:

$$E_r = -\frac{i\mathsf{k}_z}{\gamma}A_m J_{m+1}(\gamma r)\mathcal{F}_{\pm m} \qquad E_r = \frac{i\mathsf{k}_z}{\beta}\frac{J_m(\gamma a)}{K_m(\beta a)}A_m K_{m+1}(\beta r)\mathcal{F}_{\pm m}$$

$$E_\varphi = \mp\frac{\mathsf{k}_z}{\gamma}A_m J_{m+1}(\gamma r)\mathcal{F}_{\pm m} \qquad E_\varphi = \pm\frac{\mathsf{k}_z}{\beta}\frac{J_m(\gamma a)}{K_m(\beta a)}A_m K_{m+1}(\beta r)\mathcal{F}_{\pm m}$$

$$E_z = A_m J_m(\gamma r)\mathcal{F}_{\pm m} \qquad E_z = A_m\frac{J_m(\gamma a)}{K_m(\beta a)}K_m(\beta r)\mathcal{F}_{\pm m}$$

$$H_r = \pm\frac{\omega\varepsilon_1}{\gamma}A_m J_{m+1}(\gamma r)\mathcal{F}_{\pm m} \qquad H_r = \mp\frac{\omega\varepsilon_2}{\beta}\frac{J_m(\gamma a)}{K_m(\beta a)}A_m K_{m+1}(\beta r)\mathcal{F}_{\pm m}$$

$$\underbrace{\begin{aligned} H_\varphi &= -\frac{\mathrm{i}\omega\varepsilon_1}{\gamma} A_m J_{m+1}(\gamma r)\,\mathcal{F}_{\pm m} \\ H_z &= \pm\frac{\mathrm{i}\mathrm{k}_z}{\mu_0\omega} A_m J_m(\gamma r)\,\mathcal{F}_{\pm m} \end{aligned}}_{EH_{mn}\ \textit{modes}\ (r<a)} \qquad \underbrace{\begin{aligned} H_\varphi &= \frac{\mathrm{i}\omega\varepsilon_2}{\beta}\frac{J_m(\gamma a)}{K_m(\beta a)} A_m K_{m+1}(\beta r)\,\mathcal{F}_{\pm m} \\ H_z &= \pm\frac{\mathrm{i}\mathrm{k}_z}{\mu_0\omega} A_m \frac{J_m(\gamma a)}{K_m(\beta a)} K_m(\beta r)\,\mathcal{F}_{\pm m} \end{aligned}}_{EH_{mn}\ \textit{modes}\ (r<a)} \qquad [1.207]$$

$$\underbrace{\begin{aligned} E_r &= \frac{\mathrm{i}\mathrm{k}_z}{\gamma} A_m J_{m-1}(\gamma r)\,\mathcal{F}_{\pm m} \\ E_\varphi &= \mp\frac{\mathrm{k}_z}{\gamma} A_m J_{m-1}(\gamma r)\,\mathcal{F}_{\pm m} \\ E_z &= A_m J_m(\gamma r)\,\mathcal{F}_{\pm m} \\ H_r &= \pm\frac{\omega\varepsilon_1}{\gamma} A_m J_{m-1}(\gamma r)\,\mathcal{F}_{\pm m} \\ H_\varphi &= \frac{\mathrm{i}\omega\varepsilon_1}{\gamma} A_m J_{m-1}(\gamma r)\,\mathcal{F}_{\pm m} \\ H_z &= \mp\frac{\mathrm{i}\mathrm{k}_z}{\mu_0\omega} A_m J_m(\gamma r)\,\mathcal{F}_{\pm m} \end{aligned}}_{HE_{mn}\ \textit{modes}\ (r<a)} \qquad \underbrace{\begin{aligned} E_r &= \frac{\mathrm{i}\mathrm{k}_z}{\beta}\frac{J_m(\gamma a)}{K_m(\beta a)} A_m K_{m-1}(\beta r)\,\mathcal{F}_{\pm m} \\ E_\varphi &= \mp\frac{\mathrm{k}_z}{\beta}\frac{J_m(\gamma a)}{K_m(\beta a)} A_m K_{m-1}(\beta r)\,\mathcal{F}_{\pm m} \\ E_z &= A_m \frac{J_m(\gamma a)}{K_m(\beta a)} K_m(\beta r)\,\mathcal{F}_{\pm m} \\ H_r &= \pm\frac{\omega\varepsilon_2}{\beta}\frac{J_m(\gamma a)}{K_m(\beta a)} A_m K_{m-1}(\beta r)\,\mathcal{F}_{\pm m} \\ H_\varphi &= \frac{\mathrm{i}\omega\varepsilon_2}{\beta}\frac{J_m(\gamma a)}{K_m(\beta a)} A_m K_{m-1}(\beta r)\,\mathcal{F}_{\pm m} \\ E_z &= \mp\frac{\mathrm{i}\mathrm{k}_z}{\mu_0\omega} A_m \frac{J_m(\gamma a)}{K_m(\beta a)} K_m(\beta r)\,\mathcal{F}_{\pm m} \end{aligned}}_{HE_{mn}\ \textit{modes}\ (r>a)} \qquad [1.208]$$

In system $Oxyz$, we have $E_x = E_r\cos\varphi - E_\varphi\sin\varphi = \frac{1}{2}(E_r + \mathrm{i}E_\varphi)\,e^{\mathrm{i}\varphi} + \frac{1}{2}(E_r - \mathrm{i}E_\varphi)\,e^{-\mathrm{i}\varphi}$, $E_y = E_r\cos\varphi + E_\varphi\sin\varphi = \frac{1}{2}(E_\varphi - \mathrm{i}E_r)e^{\mathrm{i}\varphi} + \frac{1}{2}(E_\varphi + \mathrm{i}E_r)e^{-\mathrm{i}\varphi}$, and $E_x = E_z$ (the same expressions apply for H_x, H_y and H_z).

We thus obtain the following expressions:

$$\underbrace{\begin{aligned} E_x &= -\frac{\mathrm{i}\mathrm{k}_z}{\gamma} A_m J_{m+1}(\gamma r)\,\mathcal{F}_{\pm(m+1)} \\ E_y &= \mp\frac{\mathrm{k}_z}{\gamma} A_m J_{m+1}(\gamma r)\,\mathcal{F}_{\pm(m+1)} \\ E_z &= A_m J_m(\gamma r)\,\mathcal{F}_{\pm m} \\ H_x &= \pm\frac{\omega\varepsilon_1}{\gamma} A_m J_{m+1}(\gamma r)\,\mathcal{F}_{\pm(m+1)} \\ H_y &= -\frac{\mathrm{i}\omega\varepsilon_1}{\gamma} A_m J_{m+1}(\gamma r)\,\mathcal{F}_{\pm(m+1)} \\ H_z &= \pm\frac{\mathrm{i}\mathrm{k}_z}{\mu_0\omega} A_m J_m(\gamma r)\,\mathcal{F}_{\pm m} \end{aligned}}_{EH_{mn}\ \textit{modes}\ (r<a)} \qquad \underbrace{\begin{aligned} E_x &= \frac{\mathrm{i}\mathrm{k}_z}{\beta}\frac{J_m(\gamma a)}{K_m(\beta a)} A_m K_{m+1}(\beta r)\,\mathcal{F}_{\pm(m+1)} \\ E_y &= \pm\frac{\mathrm{k}_z}{\beta}\frac{J_m(\gamma a)}{K_m(\beta a)} A_m K_{m+1}(\beta r)\,\mathcal{F}_{\pm(m+1)} \\ E_z &= A_m \frac{J_m(\gamma a)}{K_m(\beta a)} K_m(\beta r)\,\mathcal{F}_{\pm m} \\ H_x &= \mp\frac{\omega\varepsilon_2}{\beta}\frac{J_m(\gamma a)}{K_m(\beta a)} A_m K_{m+1}(\beta r)\,\mathcal{F}_{\pm(m+1)} \\ H_y &= \frac{\mathrm{i}\omega\varepsilon_2}{\beta}\frac{J_m(\gamma a)}{K_m(\beta a)} A_m K_{m+1}(\beta r)\,\mathcal{F}_{\pm(m+1)} \\ H_z &= \pm\frac{\mathrm{i}\mathrm{k}_z}{\mu_0\omega} A_m \frac{J_m(\gamma a)}{K_m(\beta a)} K_m(\beta r)\,\mathcal{F}_{\pm m} \end{aligned}}_{EH_{mn}\ \textit{modes}\ (r>a)} \qquad [1.209]$$

$$\underbrace{\begin{aligned} E_x &= \frac{\mathrm{i}k_z}{\gamma} A_m J_{m-1}(\gamma r) \mathcal{F}_{\pm(m-1)} \\ E_y &= \mp \frac{k_z}{\gamma} A_m J_{m-1}(\gamma r) \mathcal{F}_{\pm(m-1)} \\ E_z &= A_m J_m(\gamma r) \mathcal{F}_{\pm m} \\ H_x &= \pm \frac{\omega \varepsilon_1}{\gamma} A_m J_{m-1}(\gamma r) \mathcal{F}_{\pm(m-1)} \\ H_y &= \frac{\mathrm{i}\omega \varepsilon_1}{\gamma} A_m J_{m-1}(\gamma r) \mathcal{F}_{\pm(m-1)} \\ H_z &= \mp \frac{\mathrm{i}k_z}{\mu_0 \omega} A_m J_m(\gamma r) \mathcal{F}_{\pm m} \end{aligned}}_{HE_{mn} \text{ modes } (r<a)} \qquad \underbrace{\begin{aligned} E_x &= \frac{\mathrm{i}k_z}{\beta} \frac{J_m(\gamma a)}{K_m(\beta a)} A_m K_{m-1}(\beta r) \mathcal{F}_{\pm(m-1)} \\ E_y &= \mp \frac{k_z}{\beta} \frac{J_m(\gamma a)}{K_m(\beta a)} A_m K_{m-1}(\beta r) \mathcal{F}_{\pm(m-1)} \\ E_z &= A_m \frac{J_m(\gamma a)}{K_m(\beta a)} K_m(\beta r) \mathcal{F}_{\pm m} \\ H_x &= \pm \frac{\omega \varepsilon_2}{\beta} \frac{J_m(\gamma a)}{K_m(\beta a)} A_m K_{m-1}(\beta r) \mathcal{F}_{\pm(m-1)} \\ H_y &= \frac{\mathrm{i}\omega \varepsilon_2}{\beta} \frac{J_m(\gamma a)}{K_m(\beta a)} A_m K_{m-1}(\beta r) \mathcal{F}_{\pm(m-1)} \\ H_z &= \mp \frac{\mathrm{i}k_z}{\mu_0 \omega} A_m \frac{J_m(\gamma a)}{K_m(\beta a)} K_m(\beta r) \mathcal{F}_{\pm m} \end{aligned}}_{HE_{mn} \text{ modes } (r>a)} \qquad [1.210]$$

1.4.5.4. *Construction of linearly polarized LP_{mn} modes*

When two or more modes are degenerate, it is impossible to propagate each mode individually within the guide. In practice, the degenerate modes propagate simultaneously and at the same speed. As a result, they interfere with each other. At the fiber outlet, the observed signal results from the superposition of all degenerate modes with the same propagation constant. Equations [1.209] and [1.210] (obtained under weak guiding conditions) show that the modes $EH_{m-1,n}$ and $HE_{m+1,n}$ are degenerate. Consequently, these two modes interfere, leading to a polarized mode along the Ox or Oy axis, denoted as LP_{mn}.

By adding the $EH_{m-1,n}$ and $HE_{m+1,n}$ modes (assuming $A_{m-1} = A_{m+1} = A$), we find:

$$r < a \quad \begin{cases} E_x = H_y = 0 \\ E_y = \mp \dfrac{2k_z}{\gamma} A J_m(\gamma r) \mathcal{F}_{\pm m} \\ E_z = A \left[J_{m-1}(\gamma r) e^{\mp \mathrm{i}\varphi} + J_{m+1}(\gamma r) e^{\pm \mathrm{i}\varphi} \right] \mathcal{F}_{\pm m} \\ H_x = \pm \dfrac{2\omega \varepsilon_1}{\gamma} A J_m(\gamma r) \mathcal{F}_{\pm m} \\ H_z = \pm \dfrac{\mathrm{i}k_z}{\mu_0 \omega} A \left[J_{m-1}(\gamma r) e^{\mp \mathrm{i}\varphi} - J_{m+1}(\gamma r) e^{\pm \mathrm{i}\varphi} \right] \mathcal{F}_{\pm m} \end{cases} \qquad [1.211]$$

$$r > a \left\{ \begin{aligned} E_x &= \frac{\mathrm{i}k_z}{\beta} A \left[\frac{J_{m-1}(\gamma a)}{K_{m-1}(\beta a)} + \frac{J_{m+1}(\gamma a)}{K_{m+1}(\beta a)} \right] K_m(\beta r) \mathcal{F}_{\pm m} \\ E_y &= \pm \frac{k_z}{\beta} A \left[\frac{J_{m-1}(\gamma a)}{K_{m-1}(\beta a)} - \frac{J_{m+1}(\gamma a)}{K_{m+1}(\beta a)} \right] K_m(\beta r) \mathcal{F}_{\pm m} \\ E_z &= A_m \left[\frac{J_{m-1}(\gamma a)}{K_{m-1}(\beta a)} K_{m-1}(\beta r) e^{\mp \mathrm{i}\varphi} \right. \\ &\left. + \frac{J_{m+1}(\gamma a)}{K_{m-1}(\beta a)} K_{m+1}(\beta r) e^{\pm \mathrm{i}\varphi} \right] \mathcal{F}_{\pm m} \\ H_x &= \mp \frac{\omega \varepsilon_2}{\beta} A \left[\frac{J_{m-1}(\gamma a)}{K_{m-1}(\beta a)} - \frac{J_{m+1}(\gamma a)}{K_{m+1}(\beta a)} \right] K_m(\beta r) \mathcal{F}_{\pm m} \\ H_y &= \frac{\mathrm{i}\omega \varepsilon_2}{\beta} A \left[\frac{J_{m-1}(\gamma a)}{K_{m-1}(\beta a)} + \frac{J_{m+1}(\gamma a)}{K_{m+1}(\beta a)} \right] K_m(\beta r) \mathcal{F}_{\pm m} \\ H_z &= \pm \frac{\mathrm{i}k_z}{\mu_0 \omega} A \left[\frac{J_{m-1}(\gamma a)}{K_{m-1}(\beta a)} K_{m-1}(\beta r) e^{\mp \mathrm{i}\varphi} \right. \\ &\left. - \frac{J_{m+1}(\gamma a)}{K_{m+1}(\beta a)} K_{m+1}(\beta r) e^{\pm \mathrm{i}\varphi} \right] \mathcal{F}_{\pm m} \end{aligned} \right. \quad [1.212]$$

We demonstrate from equations [1.205] and [1.206] that:

$$\frac{J_{m-1}(\gamma a)}{K_{m-1}(\gamma a)} + \frac{J_{m+1}(\gamma a)}{K_{m+1}(\gamma a)} = 0 \quad [1.213]$$

Thus, formulas [1.212] reduce to:

$$r > a \left\{ \begin{aligned} E_x &= H_y = 0 \\ E_y &= \mp \frac{2k_z}{\beta} A \frac{J_{m+1}(\gamma a)}{K_{m+1}(\beta a)} K_m(\beta r) \mathcal{F}_{\pm m} \\ E_z &= A_m \left[-K_{m-1}(\beta r) e^{\mp \mathrm{i}\varphi} + K_{m+1}(\beta r) e^{\pm \mathrm{i}\varphi} \right] \frac{J_{m+1}(\gamma a)}{K_{m+1}(\beta a)} \mathcal{F}_{\pm m} \\ H_x &= \pm \frac{2\omega \varepsilon_2}{\beta} A \frac{J_{m+1}(\gamma a)}{K_{m+1}(\beta a)} K_m(\beta r) \mathcal{F}_{\pm m} \\ H_z &= \mp \frac{\mathrm{i}k_z}{\mu_0 \omega} A \left[K_{m-1}(\beta r) e^{\mp \mathrm{i}\varphi} + K_{m+1}(\beta r) e^{\pm \mathrm{i}\varphi} \right] \frac{J_{m+1}(\gamma a)}{K_{m+1}(\beta a)} \mathcal{F}_{\pm m} \end{aligned} \right.$$

[1.214]

We clearly see that the addition of the EH$m-1,n$ and HE$m+1,n$ modes leads to a linearly polarized LPmn mode along the Oy axis (i.e. with $E_x = 0$), described by relations [1.211] and [1.214]. When subtracting the EH$m-1,n$ and HE$m+1,n$ modes, we obtain a linearly polarized mode along the Ox axis (i.e. with $E_y = 0$). In fact, the EH$m-1,n+$HE$m+1,n$ and EH$m-1,n-$HE$_{m+1,n}$ modes are identical in the sense that one can be transformed into the other simply by interchanging Ox and Oy.

In real notation, the electric and magnetic fields (of the LP_{mn} mode) are written as:

$$r < a \begin{cases} E_x = H_y = 0 \\ E_y = \mp \dfrac{2\mathsf{k}_z}{\gamma} A J_m(\gamma r) \cos(\omega t - \mathsf{k}_z z \mp m\varphi) \\ E_z = A\left[J_{m-1}(\gamma r) \cos\left[\omega t - \mathsf{k}_z z \mp (m-1)\varphi\right] + \right. \\ \qquad \left. J_{m+1}(\gamma r) \cos\left[\omega t - \mathsf{k}_z z \mp (m+1)\varphi\right]\right] \\ H_x = \pm \dfrac{2\omega\varepsilon_1}{\gamma} A J_m(\gamma r) \cos(\omega t - \mathsf{k}_z z \mp m\varphi) \\ H_z = \pm \dfrac{\mathsf{k}_z}{\mu_0 \omega} A\left[J_{m-1}(\gamma r) \sin\left[\omega t - \mathsf{k}_z z \mp (m-1)\varphi\right] - \right. \\ \qquad \left. J_{m+1}(\gamma r) \sin\left[\omega t - \mathsf{k}_z z \mp (m+1)\varphi\right]\right] \end{cases} \quad [1.215]$$

$$r > a \begin{cases} E_x = H_y = 0 \\ E_y = \mp \dfrac{2\mathsf{k}_z}{\beta} A \dfrac{J_{m+1}(\gamma a)}{K_{m+1}(\beta a)} K_m(\beta r) \cos(\omega t - \mathsf{k}_z z \mp m\varphi) \\ E_z = A_m\left[-K_{m-1}(\beta r) \cos\left[\omega t - \mathsf{k}_z z \mp (m-1)\varphi\right] + \right. \\ \qquad \left. K_{m+1}(\beta r) \cos\left[\omega t - \mathsf{k}_z z \mp (m+1)\varphi\right]\right] \dfrac{J_{m+1}(\gamma a)}{K_{m+1}(\beta a)} \\ H_x = \pm \dfrac{2\omega\varepsilon_2}{\beta} A \dfrac{J_{m+1}(\gamma a)}{K_{m+1}(\beta a)} K_m(\beta r) \cos(\omega t - \mathsf{k}_z z \mp m\varphi) \\ H_z = \mp \dfrac{\mathsf{k}_z}{\mu_0 \omega} A\left[K_{m-1}(\beta r) \sin\left[\omega t - \mathsf{k}_z z \mp (m-1)\varphi\right] + \right. \\ \qquad \left. K_{m+1}(\beta r) \sin\left[\omega t - \mathsf{k}_z z \mp (m+1)\varphi\right]\right] \dfrac{J_{m+1}(\gamma a)}{K_{m+1}(\beta a)} \end{cases} \quad [1.216]$$

The previous reasoning does not apply to the $m = 0$ case. In fact, the HE_{1n} mode cannot be associated with the $\text{EH}_{-1,n}$ mode, since there are no modes with negative m indices. The field components of the LP_{1n} mode are therefore obtained directly from those of the mode HE_{1n}, so its dispersion relation is that of the HE_{1n} mode:

$$\frac{\gamma a J_1(\gamma a)}{J_0(\gamma a)} - \frac{\beta a K_1(\beta a)}{K_0(\beta a)} = 0 \quad [1.217]$$

1.5. Resonant cavities

Instead of guiding a wave, it is also possible to trap the energy of an electromagnetic wave inside a conductor that is closed on all sides (an *electromagnetic cavity*). The boundary conditions imposed by the walls now show that monochromatic traveling waves cannot propagate inside: only standing waves can exist, with discrete, so-called *quantized frequencies*. Our aim is therefore to explicitly find the cavity's natural frequencies, as well as the configuration of the fields describing each oscillation mode.

1.5.1. *Rectangular metallic resonant cavity*

A rectangular metallic waveguide becomes a resonant cavity by adding flat walls at each end, specifically at $z = 0$ and $z = l$. For a given frequency, two progressive waves propagating in the $+Oz$ and $-Oz$ directions must overlap to form a standing wave that satisfies the boundary conditions at the borders of the walls located at $z = 0$ and $z = l$. These boundary conditions are $H_z = 0$ and $E_x = E_y = 0$ on the new walls. The interior of the cavity is assumed to be empty.

Let us first consider the TE modes. Using equations [1.84] and [1.85], we obtain the following:

$$\begin{aligned}
E_x &= \Re\left[-\frac{\mathrm{i}\omega\mu_0 \mathrm{k}_y}{\mathrm{k}_x^2+\mathrm{k}_y^2}\cos(\mathrm{k}_x x)\sin(\mathrm{k}_y y)\left(H_{mn}^+ e^{-\mathrm{i}(\omega t-\mathrm{k}_z z)} + H_{mn}^- e^{-\mathrm{i}(\omega t+\mathrm{k}_z z)}\right)\right]\\
E_y &= \Re\left[\frac{\mathrm{i}\omega\mu_0 \mathrm{k}_x}{\mathrm{k}_x^2+\mathrm{k}_y^2}\sin(\mathrm{k}_x x)\cos(\mathrm{k}_y y)\left(H_{mn}^+ e^{-\mathrm{i}(\omega t-\mathrm{k}_z z)} + H_{mn}^- e^{-\mathrm{i}(\omega t+\mathrm{k}_z z)}\right)\right]\\
E_z &= 0\\
H_x &= \Re\left[-\frac{\mathrm{i}\mathrm{k}_z \mathrm{k}_x}{\mathrm{k}_x^2+\mathrm{k}_y^2}\sin(\mathrm{k}_x x)\cos(\mathrm{k}_y y)\left(H_{mn}^+ e^{-\mathrm{i}(\omega t-\mathrm{k}_z z)} - H_{mn}^- e^{-\mathrm{i}(\omega t+\mathrm{k}_z z)}\right)\right]\\
H_y &= \Re\left[-\frac{\mathrm{i}\mathrm{k}_z \mathrm{k}_y}{\mathrm{k}_x^2+\mathrm{k}_y^2}\cos(\mathrm{k}_x x)\sin(\mathrm{k}_y y)\left(H_{mn}^+ e^{-\mathrm{i}(\omega t-\mathrm{k}_z z)} - H_{mn}^- e^{-\mathrm{i}(\omega t+\mathrm{k}_z z)}\right)\right]\\
H_z &= \Re\left[\cos(\mathrm{k}_x x)\cos(\mathrm{k}_y y)\left(H_{mn}^+ e^{-\mathrm{i}(\omega t-\mathrm{k}_z z)} + H_{mn}^- e^{-\mathrm{i}(\omega t+\mathrm{k}_z z)}\right)\right]
\end{aligned} \qquad [1.218]$$

where $\mathrm{k}_x = \frac{m\pi}{a}$ and $\mathrm{k}_y = \frac{n\pi}{b}$. H_{mn}^+ and H_{mn}^- are the amplitudes of waves propagating in the $+Oz$ and $-Oz$ directions, respectively. For H_z to cancel at $z = 0$, choose $H_{mn}^- = -H_{mn}^+$, and for it to cancel at $z = l$, quantize k_z: $\mathrm{k}_z = \frac{p\pi}{l}$, where p is a positive integer. Thus:

$$\begin{aligned}
E_x &= \frac{2\omega_{mnp}\mu_0 \mathrm{k}_y}{\mathrm{k}_x^2+\mathrm{k}_y^2} H_{mn}^+ \cos(\mathrm{k}_x x)\sin(\mathrm{k}_y y)\sin(\mathrm{k}_z z)\cos(\omega_{mnp} t)\\
E_y &= -\frac{2\omega_{mnp}\mu_0 \mathrm{k}_x}{\mathrm{k}_x^2+\mathrm{k}_y^2} H_{mn}^+ \sin(\mathrm{k}_x x)\cos(\mathrm{k}_y y)\sin(\mathrm{k}_z z)\cos(\omega_{mnp} t)\\
E_z &= 0
\end{aligned}$$

$$
\begin{aligned}
H_x &= -\frac{2\mathsf{k}_z\mathsf{k}_x}{\mathsf{k}_x^2+\mathsf{k}_y^2}H_{mn}^{+}\sin\left(\mathsf{k}_x x\right)\cos\left(\mathsf{k}_y y\right)\cos\left(\mathsf{k}_z z\right)\sin\left(\omega_{mnp}t\right)\\
H_y &= -\frac{2\mathsf{k}_z\mathsf{k}_y}{\mathsf{k}_x^2+\mathsf{k}_y^2}H_{mn}^{+}\cos\left(\mathsf{k}_x x\right)\sin\left(\mathsf{k}_y y\right)\cos\left(\mathsf{k}_z z\right)\sin\left(\omega_{mnp}t\right)\\
H_z &= 2H_{mn}^{+}\cos\left(\mathsf{k}_x x\right)\cos\left(\mathsf{k}_y y\right)\sin\left(\mathsf{k}_z z\right)\sin\left(\omega_{mnp}t\right)
\end{aligned}
\qquad [1.219]
$$

We can easily verify that E_x and E_y vanish in $z=0$ and $z=l$. This corresponds to the TE$_{mnp}$ mode. The index p must be non-null in order for H_z to be non-null as well. The indices m or n (but not both) may be zero.

For TM waves, we must take the following superposition (see [1.54] and [1.55]):

$$
\begin{aligned}
E_x &= \Re\left[-\frac{\mathsf{i}\mathsf{k}_x\mathsf{k}_z}{\mathsf{k}_x^2+\mathsf{k}_y^2}\cos\left(\mathsf{k}_x x\right)\sin\left(\mathsf{k}_y y\right)\left(E_{mn}^{+}e^{-\mathsf{i}(\omega t-\mathsf{k}_z z)}-E_{mn}^{-}e^{-\mathsf{i}(\omega t+\mathsf{k}_z z)}\right)\right]\\
E_y &= \Re\left[\frac{\mathsf{i}\mathsf{k}_y\mathsf{k}_z}{\mathsf{k}_x^2+\mathsf{k}_y^2}\sin\left(\mathsf{k}_x x\right)\cos\left(\mathsf{k}_y y\right)\left(E_{mn}^{+}e^{-\mathsf{i}(\omega t-\mathsf{k}_z z)}-E_{mn}^{-}e^{-\mathsf{i}(\omega t+\mathsf{k}_z z)}\right)\right]\\
E_z &= \Re\left[\sin\left(\mathsf{k}_x x\right)\sin\left(\mathsf{k}_y y\right)\left(E_{mn}^{+}e^{-\mathsf{i}(\omega t-\mathsf{k}_z z)}+E_{mn}^{-}e^{-\mathsf{i}(\omega t+\mathsf{k}_z z)}\right)\right]\\
H_x &= \Re\left[-\frac{\mathsf{i}\omega\mathsf{k}_y}{\mu_0c^2\left(\mathsf{k}_x^2+\mathsf{k}_y^2\right)}\sin\left(\mathsf{k}_x x\right)\cos\left(\mathsf{k}_y y\right)\right.\\
&\left.\left(E_{mn}^{+}c^{-\mathsf{i}(\omega t-\mathsf{k}_z z)}+E_{mn}^{-}e^{-\mathsf{i}(\omega t+\mathsf{k}_z z)}\right)\right]\\
H_y &= \Re\left[\frac{\mathsf{i}\omega\mathsf{k}_x}{\mu_0c^2\left(\mathsf{k}_x^2+\mathsf{k}_y^2\right)}\cos\left(\mathsf{k}_x x\right)\sin\left(\mathsf{k}_y y\right)\right.\\
&\left.\left(E_{mn}^{+}e^{-\mathsf{i}(\omega t-\mathsf{k}_z z)}+E_{mn}^{-}e^{-\mathsf{i}(\omega t+\mathsf{k}_z z)}\right)\right]\\
H_z &= 0
\end{aligned}
\qquad [1.220]
$$

To satisfy the constraints $E_x\left(z=0\right) = E_y\left(z=0\right) = 0$ and $E_x\left(z=l\right)=E_y\left(z=l\right)=0$, we must have $E_{mn}=E_{mn}^{+}$ and $\mathsf{k}_z=\frac{n\pi}{l}$ (i.e. k_z is quantized). It therefore follows that:

$$
\begin{aligned}
E_x &= -\frac{2\mathsf{k}_x\mathsf{k}_z}{\mathsf{k}_x^2+\mathsf{k}_y^2}E_{mn}^{+}\cos\left(\mathsf{k}_x x\right)\sin\left(\mathsf{k}_y y\right)\sin\left(\mathsf{k}_z z\right)\cos\left(\omega_{mnp}t\right)\\
E_y &= -\frac{2\mathsf{k}_y\mathsf{k}_z}{\mathsf{k}_x^2+\mathsf{k}_y^2}E_{mn}^{+}\sin\left(\mathsf{k}_x x\right)\cos\left(\mathsf{k}_y y\right)\sin\left(\mathsf{k}_z z\right)\cos\left(\omega_{mnp}t\right)\\
E_z &= 2E_{mn}^{+}\sin\left(\mathsf{k}_x x\right)\sin\left(\mathsf{k}_y y\right)\cos\left(\mathsf{k}_z z\right)\cos\left(\omega_{mnp}t\right)\\
H_x &= -\frac{2\omega_{mnp}\mathsf{k}_y}{\mu_0c^2\left(\mathsf{k}_x^2+\mathsf{k}_y^2\right)}E_{mn}^{+}\sin\left(\mathsf{k}_x x\right)\cos\left(\mathsf{k}_y y\right)\cos\left(\mathsf{k}_z z\right)\sin\left(\omega_{mnp}t\right)\\
H_y &= \frac{2\omega_{mnp}\mathsf{k}_x}{\mu_0c^2\left(\mathsf{k}_x^2+\mathsf{k}_y^2\right)}E_{mn}^{+}\cos\left(\mathsf{k}_x x\right)\sin\left(\mathsf{k}_y y\right)\cos\left(\mathsf{k}_z z\right)\sin\left(\omega_{mnp}t\right)\\
H_z &= 0
\end{aligned}
\qquad [1.221]
$$

where m and n must both be non-null in order for E_z to be non-null and for the solution to be non-trivial. This is the mode TM_{mnp}. In both cases, the oscillation frequencies are now quantized. The allowed values are given as:

$$\omega_{mnp} = c\sqrt{\left(\frac{m\pi}{a}\right)^2 + \left(\frac{n\pi}{b}\right)^2 + \left(\frac{p\pi}{l}\right)^2} \qquad [1.222]$$

Obviously, the distinction between the TE and TM modes is somewhat ambiguous in a rectangular cavity, since any of the three Cartesian axes can be chosen as the waveguide axis. The simplest modes are TE_{101}, TE_{011} and TM_{110}. In each case, only one component of the electric field $\vec{\mathbf{E}}$ is non-null (respectively, E_z, E_y and E_x), and the corresponding component of the magnetic field $\vec{\mathbf{H}}$ vanishes.

EXAMPLE 1.4. *TE_{101}* MODE.– In the TE_{101} mode, the fields are written as:

$$\begin{aligned}
\vec{\mathbf{E}} &= -\frac{2\sqrt{a^2+l^2}}{l} Z_0 H_{10}^+ \sin\left(\frac{\pi x}{a}\right)\sin\left(\frac{\pi z}{l}\right)\cos\left(\omega_0 t\right)\vec{\mathbf{j}} \\
\vec{\mathbf{H}} &= 2H_{10}^+\left[-\frac{a}{l}\sin\left(\frac{\pi x}{a}\right)\cos\left(\frac{\pi z}{l}\right)\vec{\mathbf{i}} + \cos\left(\frac{\pi x}{a}\right)\sin\left(\frac{\pi z}{l}\right)\vec{\mathbf{k}}\right]\sin\left(\omega_0 t\right)
\end{aligned} \qquad [1.223]$$

where $\omega_0 = \frac{c\pi\sqrt{a^2+l^2}}{al}$ is the natural frequency of the mode and $Z_0 = \sqrt{\frac{\mu_0}{\varepsilon_0}}$.

1) Since the electric field is aligned with the Oy axis, electric charges accumulate only on the faces $y = 0$ and $y = b$, with the respective surface charge densities:

$$\begin{aligned}
\sigma\left(y=0\right) &= \varepsilon_0 \vec{\mathbf{j}} \cdot \vec{\mathbf{E}}\Big|_{y=0^+} &&= -\frac{2\sqrt{a^2+l^2}}{cl} H_{10}^+ \sin\left(\frac{\pi x}{a}\right)\sin\left(\frac{\pi z}{l}\right)\cos\left(\omega_0 t\right) \\
\sigma\left(y=b\right) &= -\varepsilon_0 \vec{\mathbf{j}} \cdot \vec{\mathbf{E}}\Big|_{y=b^-} &&= -\frac{2\sqrt{a^2+l^2}}{cl} H_{10}^+ \sin\left(\frac{\pi x}{a}\right)\sin\left(\frac{\pi z}{l}\right)\cos\left(\omega_0 t\right)
\end{aligned} \qquad [1.224]$$

It is clear that the total charge is zero.

2) The discontinuity of the tangential component of $\vec{\mathbf{H}}$ at the metal – vacuum interface allows us to calculate the surface currents accumulated on the various faces of the cavity:

$$\begin{aligned}
\vec{\mathbf{K}}\Big|_{x=0} &= \vec{\mathbf{i}} \times \vec{\mathbf{H}}\Big|_{x=0^+} &&= -2H_{10}^+ \sin\left(\frac{\pi z}{l}\right)\sin\left(\omega_0 t\right)\vec{\mathbf{j}} \\
\vec{\mathbf{K}}\Big|_{x=a} &= -\vec{\mathbf{i}} \times \vec{\mathbf{H}}\Big|_{x=a^-} &&= -2H_{10}^+ \sin\left(\frac{\pi z}{l}\right)\sin\left(\omega_0 t\right)\vec{\mathbf{j}}
\end{aligned}$$

$$\begin{aligned}
&\vec{\mathbf{K}}\Big|_{y=0} = \vec{\mathbf{j}} \times \vec{\mathbf{H}}\Big|_{y=0^+} = 2H_{10}^{+}\left[\frac{a}{l}\sin\left(\frac{\pi x}{a}\right)\cos\left(\frac{\pi z}{l}\right)\vec{\mathbf{k}}\right.\\
&\left.+\cos\left(\frac{\pi x}{a}\right)\sin\left(\frac{\pi z}{l}\right)\vec{\mathbf{i}}\right]\sin(\omega_0 t)\\
&\vec{\mathbf{K}}\Big|_{y=b} = -\vec{\mathbf{j}} \times \vec{\mathbf{H}}\Big|_{y=b^-} = -2H_{10}^{+}\left[\frac{a}{l}\sin\left(\frac{\pi x}{a}\right)\cos\left(\frac{\pi z}{l}\right)\vec{\mathbf{k}}\right.\\
&\left.+\cos\left(\frac{\pi x}{a}\right)\sin\left(\frac{\pi z}{l}\right)\vec{\mathbf{i}}\right]\sin(\omega_0 t)\\
&\vec{\mathbf{K}}\Big|_{z=0} = \vec{\mathbf{k}} \times \vec{\mathbf{H}}\Big|_{z=0^+} = -2H_{10}^{+}\frac{a}{l}\sin\left(\frac{\pi x}{a}\right)\sin(\omega_0 t)\,\vec{\mathbf{j}}\\
&\vec{\mathbf{K}}\Big|_{z=l} = -\vec{\mathbf{k}} \times \vec{\mathbf{H}}\Big|_{z=l^-} = -2H_{10}^{+}\frac{a}{l}\sin\left(\frac{\pi x}{a}\right)\sin(\omega_0 t)\,\vec{\mathbf{j}}
\end{aligned} \quad [1.225]$$

3) For TE_{101} mode, the electromagnetic energy stored in the cavity is equal to:

$$\begin{aligned}
W &= \frac{1}{2}\int_{\text{cavity}}\left(\varepsilon_0\vec{\mathbf{E}}^2 + \mu_0\vec{\mathbf{H}}^2\right)dxdydz\\
&= 2\mu_0 H_{10}^{+2}\int_0^a\int_0^b\int_0^l\left\{\frac{(a^2+l^2)}{l^2}\sin^2\left(\frac{\pi x}{a}\right)\sin^2\left(\frac{\pi z}{l}\right)\cos^2(\omega_0 t)+\right.\\
&\left.\left[\frac{a^2}{l^2}\sin^2\left(\frac{\pi x}{a}\right)\cos^2\left(\frac{\pi z}{l}\right)+\cos^2\left(\frac{\pi x}{a}\right)\sin^2\left(\frac{\pi z}{l}\right)\right]\sin^2(\omega_0 t)\right\}dxdydz\\
&= \frac{\mu_0 ab\left(a^2+l^2\right)H_{10}^{+2}}{2l} = \xi_1 H_{10}^{+2}
\end{aligned} \quad [1.226]$$

where $\xi_1 = \frac{\mu_0 ab\left(a^2+l^2\right)}{2l}$ (identities [1.68] were used).

4) In reality, the metal used to make the cavity is not a perfect conductor. It has a finite conductivity γ_c. In this case, the current is not a surface current but flows within a layer of thickness δ (skin effect), which leads to energy dissipation due to Joule heating. The instantaneous power dissipated by each surface element dS of the cavity is given by the expression $dP_d = \frac{\vec{\mathbf{K}}^2}{\gamma_c\delta}dS$. Thus, the total power dissipated by the entire cavity is given as:

$$\begin{aligned}
P_d = \frac{1}{\gamma_c\delta}&\left\{\int_0^a\int_0^l\left[\vec{\mathbf{K}}^2(y=0)+\vec{\mathbf{K}}^2(y=b)\right]dxdz\right.\\
&+\int_0^b\int_0^l\left[\vec{\mathbf{K}}^2(x=0)+\right.\\
&\left.\left.\vec{\mathbf{K}}^2(x=a)\right]dydz+\int_0^a\int_0^b\left[\vec{\mathbf{K}}^2(z=0)+\vec{\mathbf{K}}^2(z=l)\right]dxdy\right\}
\end{aligned}$$

$$= \frac{2}{\gamma_c \delta} \left\{ \int_0^a \int_0^l \vec{\mathbf{K}}^2 (y=0)\, dxdz + \int_0^b \int_0^l \vec{\mathbf{K}}^2 (x=0)\, dydz \right.$$

$$\left. + \int_0^a \int_0^b \vec{\mathbf{K}}^2 (z=0)\, dxdy \right\}$$

$$= \frac{4a^2 H_{10}^{+2}}{\gamma_c \delta} \left[\frac{ab}{l^2} + \frac{bl}{a^2} + \frac{1}{2} \left(\frac{a}{l} + \frac{l}{a} \right) \right] \sin^2 (\omega_0 t) \qquad [1.227]$$

5) The power dissipated (averaged over one period) is equal to:

$$\langle P_d \rangle = \frac{4a^2 H_{10}^{+2}}{\gamma_c \delta} \left[\frac{ab}{l^2} + \frac{bl}{a^2} + \frac{1}{2} \left(\frac{a}{l} + \frac{l}{a} \right) \right] \langle \sin^2 (\omega_0 t) \rangle = \xi_2 H_{10}^{+2} \qquad [1.228]$$

where $\xi_2 = \frac{2a^2}{\gamma_c \delta} \left[\frac{ab}{l^2} + \frac{bl}{a^2} + \frac{1}{2} \left(\frac{a}{l} + \frac{l}{a} \right) \right]$ (since $\langle \sin^2 (\omega_0 t) \rangle = \frac{1}{2}$).

6) In the absence of sources compensating for the losses, the variation in total energy over time is only due to the dissipation of energy in the walls. Therefore, we write that:

$$\frac{dW_{\text{em}}}{dt} = -\langle P_d \rangle \implies \frac{dH_{10}^+}{dt} + \frac{\xi_2}{2\xi_1} H_{10}^+ = 0 \qquad [1.229]$$

7) The *quality factor* Q of a cavity is defined as the ratio of the electromagnetic energy stored in the cavity to the energy loss per oscillation cycle:

$$Q = 2\pi \frac{W_{\text{em}}}{T_0 \langle P_d \rangle} = \frac{\omega_0 W_{\text{em}}}{\langle P_d \rangle} \qquad [1.230]$$

where ω_0 is the frequency of the oscillation mode under consideration. It follows, for mode TE_{101}, that:

$$Q = \frac{\omega_0 \xi_1}{\xi_2} = \frac{\pi \gamma_c \delta Z_0}{4} \frac{2b \left(a^2 + l^2\right)^{3/2}}{al \left(a^2 + l^2\right) + 2b \left(a^3 + l^3\right)} \qquad [1.231]$$

Since both the energy W_{em} stored in the cavity and the power dissipated (averaged over one period) are proportional to the square of the field amplitudes, we see that the quality factor defined in this way is independent of the amplitude and depends only on the shape of the cavity, the oscillation mode under consideration and the conductivity of the walls.

8) Equation [1.229] admits the following solution:

$$H_{10}^+ (t) = H_{10}^+ (0)\, e^{-\frac{\xi_2 t}{2\xi_1}} \equiv H_{10}^+ (0)\, e^{-\frac{\omega_0 t}{2Q}} \qquad [1.232]$$

We can see that the variations in electric and magnetic fields [1.238] are no longer purely oscillatory over time, but that the fields decrease exponentially:

$$\begin{aligned}\vec{\mathbf{E}} &= -\frac{2\sqrt{a^2+l^2}}{l}\sqrt{\frac{\mu_0}{\varepsilon_0}}H_{10}^{+}(0)\sin\left(\frac{\pi x}{a}\right)\sin\left(\frac{\pi z}{l}\right)\cos(\omega_0 t)\,e^{-\frac{\omega_0 t}{2Q}}\vec{\mathbf{j}}\\ \vec{\mathbf{H}} &= 2H_{10}^{+}(0)\left[-\frac{a}{l}\sin\left(\frac{\pi x}{a}\right)\cos\left(\frac{\pi z}{l}\right)\vec{\mathbf{i}}\right.\\ &\left.+\cos\left(\frac{\pi x}{a}\right)\sin\left(\frac{\pi z}{l}\right)\vec{\mathbf{k}}\right]\sin(\omega_0 t)\,e^{-\frac{\omega_0 t}{2Q}}\end{aligned} \quad [1.233]$$

The higher the quality factor, the closer the damping term $e^{-\frac{\omega_0 t}{2Q}}$ is to 1.

1.5.2. *Cylindrical resonant cavity*

By closing the cylindrical waveguide with circular walls at $z = 0$ and $z = l$, a standing wave is established inside. By superimposing two traveling waves of the same frequency propagating in opposite directions along $\pm Oz$, we obtain for the TM mode:

$$\begin{aligned}E_r &= \Re\left[\frac{\mathrm{i}\mathrm{k}_z a}{\rho_{mn}}J_m'\left(\frac{\rho_{mn}r}{a}\right)e^{\pm im\varphi}\left(E_{mn}^{+}e^{\mathrm{i}\mathrm{k}_z z}-E_{mn}^{-}e^{-\mathrm{i}\mathrm{k}_z z}\right)e^{-\mathrm{i}\omega t}\right]\\ E_\varphi &= \Re\left[\mp\frac{m\mathrm{k}_z a^2}{\rho_{mn}^2 r}J_m\left(\frac{\rho_{mn}r}{a}\right)e^{\pm im\varphi}\left(E_{mn}^{+}e^{\mathrm{i}\mathrm{k}_z z}-E_{mn}^{-}e^{-\mathrm{i}\mathrm{k}_z z}\right)e^{-\mathrm{i}\omega t}\right]\\ E_z &= \Re\left[J_m\left(\frac{\rho_{mn}r}{a}\right)e^{\pm im\varphi}\left(E_{mn}^{+}e^{\mathrm{i}\mathrm{k}_z z}+E_{mn}^{-}e^{-\mathrm{i}\mathrm{k}_z z}\right)e^{-\mathrm{i}\omega t}\right]\\ H_r &= \Re\left[\pm\frac{m\omega a^2\varepsilon_0}{\rho_{mn}^2 r}J_m\left(\frac{\rho_{mn}r}{a}\right)e^{\pm im\varphi}\left(E_{mn}^{+}e^{\mathrm{i}\mathrm{k}_z z}+E_{mn}^{-}e^{-\mathrm{i}\mathrm{k}_z z}\right)e^{-\mathrm{i}\omega t}\right]\\ H_\varphi &= \Re\left[\frac{\mathrm{i}\omega a\varepsilon_0}{\rho_{mn}}J_m'\left(\frac{\rho_{mn}r}{a}\right)e^{\pm im\varphi}\left(E_{mn}^{+}e^{\mathrm{i}\mathrm{k}_z z}+E_{mn}^{-}e^{-\mathrm{i}\mathrm{k}_z z}\right)e^{-\mathrm{i}\omega t}\right]\\ H_z &= 0\end{aligned} \quad [1.234]$$

For E_r and E_φ to be canceled on the new boundaries, we must have $E_{mn}^{-} = E_{mn}^{+}$ and $\mathrm{k}_z = \frac{p\pi}{l}$, where p is an integer. It therefore follows that:

$$\begin{aligned}E_r &= -\frac{2\mathrm{k}_z a}{\rho_{mn}}J_m'\left(\frac{\rho_{mn}r}{a}\right)e^{\pm im\varphi}E_{mn}^{+}\sin(\mathrm{k}_z z)\cos(\omega_{mnp}t)\\ E_\varphi &= \mp\frac{2im\mathrm{k}_z a^2}{\rho_{mn}^2 r}J_m\left(\frac{\rho_{mn}r}{a}\right)e^{\pm im\varphi}E_{mn}^{+}\sin(\mathrm{k}_z z)\cos(\omega_{mnp}t)\\ E_z &= 2J_m\left(\frac{\rho_{mn}r}{a}\right)e^{\pm im\varphi}E_{mn}^{+}\cos(\mathrm{k}_z z)\cos(\omega_{mnp}t)\\ H_r &= \pm\frac{2m\omega_{mnp}a^2\varepsilon_0}{\rho_{mn}^2 r}J_m\left(\frac{\rho_{mn}r}{a}\right)e^{\pm im\varphi}E_{mn}^{+}\cos(\mathrm{k}_z z)\cos(\omega_{mnp}t)\\ H_\varphi &= \frac{2\omega_{mnp}a\varepsilon_0}{\rho_{mn}}J_m'\left(\frac{\rho_{mn}r}{a}\right)e^{\pm im\varphi}E_{mn}^{+}\cos(\mathrm{k}_z z)\sin(\omega_{mnp}t)\\ H_z &= 0\end{aligned} \quad [1.235]$$

Repeating the same procedure for the TE mode, we find the following:

$$\begin{aligned}
E_r &= \mp \frac{2m\omega_{mnp}a^2\mu_0}{\rho'^2_{mn}r} J_m\left(\frac{\rho'_{mn}r}{a}\right) e^{\pm im\varphi} H^+_{mn} \sin\left(\mathsf{k}_z z\right) \sin\left(\omega_{mnp}t\right) \\
E_\varphi &= \frac{2\omega_{mnp}a\mu_0}{\rho'_{mn}} J'_m\left(\frac{\rho'_{mn}r}{a}\right) e^{\pm im\varphi} H^+_{mn} \sin\left(\mathsf{k}_z z\right) \cos\left(\omega_{mnp}t\right) \\
E_z &= 0 \\
H_r &= \frac{2\mathsf{k}_z a}{\rho'_{mn}} J'_m\left(\frac{\rho'_{mn}r}{a}\right) e^{\pm im\varphi} H^+_{mn} \cos\left(\mathsf{k}_z z\right) \sin\left(\omega_{mnp}t\right) \\
H_\varphi &= \mp \frac{2m\mathsf{k}_z a^2}{\rho'^2_{mn}r} J_m\left(\frac{\rho'_{mn}r}{a}\right) e^{\pm im\varphi} H^+_{mn} \cos\left(\mathsf{k}_z z\right) \cos\left(\omega_{mnp}t\right) \\
H_z &= 2J_m\left(\frac{\rho'_{mn}r}{a}\right) e^{\pm im\varphi} H^+_{mn} \sin\left(\mathsf{k}_z z\right) \sin\left(\omega_{mnp}t\right)
\end{aligned} \qquad [1.236]$$

We clearly see that H_z, E_r and E_φ vanish at $z = 0$ and $z = l$. In both cases, the frequencies are quantized as follows:

$$\omega_{mnp} = c\sqrt{\frac{\rho^2_{mn}}{a^2} + \left(\frac{p\pi}{l}\right)^2} \qquad [1.237]$$

EXAMPLE 1.5. *TM_{010}* MODE.– In the *TM_{010}* mode, the fields are expressed as follows:

$$\begin{aligned}
\vec{\mathbf{E}} &= 2E^+_{01} J_0\left(\frac{\rho_{01}r}{a}\right) \cos\left(\omega_0 t\right) \vec{\mathbf{k}} \\
\vec{\mathbf{H}} &= 2Z_0^{-1} E^+_{01} J_0\left(\frac{\rho_{01}r}{a}\right) \sin\left(\omega_0 t\right) \vec{\mathbf{e}}_\varphi
\end{aligned} \qquad [1.238]$$

where $\omega_0 = \frac{c\rho_{01}}{a}$ is the *natural frequency of the mode* and $Z_0 = \sqrt{\frac{\mu_0}{\varepsilon_0}}$.

1) Now using the boundary conditions for a metal–vacuum interface, we give the expressions for the surface charge density σ and the surface current density $\vec{\mathbf{K}}$ on the walls of the cavity. The surface charge density on the lateral surface of the cylinder is zero for this mode, since the electric field is collinear with the Oz axis. On the flat walls of the cylinder (at $z = 0$ and $z = l$), the discontinuity of the normal component of the electric field requires that:

$$\begin{aligned}
\sigma\left(z = 0\right) &= \varepsilon_0 E_z\big|_{z=0^+} = 2\varepsilon_0 E^+_{01} J_0\left(\frac{\rho_{01}r}{a}\right) \cos\left(\omega_0 t\right) \\
\sigma\left(z = l\right) &= -\,\varepsilon_0 E_z\big|_{z=l^-} = -2\varepsilon_0 E^+_{01} J_0\left(\frac{\rho_{01}r}{a}\right) \cos\left(\omega_0 t\right)
\end{aligned} \qquad [1.239]$$

Note that the total charge sums to zero (the system is thus electrically neutral). For the calculation of the surface current density $\overrightarrow{\mathbf{K}}$, all walls carry the currents:

$$\begin{aligned}\overrightarrow{\mathbf{K}}(r=a) &= -\overrightarrow{\mathbf{e}}_r \times \overrightarrow{\mathbf{H}}\Big|_{r=a^-} = -2Z_0^{-1}J_0'(\rho_{01})\,E_{01}^{+}\sin(\omega_0 t)\,\overrightarrow{\mathbf{k}}\\ \overrightarrow{\mathbf{K}}(z=0) &= \overrightarrow{\mathbf{k}} \times \overrightarrow{\mathbf{H}}\Big|_{z=0^+} = -2Z_0^{-1}J_0'\left(\frac{\rho_{01}r}{a}\right)E_{01}^{+}\sin(\omega_0 t))\overrightarrow{\mathbf{e}}_r\\ \overrightarrow{\mathbf{K}}(z=l) &= -\overrightarrow{\mathbf{k}} \times \overrightarrow{\mathbf{H}}\Big|_{z=l^-} = 2Z_0^{-1}J_0'\left(\frac{\rho_{01}r}{a}\right)E_{01}^{+}\sin(\omega_0 t))\overrightarrow{\mathbf{e}}_r\end{aligned} \quad [1.240]$$

2) Let us calculate the electromagnetic energy stored in the cavity for this fundamental mode TM_{010}:

$$\begin{aligned}W_{\mathrm{em}} &= \int_{r=0}^{a}\int_{\varphi=0}^{2\pi}\int_{z=0}^{l}\frac{1}{2}\left(\varepsilon_0\overrightarrow{\mathbf{E}}^2+\mu_0\overrightarrow{\mathbf{H}}^2\right)rdrd\varphi dz\\ &= 4\pi\varepsilon_0 l E_{01}^2\left[\cos^2(\omega_0 t)\int_0^a J_0^2\left(\frac{\rho_{01}r}{a}\right)rdr+\sin^2(\omega_0 t)\int_0^a J_0'^2\left(\frac{\rho_{01}r}{a}\right)rdr\right]\\ &= 4\pi\varepsilon_0 l E_{01}^2\frac{a^2}{\rho_{01}^2}\left[\cos^2(\omega_0 t)\int_0^{\rho_{01}} J_0^2(x)\,xdx+\sin^2(\omega_0 t)\int_0^{\rho_{01}} J_0'^2(x)\,xdx\right]\end{aligned} \quad [1.241]$$

Since:

$$\int_0^{\rho_{01}} xJ_0^2(x)\,dx = \int_0^{\rho_{01}} xJ_0'^2(x)\,dx = I \quad [1.242]$$

(to be verified by integration by parts), we deduce that:

$$W_{\mathrm{em}} = \pi a^2 l \times 4\varepsilon_0 E_{01}^2 \times \frac{I}{\rho_{01}^2} \equiv \xi_1 E_{01}^{+2} \quad [1.243]$$

where:

$$\xi_1 = 4\pi\varepsilon_0 a^2 l\frac{I}{\rho_{01}^2} \quad [1.244]$$

3) In the case of a real conductor, the instantaneous power dissipated by the Joule effect in each surface element dS of the wall is equal to $dP_d = \frac{1}{\gamma_c\delta}\overrightarrow{\mathbf{K}}^2 dS$. At the level of the lateral wall, the dissipated power is equal to:

$$P_d\big|_{r=a} = \int_0^l\int_0^{2\pi}\frac{\overrightarrow{\mathbf{K}}^2(r=a)}{\gamma_c\delta}ad\varphi dz = \frac{8\pi a l E_{01}^{+2}}{Z_0^2\gamma_c\delta}J_0'^2(\rho_{01})\sin^2(\omega_0 t) \quad [1.245]$$

Similarly, the powers dissipated by the two flat surfaces are equal, respectively, to:

$$\begin{aligned} P_d|_{z=0} &= \int_0^a \int_0^{2\pi} \frac{\vec{\mathbf{K}}^2(z=0)}{\gamma_c \delta} r dr d\varphi = \frac{8\pi E_{01}^{+2}}{Z_0^2 \gamma_c \delta} \sin^2(\omega_0 t) \int_0^a J_0'^2\left(\frac{\rho_{01} r}{a}\right) r dr \\ &= \frac{8\pi a^2 I E_{01}^{+2}}{Z_0^2 \gamma_c \delta \rho_{01}^2} \sin^2(\omega_0 t) \\ P_d|_{z=l} &= \frac{8\pi a^2 I E_{01}^{+2}}{Z_0^2 \gamma_c \delta \rho_{01}^2} \sin^2(\omega_0 t) \end{aligned} \quad [1.246]$$

4) The powers dissipated (averaged over a period) are:

$$\begin{aligned} \langle P_d \rangle_{r=a} &= \frac{8\pi a l E_{01}^{+2}}{Z_0^2 \gamma_c \delta} J_0'^2(\rho_{01}) \langle \sin^2(\omega_0 t) \rangle = \frac{4\pi a l E_{01}^{+2}}{Z_0^2 \gamma_c \delta} J_0'^2(\rho_{01}) = \xi_2 E_{01}^{+2} \\ \langle P_d \rangle_{z=0} = \langle P_d \rangle_{z=l} &= \frac{4\pi a^2 I E_{01}^{+2}}{Z_0^2 \gamma_c \delta \rho_{01}^2} = \xi_3 E_{01}^{+2} \end{aligned} \quad [1.247]$$

where:

$$\begin{aligned} \xi_2 &= \frac{4\pi a l}{Z_0^2 \gamma_c \delta} J_0'^2(\rho_{01}) \\ \xi_3 &= \frac{4\pi a^2 I}{Z_0^2 \gamma_c \delta \rho_{01}^2} \end{aligned} \quad [1.248]$$

Thus, the power (averaged over a period) dissipated by the entire cavity is equal to:

$$\langle P_d \rangle_{\text{total}} = \langle P_d \rangle_{r=a} + \langle P_d \rangle_{z=0} + \langle P_d \rangle_{z=l} = (\xi_2 + 2\xi_3) E_{01}^{+2} \quad [1.249]$$

5) The quality factor of the cylindrical cavity for this mode is given by:

$$Q = \frac{\omega_0 W_{\text{em}}}{\langle P_d \rangle_{\text{total}}} = \frac{\omega_0 \xi_1}{\xi_2 + 2\xi_3} = \frac{l I Z_0 \gamma_c \delta}{l \rho_{01}^2 J_0'(\rho_{01}) + 2aI} \quad [1.250]$$

6) The variation in electromagnetic energy is mainly due to the dissipation of energy through the walls:

$$\frac{dW_{\text{em}}}{dt} = -\langle P_d \rangle_{\text{total}} \implies \frac{dE_{01}^+}{dt} = -\frac{\xi_2 + 2\xi_3}{2\xi_1} E_{01}^+ \quad [1.251]$$

Therefore, we obtain:

$$E_{01}^{+}(t) = E_{01}^{+}(0)\, e^{-\frac{(\xi_2+2\xi_3)t}{2\xi_1}} \equiv E_{01}^{+}(0)\, e^{-\frac{\omega_0 t}{2Q}} \qquad [1.252]$$

Also, in this case, we see that the electric and magnetic fields [1.239] no longer vary purely oscillatory over time but decrease exponentially:

$$\begin{aligned} \vec{\mathbf{E}} &= 2E_{01}^{+}(0)\, J_0\left(\frac{\rho_{01} r}{a}\right) \cos(\omega_0 t)\, e^{-\frac{\omega_0 t}{2Q}}\, \vec{\mathbf{k}} \\ \vec{\mathbf{H}} &= 2Z_0^{-1} E_{01}^{+}(0)\, J_0\left(\frac{\rho_{01} r}{a}\right) \sin(\omega_0 t)\, e^{-\frac{\omega_0 t}{2Q}}\, \vec{\mathbf{e}}_{\varphi} \end{aligned} \qquad [1.253]$$

Chapter 2

Transmission Lines

2.1. Parallel-plate transmission line

Consider two parallel conducting plates (electrodes) separated by a distance d and enclosing an LIH medium with permittivity ε_0 and permeability μ_0 (for simplicity), as illustrated in Figure 2.1. This structure is known as a *parallel plate transmission line*. We want to study the propagation of a TEM wave (where the electric and magnetic fields are perpendicular to the Oz axis) between these two electrodes. Since the spacing d is much smaller than the dimensions of the plates, the edge effects can be neglected, and we can assume that the fields depend only on the coordinate z. The perfectly conducting electrodes impose the following boundary conditions: (1) the tangential component of $\overrightarrow{\mathbf{E}}$ must vanish; (2) the normal component of $\overrightarrow{\mathbf{B}}$ (and hence of $\overrightarrow{\mathbf{H}}$) must vanish. Taking these constraints into account and neglecting the edge effects, the fields cannot depend on the coordinates x and y. They are thus of the form $\overrightarrow{\mathbf{E}} = E_x(z,t)\,\overrightarrow{\mathbf{i}}$ and $\overrightarrow{\mathbf{H}} = H_y(z,t)\,\overrightarrow{\mathbf{j}}$. Substituting these expressions into the Maxwell equations, we obtain:

$$\begin{array}{ll} \overrightarrow{\nabla} \times \overrightarrow{\mathbf{E}} = -\mu_0 \dfrac{\partial \overrightarrow{\mathbf{H}}}{\partial t} \Longrightarrow \dfrac{\partial E_x}{\partial z} = -\mu_0 \dfrac{\partial H_y}{\partial t} \text{ (i)} \\ \overrightarrow{\nabla} \times \overrightarrow{\mathbf{H}} = \varepsilon_0 \dfrac{\partial \overrightarrow{\mathbf{E}}}{\partial t} \Longrightarrow \dfrac{\partial H_y}{\partial z} = -\varepsilon_0 \dfrac{\partial E_x}{\partial t} \text{ (ii)} \end{array} \qquad [2.1]$$

These equations are the same as those established for plane waves. The wave solutions previously found are also valid here. Let us now try to introduce the voltage and current variables along the electrodes (depending on z and t). The potential difference measured between the two electrodes at position z is defined by:

$$v(z,t) = \int_0^d \overrightarrow{\mathbf{E}} \cdot d\overrightarrow{\mathbf{l}} = E_x(z,t)\,d \qquad [2.2]$$

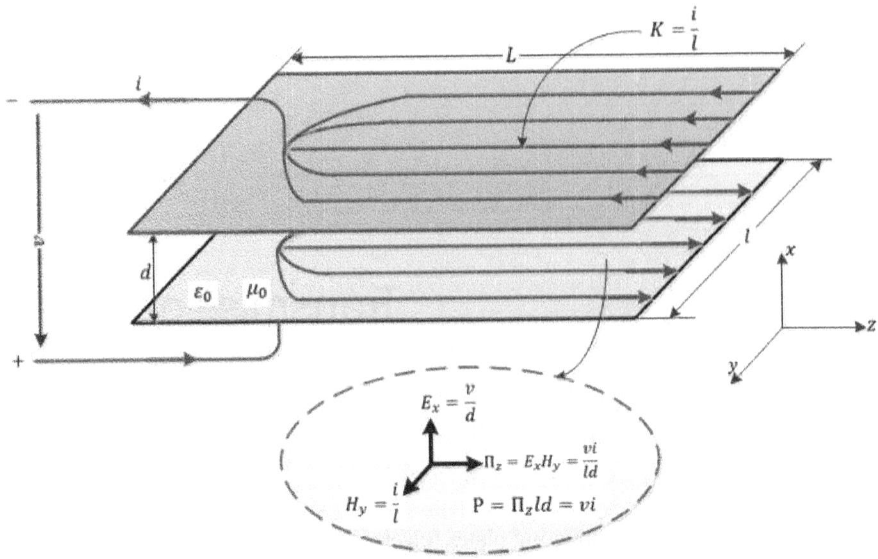

Figure 2.1. *Parallel-plate transmission line*

Since the discontinuity of the tangential component of $\vec{\mathbf{H}}$ on the surface of the plates gives us the surface current $\vec{\mathbf{K}}$, the total current $i(z,t)$ flowing along the direction Oz on the lower plate is:

$$i(z,t) = K_z(z,t)\,l = H_y(z,t)\,l \qquad [2.3]$$

By inserting equations [2.2] and [2.3] into [2.1], it comes that:

$$\begin{aligned} \frac{\partial v}{\partial z} &= -L\frac{\partial i}{\partial t} \quad \text{(i)} \\ \frac{\partial i}{\partial z} &= -C\frac{\partial v}{\partial t} \quad \text{(ii)} \end{aligned} \qquad [2.4]$$

where L and C are, respectively, the inductance and the capacitance per unit length of the parallel plate structure:

$$\begin{aligned} L &= \frac{\mu_0 d}{l} \\ C &= \frac{\varepsilon_0 l}{d} \end{aligned} \qquad [2.5]$$

These are the *characteristic equations of the transmission line* or, more simply, the *transmission equations*.

2.2. General formulation

The transmission equations [2.4] remain true for any structure with two conductors of any arbitrary shape in the Oxy plane, but whose cross-sectional area does not change along the propagation axis (here, the longitudinal axis, Oz). In general, the fields $\overrightarrow{\mathbf{E}}$ and $\overrightarrow{\mathbf{H}}$ can also depend on the transverse coordinates x and y, but always without any component lying along Oz:

$$\begin{aligned} \overrightarrow{\mathbf{E}} &= \overrightarrow{\mathbf{E}}_\perp (x,y,z,t) = E_x \overrightarrow{\mathbf{i}} + E_y \overrightarrow{\mathbf{j}} \\ \overrightarrow{\mathbf{E}} &= \overrightarrow{\mathbf{H}}_\perp (x,y,z,t) = H_x \overrightarrow{\mathbf{i}} + H_y \overrightarrow{\mathbf{j}} \end{aligned} \qquad [2.6]$$

In this case, the Maxwell equations are written as follows:

$$\begin{aligned} &\overrightarrow{\nabla}_\perp \times \overrightarrow{\mathbf{E}}_\perp + \frac{\partial}{\partial z}\left(\overrightarrow{\mathbf{k}} \times \overrightarrow{\mathbf{E}}_\perp\right) = -\mu_0 \frac{\partial \overrightarrow{\mathbf{H}}_\perp}{\partial t} \\ &\overrightarrow{\nabla}_\perp \times \overrightarrow{\mathbf{H}}_\perp + \frac{\partial}{\partial z}\left(\overrightarrow{\mathbf{k}} \times \overrightarrow{\mathbf{H}}_\perp\right) = \varepsilon_0 \frac{\partial \overrightarrow{\mathbf{E}}_\perp}{\partial t} \\ &\overrightarrow{\nabla}_\perp \cdot \overrightarrow{\mathbf{E}}_\perp = 0 \\ &\overrightarrow{\nabla}_\perp \cdot \overrightarrow{\mathbf{H}}_\perp = 0 \end{aligned} \qquad [2.7]$$

We observe the following: (1) $\overrightarrow{\nabla}_\perp \times \overrightarrow{\mathbf{E}}_\perp$ and $\overrightarrow{\nabla}_\perp \times \overrightarrow{\mathbf{H}}_\perp$ are directed along the Oz axis (longitudinal components); (2) $\overrightarrow{\mathbf{k}} \times \overrightarrow{\mathbf{E}}_\perp$ and $\overrightarrow{\mathbf{k}} \times \overrightarrow{\mathbf{H}}_\perp$ are purely in the transverse plane Oxy (transverse components). Separating the longitudinal and transverse components, we obtain:

$$\begin{aligned} &\overrightarrow{\nabla}_\perp \times \overrightarrow{\mathbf{E}}_\perp = 0 \qquad \overrightarrow{\nabla}_\perp \cdot \overrightarrow{\mathbf{E}}_\perp = 0 && \text{(i)} \\ &\overrightarrow{\nabla}_\perp \times \overrightarrow{\mathbf{H}}_\perp = 0 \qquad \overrightarrow{\nabla}_\perp \cdot \overrightarrow{\mathbf{H}}_\perp = 0 && \text{(ii)} \\ &\frac{\partial}{\partial z}\left(\overrightarrow{\mathbf{k}} \times \overrightarrow{\mathbf{E}}_\perp\right) = -\mu_0 \frac{\partial \overrightarrow{\mathbf{H}}_\perp}{\partial t} \iff \frac{\partial \overrightarrow{\mathbf{E}}_\perp}{\partial z} = \mu_0 \frac{\partial}{\partial t}\left(\overrightarrow{\mathbf{k}} \times \overrightarrow{\mathbf{H}}_\perp\right) && \text{(iii)} \\ &\frac{\partial}{\partial z}\left(\overrightarrow{\mathbf{k}} \times \overrightarrow{\mathbf{H}}_\perp\right) = \varepsilon_0 \frac{\partial \overrightarrow{\mathbf{E}}_\perp}{\partial t} && \text{(iv)} \end{aligned} \qquad [2.8]$$

The first two equations [2.8i] are those of 2D electrostatics without sources. We thus conclude that in the transverse plane Oxy (at constant z), the field $\overrightarrow{\mathbf{E}}_\perp$ derives from a scalar potential satisfying the 2D Laplace equation: $\overrightarrow{\mathbf{E}}_\perp = -\overrightarrow{\nabla}_\perp v$ and $\Delta_\perp v = 0$. The same observation holds for the magnetic excitation field $\overrightarrow{\mathbf{H}}_\perp$. Therefore, all of the tools developed to find solutions to static field problems, including 2D Laplace equation solving methods, the method of images and the method of conformal transformations, can be used here to determine $\overrightarrow{\mathbf{E}} \perp$ and $\overrightarrow{\mathbf{H}} \perp$ in the transverse plane Oxy.

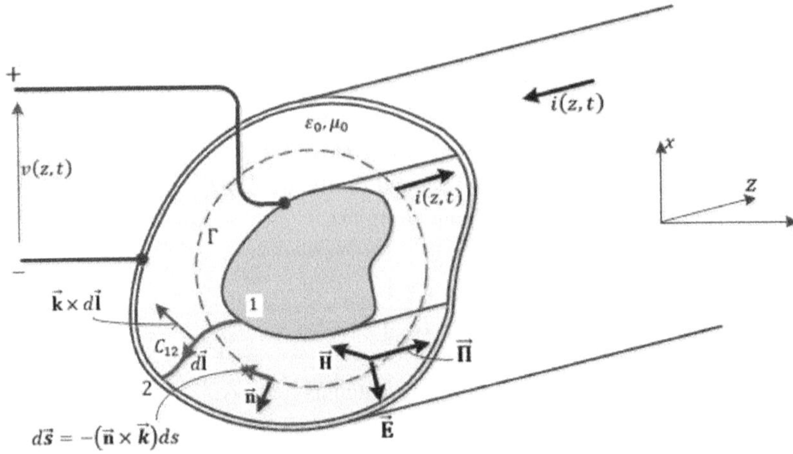

Figure 2.2. *Cross-section of an arbitrary transmission line. For a color version of this figure, see www.iste.co.uk/abdesselam/electrodynamics3.zip*

Consider a transmission line of an arbitrary shape, as shown in Figure 2.2. The transmission line consists of two perfect conductors whose cross-sectional area remains constant along the direction of Oz, but whose shape in the transverse Oxy plane is arbitrary. The electric and magnetic fields lie in the Oxy plane and behave, within this plane (i.e. with respect to x and y), like static fields. These fields can be related to the voltage and current measured along the transmission line (at the point z and at the time t) in the following way:

$$\begin{aligned} v(z,t) &= \int_{C_{12}} \vec{\mathbf{E}}_{\perp} \cdot d\vec{\mathbf{l}} \quad \text{(i)} \\ i(z,t) &= \oint_{\Gamma} \vec{\mathbf{H}}_{\perp} \cdot d\vec{\mathbf{s}} \quad \text{(ii)} \end{aligned} \qquad [2.9]$$

where C_{12} is any path located in the transverse plane (at constant z) connecting the two conductors, and Γ is any closed loop (also in the transverse plane) surrounding the inner conductor. Similarly, the charge per unit length and the flux per unit length along the transmission line are given by:

$$\begin{aligned} q(z,t) &= \varepsilon_0 \oint_{\Gamma} \vec{\mathbf{E}}_{\perp} \cdot \vec{\mathbf{n}}\, ds \quad \text{(i)} \\ \phi(z,t) &= \mu_0 \int_{C_{12}} \vec{\mathbf{H}}_{\perp} \cdot \left(\vec{\mathbf{k}} \times d\vec{\mathbf{l}}\right) \quad \text{(ii)} \end{aligned} \qquad [2.10]$$

The capacitance and inductance per unit length of the transmission line are then defined by the ratios:

$$C = \frac{q(z,t)}{v(z,t)} = \frac{\varepsilon_0 \oint_\Gamma \vec{\mathbf{E}}_\perp \cdot \vec{\mathbf{n}}\, ds}{\int_{C_{12}} \vec{\mathbf{E}}_\perp \cdot d\vec{\mathbf{l}}} \quad \text{(i)}$$

$$L = \frac{\phi(z,t)}{i(z,t)} = \frac{\mu_0 \int_{C_{12}} \vec{\mathbf{H}}_\perp \cdot \left(\vec{\mathbf{k}} \times d\vec{\mathbf{l}}\right)}{\oint_\Gamma \vec{\mathbf{H}}_\perp \cdot d\vec{\mathbf{s}}} \quad \text{(ii)} \qquad [2.11]$$

Note that C and L are constants since the geometry of the transmission line does not vary with z. Even if the fields change with z, the ratios in [2.11] do not depend on the field amplitudes.

To obtain the characteristic equations of the transmission line, take the dot product of [2.8iii] with $d\vec{\mathbf{l}}$ (which can be brought inside the derivatives since it only varies with x and y, not with z and t), and then integrate the resulting equation over a path C_{12} (at constant z) connecting the two conductors:

$$\begin{aligned}\frac{\partial}{\partial z}\left(\int_{C_{12}} \vec{\mathbf{E}}_\perp \cdot d\vec{\mathbf{l}}\right) &= \frac{\partial}{\partial t}\left(\mu_0 \int_{C_{12}} \left(\vec{\mathbf{k}} \times \vec{\mathbf{E}}_\perp\right) \cdot d\vec{\mathbf{l}}\right)\\ &= -\frac{\partial}{\partial t}\left(\mu_0 \int_{C_{12}} \vec{\mathbf{H}}_\perp \cdot \left(\vec{\mathbf{k}} \times \vec{\mathbf{l}}\right)\right)\end{aligned} \qquad [2.12]$$

The left-hand side is simply the derivative of the voltage defined in [2.9i], while the right-hand side is the time derivative (with a minus sign) of the flux per unit length defined in [2.10ii]:

$$\frac{\partial v}{\partial z} = -\frac{\partial \phi}{\partial t} = -L\frac{\partial \imath}{\partial t} \qquad [2.13]$$

This is the *first characteristic equation of the transmission line*. In the same way, let us take the dot product of [2.8iii] with $\vec{\mathbf{n}}$ (normal to Γ), and then integrate over the contour Γ (at constant z) surrounding the inner conductor:

$$\begin{aligned}\frac{\partial}{\partial z}\left(\oint_\Gamma \vec{\mathbf{E}}_\perp \cdot \vec{\mathbf{n}}\, ds\right) &= \frac{\partial}{\partial t}\left(\mu_0 \oint_\Gamma \vec{\mathbf{n}} \cdot \left(\vec{\mathbf{k}} \times \vec{\mathbf{H}}_\perp\right) ds\right)\\ &= -\frac{\partial}{\partial t}\left(\mu_0 \oint_\Gamma \vec{\mathbf{H}}_\perp \cdot d\vec{\mathbf{s}}\right)\end{aligned} \qquad [2.14]$$

The term on the left is the derivative of the charge per unit length defined in [2.10i], while the term on the right represents the time derivative of the current defined in [2.9i], that is, $\frac{1}{\varepsilon_0}\frac{\partial q}{\partial z} = -\mu_0 \frac{\partial i}{\partial t}$. By writing $q = Cv$, we obtain $C\frac{\partial v}{\partial z} = -\varepsilon_0\mu_0\frac{\partial i}{\partial t}$. Since $LC = \varepsilon_0\mu_0$, we deduce that:

$$\frac{\partial v}{\partial z} = -L\frac{\partial i}{\partial t} \tag{2.15}$$

We therefore arrive at the same equation [2.4i]. To obtain the second characteristic equation of the transmission line, we take the dot product of [2.8iv] with $d\vec{\mathbf{l}}$, and then integrate along C_{12}:

$$\begin{aligned}\frac{\partial}{\partial t}\left(\varepsilon_0 \int_{C_{12}} \vec{\mathbf{E}}_\perp \cdot d\vec{\mathbf{l}}\right) &= \frac{\partial}{\partial z}\left(\int_{C_{12}} \left(\vec{\mathbf{k}} \times \vec{\mathbf{H}}_\perp\right) \cdot d\vec{\mathbf{l}}\right) \\ &= -\frac{\partial}{\partial z}\left(\int_{C_{12}} \vec{\mathbf{H}}_\perp \cdot \left(\vec{\mathbf{k}} \times d\vec{\mathbf{l}}\right)\right)\end{aligned} \tag{2.16}$$

which yields:

$$\varepsilon_0\frac{\partial v}{\partial t} = -\frac{1}{\mu_0}\frac{\partial \phi}{\partial z} = -\frac{L}{\mu_0}\frac{\partial i}{\partial z} \implies \frac{\partial i}{\partial z} = -C\frac{\partial v}{\partial t} \tag{2.17}$$

EXAMPLE 2.1. COAXIAL CABLE.– Consider the coaxial transmission line (*coaxial cable*) shown in Figure 2.3.

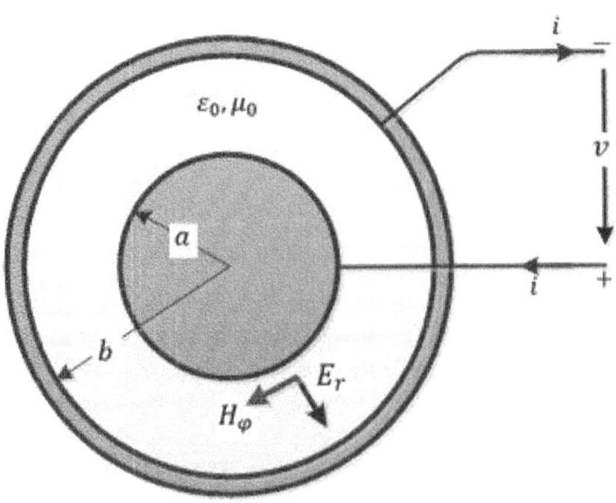

Figure 2.3. *Structure of a coaxial transmission line (coaxial cable). For a color version of this figure, see www.iste.co.uk/abdesselam/electrodynamics3.zip*

The coaxial cable consists of a cylindrical conductor (called the *core* or the *center* of the cable) with axis Oz, and radius a, surrounded by a second hollow cylindrical conductor with the same axis Oz and inner radius b $(b > a)$, called the *shield* of the cable. The space between the two conductors is typically filled with an LIH dielectric medium with permittivity ε_0 and permeability μ_0 (for simplicity). The length of the cable is much greater than b, so it can be considered infinite. We also assume that both conductors are perfect. In the transverse plane Oxy, the fields behave as if they were static, that is, independent of time. If v and i denote, respectively, the voltage difference between the two cylinders and the current carried by the inner cylinder, the fields are as follows:

$$E_r = \frac{v}{r \ln\left(\frac{b}{a}\right)} \qquad H_\varphi = \frac{i}{2\pi r} \qquad [2.18]$$

The charge per unit length q and the flux per unit length ϕ are given by:

$$q = \varepsilon_0 E_r (r = a)\, 2\pi a = \frac{2\pi\varepsilon_0 v}{\ln\left(\frac{b}{a}\right)} \qquad \phi = \int_a^b \mu_0 H_\varphi dr = \frac{\mu_0 i}{2\pi} \ln\left(\frac{b}{a}\right) \qquad [2.19]$$

Consequently, we obtain:

$$C = \frac{q}{v} = \frac{2\pi\varepsilon_0}{\ln\left(\frac{b}{a}\right)} \qquad L = \frac{\phi}{i} = \frac{\mu_0}{2\pi} \ln\left(\frac{b}{a}\right) \qquad [2.20]$$

By substituting the expressions of E_r and H_φ into [2.8], we deduce the following transmission equations:

$$\frac{\partial E_r}{\partial z} = -\mu_0 \frac{\partial H_\varphi}{\partial t} \implies \frac{\partial v}{\partial z} = -L\frac{\partial i}{\partial t} \qquad \frac{\partial H_\varphi}{\partial z} = -\varepsilon_0 \frac{\partial E_r}{\partial t} \implies \frac{\partial i}{\partial z} = -C\frac{\partial v}{\partial t} \qquad [2.21]$$

To solve the transmission equations [2.21], we proceed as follows: by combining these two equations, we deduce that the voltage v must satisfy the differential equation $\frac{\partial^2 v}{\partial z^2} = LC\frac{\partial^2 i}{\partial t^2} = \varepsilon_0\mu_0\frac{\partial^2 i}{\partial t^2}$, which can be solved by the method of separation of variables. Assuming $v(z,t) = U(z)\,T(t)$, we find $\frac{U''}{U} = \varepsilon_0\mu_0\frac{T''}{T} = -\mathsf{k}^2$, where k is a constant. This yields the following solutions: $T(t) = Ae^{-\mathsf{i}\frac{\mathsf{k}}{\sqrt{\varepsilon_0\mu_0}}t}$ and $U(t) = A'e^{\mathsf{i}\mathsf{k}z} + B'e^{-\mathsf{i}\mathsf{k}z}$. Thus, the most general solution is given by:

$$v(z,t) = \Re\left[\mathsf{F}e^{-\mathsf{i}\frac{\mathsf{k}}{\sqrt{\varepsilon_0\mu_0}}t}\left(e^{\mathsf{i}\mathsf{k}z} + \mathsf{R}e^{-\mathsf{i}\mathsf{k}z}\right)\right] \qquad [2.22]$$

where F and R are integration constants. Each term in [2.22] is easily interpreted: the first represents a forward wave propagating in the positive direction of Oz, while the second represents a backward wave in the opposite direction, resulting from a reflection of the first. The constant R thus appears here as a reflection coefficient. To obtain the expression for $i(z,t)$, we use one of the transmission equations, for example, $\frac{\partial i}{\partial t} = -\frac{1}{L}\frac{\partial v}{\partial z} = -\frac{\mathrm{ikF}}{L} e^{-\mathrm{i}\frac{\mathrm{k}}{\sqrt{\varepsilon_0\mu_0}}t}\left(e^{\mathrm{ik}z} - \mathrm{R}e^{-\mathrm{ik}z}\right)$. After integration, this yields:

$$i(z,t) = \Re\left[\frac{\sqrt{\varepsilon_0\mu_0}\mathrm{F}}{L} e^{-\mathrm{i}\frac{\mathrm{k}}{\sqrt{\varepsilon_0\mu_0}}t}\left(e^{\mathrm{ik}z} - \mathrm{R}e^{-ikz}\right)\right] \quad [2.23]$$

The constant F can be determined by specifying the voltage at $z = 0$. Taking $v(0,t) = v_0 e^{-i\omega t}$, we find $\mathrm{k} = \sqrt{\varepsilon_0\mu_0}\omega$ and $\mathrm{F} = \frac{v_0}{1+\mathrm{R}}$. Given these results, the expressions for voltage $v(z,t)$ and current intensity $i(z,t)$ are as follows:

$$\begin{aligned} v(z,t) &= \Re\left[\frac{v_0}{1+\mathrm{R}} e^{-\mathrm{i}\omega t}\left(e^{\mathrm{ik}z} + \mathrm{R}e^{-\mathrm{ik}z}\right)\right] \\ i(z,t) &= \Re\left[\frac{1}{\mathrm{Z}_0}\frac{v_0}{1+\mathrm{R}} e^{-\mathrm{i}\omega t}\left(e^{\mathrm{ik}z} - \mathrm{R}e^{-\mathrm{ik}z}\right)\right] \end{aligned} \quad [2.24]$$

where:

$$\mathrm{Z}_0 = \frac{1}{2\pi}\sqrt{\frac{\mu_0}{\varepsilon_0}}\ln\left(\frac{b}{a}\right) \quad [2.25]$$

Z_0 is called the *characteristic impedance of the cable*. The ratio is given as:

$$\hat{\mathrm{Z}}(z) = \frac{\hat{v}(z)}{\hat{i}(z)} = Z_c\frac{e^{\mathrm{ik}z} + \mathrm{R}e^{-\mathrm{ik}z}}{e^{\mathrm{ik}z} - \mathrm{R}e^{-\mathrm{ik}z}} \quad [2.26]$$

which defines a complex impedance at position z.

2.3. Modeling of lossless transmission lines

Transmission equations [2.4] can also be derived using an infinitesimal electric quadrupole uniformly distributed (one after the other) in the Oz direction of the line, without resorting to the Maxwell equations. Each quadrupole consists of an inductor with inductance coefficient Ldz and a capacitor with capacitance Cdz, connected in parallel with the inductor, as illustrated in Figure 2.4. The current intensity and voltage at the input of the quadrupole at position z are denoted by $i(z,t)$ and $v(z,t)$, respectively. At the output of the quadrupole located at $z + dz$, their respective values are $i(z+dz,t)$ and $v(z+dz,t)$. The voltage is the potential difference between the central conductor and the outer conductor.

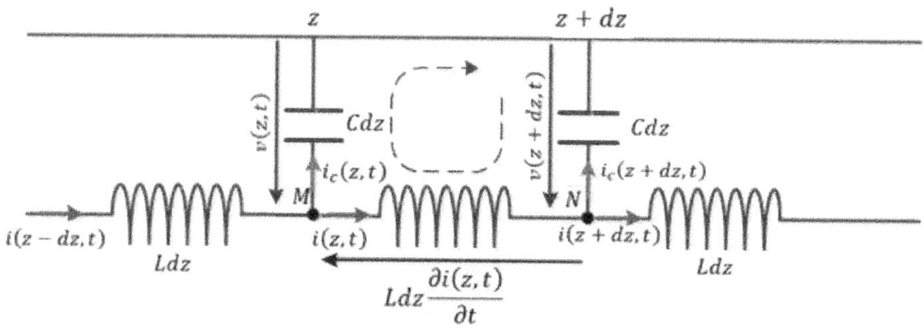

Figure 2.4. *Modeling of a lossless transmission line. For a color version of this figure, see www.iste.co.uk/abdesselam/electrodynamics3.zip*

Voltage–current relationships can be easily established by applying the first and second Kirchhoff laws. At node M, we obtain:

$$i(z-dz,t) = i(z,t) + i_c(z,t) \qquad [2.27]$$

where $i_c(z,t)$ is the current entering the capacitor. Since the charge on the capacitor is $Q = Cdzv(z,t)$, we write $i_c(z,t) = \frac{\partial Q}{\partial t} = Cdz\frac{\partial v}{\partial t}$. This leads to $Cdz\frac{\partial v}{\partial t} = -[i(z,t) - i(z-dz,t)] = -\frac{\partial i}{\partial z}dz$, and consequently:

$$\frac{\partial i}{\partial z} = -C\frac{\partial v}{\partial t} \qquad [2.28]$$

This is the *first transmission equation*. Similarly, applying Kirchhoff's loop law to the circuit between z and $z+dz$ leads to the second transmission equation $v(z,t) = v(z+dz,t) + Ldz\frac{\partial i}{\partial t}$. We can then conclude that:

$$\frac{\partial v}{\partial z} = -L\frac{\partial i}{\partial t} \qquad [2.29]$$

2.4. Modeling of a transmission line with losses

The modeling of a transmission line as a succession of quadrupoles can be easily extended to include the losses observed in conductors (due to the Joule effect in real conductors) and in dielectrics. To take these losses into account, the structure of the elementary quadrupole of the line must be modified. Thus, to consider losses in the conducting wires, a resistance Rdz is added in series with the inductance Ldz, where R is the total resistance of the two conductors per unit length of the line. For dielectric losses, a resistance is added in parallel with the capacitance Cdz. The corresponding conductance is Gdz, where G is the conductance per unit length of the line associated with dielectric losses. The lossy transmission line is therefore modeled as illustrated in Figure 2.5.

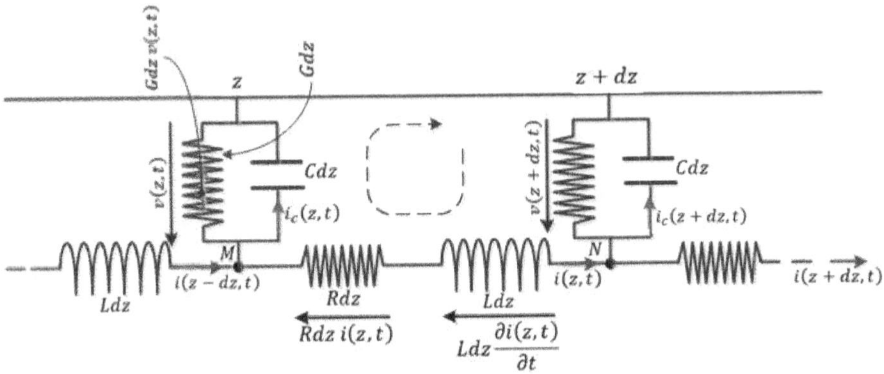

Figure 2.5. *Modeling a transmission line with losses. For a color version of this figure, see www.iste.co.uk/abdesselam/electrodynamics3.zip*

The equations previously established at node M and in the loop (located between M and N) must be modified as follows:

$$\begin{aligned} i\left(z-dz,t\right) &= i\left(z,t\right)+Cdz\frac{\partial v}{\partial t}+Gdzv \\ v\left(z,t\right) &= v\left(z+dz,t\right)+Ldz\frac{\partial i}{\partial t}+Rdzi \end{aligned} \qquad [2.30]$$

which gives:

$$\begin{aligned} \frac{\partial i}{\partial z} &= -C\frac{\partial v}{\partial t}-Gv \quad \text{(i)} \\ \frac{\partial v}{\partial z} &= -L\frac{\partial i}{\partial t}-Ri \quad \text{(ii)} \end{aligned} \qquad [2.31]$$

These are the *transmission equations with losses*. By setting $R = G = 0$, we recover the lossless transmission equations [2.4].

By multiplying [2.31i] by v and [2.31ii] by i, and then summing the results, we obtain a form equivalent to the Poynting theorem:

$$\frac{\partial}{\partial z}\left(vi\right)=-\frac{\partial}{\partial t}\left(\frac{1}{2}Cv^2+\frac{1}{2}Li^2\right)-Gv^2-Ri^2 \qquad [2.32]$$

The power vi flowing along the transmission line is partly stored in the inductance and capacitance, while the other part is dissipated in the resistance and conductance through the Joule effect.

In terms of fields, the total electromagnetic power flowing through the transmission line at position z is equal to:

$$P(z,t) = \int_S \left(\vec{\mathbf{E}}_\perp \times \vec{\mathbf{H}}_\perp\right) \cdot \vec{\mathbf{k}}\, dS = \int_S \vec{\mathbf{E}}_\perp \cdot \left(\vec{\mathbf{H}}_\perp \times \vec{\mathbf{k}}\right) dS \qquad [2.33]$$

where S is the cross-sectional area bounded between the two conductors (at constant z). Since $\vec{\nabla}_\perp \times \vec{\mathbf{E}}_\perp = 0$, we write $\vec{\mathbf{E}}_\perp = -\vec{\nabla}_\perp V$, and consequently:

$$P(z,t) = \int_S \left(\vec{\mathbf{k}} \times \vec{\mathbf{H}}_\perp\right) \cdot \vec{\nabla}_\perp V\, dS \qquad [2.34]$$

Noting that $\vec{\nabla}_\perp \cdot \left[V\left(\vec{\mathbf{k}} \times \vec{\mathbf{H}}_\perp\right)\right] = \left(\vec{\mathbf{k}} \times \vec{\mathbf{H}}_\perp\right) \cdot \vec{\nabla}_\perp V + V \vec{\nabla}_\perp \cdot \left(\vec{\mathbf{k}} \times \vec{\mathbf{H}}_\perp\right) = \left(\vec{\mathbf{k}} \times \vec{\mathbf{H}}_\perp\right) \cdot \vec{\nabla}_\perp V - V \vec{\mathbf{k}} \cdot \left(\vec{\mathbf{k}} \times \vec{\mathbf{H}}_\perp\right) = \left(\vec{\mathbf{k}} \times \vec{\mathbf{H}}_\perp\right) \cdot \vec{\nabla}_\perp V$, since $\vec{\nabla}_\perp \times \vec{\mathbf{H}}_\perp = 0$, the integral [2.34] can now be converted to a curvilinear integral using the divergence theorem in two dimensions:

$$P(z,t) = \int_S \vec{\nabla}_\perp \cdot \left[V\left(\vec{\mathbf{k}} \times \vec{\mathbf{H}}_\perp\right)\right] dS = \oint_{\Gamma_{\text{ext}}} V\left(\vec{\mathbf{k}} \times \vec{\mathbf{H}}_\perp\right) \cdot \vec{\mathbf{n}}\, ds + \oint_{\Gamma_{\text{int}}} V\left(\vec{\mathbf{k}} \times \vec{\mathbf{H}}_\perp\right) \cdot \left(-\vec{\mathbf{n}}\right) ds \qquad [2.35]$$

where Γ_{ext} and Γ_{int} are, respectively, the contours of the outer and inner conductors in the transverse plane Oxy. The minus sign appears in the second integral because $\vec{\mathbf{n}}$ is defined inward with respect to the surface S in Figure 2.2 (when traversing the contour Γ_{int}). By choosing $V = 0$ on the outer conductor (reference level), the integral on the outer contour must be zero. Only the integral on the inner contour is non-zero, where $V = v$:

$$P(z,t) = -\oint_{\Gamma_{\text{int}}} v\left(\vec{\mathbf{k}} \times \vec{\mathbf{H}}_\perp\right) \cdot \vec{\mathbf{n}}\, ds = v \oint_{\Gamma_{\text{int}}} \vec{\mathbf{H}}_\perp \cdot \left(\vec{\mathbf{k}} \times \vec{\mathbf{n}}\right) ds = v \oint_{\Gamma_{\text{int}}} \vec{\mathbf{H}}_\perp \cdot d\vec{\mathbf{s}} = vi \qquad [2.36]$$

The electromagnetic power flowing through the transmission line is equal to the power of the circuit.

2.5. Wave equations and sinusoidal solutions

Restricting ourselves now to lossless transmission lines ($R = G = 0$), the two coupled transmission equations [2.31] can be reduced to two wave equations in v and i:

$$\begin{aligned}\frac{\partial^2 v}{\partial t^2} &= c^2 \frac{\partial^2 v}{\partial z^2}\\ \frac{\partial^2 i}{\partial t^2} &= c^2 \frac{\partial^2 i}{\partial z^2}\end{aligned} \qquad [2.37]$$

where $c = \frac{1}{\sqrt{LC}} = \frac{1}{\sqrt{\varepsilon_0 \mu_0}}$ is the wave velocity. The solutions of equations [2.37] are of the form:

$$\begin{aligned}v(z,t) &= V_+\left(t - \frac{z}{c}\right) + V_-\left(t + \frac{z}{c}\right)\\ i(z,t) &= I_+\left(t - \frac{z}{c}\right) + I_-\left(t + \frac{z}{c}\right)\end{aligned} \qquad [2.38]$$

where V_+, V_-, I_+ and I_- are determined by the boundary conditions imposed at the ends of the transmission line. By substituting these solutions into [2.4], we obtain:

$$\begin{aligned}V_+\left(t - \frac{z}{c}\right) &= \mathsf{Z}_0 I_+\left(t - \frac{z}{c}\right)\\ V_-\left(t + \frac{z}{c}\right) &= -\mathsf{Z}_0 I_-\left(t + \frac{z}{c}\right)\end{aligned} \qquad [2.39]$$

where $\mathsf{Z}_0 = \sqrt{\frac{L}{C}}$. Z_0 is called the *characteristic impedance of the lossless line*. Its inverse, $\mathsf{Y}_0 = \frac{1}{\mathsf{Z}_0}$, is also called the *characteristic admittance of the line*. Transmission lines are often driven by sources that vary sinusoidally. This implies that the voltage and current along the line also vary sinusoidally with time. Thus, we take $V_+\left(t - \frac{z}{c}\right) = \Re\left[\hat{V}_+ e^{-\mathrm{i}(\omega t - \mathsf{k}z)}\right]$ and $V_-\left(t + \frac{z}{c}\right) = \Re\left[\hat{V}_- e^{-\mathrm{i}(\omega t + \mathsf{k}z)}\right]$, with $\mathsf{k} = \frac{\omega}{c}$. Here, $\hat{V}_+$ and $\hat{V}_-$ are complex integration constants to be determined by boundary conditions. Consequently, it follows that:

$$\begin{aligned}v(z,t) &= \Re\left[\left(\hat{V}_+ e^{\mathrm{i}\mathsf{k}z} + \hat{V}_- e^{-\mathrm{i}\mathsf{k}z}\right) e^{-\mathrm{i}\omega t}\right] = \Re\left[\hat{v}(z)\, e^{-\mathrm{i}\omega t}\right]\\ i(z,t) &= \Re\left[\mathsf{Y}_0\left(\hat{V}_+ e^{\mathrm{i}\mathsf{k}z} - \hat{V}_- e^{-\mathrm{i}\mathsf{k}z}\right) e^{-\mathrm{i}\omega t}\right] = \Re\left[\hat{i}(z)\, e^{-\mathrm{i}\omega t}\right]\end{aligned} \qquad [2.40]$$

where $\hat{v}(z)$ and $\hat{i}(z)$ are the complex amplitudes of the waves:

$$\begin{aligned}\hat{v}(z) &= \hat{V}_+ e^{\mathrm{i}\mathsf{k}z} + \hat{V}_- e^{-\mathrm{i}\mathsf{k}z}\\ \hat{i}(z) &= \mathsf{Y}_0\left(\hat{V}_+ e^{\mathrm{i}\mathsf{k}z} - \hat{V}_- e^{-\mathrm{i}\mathsf{k}z}\right)\end{aligned} \qquad [2.41]$$

The ratio is given is:

$$\hat{Z}(z) = \frac{\hat{v}(z)}{\hat{i}(z)} = Z_0 \left(\frac{\hat{V}_+ e^{ikz} + \hat{V}_- e^{-ikz}}{\hat{V}_+ e^{ikz} - \hat{V}_- e^{-ikz}} \right) \qquad [2.42]$$

is called the *complex impedance of the transmission line at position z.*

EXAMPLE 2.2. SHORT-CIRCUITED LINE.– The transmission line depicted in Figure 2.6 is excited by a sinusoidal voltage source at $z = 0$, which imposes the following boundary condition:

$$v(z = 0, t) = V_0 \cos \omega t = \Re \left[V_0 e^{-i\omega t} \right] \implies \hat{v}(z = 0) = V_0 = \hat{V}_+ + \hat{V}_- \qquad [2.43]$$

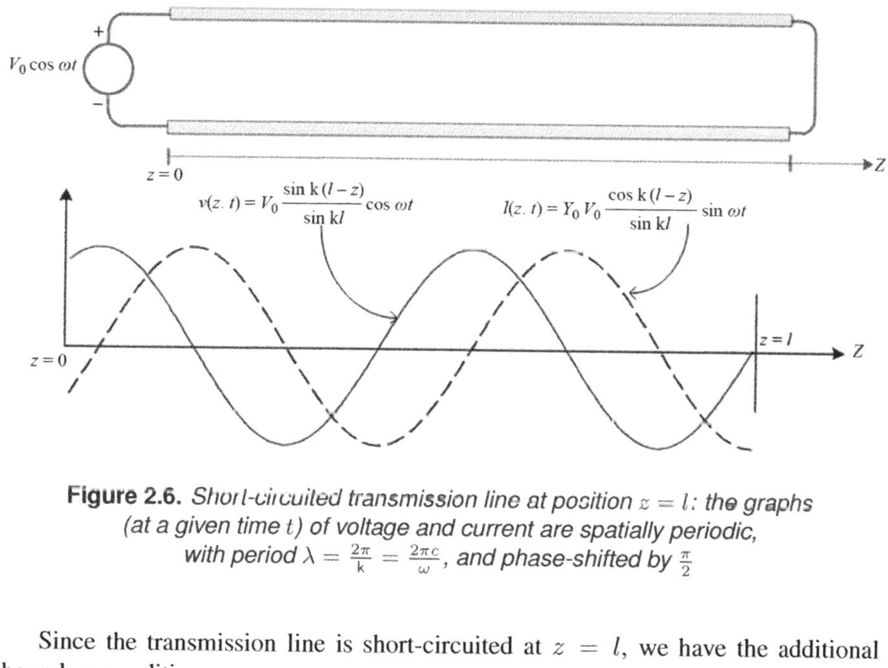

Figure 2.6. *Short-circuited transmission line at position $z = l$: the graphs (at a given time t) of voltage and current are spatially periodic, with period $\lambda = \frac{2\pi}{k} = \frac{2\pi c}{\omega}$, and phase-shifted by $\frac{\pi}{2}$*

Since the transmission line is short-circuited at $z = l$, we have the additional boundary condition:

$$v(z = l, t) = 0 \implies \hat{v}(z = l) = 0 = \hat{V}_+ e^{ikl} + \hat{V}_- e^{-ikl} \qquad [2.44]$$

By simultaneously solving [2.43] and [2.44], we find:

$$\begin{aligned} \hat{V}_+ &= -V_0 \frac{e^{-ikl}}{2i \sin kl} \\ \hat{V}_- &= V_0 \frac{e^{ikl}}{2i \sin kl} \end{aligned} \qquad [2.45]$$

and, consequently:

$$\hat{v}(z) = V_0 \frac{\sin \mathsf{k}(l - z)}{\sin \mathsf{k}l}$$
$$\hat{i}(z) = \mathrm{i}\mathsf{Y}_0 V_0 \frac{\cos \mathsf{k}(l - z)}{\sin \mathsf{k}l} \qquad [2.46]$$

Therefore, the voltage and current as functions of z and t (i.e. along the transmission line) are:

$$v(z,t) = \Re\left[\hat{v}(z)\, e^{-\mathrm{i}\omega t}\right] = V_0 \frac{\sin \mathsf{k}(l - z)}{\sin \mathsf{k}l} \cos \omega t$$
$$i(z,t) = \Re\left[\hat{i}(z)\, e^{-\mathrm{i}\omega t}\right] = \mathsf{Y}_0 V_0 \frac{\cos \mathsf{k}(l - z)}{\sin \mathsf{k}l} \sin \omega t \qquad [2.47]$$

The complex impedance at any point z is equal to:

$$\hat{\mathsf{Z}}(z) = \frac{\hat{v}(z)}{\hat{i}(z)} = -\mathrm{i}\mathsf{Z}_0 \tan \mathsf{k}(l - z) \qquad [2.48]$$

In particular, at $z = 0$, the transmission line appears to the generator as an impedance with the value $\hat{Z}(0) = -\mathrm{i}\mathsf{Z}_0 \tan \mathsf{k}l$. The following can be observed:

1) When $\mathsf{k}l = n\pi$, with $n \in \mathbb{N}^*$, the impedance at point $z = 0$ is zero, and the transmission line behaves like a short circuit.

2) When $\mathsf{k}l = \frac{(2n-1)\pi}{2}$, with $n \in \mathbb{N}^*$, the impedance at point $z = 0$ is infinite, and the transmission line behaves like an open circuit.

3) When $l \ll \lambda$, the following approximations can be made: $\sin \mathsf{k}z \simeq \mathsf{k}z$ and $\cos \mathsf{k}z \simeq 1$. Thus, equations [2.47] reduce to:

$$v(z,t) = V_0\left(1 - \frac{z}{l}\right)\cos \omega t \implies v(0,t) = V_0 \cos \omega t$$
$$i(z,t) = \frac{Y_0 V_0}{\mathsf{k}l} \sin \omega t \implies i(0,t) = \frac{Y_0 V_0}{\mathsf{k}l} \sin \omega t \qquad [2.49]$$

and, consequently:

$$v(0,t) = Ll \frac{di(0,t)}{dt} \qquad [2.50]$$

We see that the short-circuited transmission line behaves like an inductor with inductance Ll.

EXAMPLE 2.3. OPEN-CIRCUITED LINE.– If the short circuit at the end $z = l$ is removed, as shown in Figure 2.7, the boundary conditions are modified as follows:

$$i(z = l, t) = 0 \implies \hat{V}_+ e^{\mathrm{i}\mathsf{k}l} - \hat{V}_- e^{-\mathrm{i}\mathsf{k}l} = 0$$
$$v(z = 0, t) = \Re\left[V_0 e^{-\mathrm{i}\omega t}\right] \implies \hat{V}_+ + \hat{V}_- = V_0 \qquad [2.51]$$

which gives:

$$\hat{V}_+ = V_0 \frac{e^{-ikl}}{2\cos kl}$$
$$\hat{V}_- = V_0 \frac{e^{ikl}}{2\cos kl} \quad [2.52]$$

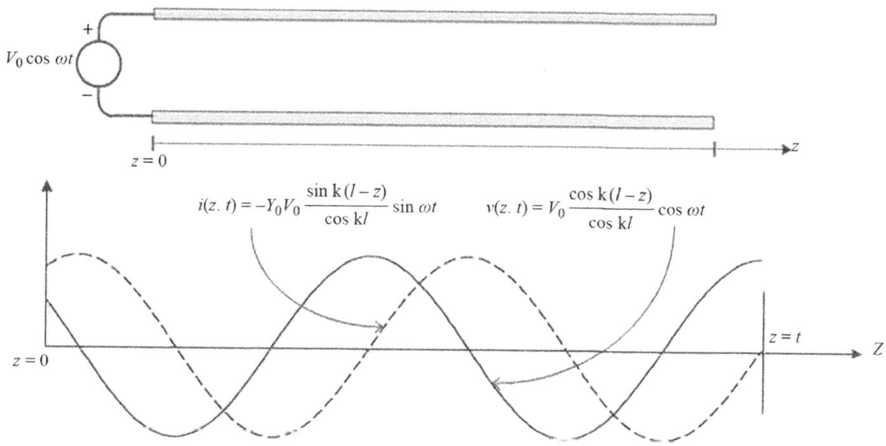

Figure 2.7. *Open transmission line at $z = l$: the graphs (at a given instant t) of the voltage and current are phase-shifted by $\frac{\pi}{2}$*

It follows that:

$$v(z,t) = V_0 \frac{\cos k(l-z)}{\cos kl} \cos \omega t$$
$$i(z,t) = -Y_0 V_0 \frac{\sin k(l-z)}{\cos kl} \sin \omega t \quad [2.53]$$

The complex impedance at any point z is given by:

$$\hat{Z}(z) = \frac{\hat{v}(z)}{\hat{i}(z)} = -\frac{iZ_0}{\tan k(l-z)} \quad [2.54]$$

At the end $z = 0$, the transmission line appears to the generator as an impedance of value $\hat{Z}(0) = -\frac{iZ_0}{\tan kl}$:

1) When $kl = n\pi$, with $n \in \mathbb{N}^*$, the impedance is infinite at point $z = 0$ and the transmission line behaves like an open circuit.

2) When $kl = \frac{(2n-1)\pi}{2}$, with $n \in \mathbb{N}^*$, the impedance is zero at point $z = 0$ and the transmission line behaves like a short circuit.

3) When $l \ll \lambda$, equations [2.53] can be approximated as follows:

$$\begin{aligned} v(z,t) &= V_0 \cos\omega t \implies v(0,t) = V_0 \cos\omega t \\ i(z,t) &= -\mathsf{Y}_0 V_0 \mathsf{k}(l-z)\sin\omega t \implies i(0,t) = -\mathsf{Y}_0 V_0 \mathsf{k} l \sin\omega t \end{aligned} \quad [2.55]$$

and, consequently:

$$i(0,t) = Cl\frac{dv(0,t)}{dt} \quad [2.56]$$

We see that the open transmission line behaves like a capacitor with a value of Cl.

2.6. Effects of losses in transmission lines

Taking into account ohmic losses in the transmission line, the voltage and current waves must decrease as they propagate. Since equations [2.31] are linear with constant coefficients, let us try sinusoidal solutions of the following form (generally, the generator feeds the line with sinusoidal voltage):

$$\begin{aligned} v(z,t) &= \Re\left[\hat{V}_+ e^{-\mathrm{i}(\omega t - \mathsf{k}z)} + \hat{V}_- e^{-\mathrm{i}(\omega t + \mathsf{k}z)}\right] \\ i(z,t) &= \Re\left[\hat{I}_+ e^{-\mathrm{i}(\omega t - \mathsf{k}z)} + \hat{I}_- e^{-\mathrm{i}(\omega t + \mathsf{k}z)}\right] \end{aligned} \quad [2.57]$$

The plus sign is used for waves propagating in the positive direction of the Oz axis, and the minus sign for waves propagating in the opposite direction. Substituting [2.57] into [2.31], we obtain:

$$\begin{aligned} \mathsf{ik}\left(\hat{I}_+ e^{\mathsf{ik}z} - \hat{I}_- e^{-\mathsf{ik}z}\right) &= (\mathrm{i}C\omega - G)\left(\hat{V}_+ e^{\mathsf{ik}z} + \hat{V}_+ e^{\mathsf{ik}z}\right) \\ \mathsf{ik}\left(\hat{V}_+ e^{\mathsf{ik}z} - \hat{V}_+ e^{\mathsf{ik}z}\right) &= (\mathrm{i}L\omega - R)\left(\hat{I}_+ e^{\mathsf{ik}z} + \hat{I}_+ e^{\mathsf{ik}z}\right) \end{aligned} \quad [2.58]$$

It follows that:

$$\frac{\hat{V}_\pm}{\hat{I}_\pm} = \pm\frac{\mathsf{ik}}{\mathrm{i}C\omega - G} = \pm\frac{\mathrm{i}L\omega - R}{\mathsf{ik}} \quad [2.59]$$

To obtain the relationship between k and ω, we begin by expanding equation [2.59]:

$$\mathsf{k}^2 = LC\omega^2 + \mathrm{i}\omega(LG + CR) - RG \quad [2.60]$$

We see that the wavenumber k is complex. By setting $\mathsf{k} = \mathsf{k}_r + \mathrm{i}\mathsf{k}_i$, it comes that:

$$\begin{aligned} \mathsf{k}_r^2 - \mathsf{k}_i^2 &= LC\omega^2 - RG \\ 2\mathsf{k}_r\mathsf{k}_i &= \omega(LG + CR) \end{aligned} \quad [2.61]$$

and, consequently:

$$\begin{aligned} \mathsf{k}_r &= \sqrt{\frac{1}{2}\left[\sqrt{(R^2+L^2\omega^2)\,(G^2+C^2\omega^2)}+LC\omega^2-RG\right]} \\ \mathsf{k}_i &= \sqrt{\frac{1}{2}\left[\sqrt{(R^2+L^2\omega^2)\,(G^2+C^2\omega^2)}-LC\omega^2+RG\right]} \end{aligned} \qquad [2.62]$$

2.6.1. *Low-loss limit*

In the low-loss limit ($R, G \longrightarrow 0$), the (second order) terms RG and k_i^2 are negligible compared to k_r^2 (the attenuation wavenumber k_i is small compared to the propagation wavenumber k_r) in equation [2.61]. In this limit, the approximate solutions are given as:

$$\begin{aligned} \mathsf{k}_r &\simeq \sqrt{LC}\omega = \frac{\omega}{c} \\ \mathsf{k}_i &= \frac{\omega\,(LG+CR)}{2\mathsf{k}_r} = \frac{1}{2}\left[R\sqrt{\frac{C}{L}}+G\sqrt{\frac{L}{C}}\right] = \frac{1}{2}\left[R\mathsf{Y}_0+G\mathsf{Z}_0\right] \end{aligned} \qquad [2.63]$$

If $R = 0$, we can directly compare the solutions [2.63] to those found for plane waves propagating in a dielectric medium with permittivity ε_0, permeability μ_0 and low conductivity γ. In such a medium, the wavenumber is complex:

$$\mathsf{k} \simeq \frac{\omega}{c} + \mathrm{i}\frac{\gamma}{2}\sqrt{\frac{\mu_0}{\varepsilon_0}} \qquad [2.64]$$

Comparing equations [2.63] and [2.64] for $R = 0$, we deduce $G\mathsf{Z}_0 = \gamma\sqrt{\frac{\mu_0}{\varepsilon_0}}$.

2.6.2. *Distortionless lines*

From [2.59], we can also directly derive the value of k:

$$\mathrm{i}\mathsf{k} = \sqrt{(\mathrm{i}L\omega - R)\,(\mathrm{i}C\omega - G)} \qquad [2.65]$$

and, consequently:

$$\frac{\hat{V}_\pm}{\hat{I}_\pm} = \pm\sqrt{\frac{\mathrm{i}L\omega - R}{\mathrm{i}C\omega - G}} = \pm\sqrt{\frac{L}{C}}\sqrt{\frac{\mathrm{i}\omega - \frac{R}{L}}{\mathrm{i}\omega - \frac{G}{C}}} = \pm\hat{\mathsf{Z}}_c \qquad [2.66]$$

If the line parameters are adjusted such that:

$$\frac{R}{L} = \frac{G}{C} \qquad [2.67]$$

the ratios $\frac{\hat{V}_{\pm}}{\hat{I}_{\pm}}$ become independent of frequency ω and equal to the lossless line impedance Z_0. Condition [2.67] is called the *Heaviside condition*. When studying the variation of k_i with respect to L, we find that k_i reaches its minimum and becomes independent of frequency if the Heaviside condition is satisfied. In fact, it can be verified that $\frac{\partial \mathsf{k}_i^2}{\partial L} = \frac{\omega^2}{2}\left[L\sqrt{\frac{G^2+C^2\omega^2}{R^2+L^2\omega^2}} - C\right] = 0$ for $LG = RC$. In this case, the complex wavenumber is written as $\mathsf{ik} = \sqrt{LC}\left(\mathsf{i}\omega - \frac{R}{L}\right))$ and, therefore:

$$\begin{aligned} \mathsf{k}_r &= \omega\sqrt{LC} \\ \mathsf{k}_i &= \sqrt{RG} \end{aligned} \qquad [2.68]$$

Although the waves are attenuated, all frequencies propagate with the same phase and group velocities (as in a lossless line):

$$\begin{aligned} v_\varphi &= \frac{\omega}{\mathsf{k}_r} = \frac{1}{\sqrt{LC}} \\ v_g &= \frac{d\omega}{d\mathsf{k}_r} = \frac{1}{\sqrt{LC}} \end{aligned} \qquad [2.69]$$

Since all Fourier components of the excitation signal travel at the same speed, the shape of the signal remains unchanged as it propagates along the line. Such lines are called *distortionless transmission lines*.

2.6.3. *Solutions with arbitrary impedance at the line output*

In any transmission line excited by a sinusoidal voltage source, the voltage and current waves are each the sum of two traveling plane waves, one propagating in the positive direction of Oz and the other in the opposite direction: $v(z,t) = \Re\left[\left(\hat{V}_+ e^{\mathsf{ik}z} + \hat{V}_- e^{-\mathsf{ik}z}\right)e^{-\mathsf{i}\omega t}\right]$ and $i(z,t) = \Re\left[\left(\hat{I}_+ e^{\mathsf{ik}z} + \hat{I}_- e^{-\mathsf{ik}z}\right)e^{-\mathsf{i}\omega t}\right]$, where $\mathsf{ik} = \sqrt{(\mathsf{i}L\omega - R)(\mathsf{i}C\omega - G)}$ and $\hat{Z}_c = \frac{\hat{V}_+}{\hat{I}_+} = -\frac{\hat{V}_-}{\hat{I}_-} = \sqrt{\frac{\mathsf{i}L\omega - R}{\mathsf{i}C\omega - G}}$. Here, $\mathsf{k} = \mathsf{k}_r + \mathsf{ik}_i$ is the complex wavenumber (k_r is the propagation constant, and k_i is the attenuation coefficient). In the case of a lossless line ($R = G = 0$), we obtain $\mathsf{k}_r = \sqrt{LC}\omega$ and $\mathsf{k}_i = 0$: the wave does not weaken over the course of its propagation, which occurs at the speed corresponding to the propagation of electromagnetic waves in the dielectric medium used.

The constants $\hat{V}_+$ and $\hat{V}_-$ (and consequently $\hat{I}_+$ and $\hat{I}_-$) can be determined by specifying the impedance values $\hat{Z}(z) = \frac{\hat{V}(z)}{\hat{I}(z)}$ at the input and output of the line. The functions $u(z,t)$ and $i(z,t)$ can also be expressed in terms of the reflection coefficients at the input and output of the line:

$$\hat{\mathsf{r}} = \frac{\hat{V}_-}{\hat{V}_+}$$
$$\hat{\rho} = \frac{\hat{V}_-}{\hat{V}_+} e^{-2\mathrm{i}kl} \tag{2.70}$$

In terms of $\hat{\mathsf{r}}$ and $\hat{\rho}$, the voltage and current are written as:

$$\begin{aligned}
v(z,t) &= \Re\left[\hat{V}(0)\left(\tfrac{e^{\mathrm{i}kz}+\hat{\mathsf{r}}e^{-\mathrm{i}kz}}{1+\hat{\mathsf{r}}}\right)e^{-\mathrm{i}\omega t}\right] \\
&= \Re\left[\hat{V}(0)\left(\frac{e^{-\mathrm{i}k(l-z)}+\hat{\rho}e^{\mathrm{i}k(l-z)}}{e^{-\mathrm{i}kl}+\hat{\rho}e^{\mathrm{i}kl}}\right)e^{-\mathrm{i}\omega t}\right] \\
i(z,t) &= \Re\left[\tfrac{\hat{V}(0)}{\hat{Z}_c}\left(\tfrac{e^{\mathrm{i}kz}-\hat{\mathsf{r}}e^{-\mathrm{i}kz}}{1+\hat{\mathsf{r}}}\right)e^{-\mathrm{i}\omega t}\right] \\
&= \Re\left[\frac{\hat{V}(0)}{\hat{Z}_c}\left(\frac{e^{-\mathrm{i}k(l-z)}-\hat{\rho}e^{\mathrm{i}k(l-z)}}{e^{-\mathrm{i}kl}+\hat{\rho}e^{\mathrm{i}kl}}\right)e^{-\mathrm{i}\omega t}\right]
\end{aligned} \tag{2.71}$$

where $\hat{V}(0)$ is the complex amplitude of the voltage applied at the input of the line.

If the line is terminated with an output impedance $Z(l) = Z_R$ (at $z = l$), called the *charge impedance*, we can always find the impedance at any given position z using the expression:

$$\hat{Z}(z) = \frac{\hat{V}(z)}{\hat{I}(z)} = Z_c\left(\frac{\hat{V}_+e^{\mathrm{i}kz}+\hat{V}_-e^{-\mathrm{i}kz}}{\hat{V}_+e^{\mathrm{i}kz}-\hat{V}_-e^{-\mathrm{i}kz}}\right) = Z_c\left(\frac{\hat{I}_+e^{\mathrm{i}kz}-\hat{I}_-e^{-\mathrm{i}kz}}{\hat{I}_+e^{\mathrm{i}kz}+\hat{I}_-e^{-\mathrm{i}kz}}\right) \tag{2.72}$$

At the input of the line ($z = 0$), we write:

$$\begin{cases} \hat{I}_+ + \hat{I}_- = \hat{I}(0) \\ Z_c\hat{I}_+ - Z_c\hat{I}_- = \hat{V}(0) \end{cases} \Longrightarrow \begin{cases} \hat{I}_+ = \dfrac{1}{2}\left(\hat{I}(0) + \dfrac{\hat{V}(0)}{\hat{Z}_c}\right) \\ \hat{I}_- = \dfrac{1}{2}\left(\hat{I}(0) - \dfrac{\hat{V}(0)}{\hat{Z}_c}\right) \end{cases}$$
$$\Longrightarrow \begin{cases} \hat{V}_+ = \dfrac{1}{2}\left(\hat{V}(0) + \hat{I}(0)\hat{Z}_c\right) \\ \hat{V}_- = \dfrac{1}{2}\left(\hat{V}(0) - \hat{I}(0)\hat{Z}_c\right) \end{cases} \tag{2.73}$$

which gives (in terms of the input impedance):

$$\begin{aligned}
\hat{I}(z) &= \hat{I}(0)\cos kz + \mathrm{i}\frac{\hat{V}(0)}{\hat{Z}_c}\sin kz = \frac{\hat{V}(0)}{\hat{Z}_c\hat{Z}(0)}\left[\hat{Z}_c\cos kz + \mathrm{i}\hat{Z}(0)\sin kz\right] \\
\hat{V}(z) &= \hat{V}(0)\cos kz + \mathrm{i}\hat{I}(0)\hat{Z}_c\sin kz = \frac{\hat{V}(0)}{\hat{Z}(0)}\left[\hat{Z}(0)\cos kz + \mathrm{i}\hat{Z}_c\sin kz\right]
\end{aligned} \tag{2.74}$$

and, consequently:

$$\hat{Z}(z) = \hat{Z}_c\left(\frac{\hat{V}(0)\cos \mathsf{k}z + \mathrm{i}\hat{I}(0)\hat{Z}_c \sin \mathsf{k}z}{\hat{Z}_c\hat{I}(0)\cos \mathsf{k}z + \mathrm{i}\hat{V}(0)\sin \mathsf{k}z}\right) = \hat{Z}_c\left(\frac{\hat{Z}(0) + \mathrm{i}\hat{Z}_c \tan \mathsf{k}z}{\hat{Z}_c + \mathrm{i}\hat{Z}(0)\tan \mathsf{k}z}\right) \quad [2.75]$$

From the previous expression, we can compute the impedance at the end of the line ($z = l$):

$$\hat{Z}_R = Z(l) = \hat{Z}_c\left(\frac{\hat{Z}(0) + \mathrm{i}\hat{Z}_c \tan \mathsf{k}l}{\hat{Z}_c + \mathrm{i}\hat{Z}(0)\tan \mathsf{k}l}\right) \iff \hat{Z}(0) = \hat{Z}_c\left(\frac{\hat{Z}_R - \mathrm{i}\hat{Z}_c \tan \mathsf{k}l}{\hat{Z}_c - \mathrm{i}\hat{Z}_R \tan \mathsf{k}l}\right) \quad [2.76]$$

These two relations show that the input and output impedances are not independent. In terms of the output impedance $\hat{Z}_R$, we find:

$$\begin{aligned} \hat{I}(z) &= \frac{\hat{V}(0)}{\hat{Z}_c}\frac{\hat{Z}_c \cos \mathsf{k}(l-z) - \mathrm{i}\hat{Z}_R \sin \mathsf{k}(l-z)}{\hat{Z}_R \cos \mathsf{k}l - \mathrm{i}\hat{Z}_c \sin \mathsf{k}l} \\ \hat{V}(z) &= \hat{V}(0)\frac{\hat{Z}_R \cos \mathsf{k}(l-z) - \mathrm{i}\hat{Z}_c \sin \mathsf{k}(l-z)}{\hat{Z}_R \cos \mathsf{k}l - \mathrm{i}\hat{Z}_c \sin \mathsf{k}l} \end{aligned} \quad [2.77]$$

Two output configurations are of particular interest. When the line is open at the output, $\hat{Z}_R$ must be considered infinite and, consequently, the input impedance $\hat{Z}(0)$ is:

$$\hat{Z}(0) = \mathrm{i}\frac{\hat{Z}_c}{\tan \mathsf{k}l} \equiv \hat{Z}_{\text{open}} \quad [2.78]$$

When the line is short-circuited in the output, we must take $\hat{Z}_R = 0$. Thus, the input impedance is equal to:

$$\hat{Z}(0) = -\mathrm{i}\hat{Z}_c \tan \mathsf{k}l \equiv \hat{Z}_{\text{SC}} \quad [2.79]$$

Thus, we deduce the important relation $\hat{Z}_{\text{open}}\hat{Z}_{\text{CC}} = \hat{Z}_c^2$.

For $z = l$, we have:

$$\hat{I}_R = \hat{I}(l) = \frac{\hat{V}(0)}{\hat{Z}_R \cos \mathsf{k}l - \mathrm{i}\hat{Z}_c \sin \mathsf{k}l} \implies \hat{Z}_{tR} = \frac{\hat{V}(0)}{\hat{I}_R} = \hat{Z}_R \cos \mathsf{k}l - \mathrm{i}\hat{Z}_c \sin \mathsf{k}l \quad [2.80]$$

$\hat{Z}_{tR}$ is called the *transfer impedance*.

The reflection coefficients at the input and output of the line are also written as:

$$\hat{\mathsf{r}} = \frac{\hat{Z}(0) - \hat{Z}_c}{\hat{Z}(0) + \hat{Z}_c} = \frac{\hat{Z}_R - \hat{Z}_c}{\hat{Z}_R + \hat{Z}_c} e^{2\mathsf{i}kl}$$

$$\hat{\rho} = \frac{\hat{Z}(0) - \hat{Z}_c}{\hat{Z}(0) + \hat{Z}_c} e^{-2\mathsf{i}kl} = \frac{\hat{Z}_R - \hat{Z}_c}{\hat{Z}_R + \hat{Z}_c} \qquad [2.81]$$

We clearly see that when $\hat{Z}_R = \hat{Z}_c$, the reflection coefficient is zero. In this case, the impedance $\hat{Z}(z)$ is equal to $\hat{Z}_c$ along the line and, in particular, at the input: $\hat{Z}(0) = \hat{Z}_c$. When $\hat{Z}_R \neq \hat{Z}_c$, the reflection coefficient $\hat{\rho}$ is non-zero. As a result, a reflected wave is generated at the end of the line, which interferes with the incident wave to produce a standing wave. As a result, we observe along the transmission line an alternation of maxima (antinodes) and minima (nodes) in the signal amplitude. According to equation [2.71], we have:

$$\left|\hat{V}(z)\right|^2 = \left|\hat{V}(0)\right|^2 \frac{e^{2\mathsf{k}_i(l-z)} + |\hat{\rho}|^2 e^{-2\mathsf{k}_i(l-z)} + \hat{\rho} e^{2\mathsf{i}\mathsf{k}_r(l-z)} + \hat{\rho}^\star e^{-2\mathsf{i}\mathsf{k}_r(l-z)}}{e^{2\mathsf{k}_i l} + |\hat{\rho}|^2 e^{-2\mathsf{k}_i l} + \hat{\rho} e^{2\mathsf{i}\mathsf{k}_r l} + \hat{\rho}^\star e^{-2\mathsf{i}\mathsf{k}_r l}}$$

$$= \left|\hat{V}(0)\right|^2 \frac{e^{2\mathsf{k}_i(l-z)} + |\hat{\rho}|^2 e^{-2\mathsf{k}_i(l-z)} + 2|\hat{\rho}|\cos[2\mathsf{k}_r(l-z) + \arg\hat{\rho}]}{e^{2\mathsf{k}_i l} + |\hat{\rho}|^2 e^{-2\mathsf{k}_i l} + 2|\hat{\rho}|\cos[2\mathsf{k}_r l + \arg\hat{\rho}]} \qquad [2.82]$$

The third term in [2.82] is the interference term between the incident and reflected waves. If we limit ourselves to transmission lines with very low losses, we can make the following approximation:

$$\left|\hat{V}(z)\right|^2 = \left|\hat{V}(0)\right|^2 \frac{1 + |\hat{\rho}|^2 + 2|\hat{\rho}|\cos[2\mathsf{k}_r(l-z) + \arg\hat{\rho}]}{1 + |\hat{\rho}|^2 + 2|\hat{\rho}|\cos[2\mathsf{k}_r l + \arg\hat{\rho}]} \qquad [2.83]$$

The voltage amplitude reaches a maximum value proportional to $1 + |\hat{\rho}|$ when $2\mathsf{k}_r(l-z) + \arg\hat{\rho} = 2\pi n$, where n is an integer. Since $0 \leq z \leq l$, we deduce that:

$$\frac{\arg\hat{\rho}}{2\pi} \leq n \leq \frac{\mathsf{k}_r l}{\pi} + \frac{\arg\hat{\rho}}{2\pi} \qquad [2.84]$$

Similarly, the voltage amplitude reaches a minimum value proportional to $1 - |\hat{\rho}|$ when $2\mathsf{k}_r(l-z) + \arg\hat{\rho} = (2n+1)\pi$, where n is an integer such that:

$$\frac{\arg\hat{\rho}}{2\pi} - \frac{1}{2} \leq n \leq \frac{\mathsf{k}_r l}{\pi} + \frac{\arg\hat{\rho}}{2\pi} - \frac{1}{2} \qquad [2.85]$$

Therefore, the antinodes and nodes are located at the positions $z_n = l + \frac{\arg\hat{\rho}}{2\mathsf{k}_r} - \frac{n\pi}{\mathsf{k}_r}$ and $z'_n = l + \frac{\arg\hat{\rho}}{2\mathsf{k}_r} - \frac{n\pi}{\mathsf{k}_r} - \frac{\pi}{2\mathsf{k}_r} = l + \frac{\arg\hat{\rho}}{2\mathsf{k}_r} - \frac{n\pi}{\mathsf{k}_r} - \frac{\lambda}{4}$, respectively. We thus see that the antinodes and nodes are separated by a quarter of the wavelength.

In general, the appearance of standing waves is characterized by a coefficient called the *standing wave ratio* (SWR), defined as follows:

$$S = \frac{\left|\hat{V}\right|_{\max}}{\left|\hat{V}\right|_{\min}} \qquad [2.86]$$

In the case of a transmission line with very low losses, we have approximately the following relationship:

$$S = \frac{1+|\hat{\rho}|}{1-|\hat{\rho}|} \qquad [2.87]$$

The value of S varies between 1 ($\left|\hat{V}\right|_{\max} = \left|\hat{V}\right|_{\min}$ when $\hat{Z}_R = \hat{Z}_c$) and $+\infty$ ($\left|\hat{V}\right|_{\min} = 0$). If S increases, the voltage and current amplitudes also increase. This results in an increase in the energy dissipated in the line.

EXAMPLE 2.4. OPEN-CIRCUITED LINE.–

1) When $\hat{Z}_R$ is infinite (open transmission line at the output), the reflection coefficient $\hat{\rho}$ equals 1 at the end of the line. According to [2.77], we have:

$$\hat{I}(z) = \frac{\hat{V}(0)}{\hat{Z}_c}\frac{-\mathrm{i}\sin \mathsf{k}(l-z)}{\cos \mathsf{k}l}$$

$$\hat{V}(z) = \hat{V}(0)\frac{\cos \mathsf{k}(l-z)}{\cos \mathsf{k}l} \qquad [2.88]$$

Since $\sin \mathsf{k}h = \sin \mathsf{k}_r h \cosh \mathsf{k}_i h + \mathrm{i}\cos \mathsf{k}_r h \sinh \mathsf{k}_i h$, $\cos \mathsf{k}h = \cos \mathsf{k}_r h \cosh \mathsf{k}_i h$ $-\mathrm{i}\sin \mathsf{k}_r h \sinh \mathsf{k}_i h$ and $\cos \mathsf{k}l = \cos \mathsf{k}_r l \cosh \mathsf{k}_i l - \mathrm{i}\sin \mathsf{k}_r l \sinh \mathsf{k}_i l$, where $h = l - z$, it can be derived that:

$$|\hat{I}(z)|^2 = \frac{|\hat{V}(0)|^2}{|\hat{Z}_c|^2}\frac{\sin^2 \mathsf{k}_r(l-z)\cosh^2 \mathsf{k}_i(l-z) + \cos^2 \mathsf{k}_r(l-z)\sinh^2 \mathsf{k}_i(l-z)}{\cos^2 \mathsf{k}_r l \cosh^2 \mathsf{k}_i l + \sin^2 \mathsf{k}_r l \sinh^2 \mathsf{k}_i l}$$

$$= \frac{|\hat{V}(0)|^2}{|\hat{Z}_c|^2}\frac{\sin^2 \mathsf{k}_r(l-z) + \sinh^2 \mathsf{k}_i(l-z)}{\cos^2 \mathsf{k}_r l + \sinh^2 \mathsf{k}_i l}$$

$$|\hat{V}(z)|^2 = |\hat{V}(0)|^2\frac{\cos^2 \mathsf{k}_r(l-z)\cosh^2 \mathsf{k}_i(l-z) + \sin^2 \mathsf{k}_r(l-z)\sinh^2 \mathsf{k}_i(l-z)}{\cos^2 \mathsf{k}_r l \cosh^2 \mathsf{k}_i l + \sin^2 \mathsf{k}_r l \sinh^2 \mathsf{k}_i l}$$

$$= |\hat{V}(0)|^2\frac{\cos^2 \mathsf{k}_r(l-z) + \sinh^2 \mathsf{k}_i(l-z)}{\cos^2 \mathsf{k}_r l + \sinh^2 \mathsf{k}_i l} \qquad [2.89]$$

If k_i is small, the variations of $\left|\hat{V}(z)\right|^2$ and $\left|\hat{I}(z)\right|^2$ are governed, respectively, by those of $\cos^2 k_r(l-z)$ and $\sin^2 k_r(l-z)$ (the rapidly varying functions) because the term $\sinh^2 k_i(l-z)$ changes very little over the distance l (a slowly varying function - $|k_r l| \ll 1$). The voltage amplitude therefore reaches its maximum values at distances $z_n = l - \frac{n\pi}{k_r} = l - \frac{n\lambda}{2}$ and its minimum values at distances $z'_n = l - \left(n+\frac{1}{2}\right)\frac{\pi}{k_r} = l - \left(n+\frac{1}{2}\right)\frac{\lambda}{2}$, where n is an integer that satisfies the condition $0 \leq n \leq \frac{k_r l}{\pi}$. For the current, it follows that $\left|\hat{I}(z)\right|^2$ is minimum when $\left|\hat{V}(z)\right|^2$ is maximum and vice versa. The antinodes (maxima) and nodes (minima) are spaced by a distance of $\frac{\lambda}{4}$. The SWR is approximately written as:

$$S(n) = \frac{\left|\hat{V}\right|_{\max}}{\left|\hat{V}\right|_{\min}} = \sqrt{\frac{1+\sinh^2\left[\frac{n\pi}{2Q}\right]}{\sinh^2\left[\frac{\left(n+\frac{1}{2}\right)\pi}{2Q}\right]}} \approx \frac{4Q}{(2n+1)\pi} \qquad [2.90]$$

where $Q = \frac{k_r}{2k_i}$. $S(n)$ increases as we approach the line input: it reaches its minimum value at the output and is approximately given by $S \approx \frac{1}{k_i l}$.

2) For more precision, the extrema of $\left|\hat{V}(z)\right|^2$ are found by canceling the derivative: $\frac{d\left|\hat{V}(z)\right|^2}{dz} = k_r \sin 2k_r(l-z) - k_i \sinh 2k_i(l-z) = 0$. This gives:

$$\frac{\sin 2k_r(l-z)}{2k_r(l-z)} = \frac{k_i \sinh 2k_i(l-z)}{2k_r^2(l-z)} \simeq \frac{k_i^2}{k_r^2} \qquad [2.91]$$

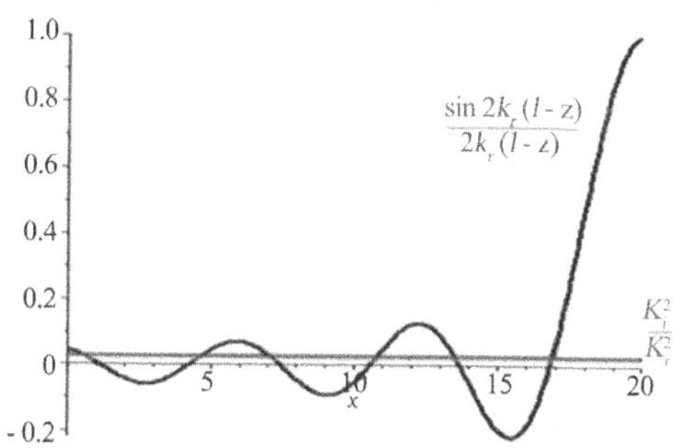

Figure 2.8. *Graphical resolution of the equation $\frac{\sin 2k_r(l-z)}{2k_r(l-z)} \simeq \frac{k_i^2}{k_r^2}$ for low losses. For a color version of this figure, see www.iste.co.uk/abdesselam/electrodynamics3.zip*

The solutions correspond to the intersection points between the curve $\frac{\sin 2k_r(l-z)}{2k_r(l-z)}$ and the horizontal line $\frac{k_i^2}{k_r^2}$, as illustrated in Figure 2.8. We can see, graphically, that the solutions are located in proximity to the zeros of the function $\sin 2\mathsf{k}_r\,(l-z)$, that is, at the points $l-\frac{n\pi}{2\mathsf{k}_r}$, where n is an integer satisfying the condition $0\le n\le\frac{2\mathsf{k}_r l}{\pi}$. By performing a first-order Taylor expansion near the point $l-\frac{n\pi}{2\mathsf{k}_r}$, we obtain:

$$-\left.\frac{\cos 2\mathsf{k}_r\,(l-z)}{(l-z)}\right|_{z=l-\frac{n\pi}{2\mathsf{k}_r}}\left(z-l+\frac{n\pi}{2\mathsf{k}_r}\right)\simeq\frac{\mathsf{k}_i^2}{\mathsf{k}_r^2}$$

$$\Longrightarrow\; z_n=l-\frac{n\pi}{2\mathsf{k}_r}\left(1+\frac{(-1)^n\,\mathsf{k}_i^2}{\mathsf{k}_r^2}\right)\qquad[2.92]$$

The even values of n correspond to the maxima and the odd values of n to the minima. We see that the deviations from the lossless case are on the order of $\frac{\mathsf{k}_i^2}{\mathsf{k}_r^2}$.

EXAMPLE 2.5. SHORT-CIRCUITED LINE.– When $\hat{Z}_R$ is zero (i.e. when the transmission line is short-circuited at the output), the reflection coefficient $\hat{\rho}$ is equal to -1 at the end of the line. According to [2.77], we have:

$$\begin{cases}\hat{I}\,(z)=\dfrac{\hat{V}\,(0)}{\hat{Z}_c}\dfrac{\mathrm{i}\cos\mathsf{k}\,(l-z)}{\sin\mathsf{k}l}\\ \hat{V}\,(z)=\hat{V}\,(0)\,\dfrac{\sin\mathsf{k}\,(l-z)}{\sin\mathsf{k}l}\end{cases}$$

$$\Longrightarrow\begin{cases}\left|\hat{I}\,(z)\right|^2=\dfrac{\left|\hat{V}\,(0)\right|^2}{\left|\hat{Z}_c\right|^2}\dfrac{\cos^2\mathsf{k}_r\,(l-z)+\sinh^2\mathsf{k}_i\,(l-z)}{\sin^2\mathsf{k}_r l+\sinh^2\mathsf{k}_i l}\\ \left|\hat{V}\,(z)\right|^2=\left|\hat{V}\,(0)\right|^2\dfrac{\sin^2\mathsf{k}_r\,(l-z)+\sinh^2\mathsf{k}_i\,(l-z)}{\sin^2\mathsf{k}_r l+\sinh^2\mathsf{k}_i l}\end{cases}\qquad[2.93]$$

The same results developed in example 2.4 also remain valid here. The extrema of voltage and current still occur at the distances given by [2.92]: even values of n correspond to voltage minima and current maxima, while odd values of n correspond to voltage maxima and current minima.

2.7. Resonance effect in a transmission line

Resonance effects can be observed in transmission lines. These effects tend to increase the maximum voltage or current values. For simplicity, we limit ourselves to an open line. When the length of the open line is equal to $\frac{(2n+1)\lambda}{4}$, the amplitudes of the voltage or current maxima can significantly increase. In this case, we say that a *resonance effect* occurs. From the expression of the voltage at the end of the line $\left|\hat{V}_R\right|^2=\left|\hat{V}\,(l)\right|^2=\frac{\left|\hat{V}(0)\right|^2}{\cos^2\mathsf{k}_r l+\sinh^2\mathsf{k}_i l}$, we see that a resonance occurs when

$\cos \mathsf{k}_r l = 0$. In this case, we obtain $\frac{|\hat{V}_R|^2}{|\hat{V}(0)|^2} = \frac{1}{\sinh^2 \mathsf{k}_i l} \simeq \frac{1}{\mathsf{k}_i^2 l^2}$. The voltage amplitude at the output can become very large when $\mathsf{k}_i l$ is small – this corresponds to an overvoltage at the output. We therefore have resonance if $\mathsf{k}_r l = \frac{(2n+1)\pi}{2} \Longrightarrow l = \frac{(2n+1)\pi}{2\mathsf{k}_r} = \frac{(2n+1)\lambda}{4}$, $n = 0, 1, 2, \ldots$. Hence, we conclude that:

$$\frac{\left|\hat{V}_R\right|}{\left|\hat{V}(0)\right|} = \frac{1}{\mathsf{k}_i l} = \frac{4Q}{(2n+1)\pi} \quad [2.94]$$

where $Q = \frac{\mathsf{k}_r}{2\mathsf{k}_i}$ is the *overvoltage coefficient of the line* (also known as the *Q-factor of the line*). At the input of the line, the impedance is $\left|\hat{Z}(0)\right| = \frac{|\hat{Z}_c|}{|\tan \mathsf{k}l|} = \frac{|\hat{Z}_c| \sinh \mathsf{k}_i l}{\cosh \mathsf{k}_i l} \simeq \left|\hat{Z}_c\right| \mathsf{k}_i l$. The input impedance is therefore very small and, consequently, the current $\left|\hat{I}(0)\right| = \frac{|\hat{V}(0)|}{|\hat{Z}(0)|}$ is very large.

2.8. Energy aspect in a transmission line

We will now focus on the energy aspect in a transmission line. We still limit ourselves to a low-loss open line. Before starting our discussion, let us make a simple remark: the complex voltage is expressed as a complex number $\hat{V}(z)$ multiplied by the temporal factor $e^{-i\omega t}$, and consequently, the corresponding physical measured voltage is $v(z,t) = \Re\left[\hat{V}(z) e^{-i\omega t}\right] = \Re\left[\hat{V}(z)\right] \cos \omega t + \Im\left[\hat{V}(z)\right] \sin \omega t$. The time average of v^2 is therefore $\langle v^2 \rangle = \frac{1}{2}\left(\Re\left[\hat{V}(z)\right]^2 + \Im\left[\hat{V}(z)\right]^2\right) = \frac{1}{2}\left|\hat{V}(z)\right|^2$. Similarly, we verify that $\langle i^2 \rangle = \frac{1}{2}\left(\Re\left[\hat{I}(z)\right]^2 + \Im\left[\hat{I}(z)\right]^2\right) = \frac{1}{2}\left|\hat{I}(z)\right|^2$. The following are the essential results:

– The time averages of the real voltage and current along an open transmission line are given by:

$$\begin{aligned} \langle i^2 \rangle &= \frac{\left|\hat{V}(0)\right|^2}{2\left|\hat{Z}_c\right|^2} \frac{|\sin \mathsf{k}(l-z)|^2}{|\cos \mathsf{k}l|^2} \\ \langle v^2 \rangle &= \frac{\left|\hat{V}(0)\right|^2}{2} \frac{|\cos \mathsf{k}(l-z)|^2}{|\cos \mathsf{k}l|^2} \end{aligned} \quad [2.95]$$

– The average electromagnetic energy stored per unit length is given by $E = \frac{C}{2}\langle v^2\rangle + \frac{L}{2}\langle i^2\rangle$. If k_i is small, then $\hat{Z}_c$ is approximately equal to $\sqrt{\frac{L}{C}}$, and:

$$E \simeq \frac{C\left|\hat{V}(0)\right|^2}{4}\frac{\left|\cos \mathsf{k}(l-z)\right|^2 + \left|\sin \mathsf{k}(l-z)\right|^2}{\left|\cos \mathsf{k}l\right|^2}$$

$$= \frac{C\left|\hat{V}(0)\right|^2}{4}\frac{\left|\cosh \mathsf{k}_i(l-z)\right|^2 + \left|\sinh \mathsf{k}_i(l-z)\right|^2}{\left|\cos \mathsf{k}l\right|^2} \simeq \frac{C\left|\hat{V}(0)\right|^2}{4}\frac{1}{\left|\cos \mathsf{k}l\right|^2} \quad [2.96]$$

Thus, the total average electromagnetic energy stored in the entire line is equal to:

$$E_{\text{stored}} = El = \frac{C\left|\hat{V}(0)\right|^2}{4}\frac{l}{\left|\cos \mathsf{k}l\right|^2} \quad [2.97]$$

– The average power dissipated per unit length is:

$$\langle P_{dl}\rangle = R\langle i^2\rangle + G\langle v^2\rangle \simeq \frac{\left|\hat{V}(0)\right|^2}{2\left|\cos \mathsf{k}l\right|^2}\left[\frac{RC}{L}\left|\sin \mathsf{k}(l-z)\right|^2 + G\left|\cos \mathsf{k}(l-z)\right|^2\right]$$

$$\simeq \frac{\left|\hat{V}(0)\right|^2}{2\left|\cos \mathsf{k}l\right|^2}\left[\frac{RC}{L}\sin^2 \mathsf{k}_r(l-z) + G\cos^2 \mathsf{k}_r(l-z)\right] \quad [2.98]$$

– The average power dissipated by the whole line is:

$$\langle P_d\rangle = \frac{\left|\hat{V}(0)\right|^2}{2\left|\cos \mathsf{k}l\right|^2}\left[\frac{RC}{L}\int_0^l \sin^2 \mathsf{k}_r(l-z)\,dz + G\int_0^l \cos^2 \mathsf{k}_r(l-z)\,dz\right]$$

$$= \frac{\left|\hat{V}(0)\right|^2}{2\left|\cos \mathsf{k}l\right|^2}\left[\left(\frac{RC}{L}+G\right)\frac{l}{2} + \left(G-\frac{RC}{L}\right)\frac{\sin 2\mathsf{k}_r l}{4\mathsf{k}_r l}\right] \quad [2.99]$$

– If there is a resonance, the second term of [2.99] vanishes, since $\mathsf{k}_r l = \frac{(2n+1)\pi}{2}$. It therefore follows that:

$$\langle P_d\rangle = \frac{\left|\hat{V}(0)\right|^2}{2\left|\cos \mathsf{k}l\right|^2}\left(\frac{RC}{L}+G\right)\frac{l}{2} = \frac{\left|\hat{V}(0)\right|^2 \mathsf{k}_i l}{2\left|\cos \mathsf{k}l\right|^2}\sqrt{\frac{C}{L}} \quad [2.100]$$

– The average energy dissipated during one period (in the case of resonance) is:

$$E_{\text{dissipated}} = \langle P_d\rangle T = \frac{\left|\hat{V}(0)\right|^2 \mathsf{k}_i l}{2\left|\cos \mathsf{k}l\right|^2}\sqrt{\frac{C}{L}}\frac{2\pi}{\omega} \quad [2.101]$$

Thus, we deduce that:

$$\frac{E_{\text{stored}}}{E_{\text{dissipated}}} = \frac{\sqrt{LC}\omega}{4\pi \mathsf{k}_i} = \frac{\mathsf{k}_r}{4\pi \mathsf{k}_i} = \frac{Q}{2\pi} \quad [2.102]$$

We see that here Q has another meaning and can therefore be interpreted as a quality factor.

Chapter 3

Electromagnetic Radiation

3.1. Retarded potentials

Any time-dependent distribution of charges or currents radiates an electromagnetic field. For simplicity, we will assume that the radiation source is a charge or current density undergoing sinusoidal motion at frequency ω. Solving a radiation problem involves calculating the electric and magnetic fields for a given distribution. Using the vector and scalar potentials in these calculations greatly simplifies the problem.

3.1.1. *Inhomogeneous wave equations*

The Maxwell equations can be written as:

$$\begin{aligned} &\vec{\nabla} \times \vec{\mathbf{E}} = -\frac{\partial \vec{\mathbf{B}}}{\partial t} && \text{(i)} \\ &\vec{\nabla} \times \vec{\mathbf{H}} = \vec{\mathbf{J}} + \frac{\partial \vec{\mathbf{D}}}{\partial t} && \text{(ii)} \\ &\vec{\nabla} \cdot \vec{\mathbf{B}} = 0 && \text{(iii)} \\ &\vec{\nabla} \times \vec{\mathbf{D}} = \rho && \text{(iv)} \end{aligned} \qquad [3.1]$$

where $\rho \equiv \rho\left(\vec{\mathbf{r}}, t\right)$ and $\vec{\mathbf{J}} \equiv \vec{\mathbf{J}}\left(\vec{\mathbf{r}}, t\right)$ are, respectively, the charge and current densities.

From equation [3.1iii], it follows that the magnetic field $\vec{\mathbf{B}}$ derives from a *vector potential* $\vec{\mathbf{A}}$:

$$\vec{\mathbf{B}} = \vec{\nabla} \times \vec{\mathbf{A}} \qquad [3.2]$$

Similarly, from equation [3.1i], it follows that $\vec{\nabla}\times\left(\vec{\mathbf{E}}+\frac{\partial\vec{\mathbf{A}}}{\partial t}\right)=0$. This results in the fact that $\vec{\mathbf{E}}+\frac{\partial\vec{\mathbf{A}}}{\partial t}$ derives from a *scalar electric potential*, that is, $\vec{\mathbf{E}}+\frac{\partial\vec{\mathbf{A}}}{\partial t}=-\vec{\nabla}V$ (the minus sign being conventional). Thus, we obtain:

$$\vec{\mathbf{E}}=-\vec{\nabla}V-\frac{\partial\vec{\mathbf{A}}}{\partial t}\qquad [3.3]$$

Substituting the expressions [3.2] and [3.3] into [3.1ii] and taking into account that $\vec{\mathbf{D}}=\varepsilon_0\vec{\mathbf{E}}$ and $\vec{\mathbf{B}}=\mu_0\vec{\mathbf{H}}$ (LIH medium), we get $\vec{\nabla}\times\left(\vec{\nabla}\times\vec{\mathbf{A}}\right)=\mu_0\vec{\mathbf{J}}+\frac{1}{c^2}\left[-\vec{\nabla}\left(\frac{\partial V}{\partial t}\right)-\frac{\partial^2\vec{\mathbf{A}}}{\partial t^2}\right]$ and, consequently:

$$\Delta\vec{\mathbf{A}}-\vec{\nabla}\left[\vec{\nabla}\cdot\vec{\mathbf{A}}+\frac{1}{c^2}\frac{\partial V}{\partial t}\right]-\frac{1}{c^2}\frac{\partial^2\vec{\mathbf{A}}}{\partial t^2}=-\mu_0\vec{\mathbf{J}}\qquad [3.4]$$

where $\varepsilon_0\mu_0c^2=1$ (c is the speed of light in vacuum).

So far, we have only specified the curl of $\vec{\mathbf{A}}$ in equation [3.2], and its divergence remains undetermined. This means that $\vec{\mathbf{A}}$ is, in fact, defined up to an additional term of the form $\vec{\nabla}f$, where f is a scalar field. The Helmholtz–Hodge theorem indicates that to uniquely specify the vector potential $\vec{\mathbf{A}}$, we must also specify the divergence of $\vec{\mathbf{A}}$. This is called *fixing the gauge*. It can be seen that by fixing:

$$\vec{\nabla}\cdot\vec{\mathbf{A}}=-\frac{1}{c^2}\frac{\partial V}{\partial t}\qquad [3.5]$$

the term in the middle on the left side of equation [3.4] becomes zero, and consequently, the equation reduces to:

$$\Delta\vec{\mathbf{A}}-\frac{1}{c^2}\frac{\partial^2\vec{\mathbf{A}}}{\partial t^2}=-\mu_0\vec{\mathbf{J}}\qquad [3.6]$$

We thus conclude that $\vec{\mathbf{A}}$ and $\vec{\mathbf{J}}$ are related by an inhomogeneous vector wave equation. The gauge condition [3.5] is called *the Lorenz gauge*. When $\vec{\mathbf{J}}$ is known, equation [3.6] can be solved to obtain $\vec{\mathbf{A}}$, and then [3.5] can be used to determine V. The scalar potential can also be directly found by substituting [3.3] into the Maxwell–Gauss equation [3.1iv]:

$$\Delta V+\frac{\partial\left(\vec{\nabla}\cdot\vec{\mathbf{A}}\right)}{\partial t}=-\frac{\rho}{\varepsilon_0}\qquad [3.7]$$

The second term on the left can be expressed in terms of V using the Lorenz gauge condition [3.5], which gives the inhomogeneous scalar wave equation:

$$\Delta V - \frac{1}{c^2}\frac{\partial^2 V}{\partial t^2} = -\frac{\rho}{\varepsilon_0} \qquad [3.8]$$

3.1.2. *Solutions of the inhomogeneous wave equations*

We now aim to find the general solutions of equations [3.6] and [3.8]. We observe that the three scalar equations from equation [3.6] (one for each component) and equation [3.8] share the same form. Therefore, it suffices to find the general solution of one of these equations and then apply the result to the others, each time respecting the boundary conditions specific to each case. As was done earlier, we first solve equation [3.8] for a point charge source and then extend the result to an arbitrary distribution by superimposing all of the solutions associated with the elementary charges forming the distribution. To this end, consider a time-varying point charge $Q(t)$ located at $r = 0$. Due to symmetry, the resulting potential V depends only on r. Hence, equation [3.8] becomes:

$$\frac{1}{r^2}\frac{\partial}{\partial r}\left(r^2\frac{\partial V}{\partial r}\right) - \frac{1}{c^2}\frac{\partial^2 V}{\partial t^2} = 0 \qquad [3.9]$$

for $r > 0$. The right-hand side is zero because the charge density is zero everywhere except at $r = 0$. Multiplying equation [3.9] by r and noting that $\frac{1}{r}\frac{\partial}{\partial r}\left(r^2\frac{\partial V}{\partial r}\right) = \frac{\partial^2}{\partial r^2}(rV)$, we also obtain:

$$\frac{\partial^2}{\partial r^2}(rV) - \frac{1}{c^2}\frac{\partial^2}{\partial t^2}(rV) = 0 \qquad [3.10]$$

This equation admits the following solution:

$$rV = f_+\left(t - \frac{r}{c}\right) + f_-\left(t + \frac{r}{c}\right) \qquad [3.11]$$

The term $f_-\left(t + \frac{r}{c}\right)$ must be eliminated because there are no sources or obstacles (to produce reflected waves) outside $r = 0$. As a result, all waves radiate outward radially from the point charge. The arbitrary function f_+ is determined by recognizing that the retardation effect cannot be observed as $r \longrightarrow 0$, that is, in the neighborhood of the charge. In this region, the potential should approach the Coulomb potential of a point charge, that is, $\lim_{r\longrightarrow 0} V = \frac{Q(t)}{4\pi\varepsilon_0 r}$. Thus, we obtain:

$$f_+(t) = \frac{Q(t)}{4\pi\varepsilon_0} \qquad [3.12]$$

The potential due to a point charge is then obtained from equations [3.11] and [3.12] by replacing the time t with the retarded time $t - \frac{r}{c}$:

$$V(r,t) = \frac{Q\left(t - \frac{r}{c}\right)}{4\pi\varepsilon_0 r} \qquad [3.13]$$

The potential at time t does not depend on the current value of the charge $Q(t)$, but on its value at the earlier time $t - \frac{r}{c}$ (the moment when the received wave was emitted). The potential due to an arbitrary charge distribution $\rho(t)$ is obtained by replacing $Q(t)$ with the differential charge $\rho(t)\,d\tau'$ and then integrating over the entire volume τ' occupied by the distribution:

$$V\left(\vec{\mathbf{r}},t\right) = \int_{\tau'} \frac{Q\left(t - \frac{\left|\vec{\mathbf{r}} - \vec{\mathbf{r}}'\right|}{c}\right)}{4\pi\varepsilon_0\left|\vec{\mathbf{r}} - \vec{\mathbf{r}}'\right|} d\tau \qquad [3.14]$$

where $\left|\vec{\mathbf{r}} - \vec{\mathbf{r}}'\right|$ is the distance measured between the differential charge element and the field point.

The vector potential $\vec{\mathbf{A}}$ from equation [3.6] is obtained from an integral equation similar to equation [3.14], since each component of $\vec{\mathbf{A}}$ satisfies an equation of the same form as equation [3.8], provided that $\frac{\rho}{\varepsilon_0}$ is replaced by $\mu_0 \vec{\mathbf{J}}$:

$$\vec{\mathbf{A}}\left(\vec{\mathbf{r}},t\right) = \int_{\tau'} \frac{\mu_0 \vec{\mathbf{J}}\left(t - \frac{\left|\vec{\mathbf{r}} - \vec{\mathbf{r}}'\right|}{c}\right)}{4\pi\left|\vec{\mathbf{r}} - \vec{\mathbf{r}}'\right|} d\tau \qquad [3.15]$$

3.1.3. *Solutions by Fourier analysis*

There is a more refined method to solve inhomogeneous wave equations [3.6] and [3.8]. Let us begin by writing them in a more general form, in terms of an amplitude $\psi\left(\vec{\mathbf{r}},t\right)$, which may represent V or any component of $\vec{\mathbf{A}}$:

$$\Delta\psi - \frac{1}{c^2}\frac{\partial^2\psi}{\partial t^2} = -f\left(\vec{\mathbf{r}},t\right) \qquad [3.16]$$

where $f\left(\vec{\mathbf{r}},t\right)$ is an arbitrary function of space and time (source). To eliminate the time part of this equation, let us introduce the Fourier transforms:

$$\begin{cases} \psi\left(\overrightarrow{\mathbf{r}},t\right) = \dfrac{1}{2\pi}\displaystyle\int_{-\infty}^{+\infty} \Psi\left(\overrightarrow{\mathbf{r}},\omega\right) e^{-\mathrm{i}\omega t} d\omega \\ \Psi\left(\overrightarrow{\mathbf{r}},\omega\right) = \displaystyle\int_{-\infty}^{+\infty} \psi\left(\overrightarrow{\mathbf{r}},t\right) e^{\mathrm{i}\omega t} dt \end{cases}$$
$$\begin{cases} f\left(\overrightarrow{\mathbf{r}},t\right) = \dfrac{1}{2\pi}\displaystyle\int_{-\infty}^{+\infty} F\left(\overrightarrow{\mathbf{r}},\omega\right) e^{-\mathrm{i}\omega t} d\omega \\ F\left(\overrightarrow{\mathbf{r}},\omega\right) = \displaystyle\int_{-\infty}^{+\infty} f\left(\overrightarrow{\mathbf{r}},t\right) e^{\mathrm{i}\omega t} dt \end{cases} \qquad [3.17]$$

By substituting [3.17] into [3.16], we directly obtain the Helmholtz equation:

$$\Delta\Psi\left(\overrightarrow{\mathbf{r}},\omega\right) + \mathsf{k}^2\Psi\left(\overrightarrow{\mathbf{r}},\omega\right) = -F\left(\overrightarrow{\mathbf{r}},\omega\right) \qquad [3.18]$$

where $\mathsf{k} = \frac{\omega}{v}$. The same result is obtained by assuming that the source f and the wave ψ oscillate at a well-defined frequency ω.

We will solve the Helmholtz equation [3.18] using the Green function method. The Green function $G\left(\overrightarrow{\mathbf{r}}, \overrightarrow{\mathbf{r}}'\right)$ is defined as the solution to the Helmholtz equation for a point source located at $\overrightarrow{\mathbf{r}}'$:

$$\left(\Delta + \mathsf{k}^2\right) G\left(\overrightarrow{\mathbf{r}}, \overrightarrow{\mathbf{r}}'\right) = -\delta^3\left(\overrightarrow{\mathbf{r}} - \overrightarrow{\mathbf{r}}'\right) \qquad [3.19]$$

The solution $G\left(\overrightarrow{\mathbf{r}}, \overrightarrow{\mathbf{r}}'\right)$ depends on the boundary conditions imposed on the problem. The function $G\left(\overrightarrow{\mathbf{r}}, \overrightarrow{\mathbf{r}}'\right)$ has two arguments: the observation point $\overrightarrow{\mathbf{r}}$ and the source point $\overrightarrow{\mathbf{r}}'$. It is interpreted as the amplitude, at the point $\overrightarrow{\mathbf{r}}$, of the wave generated by a monochromatic source located at $\overrightarrow{\mathbf{r}}'$. Once the Green function $G\left(\overrightarrow{\mathbf{r}}, \overrightarrow{\mathbf{r}}'\right)$ is determined, the general solution of the Helmholtz equation [3.18] can be calculated using the principle of superposition:

$$\Psi\left(\overrightarrow{\mathbf{r}},\omega\right) = \int_{\tau'} d\tau' G\left(\overrightarrow{\mathbf{r}}, \overrightarrow{\mathbf{r}}'\right) F\left(\overrightarrow{\mathbf{r}}',\omega\right) \qquad [3.20]$$

In fact, let us apply the operator $\Delta + \mathsf{k}^2$ to this expression:

$$\begin{aligned} \left(\Delta + \mathsf{k}^2\right)\Psi\left(\overrightarrow{\mathbf{r}},\omega\right) &= \int_{\tau'} d\tau' \left(\Delta + \mathsf{k}^2\right) G\left(\overrightarrow{\mathbf{r}}, \overrightarrow{\mathbf{r}}'\right) F\left(\overrightarrow{\mathbf{r}}',\omega\right) \\ &= -\int_{\tau'} d\tau' \delta^3\left(\overrightarrow{\mathbf{r}} - \overrightarrow{\mathbf{r}}'\right) F\left(\overrightarrow{\mathbf{r}}',\omega\right) = -F\left(\overrightarrow{\mathbf{r}},\omega\right) \end{aligned} \qquad [3.21]$$

which is in agreement with equation [3.18].

Let us now find the Green function $G(\vec{\mathbf{r}}, \vec{\mathbf{r}}')$ at the point $\vec{\mathbf{r}}$ due to a unit point source located at $\vec{\mathbf{r}}'$ in the absence of any obstacles. The translational invariance of the system means that the Green function $G(\vec{\mathbf{r}}, \vec{\mathbf{r}}')$ depends only on the relative position $\vec{\mathbf{r}} - \vec{\mathbf{r}}'$ of the observation point with respect to the source, and rotational invariance means that $G(\vec{\mathbf{r}}, \vec{\mathbf{r}}')$ depends only on the relative distance $|\vec{\mathbf{r}} - \vec{\mathbf{r}}'|$ between the observation point and the source. Setting $R = |\vec{\mathbf{r}} - \vec{\mathbf{r}}'|$, we see that the function G must satisfy $\frac{1}{R}\frac{d^2}{R^2}(RG) + \mathrm{k}^2 G = 0$ or equivalently:

$$\frac{d^2}{R^2}(RG) + \mathrm{k}^2 (RG) = 0 \qquad [3.22]$$

for $R \neq 0$. This last equation is homogeneous and admits the solution:

$$G = A\frac{e^{\mathrm{ik}R}}{R} + B\frac{e^{-\mathrm{ik}R}}{R} \qquad [3.23]$$

where A and B are constants. As the wave approaches the source, the second term must be discarded. There is no reflected wave (moving away from the source) due to the absence of obstacles (which would produce reflected waves). To determine A, substitute the solution [3.23] into [3.19] and then integrate the result inside a sphere with radius a centered at $\vec{\mathbf{r}}'$:

$$\begin{aligned}
&\int_{\text{V.Sphere}} (\Delta + \mathrm{k}^2)\, G(R)\, d\tau = \oint_{\text{S.Sphere}} d\vec{\mathbf{S}} \cdot \vec{\nabla} G + 4\pi \mathrm{k}^2 \int_0^a G(R)\, R^2 dR \\
&= 4\pi R^2 \frac{d}{dR}\left(A\frac{e^{\mathrm{ik}R}}{R}\right)\bigg|_{R=a} + 4\pi \mathrm{k}^2 A \int_0^a e^{\mathrm{ik}R} R dR \\
&= 4\pi A(\mathrm{ik}a - 1)\, e^{\mathrm{ik}a} + 4\pi A\left[(1 - \mathrm{ik}a)\, e^{\mathrm{ik}a} - 1\right] = -4\pi A
\end{aligned} \qquad [3.24]$$

On the other hand, the integral of equation [3.19] over this sphere equals -1 due to the Dirac delta function on the right-hand side. We therefore conclude that $A = \frac{1}{4\pi}$. Thus, the desired Green function is:

$$G(\vec{\mathbf{r}}, \vec{\mathbf{r}}') = \frac{e^{\mathrm{ik}|\vec{\mathbf{r}} - \vec{\mathbf{r}}'|}}{4\pi |\vec{\mathbf{r}} - \vec{\mathbf{r}}'|} \qquad [3.25]$$

The general solution of equation [3.18] is therefore:

$$\Psi(\vec{\mathbf{r}}, \omega) = \int_{\tau'} F(\vec{\mathbf{r}}', \omega) \frac{e^{\mathrm{ik}|\vec{\mathbf{r}} - \vec{\mathbf{r}}'|}}{4\pi |\vec{\mathbf{r}} - \vec{\mathbf{r}}'|} d\tau' \qquad [3.26]$$

To find the function $\psi\left(\vec{\mathbf{r}},t\right)$ now, we simply calculate the inverse Fourier transform:

$$\psi\left(\vec{\mathbf{r}},t\right)=\frac{1}{4\pi}\int\frac{d\omega}{2\pi}e^{-\mathrm{i}\omega t}\int_{\tau'}F\left(\vec{\mathbf{r}}',\omega\right)\frac{e^{\mathrm{i}\omega\frac{\left|\vec{\mathbf{r}}-\vec{\mathbf{r}}'\right|}{c}}}{\left|\vec{\mathbf{r}}-\vec{\mathbf{r}}'\right|}d\tau'=\frac{1}{4\pi}\int_{\tau'}\frac{d\tau'}{\left|\vec{\mathbf{r}}-\vec{\mathbf{r}}'\right|}\times$$

$$\int\frac{d\omega}{2\pi}F\left(\vec{\mathbf{r}}',\omega\right)e^{-\mathrm{i}\omega\left(t-\frac{\left|\vec{\mathbf{r}}-\vec{\mathbf{r}}'\right|}{c}\right)}=\frac{1}{4\pi}\int_{\tau'}\frac{f\left(\vec{\mathbf{r}}',t-\frac{\left|\vec{\mathbf{r}}-\vec{\mathbf{r}}'\right|}{c}\right)}{\left|\vec{\mathbf{r}}-\vec{\mathbf{r}}'\right|}d\tau'$$

[3.27]

Let us now apply this result to electromagnetic potentials. For V, we take $f=\frac{\rho}{\varepsilon_0}$, while for the components A_x, A_y and A_z of $\vec{\mathbf{A}}$, we take, respectively, $f=\mu_0 J_x$, $f=\mu_0 J_y$ and $f=\mu_0 J_z$. This yields the solutions [3.14] and [3.15]. The resulting potentials are called *retarded potentials*. They have the same form as in electrostatics and magnetostatics, except that the sources are not taken at the observation time t, but at the earlier time $t-\frac{\left|\vec{\mathbf{r}}-\vec{\mathbf{r}}'\right|}{c}$. The difference between the two lies in the time required for the interaction to propagate from $\vec{\mathbf{r}}'$ to $\vec{\mathbf{r}}$.

It is also possible to directly introduce a Green function for the inhomogeneous wave equation [3.16]. Since time is now involved, the Green function for the problem is introduced as a solution to the equation:

$$\Delta G\left(\vec{\mathbf{r}}-\vec{\mathbf{r}}',t-t'\right)-\frac{1}{c^2}\frac{\partial^2}{\partial t^2}G\left(\vec{\mathbf{r}}-\vec{\mathbf{r}}',t-t'\right)=-\delta^3\left(\vec{\mathbf{r}}-\vec{\mathbf{r}}'\right)\delta\left(t-t'\right)$$

[3.28]

with the same boundary conditions as those of the original problem. The Green function is an impulse response (i.e. the response to a point-like and infinitely brief source term). Once this impulse response is known, the solution for any arbitrary source term can be obtained directly by linear superposition. The general solution of the problem with the source term $f\left(\vec{\mathbf{r}}',t'\right)$ is written as:

$$\psi\left(\vec{\mathbf{r}},t\right)=\int G\left(\vec{\mathbf{r}}-\vec{\mathbf{r}}',t-t'\right)f\left(\vec{\mathbf{r}}',t'\right)d\tau'dt' \quad [3.29]$$

Defining $\mathsf{G}\left(\vec{\mathbf{r}}-\vec{\mathbf{r}}',\omega\right)$ as the Fourier transform with respect to the variable $\xi=t-t'$:

$$\mathsf{G}\left(\vec{\mathbf{r}}-\vec{\mathbf{r}}',\omega\right)=\int G\left(\vec{\mathbf{r}}-\vec{\mathbf{r}}',u\right)e^{\mathrm{i}\omega\xi}d\xi \quad [3.30]$$

Equation [3.28] becomes:

$$\Delta \mathsf{G}\left(\vec{\mathbf{r}} - \vec{\mathbf{r}}', \omega\right) + \mathsf{k}^2 \mathsf{G}\left(\vec{\mathbf{r}} - \vec{\mathbf{r}}', \omega\right) = -\delta^3\left(\vec{\mathbf{r}} - \vec{\mathbf{r}}'\right) \tag{3.31}$$

where $\mathsf{k} = \frac{\omega}{c}$. This is the Helmholtz equation that describes a monochromatic wave generated by a point source located at the point $\vec{\mathbf{r}}'$. The solution to this equation (outgoing wave) is:

$$\mathsf{G}\left(\vec{\mathbf{r}} - \vec{\mathbf{r}}', t - t'\right) = \frac{1}{4\pi\left|\vec{\mathbf{r}} - \vec{\mathbf{r}}'\right|}\delta\left(t - t' - \frac{\left|\vec{\mathbf{r}} - \vec{\mathbf{r}}'\right|}{c}\right) \tag{3.32}$$

Using the Green function [3.32], we can directly write the solutions to equation [3.16] as:

$$\begin{aligned} \psi\left(\vec{\mathbf{r}}, t\right) &= \int d\tau' \int dt' \frac{\delta\left(t - t' - \frac{\left|\vec{\mathbf{r}} - \vec{\mathbf{r}}'\right|}{c}\right)}{4\pi\left|\vec{\mathbf{r}} - \vec{\mathbf{r}}'\right|} f\left(\vec{\mathbf{r}}', t'\right) \\ &= \frac{1}{4\pi}\int \frac{f\left(\vec{\mathbf{r}}', t - \frac{\left|\vec{\mathbf{r}} - \vec{\mathbf{r}}'\right|}{c}\right)}{\left|\vec{\mathbf{r}} - \vec{\mathbf{r}}'\right|} d\tau' \end{aligned} \tag{3.33}$$

3.2. Radiation from a localized oscillating source

Let us now focus on radiation from a monochromatic source. To simplify, we consider a distribution of charges and current whose respective densities, $\rho\left(\vec{\mathbf{r}}, t\right)$ and $\vec{\mathbf{J}}\left(\vec{\mathbf{r}}, t\right)$, are written as follows (harmonic time dependence):

$$\begin{aligned} \rho\left(\vec{\mathbf{r}}, t\right) &= \Re\left[\hat{\rho}\left(\vec{\mathbf{r}}\right) e^{-\mathrm{i}\omega t}\right] \\ \vec{\mathbf{J}}\left(\vec{\mathbf{r}}, t\right) &= \Re\left[\vec{\hat{\mathbf{J}}}\left(\vec{\mathbf{r}}\right) e^{-\mathrm{i}\omega t}\right] \end{aligned} \tag{3.34}$$

where $\vec{\hat{\mathbf{J}}}$ and $\hat{\rho}$ must satisfy the continuity equation $\vec{\nabla} \cdot \vec{\hat{\mathbf{J}}} = \mathrm{i}\omega\hat{\rho}$. If the densities are not sinusoidal, they can still be expressed as a Fourier transform in time, so that each component becomes monochromatic. In that case, it is sufficient to determine the radiation for the monochromatic source and then apply the inverse Fourier transform to obtain the radiation produced by the original source. From equation [3.15], we then deduce the expression for the vector potential:

$$\vec{\mathbf{A}}\left(\vec{\mathbf{r}}, t\right) = \Re\left[\frac{\mu_0}{4\pi}\int_{\tau'} \frac{\vec{\hat{\mathbf{J}}}\left(\vec{\mathbf{r}}'\right) e^{-\mathrm{i}\omega\left(t - \frac{\left|\vec{\mathbf{r}} - \vec{\mathbf{r}}'\right|}{v}\right)}}{\left|\vec{\mathbf{r}} - \vec{\mathbf{r}}'\right|} d\tau'\right] = \Re\left[\vec{\hat{\mathbf{A}}}\left(\vec{\mathbf{r}}\right) e^{-\mathrm{i}\omega t}\right] \tag{3.35}$$

where $\vec{\hat{\mathbf{A}}}(\vec{\mathbf{r}})$ is the complex amplitude:

$$\vec{\hat{\mathbf{A}}}(\vec{\mathbf{r}}) = \frac{\mu_0}{4\pi}\int_{\tau'} \frac{\vec{\hat{\mathbf{J}}}(\vec{\mathbf{r}}')\, e^{\mathrm{i}k|\vec{\mathbf{r}}-\vec{\mathbf{r}}'|}}{|\vec{\mathbf{r}}-\vec{\mathbf{r}}'|}d\tau' \qquad [3.36]$$

The integration is performed over the region containing the moving charges. The integral equation [3.35] represents the expression for the retarded vector potential in the monochromatic regime. We observe that each point $\vec{\mathbf{r}}'$ of the source radiates an outgoing spherical wave with an amplitude proportional to $\vec{\hat{\mathbf{J}}}(\vec{\mathbf{r}}')$. The superposition of all of these spherical waves (which interfere with each other) constructs the vector potential. Similarly, the scalar potential is written as:

$$V(\vec{\mathbf{r}}, t) = \Re\left[\hat{V}(\vec{\mathbf{r}})\, e^{-\mathrm{i}\omega t}\right] \qquad [3.37]$$

where:

$$\hat{V}(\vec{\mathbf{r}}) = \frac{1}{4\pi\varepsilon_0}\int_{\tau'} \frac{\hat{\rho}(\vec{\mathbf{r}}')\, e^{\mathrm{i}k|\vec{\mathbf{r}}-\vec{\mathbf{r}}'|}}{|\vec{\mathbf{r}}-\vec{\mathbf{r}}'|}d\tau' \qquad [3.38]$$

Since the (radiation) fields are observed at a distance much greater than the size of the source, it is more practical to place the source approximately around the origin and to express the potentials in terms of the observer's spherical coordinates. Consider a distribution that occupies a volume τ', which does not extend beyond a distance r'_{max} from the origin of the coordinates. The observer is located at a distance $r \gg r'_{\mathrm{max}}$ from the origin. It is also assumed that the radius r'_{max} of the radiating system is much smaller than the radiation wavelength λ, that is, $\lambda \gg r'_{\mathrm{max}}$. In general, it is very difficult to evaluate the integrals [3.36] and [3.38]. Consequently, various approximations must be used to compute them. Since the problem involves two characteristic length scales (the distance r and the radiation wavelength $\lambda = \frac{2\pi}{\mathrm{k}}$), we will divide the space into three zones:

1) Static zone

This is defined by the condition $r \ll \lambda$, which implies that $e^{\mathrm{i}k|\vec{\mathbf{r}}-\vec{\mathbf{r}}'|} \approx 1$. In this region, we see that the effect of retardation is negligible, and we obtain the same expression for $\vec{\mathbf{A}}$ as in the static case:

$$\vec{\hat{\mathbf{A}}}(\vec{\mathbf{r}}) = \frac{\mu_0}{4\pi}\int_{\tau'} \frac{\vec{\hat{\mathbf{J}}}(\vec{\mathbf{r}}')}{|\vec{\mathbf{r}}-\vec{\mathbf{r}}'|}d\tau' \qquad [3.39]$$

Using the expansion (in terms of spherical harmonic functions):

$$\frac{1}{|\vec{\mathbf{r}}-\vec{\mathbf{r}}'|} = \sum_{l,m} \frac{4\pi}{2l+1}\frac{r_<^l}{r_>^{l+1}} Y_l^m(\theta,\varphi)\, Y_l^{m\star}(\theta',\varphi') \qquad [3.40]$$

the potential [3.39] can be expressed, when $r > r'$, as follows:

$$\vec{\hat{\mathbf{A}}}\left(\vec{\mathbf{r}}\right) = \mu_0 \sum_{l,m} \frac{1}{2l+1} \frac{Y_l^m(\theta,\varphi)}{r^{l+1}} \int_{\tau'} \vec{\hat{\mathbf{J}}}\left(\vec{\mathbf{r}}'\right) r'^l Y_l^{m\star}(\theta',\varphi')\, d\tau' \qquad [3.41]$$

For the scalar potential, we obtain:

$$\begin{aligned} \hat{V}\left(\vec{\mathbf{r}}\right) &= \int_{\tau'} \frac{\hat{\rho}\left(\vec{\mathbf{r}}'\right)}{4\pi\varepsilon_0 \left|\vec{\mathbf{r}} - \vec{\mathbf{r}}'\right|} d\tau' \\ &= \sum_{l,m} \frac{1}{2l+1} \frac{Y_l^m(\theta,\varphi)}{\varepsilon_0 r^{l+1}} \int_{\tau'} \hat{\rho}\left(\vec{\mathbf{r}}'\right) r'^l Y_l^{m\star}(\theta',\varphi')\, d\tau' \qquad [3.42] \end{aligned}$$

2) Radiation zone

This zone is defined by $r \gg \lambda$. When $r \gg \lambda$, a Taylor expansion in terms of $\frac{r'}{r}$ can be performed in the expression $\frac{e^{\mathrm{i}k\left|\vec{\mathbf{r}} - \vec{\mathbf{r}}'\right|}}{\left|\vec{\mathbf{r}} - \vec{\mathbf{r}}'\right|}$. Thus, we obtain:

$$\begin{aligned} \frac{e^{\mathrm{i}k\left|\vec{\mathbf{r}} - \vec{\mathbf{r}}'\right|}}{\left|\vec{\mathbf{r}} - \vec{\mathbf{r}}'\right|} &\approx \frac{e^{\mathrm{i}k\left|\vec{\mathbf{r}} - \vec{\mathbf{r}}'\right|}}{r} = \frac{e^{\mathrm{i}k\sqrt{r^2 - 2\vec{\mathbf{r}}\cdot\vec{\mathbf{r}}' + r'^2}}}{r} = \frac{e^{\mathrm{i}kr\sqrt{1 - 2\frac{\vec{\mathbf{e}}_r\cdot\vec{\mathbf{r}}'}{r} + \frac{r'^2}{r^2}}}}{r} \\ &\approx \frac{e^{\mathrm{i}kr\left[1 - \frac{\vec{\mathbf{e}}_r\cdot\vec{\mathbf{r}}'}{r} + \cdots\right]}}{r} \qquad [3.43] \end{aligned}$$

where $\vec{\mathbf{e}}_r = \frac{\vec{\mathbf{r}}}{r}$ is the unit vector in the direction of the observer. We have used the fact that $r \gg r'_{\mathrm{max}}$. Now, for this expansion to be valid, the second-order term in the expansion [3.43] must also be negligible in the phase term. This means that $\frac{kr'^2_{\mathrm{max}}}{r^2} \ll 2\pi$, or equivalently, $r \gg \frac{r'^2_{\mathrm{max}}}{\lambda}$. We therefore conclude that the far-field conditions are $r \gg r'_{\mathrm{max}}$ and $r \gg \frac{r'^2_{\mathrm{max}}}{\lambda}$. This region is called the *radiation zone*. In this zone, the term $\frac{e^{\mathrm{i}k\left|\vec{\mathbf{r}} - \vec{\mathbf{r}}'\right|}}{\left|\vec{\mathbf{r}} - \vec{\mathbf{r}}'\right|}$ appearing in equation [3.36] can therefore be simplified as follows:

$$\frac{e^{\mathrm{i}k\left|\vec{\mathbf{r}} - \vec{\mathbf{r}}'\right|}}{\left|\vec{\mathbf{r}} - \vec{\mathbf{r}}'\right|} \simeq \frac{e^{\mathrm{i}kr}}{r} e^{-\mathrm{i}k\vec{\mathbf{e}}_r\cdot\vec{\mathbf{r}}'} \qquad [3.44]$$

and, consequently, the vector potential [3.36] is written as:

$$\vec{\hat{\mathbf{A}}}\left(\vec{\mathbf{r}}\right) = \frac{\mu_0}{4\pi} \frac{e^{\mathrm{i}kr}}{r} \int_{\tau'} \vec{\hat{\mathbf{J}}}\left(\vec{\mathbf{r}}'\right) e^{-\mathrm{i}k\vec{\mathbf{e}}_r\cdot\vec{\mathbf{r}}'} d\tau' \equiv \frac{\mu_0}{4\pi} \frac{e^{\mathrm{i}kr}}{r} \int_{\tau'} \vec{\hat{\mathbf{J}}}\left(\vec{\mathbf{r}}'\right) e^{-\mathrm{i}\vec{\mathrm{k}}\cdot\vec{\mathbf{r}}'} d\tau' \qquad [3.45]$$

with $\overrightarrow{\mathsf{k}} = \mathsf{k}\overrightarrow{\mathbf{e}}_r$. For the scalar potential, we find:

$$\hat{V}\left(\overrightarrow{\mathbf{r}}\right) = \frac{1}{4\pi\varepsilon_0}\frac{e^{\mathsf{i}kr}}{r}\int_{\tau'}\hat{\rho}\left(\overrightarrow{\mathbf{r}}'\right)e^{-\mathsf{i}\overrightarrow{\mathsf{k}}\cdot\overrightarrow{\mathbf{r}}'}d\tau' \qquad [3.46]$$

Here are the main characteristics of radiation in the region of far-field:

– The radiated vector potential takes the form of a spherical wave whose amplitude is modulated by an interference term that accounts for the phase shifts between the different points of the source.

– Since $r'_{\max} \ll \lambda$ (i.e. $\mathsf{k}r' \ll 1$), we can expand the exponential inside the integral [3.45] to obtain:

$$\overrightarrow{\hat{\mathbf{A}}}\left(\overrightarrow{\mathbf{r}}\right) = \frac{\mu_0}{4\pi}\frac{e^{\mathsf{i}kr}}{r}\sum_l\frac{(-\mathsf{ik})^l}{l!}\int_{\tau'}\overrightarrow{\hat{\mathbf{J}}}\left(\overrightarrow{\mathbf{r}}'\right)\left(\overrightarrow{\mathsf{k}}\cdot\overrightarrow{\mathbf{r}}'\right)^l d\tau' \qquad [3.47]$$

– Mathematically, the integral in equation [3.45] is the three-dimensional spatial Fourier transform of the current density $\overrightarrow{\hat{\mathbf{J}}}\left(\overrightarrow{\mathbf{r}}'\right)$. This relation describes the connection between the angular distribution of the radiated field (dependence on $\overrightarrow{\mathbf{e}}_r$) and the spatial distribution of the current in the source. To this end, let us define:

$$\overrightarrow{\hat{\mathsf{N}}}\left(\overrightarrow{\mathsf{k}}\right) = \int_{\tau'}\overrightarrow{\hat{\mathbf{J}}}\left(\overrightarrow{\mathbf{r}}'\right)e^{-\mathsf{i}\overrightarrow{\mathsf{k}}\cdot\overrightarrow{\mathbf{r}}'}d\tau' \qquad [3.48]$$

The vector $\overrightarrow{\hat{\mathsf{N}}}\left(\overrightarrow{\mathsf{k}}\right)$, which depends only on the direction of the observation point, is called the *radiation vector*. If it is a surface or line current distribution, equation [3.48] can be, respectively, replaced by:

$$\overrightarrow{\hat{\mathsf{N}}}\left(\overrightarrow{\mathsf{k}}\right) = \begin{cases}\displaystyle\int_{S'}\overrightarrow{\hat{\mathbf{K}}}\left(\overrightarrow{\mathbf{r}}'\right)e^{-\mathsf{i}\overrightarrow{\mathsf{k}}\cdot\overrightarrow{\mathbf{r}}'}dS' & \text{surfacedist.}\\ \displaystyle\int_{L'}\overrightarrow{\hat{\mathbf{I}}}\left(\overrightarrow{\mathbf{r}}'\right)e^{-\mathsf{i}\overrightarrow{\mathsf{k}}\cdot\overrightarrow{\mathbf{r}}'}dl' \equiv \hat{I}\int_{L'}d\overrightarrow{\mathbf{l}}\left(\overrightarrow{\mathbf{r}}'\right)e^{-\mathsf{i}\overrightarrow{\mathsf{k}}\cdot\overrightarrow{\mathbf{r}}'} & \text{linedist.}\end{cases} \qquad [3.49]$$

where $\overrightarrow{\hat{\mathbf{K}}}$ and $\hat{I}$ are, respectively, the surface current and the current carried by the wire. Thus, equation [3.45] becomes:

$$\overrightarrow{\hat{\mathbf{A}}}\left(\overrightarrow{\mathbf{r}}\right) = \frac{\mu_0}{4\pi}\frac{e^{\mathsf{i}kr}}{r}\overrightarrow{\hat{\mathsf{N}}}\left(\overrightarrow{\mathsf{k}}\right) \qquad [3.50]$$

– A similar expression exists for the scalar potential:

$$\hat{V}\left(\overrightarrow{\mathbf{r}}\right) = \frac{e^{\mathsf{i}kr}}{4\pi\varepsilon_0 r}\hat{\mathrm{P}}\left(\overrightarrow{\mathsf{k}}\right) \qquad [3.51]$$

with:

$$\hat{\mathrm{P}}\left(\overrightarrow{\mathrm{k}}\right) = \int_{\tau'} \hat{\rho}\left(\overrightarrow{\mathbf{r}}'\right) e^{-\mathrm{i}\overrightarrow{\mathrm{k}}\cdot\overrightarrow{\mathbf{r}}'} d\tau' \quad [3.52]$$

It is possible to express $\hat{\mathrm{P}}\left(\overrightarrow{\mathrm{k}}\right)$ in terms of $\overrightarrow{\hat{\mathrm{N}}}\left(\overrightarrow{\mathrm{k}}\right)$ by using the continuity equation. Indeed, we have: $\overrightarrow{\nabla}\cdot\overrightarrow{\mathbf{J}} + \frac{\partial\rho}{\partial t} = 0$, which gives $\hat{\rho} = \frac{-\mathrm{i}}{\omega}\overrightarrow{\nabla}\cdot\overrightarrow{\hat{\mathbf{J}}}$, and, consequently:

$$\begin{aligned}\hat{\mathrm{P}}\left(\overrightarrow{\mathrm{k}}\right) &= \frac{-\mathrm{i}}{\omega}\int_{\tau'} \overrightarrow{\nabla}\cdot\overrightarrow{\hat{\mathbf{J}}}\left(\overrightarrow{\mathbf{r}}'\right) e^{-\mathrm{i}\overrightarrow{\mathrm{k}}\cdot\overrightarrow{\mathbf{r}}'} d\tau' \\ &= \frac{-\mathrm{i}}{\omega}\int_{\tau'}\left[\overrightarrow{\nabla}\cdot\left(\overrightarrow{\hat{\mathbf{J}}}\left(\overrightarrow{\mathbf{r}}'\right) e^{-\mathrm{i}\overrightarrow{\mathrm{k}}\cdot\overrightarrow{\mathbf{r}}'}\right) - \overrightarrow{\hat{\mathbf{J}}}\left(\overrightarrow{\mathbf{r}}'\right)\cdot\overrightarrow{\nabla} e^{-\mathrm{i}\overrightarrow{\mathrm{k}}\cdot\overrightarrow{\mathbf{r}}'}\right] d\tau' \\ &= \frac{-\mathrm{i}}{\omega}\oint_{S'} e^{-\mathrm{i}\overrightarrow{\mathrm{k}}\cdot\overrightarrow{\mathbf{r}}'}\overrightarrow{\hat{\mathbf{J}}}\left(\overrightarrow{\mathbf{r}}'\right)\cdot d\overrightarrow{\mathbf{S}}' + \frac{\mathrm{i}}{\omega}\int_{\tau'}\overrightarrow{\hat{\mathbf{J}}}\left(\overrightarrow{\mathbf{r}}'\right)\cdot\overrightarrow{\nabla} e^{-\mathrm{i}\overrightarrow{\mathrm{k}}\cdot\overrightarrow{\mathbf{r}}'} d\tau' \quad [3.53]\end{aligned}$$

The first integral is zero, since no current can escape from the surface S'. Thus, we are left with:

$$\hat{\mathrm{P}}\left(\overrightarrow{\mathrm{k}}\right) = \frac{1}{\omega}\overrightarrow{\mathrm{k}}\cdot\int_{\tau'}\overrightarrow{\hat{\mathbf{J}}}\left(\overrightarrow{\mathbf{r}}'\right) e^{-\mathrm{i}\overrightarrow{\mathrm{k}}\cdot\overrightarrow{\mathbf{r}}'} d\tau' = \frac{1}{c}\overrightarrow{\mathbf{e}}_r\cdot\overrightarrow{\hat{\mathbf{N}}}\left(\overrightarrow{\mathrm{k}}\right) \quad [3.54]$$

– Magnetic field: starting from the exact expression of $\overrightarrow{\hat{\mathbf{A}}}$, we can determine $\overrightarrow{\nabla}\times\overrightarrow{\hat{\mathbf{A}}}$, and then apply the far-field approximation:

$$\begin{aligned}\overrightarrow{\nabla}\times\overrightarrow{\hat{\mathbf{A}}} &= \frac{\mu_0}{4\pi}\int_{\tau'}\overrightarrow{\nabla}\times\left[\overrightarrow{\hat{\mathbf{J}}}'\frac{e^{\mathrm{ik}\left|\overrightarrow{\mathbf{r}}-\overrightarrow{\mathbf{r}}'\right|}}{\left|\overrightarrow{\mathbf{r}}-\overrightarrow{\mathbf{r}}'\right|}\right] d\tau' = \frac{\mu_0}{4\pi}\int_{\tau'}\overrightarrow{\nabla}\left(\frac{e^{\mathrm{ik}\left|\overrightarrow{\mathbf{r}}-\overrightarrow{\mathbf{r}}'\right|}}{\left|\overrightarrow{\mathbf{r}}-\overrightarrow{\mathbf{r}}'\right|}\right)\times\overrightarrow{\hat{\mathbf{J}}}' d\tau' \\ &= \frac{\mu_0}{4\pi}\int_{\tau'}\left(\mathrm{ik} - \frac{1}{\left|\overrightarrow{\mathbf{r}}-\overrightarrow{\mathbf{r}}'\right|}\right)\frac{e^{\mathrm{ik}\left|\overrightarrow{\mathbf{r}}-\overrightarrow{\mathbf{r}}'\right|}}{\left|\overrightarrow{\mathbf{r}}-\overrightarrow{\mathbf{r}}'\right|}\overrightarrow{\nabla}\left(\left|\overrightarrow{\mathbf{r}}-\overrightarrow{\mathbf{r}}'\right|\right)\times\overrightarrow{\hat{\mathbf{J}}}' d\tau' \quad [3.55]\end{aligned}$$

where $\overrightarrow{\hat{\mathbf{J}}}' = \overrightarrow{\hat{\mathbf{J}}}\left(\overrightarrow{\mathbf{r}}'\right)$. Since $r \gg r'_{\max}$, we use the approximation

$$\overrightarrow{\nabla}\left(\left|\overrightarrow{\mathbf{r}}-\overrightarrow{\mathbf{r}}'\right|\right) = \frac{\overrightarrow{\mathbf{r}}-\overrightarrow{\mathbf{r}}'}{\left|\overrightarrow{\mathbf{r}}-\overrightarrow{\mathbf{r}}'\right|} \approx \overrightarrow{\mathbf{e}}_r$$

We finally obtain:

$$\begin{aligned}
\vec{\nabla} \times \vec{\hat{\mathbf{A}}} &\simeq \frac{\mathrm{i}\mu_0 \mathrm{k}}{4\pi} \vec{\mathbf{e}}_r \times \int_{\tau'} \vec{\hat{\mathbf{J}}}' \frac{e^{\mathrm{ik}\left|\vec{\mathbf{r}} - \vec{\mathbf{r}}'\right|}}{\left|\vec{\mathbf{r}} - \vec{\mathbf{r}}'\right|} d\tau' \\
&\simeq \frac{\mathrm{i}\mu_0 \mathrm{k}}{4\pi} \vec{\mathbf{e}}_r \times \left[\frac{e^{\mathrm{ik}r}}{r} \int_{\tau'} \vec{\hat{\mathbf{J}}}' e^{-\mathrm{ik}\vec{\mathbf{e}}_r \cdot \vec{\mathbf{r}}'} d\tau' \right] \\
&= \frac{\mathrm{i}\mu_0 \mathrm{k}}{4\pi} \frac{e^{\mathrm{ik}r}}{r} \vec{\mathbf{e}}_r \times \vec{\hat{\mathbf{N}}}\left(\vec{\mathrm{k}}\right) \qquad [3.56]
\end{aligned}$$

The expression for the complex amplitude of the magnetic field, $\vec{\hat{\mathbf{B}}} = \vec{\nabla} \times \vec{\hat{\mathbf{A}}}$, is therefore:

$$\vec{\hat{\mathbf{B}}} = \frac{\mathrm{i}\mu_0 \mathrm{k}}{4\pi} \frac{e^{\mathrm{ik}r}}{r} \vec{\mathbf{e}}_r \times \vec{\hat{\mathbf{N}}}\left(\vec{\mathrm{k}}\right) \iff \vec{\hat{\mathbf{H}}} = \frac{\mathrm{ik}}{4\pi} \frac{e^{\mathrm{ik}r}}{r} \vec{\mathbf{e}}_r \times \vec{\hat{\mathbf{N}}}\left(\vec{\mathrm{k}}\right) \qquad [3.57]$$

– Electric field: the electric field is obtained in a similar manner:

$$\begin{aligned}
\vec{\hat{\mathbf{E}}} &= -\vec{\nabla}\hat{V} + \mathrm{i}\omega \vec{\hat{\mathbf{A}}} \\
&= -\frac{1}{4\pi\varepsilon_0} \int_{\tau'} \hat{\rho}' \vec{\nabla} \left(\frac{e^{\mathrm{ik}\left|\vec{\mathbf{r}} - \vec{\mathbf{r}}'\right|}}{\left|\vec{\mathbf{r}} - \vec{\mathbf{r}}'\right|} \right) d\tau' + \frac{\mathrm{i}\omega\mu_0}{4\pi} \int_{\tau'} \vec{\hat{\mathbf{J}}}' \frac{e^{\mathrm{ik}\left|\vec{\mathbf{r}} - \vec{\mathbf{r}}'\right|}}{\left|\vec{\mathbf{r}} - \vec{\mathbf{r}}'\right|} d\tau' \\
&\simeq -\frac{\mathrm{ik}}{4\pi\varepsilon_0} \frac{e^{\mathrm{ik}r}}{r} \int_{\tau'} \hat{\rho}' e^{-\mathrm{i}\vec{\mathrm{k}} \cdot \vec{\mathbf{r}}} d\tau' + \frac{\mathrm{i}\omega\mu_0}{4\pi} \frac{e^{\mathrm{ik}r}}{r} \int_{\tau'} \vec{\hat{\mathbf{J}}}' e^{-\mathrm{i}\vec{\mathrm{k}} \cdot \vec{\mathbf{r}}} d\tau' \\
&= -\frac{\mathrm{ik}}{4\pi\varepsilon_0} \frac{e^{\mathrm{ik}r}}{r} \left[\hat{\mathrm{P}}\left(\vec{\mathrm{k}}\right) \vec{\mathbf{u}} - \frac{1}{c} \vec{\hat{\mathbf{N}}}\left(\vec{\mathrm{k}}\right) \right] \\
&= -\frac{\mathrm{ik}}{4\pi\varepsilon_0 c} \frac{e^{\mathrm{ik}r}}{r} \left[\left(\vec{\mathbf{e}}_r \cdot \vec{\hat{\mathbf{N}}}\left(\vec{\mathrm{k}}\right) \right) \vec{\mathbf{e}}_r - \vec{\hat{\mathbf{N}}}\left(\vec{\mathrm{k}}\right) \right] \\
&= \frac{\mathrm{ik}}{4\pi\varepsilon_0 c} \frac{e^{\mathrm{ik}r}}{r} \left(\vec{\mathbf{e}}_r \times \vec{\hat{\mathbf{N}}}\left(\vec{\mathrm{k}}\right) \right) \times \vec{\mathbf{e}}_r = \sqrt{\frac{\mu_0}{\varepsilon_0}} \vec{\hat{\mathbf{H}}} \times \vec{\mathbf{e}}_r \qquad [3.58]
\end{aligned}$$

Relations [3.57] and [3.58] are the same as those for a plane wave. In the far-field region, the spherical wave effectively becomes a plane wave propagating in the direction $\vec{\mathbf{e}}_r$.

– Radiated power: to determine the power radiated by an arbitrary source in the far-field region, we must first compute the time-averaged value of the Poynting vector:

$$\langle\vec{\mathbf{\Pi}}\rangle = \frac{1}{2}\left[\vec{\hat{\mathbf{E}}} \times \vec{\hat{\mathbf{H}}}^{\star}\right] = \frac{1}{2}\sqrt{\frac{\mu_0}{\varepsilon_0}}\left|\vec{\hat{\mathbf{H}}}\right|^2 \vec{\mathbf{e}}_r = \sqrt{\frac{\mu_0}{\varepsilon_0}}\frac{\mathsf{k}^2}{32\pi^2 r^2}\left|\vec{\mathbf{e}}_r \times \vec{\hat{\mathbf{N}}}\right|^2 \vec{\mathbf{e}}_r \qquad [3.59]$$

Thus, the power radiated in a given direction $\vec{\mathbf{e}}_r$, in the elementary solid angle $d\Omega$, is written as:

$$dP = \langle\vec{\mathbf{\Pi}} \cdot \vec{\mathbf{e}}_r\rangle dS = \langle\vec{\mathbf{\Pi}} \cdot \vec{\mathbf{e}}_r\rangle r^2 d\Omega \qquad [3.60]$$

which gives:

$$\frac{dP}{d\Omega} = r^2\langle\vec{\mathbf{\Pi}} \cdot \vec{\mathbf{e}}_r\rangle = \sqrt{\frac{\mu_0}{\varepsilon_0}}\frac{\mathsf{k}^2}{32\pi^2}\left|\vec{\mathbf{e}}_r \times \vec{\hat{\mathbf{N}}}\right|^2 \qquad [3.61]$$

$\frac{dP}{d\Omega}$ represents the power radiated per unit of the solid angle. If $\vec{\hat{\mathbf{N}}}$ is real (the different components of $\vec{\hat{\mathbf{N}}}$ have the same phase), we obtain:

$$\frac{dP}{d\Omega} = \sqrt{\frac{\mu_0}{\varepsilon_0}}\frac{\mathsf{k}^2}{32\pi^2}\left|\vec{\hat{\mathbf{N}}}\right|^2 \sin^2\gamma \qquad [3.62]$$

where γ is the angle between $\vec{\hat{\mathbf{N}}}$ and $\vec{\mathbf{e}}_r$. The total power radiated into all space is obtained by integration:

$$P_{\text{radiated}} = \int_{4\pi} \frac{dP}{d\Omega} d\Omega \qquad [3.63]$$

– By writing $\vec{\hat{\mathbf{N}}} = \hat{N}_r \vec{\mathbf{e}}_r + \hat{N}_\theta \vec{\mathbf{e}}_\theta + \hat{N}_\varphi \vec{\mathbf{e}}_\varphi$, we immediately see that:

$$\begin{aligned} \vec{\hat{\mathbf{H}}} &= \frac{\mathsf{ik}}{4\pi}\frac{e^{\mathsf{ik}r}}{r}\left(\hat{N}_\theta \vec{\mathbf{e}}_\varphi - \hat{N}_\varphi \vec{\mathbf{e}}_\theta\right) && \text{(i)} \\ \vec{\hat{\mathbf{E}}} &= \sqrt{\frac{\mu_0}{\varepsilon_0}}\frac{\mathsf{ik}}{4\pi}\frac{e^{\mathsf{ik}r}}{r}\left(\hat{N}_\theta \vec{\mathbf{e}}_\theta + \hat{N}_\varphi \vec{\mathbf{e}}_\varphi\right) && \text{(ii)} \\ \frac{dP}{d\Omega} &= \sqrt{\frac{\mu_0}{\varepsilon_0}}\frac{\mathsf{k}^2}{32\pi^2}\left[\left|\hat{N}_\theta\right|^2 + \left|\hat{N}_\varphi\right|^2\right] && \text{(iii)} \end{aligned} \qquad [3.64]$$

3) Intermediate zone

In this region, r and λ are of the same order of magnitude. We must use the exact expansion of the term $\frac{e^{\mathsf{ik}|\vec{\mathbf{r}}-\vec{\mathbf{r}}'|}}{|\vec{\mathbf{r}}-\vec{\mathbf{r}}'|}$:

$$\frac{e^{\mathsf{ik}|\vec{\mathbf{r}}-\vec{\mathbf{r}}'|}}{|\vec{\mathbf{r}}-\vec{\mathbf{r}}'|} = 4\pi\mathsf{ik}\sum_{l,m} h_l^{(1)}(\mathsf{k}r_>)\, j_l(\mathsf{k}r_<)\, Y_l^m(\theta,\varphi)\, Y_l^{m\star}(\theta',\varphi') \qquad [3.65]$$

where $h_l^{(1)} = j_l + in_l$ is the *spherical Hankel function of the first kind.* Using equation [3.65], the vector potential becomes:

$$\vec{\hat{\mathbf{A}}}(\vec{\mathbf{r}}) = \mathrm{i}\mu_0 \mathrm{k} \sum_{l,m} h_l^{(1)}(\mathrm{k}r)\, Y_l^m(\theta,\varphi) \int_{\tau'} \vec{\hat{\mathbf{J}}}(\vec{\mathbf{r}}')\, j_l(\mathrm{k}r')\, Y_l^{m\star}(\theta',\varphi')\, d\tau' \quad [3.66]$$

Equation [3.66] is valid for an arbitrary source (whether small or large). At this point, no approximation regarding the source size has been made in this equation. Since $r'_{\max} \ll \lambda$ (i.e. $\mathrm{k}r' \ll 1$), we can take the approximation $j_l(\mathrm{k}r') \approx \frac{(\mathrm{k}r')^l}{(2l+1)!!}$, and consequently, the vector potential $\vec{\hat{\mathbf{A}}}$ becomes:

$$\vec{\hat{\mathbf{A}}}(\vec{\mathbf{r}}) = \mathrm{i}\mu_0 \mathrm{k} \sum_{l,m} h_l^{(1)}(\mathrm{k}r)\, Y_l^m(\theta,\varphi) \int_{\tau'} \frac{(\mathrm{k}r')^l}{(2l+1)!!} \vec{\hat{\mathbf{J}}}(\vec{\mathbf{r}}')\, Y_l^{m\star}(\theta',\varphi')\, d\tau' [3.67]$$

In the same way, we find:

$$\hat{V}(\vec{\mathbf{r}}) = \frac{\mathrm{ik}}{\varepsilon_0} \sum_{l,m} h_l^{(1)}(\mathrm{k}r)\, Y_l^m(\theta,\varphi) \int_{\tau'} \frac{(\mathrm{k}r')^l}{(2l+1)!!} \hat{\rho}(\vec{\mathbf{r}}')\, Y_l^{m\star}(\theta',\varphi')\, d\tau' \quad [3.68]$$

3.2.1. *Radiation from an electric dipole*

We saw earlier that if the radiating system is spatially small compared to the radiated wave (i.e. the source dimensions are much smaller than the wavelength of emission, $r'_{\max} \ll \lambda$), the expressions for the electromagnetic field simplify considerably. Far from the source ($r \gg r'_{\max}$), the radiated field becomes equivalent to the field produced by a single electric dipole. This is known as the *electric dipole approximation*. By retaining only the first term ($l = 0$) in equation [3.67] and using the facts that $h_0^{(1)} = \frac{-\mathrm{i}e^{\mathrm{i}x}}{x}$ and $Y_0^0 = \frac{1}{\sqrt{2\pi}}$, the vector potential $\vec{\hat{\mathbf{A}}}$ becomes:

$$\vec{\hat{\mathbf{A}}}(\vec{\mathbf{r}}) = \frac{\mu_0}{4\pi} \frac{e^{\mathrm{ik}r}}{r} \int_{\tau'} \vec{\hat{\mathbf{J}}}(\vec{\mathbf{r}}')\, d\tau' \quad [3.69]$$

This expression can also be obtained from equation [3.47], which is valid in the radiation zone, by retaining only the zero-order term in the expansion.

To express the integral given in equation [3.69] as a function of charge density ρ, it suffices to note that

$$\begin{aligned}\vec{\nabla}' \cdot \left[x' \vec{\hat{\mathbf{J}}}'\right] &= \left(\vec{\nabla}' x'\right) \cdot \vec{\hat{\mathbf{J}}}' + x' \left(\vec{\nabla}' \cdot \vec{\hat{\mathbf{J}}}'\right) \\ &= \vec{\mathbf{i}} \cdot \underline{\mathrm{f}} \vec{J}' + x' \left(\vec{\nabla}' \cdot \vec{\hat{\mathbf{J}}}'\right) = \hat{J}'_x + x' \left(\vec{\nabla}' \cdot \vec{\hat{\mathbf{J}}}'\right)\end{aligned}$$

From this, it follows that $\hat{J}'_x = \vec{\nabla}' \cdot \left[x' \vec{\hat{\mathbf{J}}}'\right] - x' \left(\vec{\nabla}' \cdot \vec{\hat{\mathbf{J}}}'\right)$. Integrating $\hat{J}'_x$ over a volume large enough to contain all the current, we find:

$$\int_{\tau'} \hat{J}'_x d\tau' = \int_{\tau'} \vec{\nabla}' \cdot \left[x' \vec{\hat{\mathbf{J}}}'\right] d\tau' - \int_{\tau'} x' \left(\vec{\nabla}' \cdot \vec{\hat{\mathbf{J}}}'\right) d\tau'$$

$$= \oint_{S'} x' \vec{\hat{\mathbf{J}}}' \cdot d\vec{\mathbf{S}}' + \int_{\tau'} x' \frac{\partial \hat{\rho}'}{\partial t} d\tau' = -\mathrm{i}\omega \int_{\tau'} x' \hat{\rho}' d\tau' \quad [3.70]$$

The surface integral is zero, as no current can leave S'. Repeating the same for the other two components, we obtain:

$$\int_{\tau'} \vec{\hat{\mathbf{J}}}' d\tau' = -\mathrm{i}\omega \int_{\tau'} \vec{\mathbf{r}}' \hat{\rho}' d\tau' = -\mathrm{i}\omega \vec{\hat{\mathbf{p}}}_0 \quad [3.71]$$

where $\vec{\hat{\mathbf{p}}}_0$ is the complex amplitude of the electric dipole moment of the charge distribution $\left(\vec{\mathbf{p}} = \vec{\hat{\mathbf{p}}}_0 e^{-\mathrm{i}\omega t}\right)$. Thus, equation [3.69] is written as:

$$\vec{\hat{\mathbf{A}}}\left(\vec{\mathbf{r}}\right) = \frac{-\mathrm{i}\omega\mu_0}{4\pi} \frac{e^{ikr}}{r} \vec{\hat{\mathbf{p}}}_0 \quad [3.72]$$

Now, let us calculate the fields. To determine the magnetic field of the dipole, we take the curl of the potential vector:

$$\vec{\hat{\mathbf{B}}} = \vec{\nabla} \times \vec{\hat{\mathbf{A}}} = \frac{-\mathrm{i}\omega\mu_0}{4\pi} \vec{\nabla}\left(\frac{e^{ikr}}{r}\right) \times \vec{\hat{\mathbf{p}}}_0 = \frac{\omega \mathrm{k}\mu_0}{4\pi}\left(1 - \frac{1}{\mathrm{i}\mathrm{k}r}\right) \frac{e^{ikr}}{r} \vec{\mathbf{e}}_r \times \vec{\hat{\mathbf{p}}}_0 \quad [3.73]$$

We thus see that the field $\vec{\hat{\mathbf{B}}}$ naturally separates into two terms: the first varies as $\frac{1}{r}$ and dominates in the radiation zone ($\mathrm{k}r \gg 1$), while the second varies as $\frac{1}{r^2}$ and dominates in the static zone ($\mathrm{k}r \ll 1$). In the radiation zone, we take the following approximation:

$$\vec{\hat{\mathbf{B}}} \simeq \frac{\omega \mathrm{k}\mu_0}{4\pi} \frac{e^{ikr}}{r} \vec{\mathbf{e}}_r \times \vec{\hat{\mathbf{p}}}_0 \quad [3.74]$$

In the static zone, we retain the term:

$$\vec{\hat{\mathbf{B}}} \simeq \frac{\mathrm{i}\omega\mu_0}{4\pi} \frac{e^{\mathrm{ik}r}}{r^2} \vec{\mathbf{e}}_r \times \vec{\hat{\mathbf{p}}}_0 \implies \vec{\mathbf{B}} \simeq \Re\left[\frac{\mathrm{i}\omega\mu_0}{4\pi r^2} \vec{\mathbf{e}}_r \times \vec{\hat{\mathbf{p}}}_0 e^{-\mathrm{i}\omega\left(t - \frac{r}{c}\right)}\right] \quad [3.75]$$

The magnetic field here is nothing other than the magnetostatic field created by a dipolar current, multiplied by the retardation factor $e^{\mathrm{ik}r}$. We thus observe that when we are sufficiently close to the source, that is, at a small fraction of the wavelength, the fields are merely static fields (resulting from the charge distribution) delayed by $\frac{r}{c}$.

For the field $\overrightarrow{\hat{\mathbf{E}}}$, we would normally compute $\hat{V}$ and then use $\overrightarrow{\hat{\mathbf{E}}} = -\overrightarrow{\nabla}\hat{V} + \mathrm{i}\omega\overrightarrow{\hat{\mathbf{A}}}$. However, we will proceed differently here for the sake of simplicity. Outside the source region (where $\overrightarrow{\hat{\mathbf{J}}} = 0$), it is easier to obtain $\overrightarrow{\hat{\mathbf{E}}}$ from the relation $\overrightarrow{\nabla} \times \overrightarrow{\hat{\mathbf{B}}} = \mu_0\varepsilon_0\frac{\partial\overrightarrow{\hat{\mathbf{E}}}}{\partial t} = -\mathrm{i}\frac{\omega}{c^2}\overrightarrow{\hat{\mathbf{E}}} = -\frac{\mathrm{ik}}{c}\overrightarrow{\hat{\mathbf{E}}}$:

$$\begin{aligned}\overrightarrow{\hat{\mathbf{E}}} &= \frac{\mathrm{i}c}{\mathrm{k}}\overrightarrow{\nabla} \times \overrightarrow{\hat{\mathbf{B}}} = \frac{\mathrm{i}c\omega\mu_0}{4\pi}\left[\overrightarrow{\nabla}\left(\frac{e^{ikr}}{r} - \frac{e^{ikr}}{\mathrm{i}\mathrm{k}r^2}\right) \times \left(\overrightarrow{\mathbf{e}}_r \times \overrightarrow{\hat{\mathbf{p}}}_0\right)\right. \\ &\left.+\left(\frac{e^{ikr}}{r} - \frac{e^{ikr}}{\mathrm{i}\mathrm{k}r^2}\right)\overrightarrow{\nabla} \times \left(\overrightarrow{\mathbf{e}}_r \times \overrightarrow{\hat{\mathbf{p}}}_0\right)\right] \\ &\frac{\mathrm{i}c\omega\mu_0}{4\pi}\left[\mathrm{ik}\frac{e^{ikr}}{r}\overrightarrow{\mathbf{e}}_r \times \left(\overrightarrow{\mathbf{e}}_r \times \overrightarrow{\hat{\mathbf{p}}}_0\right) - 2\frac{e^{ikr}}{r^2}\overrightarrow{\mathbf{e}}_r \times \left(\overrightarrow{\mathbf{e}}_r \times \overrightarrow{\hat{\mathbf{p}}}_0\right) + \right. \\ &\left.\frac{2e^{ikr}}{\mathrm{i}\mathrm{k}r^3}\overrightarrow{\mathbf{e}}_r \times \left(\overrightarrow{\mathbf{e}}_r \times \overrightarrow{\hat{\mathbf{p}}}_0\right) + \frac{e^{ikr}}{\mathrm{i}\mathrm{k}r^2}(\mathrm{i}\mathrm{k}r - 1)\,\overrightarrow{\nabla} \times \left(\overrightarrow{\mathbf{e}}_r \times \overrightarrow{\hat{\mathbf{p}}}_0\right)\right]\end{aligned} \quad [3.76]$$

Since:

$$\begin{aligned}&\overrightarrow{\mathbf{e}}_r \times \left(\overrightarrow{\mathbf{e}}_r \times \overrightarrow{\hat{\mathbf{p}}}_0\right) = \left(\overrightarrow{\mathbf{e}}_r \cdot \overrightarrow{\hat{\mathbf{p}}}_0\right)\overrightarrow{\mathbf{e}}_r - \overrightarrow{\hat{\mathbf{p}}}_0 \\ &\overrightarrow{\nabla} \times \left(\overrightarrow{\mathbf{e}}_r \times \overrightarrow{\hat{\mathbf{p}}}_0\right) = -\overrightarrow{\hat{\mathbf{p}}}_0\left(\overrightarrow{\nabla} \cdot \overrightarrow{\mathbf{e}}_r\right) + \left(\overrightarrow{\hat{\mathbf{p}}}_0 \cdot \overrightarrow{\nabla}\right)\overrightarrow{\mathbf{e}}_r \\ &= -\overrightarrow{\hat{\mathbf{p}}}_0\left(\overrightarrow{\nabla} \cdot \left[\frac{\overrightarrow{\mathbf{r}}}{r}\right]\right) + \hat{p}_{0x}\frac{\partial}{\partial x}\left[\frac{\overrightarrow{\mathbf{r}}}{r}\right] + \hat{p}_{0y}\frac{\partial}{\partial y}\left[\frac{\overrightarrow{\mathbf{r}}}{r}\right] + \hat{p}_{0z}\frac{\partial}{\partial z}\left[\frac{\overrightarrow{\mathbf{r}}}{r}\right] \\ &= -\overrightarrow{\hat{\mathbf{p}}}_0\left[\overrightarrow{\nabla}\left(\frac{1}{r}\right) \cdot \overrightarrow{\mathbf{r}} + \frac{\overrightarrow{\nabla} \cdot \overrightarrow{\mathbf{r}}}{r}\right] + \hat{p}_{0x}\left[\frac{\overrightarrow{\mathbf{i}}}{r} - \frac{x\overrightarrow{\mathbf{r}}}{r^3}\right] + \\ &\hat{p}_{0y}\left[\frac{\overrightarrow{\mathbf{j}}}{r} - \frac{y\overrightarrow{\mathbf{r}}}{r^3}\right] + \hat{p}_{0z}\left[\frac{\overrightarrow{\mathbf{k}}}{r} - \frac{z\overrightarrow{\mathbf{r}}}{r^3}\right] = -\overrightarrow{\hat{\mathbf{p}}}_0\left[-\frac{\overrightarrow{\mathbf{r}}}{r^3} \cdot \overrightarrow{\mathbf{r}} + \frac{3}{r}\right] \\ &+\left[\frac{\overrightarrow{\hat{\mathbf{p}}}_0}{r} - \frac{\overrightarrow{\hat{\mathbf{p}}}_0 \cdot \overrightarrow{\mathbf{r}}}{r^3}\right] = -\frac{\overrightarrow{\hat{\mathbf{p}}}_0}{r} - \frac{\overrightarrow{\hat{\mathbf{p}}}_0 \cdot \overrightarrow{\mathbf{r}}}{r^3}\end{aligned} \quad [3.77]$$

it finally comes that:

$$\vec{\hat{\mathbf{E}}} = \frac{c\omega \mathrm{k}\mu_0}{4\pi}\frac{e^{ikr}}{r}\vec{\mathbf{e}}_r \times \left(\vec{\hat{\mathbf{p}}}_0 \times \vec{\mathbf{e}}_r\right) + \frac{c\omega\mu_0}{4\pi \mathrm{k}}\frac{(\mathrm{i}\mathrm{k}r-1)\,e^{ikr}}{r^2}$$
$$\times\left[-\frac{2}{r}\vec{\mathbf{e}}_r \times \left(\vec{\mathbf{e}}_r \times \vec{\hat{\mathbf{p}}}_0\right) + \vec{\nabla} \times \left(\vec{\mathbf{e}}_r \times \vec{\hat{\mathbf{p}}}_0\right)\right]$$
$$= \frac{c\omega \mathrm{k}\mu_0}{4\pi}\frac{e^{ikr}}{r}\vec{\mathbf{e}}_r \times \left(\vec{\hat{\mathbf{p}}}_0 \times \vec{\mathbf{e}}_r\right) + \frac{(\mathrm{i}\mathrm{k}r-1)\,e^{ikr}}{4\pi\varepsilon_0 r^3}$$
$$\times\left[\vec{\hat{\mathbf{p}}}_0 - 3\left(\vec{\mathbf{e}}_r \cdot \vec{\hat{\mathbf{p}}}_0\right)\vec{\mathbf{e}}_r\right] \quad [3.78]$$

Similarly, two limiting cases appear:

1) *Radiation zone* ($\mathrm{k}r \gg 1$): in this region, the expression for the electric field is dominated by the term in $\frac{1}{r}$. Thus, $\vec{\hat{\mathbf{E}}}$ will approximately take the form:

$$\vec{\hat{\mathbf{E}}} \simeq \frac{c\omega \mathrm{k}\mu_0}{4\pi}\frac{e^{ikr}}{r}\vec{\mathbf{e}}_r \times \left(\vec{\hat{\mathbf{p}}}_0 \times \vec{\mathbf{e}}_r\right) = c\vec{\hat{\mathbf{B}}} \times \vec{\mathbf{e}}_r \quad [3.79]$$

2) *Static zone* ($\mathrm{k}r \ll 1$): near the source, we retain only the term in $\frac{1}{r^3}$ in the expression for $\vec{\hat{\mathbf{E}}}$. In this case, the electric field is approximately:

$$\vec{\hat{\mathbf{E}}} \simeq \frac{3\left(\vec{\mathbf{e}}_r \cdot \vec{\hat{\mathbf{p}}}_0\right)\vec{\mathbf{e}}_r - \vec{\hat{\mathbf{p}}}_0}{4\pi\varepsilon_0 r^3}e^{\mathrm{i}\mathrm{k}r}$$
$$\vec{\mathbf{E}} \simeq \Re\left[\frac{3\left(\vec{\mathbf{e}}_r \cdot \vec{\hat{\mathbf{p}}}_0\right)\vec{\mathbf{e}}_r - \vec{\hat{\mathbf{p}}}_0}{4\pi\varepsilon_0 r^3}e^{\mathrm{i}\omega\left(t-\frac{r}{c}\right)}\right] \quad [3.80]$$

This is exactly the electrostatic field created by an electric dipole seen in an earlier instant $t - \frac{r}{c}$.

It is important to note the following:

– The magnetic field $\vec{\hat{\mathbf{B}}}$ radiated by the electric dipole is transverse everywhere in space.

– The electric field $\vec{\hat{\mathbf{E}}}$ has a component along $\vec{\mathbf{e}}_r$. The nonzero longitudinal component (generated by the second term in equation [3.78]) of the electric field turns out to be a general feature of multipolar electric radiation. Very far from the source, the electric field becomes purely transverse (the longitudinal part is negligible).

We thus conclude that an oscillating electric dipole generates TM waves. Later, we will see that the radiation from a magnetic dipole has a purely transverse electric field. In contrast, $\vec{\mathbf{B}}$ has a longitudinal component.

The time-averaged value of the complex Poynting vector is given as:

$$\begin{aligned}\langle\vec{\mathbf{\Pi}}\rangle &= \frac{1}{2\mu_0}\Re\left[\vec{\mathbf{E}}\times\vec{\mathbf{B}}^\star\right] = \Re\left\{\frac{c\omega^2\mathsf{k}^2\mu_0}{32\pi^2r^2}\frac{1+\mathsf{i}\mathsf{k}r}{\mathsf{i}\mathsf{k}r}\left[\vec{\mathbf{e}}_r\times\left(\hat{\vec{\mathbf{p}}}_0\times\vec{\mathbf{e}}_r\right)\right]\right.\\ &\times\left(\vec{\mathbf{e}}_r\times\hat{\vec{\mathbf{p}}}_0\right) - \frac{\mathsf{i}\left(1+\mathsf{k}^2r^2\right)\omega}{32\pi^2\varepsilon_0 r^5}\left[\hat{\vec{\mathbf{p}}}_0\times\left(\vec{\mathbf{e}}_r\times\hat{\vec{\mathbf{p}}}_0\right) -\right.\\ &\left.\left. 3\left(\hat{\vec{\mathbf{p}}}_0\cdot\vec{\mathbf{e}}_r\right)\times\left[\vec{\mathbf{e}}_r\times\left(\vec{\mathbf{e}}_r\times\hat{\vec{\mathbf{p}}}_0\right)\right]\right]\right\}\end{aligned} \quad [3.81]$$

By eliminating all of the imaginary terms, we are left with:

$$\begin{aligned}\langle\vec{\mathbf{\Pi}}\rangle &= \frac{c\omega^2\mathsf{k}^2\mu_0}{32\pi^2r^2}\left[\vec{\mathbf{e}}_r\times\left(\hat{\vec{\mathbf{p}}}_0\times\vec{\mathbf{e}}_r\right)\right]\times\left(\vec{\mathbf{e}}_r\times\hat{\vec{\mathbf{p}}}_0\right)\\ &= \frac{\omega^4\mu_0}{32\pi^2r^2c}\left|\vec{\mathbf{e}}_r\times\hat{\vec{\mathbf{p}}}_0\right|^2\vec{\mathbf{e}}_r = \frac{\omega^4\mu_0}{32\pi^2r^2c}\left|\hat{\vec{\mathbf{p}}}_0\right|^2\sin^2\theta\,\vec{\mathbf{e}}_r\end{aligned} \quad [3.82]$$

where θ is the angle measured between the dipole axis and the direction of observation. The power radiated in the direction $\vec{\mathbf{e}}_r$, within the solid angle $d\Omega$, is given by:

$$\frac{dP}{d\Omega} = \langle\vec{\mathbf{\Pi}}\rangle\cdot\vec{\mathbf{e}}_r r^2 = \frac{\mu_0\omega^4\left|\hat{\vec{\mathbf{p}}}_0\right|^2}{32\pi^2c}\sin^2\theta \quad [3.83]$$

The angular distribution is illustrated in Figure 3.1.

The radiated power is zero in the direction of the dipole axis (i.e. when $\theta = 0$). The total power P radiated by the dipole in all space is obtained by integrating equation [3.83] in all directions (i.e. in the full solid angle 4π):

$$\begin{aligned}\langle P\rangle &= \frac{\mu_0\omega^4\left|\hat{\vec{\mathbf{p}}}_0\right|^2}{32\pi^2c}\int_{4\pi}\sin^2\theta d\Omega = \frac{\mu_0\omega^4\left|\hat{\vec{\mathbf{p}}}_0\right|^2}{32\pi^2c}\int_0^{2\pi}d\varphi\int_0^{\pi}\sin^2\theta d\theta\\ &= \frac{\mu_0\omega^4\left|\hat{\vec{\mathbf{p}}}_0\right|^2}{32\pi^2c}\times\frac{8\pi}{3} = \frac{\mu_0\omega^4\left|\hat{\vec{\mathbf{p}}}_0\right|^2}{12\pi c}\end{aligned} \quad [3.84]$$

EXAMPLE 3.1. ROTATING WIRE.– Let us find the radiated power from a short wire of length l, initially positioned along the Ox axis, centered at the origin and carrying a linear charge density $\lambda_0 x'$ along its length. The wire rotates in the Oxy plane around the Oz axis with an angular frequency ω, as illustrated in Figure 3.2(a). As the wire rotates, its dipole moment at time t is given by:

$$\vec{\mathbf{p}}(t) = \int_{-\frac{l}{2}}^{\frac{l}{2}} \lambda_0 x'^2 dx' \left(\cos\omega t\,\vec{\mathbf{i}} + \sin\omega t\,\vec{\mathbf{j}}\right) = \frac{\lambda_0 l^3}{12}\left(\cos\omega t\,\vec{\mathbf{i}} + \sin\omega t\,\vec{\mathbf{j}}\right)$$
$$= \Re\left[\frac{\lambda_0 l^3}{12}\left(\vec{\mathbf{i}} + \mathrm{i}\,\vec{\mathbf{j}}\right) e^{-\mathrm{i}\omega t}\right] \quad [3.85]$$

and consequently:

$$\vec{\hat{\mathbf{p}}}_0 = \frac{\lambda_0 l^3}{12}\left(\vec{\mathbf{i}} + \mathrm{i}\,\vec{\mathbf{j}}\right) \quad [3.86]$$

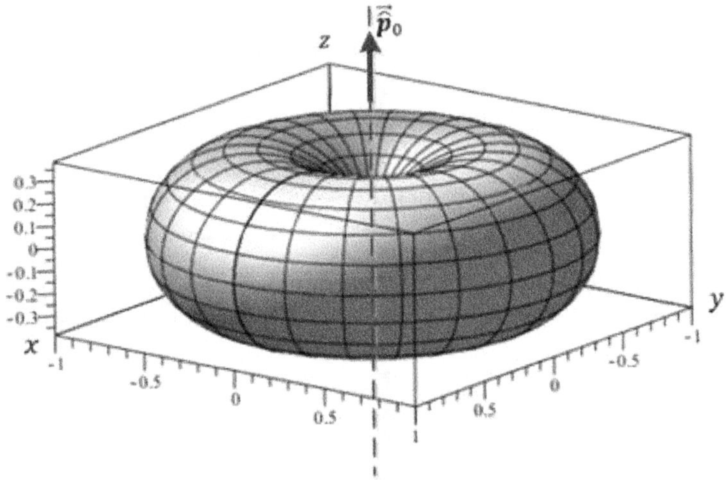

Figure 3.1. *Angular distribution of the power radiated in space. For a color version of this figure, see www.iste.co.uk/abdesselam/electrodynamics3.zip*

The expression [3.85] can be interpreted as two linear oscillators performing harmonic oscillations that are phase-shifted by $\frac{\pi}{2}$ with respect to each other. Let us begin by evaluating $\vec{\mathbf{e}}_r \times \vec{\hat{\mathbf{p}}}_0$:

$$\vec{\mathbf{e}}_r \times \vec{\hat{\mathbf{p}}}_0 = \begin{vmatrix} \vec{\mathbf{i}} & \vec{\mathbf{j}} & \vec{\mathbf{k}} \\ \sin\theta\cos\varphi & \sin\theta\sin\varphi & \cos\theta \\ \frac{\lambda_0 l^3}{12} & \frac{\mathrm{i}\lambda_0 l^3}{12} & 0 \end{vmatrix}$$
$$= \frac{\lambda_0 l^3}{12}\left[-\mathrm{i}\cos\theta\,\vec{\mathbf{i}} + \cos\theta\,\vec{\mathbf{j}} + \mathrm{i}\sin\theta e^{\mathrm{i}\varphi}\,\vec{\mathbf{k}}\right] \quad [3.87]$$

which yields:

$$\left|\overrightarrow{\mathbf{e}}_r \times \overrightarrow{\hat{\mathbf{p}}}_0\right|^2 = \frac{\lambda_0^2 l^6}{144}\left[2\cos^2\theta + \sin^2\theta\right] \qquad [3.88]$$

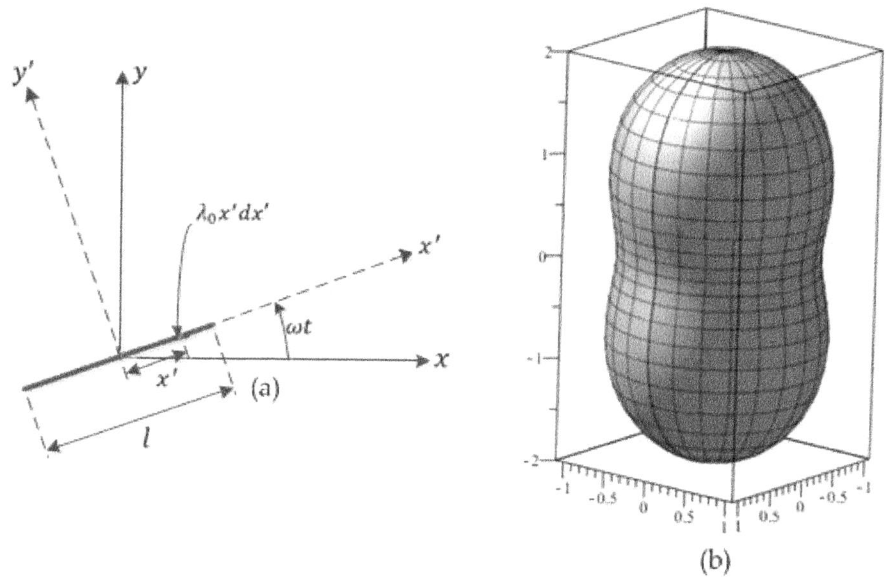

Figure 3.2. *(a) A wire of length l carrying a linear charge density $\lambda_0 x'$ in uniform rotation around the Oz axis. (b) Spatial distribution of the power radiated by the dipole. For a color version of this figure, see www.iste.co.uk/abdesselam/electrodynamics3.zip*

The power radiated, per unit of the solid angle, is then:

$$\frac{dP}{d\Omega} = \frac{\mu_0 \omega^4 \lambda_0^2 l^6 \left(2\cos^2\theta + \sin^2\theta\right)}{4608\pi^2 c} \qquad [3.89]$$

The power is maximal in the directions $\pm Oz$, as illustrated in Figure 3.2(b). The total power radiated in all space is equal to:

$$P = \frac{\mu_0 \omega^4 \lambda_0^2 l^6}{4608\pi^2 c}\int_0^{\pi}\int_0^{2\pi}\left(1+\cos^2\theta\right)\sin\theta d\theta d\varphi = \frac{\mu_0 \omega^4 \lambda_0^2 l^6}{864\pi c} \qquad [3.90]$$

3.2.2. *Radiations from the magnetic moment and the electric quadrupole*

If the first term $(l = 0)$ of the vector potential [3.66] vanishes, or if the dimensions of the source are not sufficiently small, the next term $(l = 1)$ will contribute significantly to the field. As we shall see in what follows, this term gives rise to electric quadrupole radiation and magnetic dipole radiation. Let us return to formula [3.66]. Using the identity:

$$\sum_{m=-l}^{l} Y_l^m(\theta,\varphi)\,Y_l^{m\star}(\theta',\varphi') = \frac{2l+1}{4\pi}P_l(\cos\gamma) = \frac{2l+1}{4\pi}P_l\left(\vec{\mathbf{e}}_r\cdot\vec{\mathbf{e}}'_r\right) \qquad [3.91]$$

(γ is the angle measured between $\vec{\mathbf{r}}$ and $\vec{\mathbf{r}}'$), equation [3.66] is written as:

$$\vec{\hat{\mathbf{A}}}\left(\vec{\mathbf{r}}\right) = \mathrm{i}\mu_0\mathsf{k}\sum_l \frac{2l+1}{4\pi}h_l^{(1)}(\mathsf{k}r)\int_{\tau'}\vec{\hat{\mathbf{J}}}\left(\vec{\mathbf{r}}'\right)j_l(\mathsf{k}r')\,P_l\left(\vec{\mathbf{e}}_r\cdot\vec{\mathbf{e}}'_r\right)d\tau' \qquad [3.92]$$

Since $P_1(x) = x$, the term $l = 1$ reduces to:

$$\vec{\hat{\mathbf{A}}}_1\left(\vec{\mathbf{r}}\right) = \frac{3\mathrm{i}\mu_0\mathsf{k}}{4\pi}h_1^{(1)}(\mathsf{k}r)\int_{\tau'}\vec{\hat{\mathbf{J}}}\left(\vec{\mathbf{r}}'\right)j_1(\mathsf{k}r')\left(\vec{\mathbf{e}}_r\cdot\vec{\mathbf{e}}'_r\right)d\tau' \qquad [3.93]$$

In particular, when the dimensions of the source are very small compared to the wavelength ($r'_{\mathrm{max}} \ll \lambda \Longrightarrow \mathsf{k}r \ll 1$), we can replace $j_1(\mathsf{k}r')$ with $\frac{\mathsf{k}r'}{3}$. Thus, the term $l = 1$ can be approximated by the expression:

$$\begin{aligned}\vec{\hat{\mathbf{A}}}_1\left(\vec{\mathbf{r}}\right) &= \frac{\mathrm{i}\mu_0\mathsf{k}}{4\pi}h_1^{(1)}(\mathsf{k}r)\int_{\tau'}\vec{\hat{\mathbf{J}}}\left(\vec{\mathbf{r}}'\right)\mathsf{k}r'\left(\vec{\mathbf{e}}_r\cdot\vec{\mathbf{e}}'_r\right)d\tau'\\ &= -\frac{\mathrm{i}\mu_0\mathsf{k}^2}{4\pi}\frac{e^{\mathrm{i}\mathsf{k}r}}{\mathsf{k}r}\left(1+\frac{\mathrm{i}}{\mathsf{k}r}\right)\int_{\tau'}\vec{\hat{\mathbf{J}}}\left(\vec{\mathbf{r}}'\right)\left(\vec{\mathbf{e}}_r\cdot\vec{\mathbf{r}}'\right)d\tau'\\ &= -\frac{\mathrm{i}\mu_0\mathsf{k}e^{\mathrm{i}\mathsf{k}r}}{4\pi r^2}\left(1-\frac{1}{\mathrm{i}\mathsf{k}r}\right)\int_{\tau'}\vec{\hat{\mathbf{J}}}\left(\vec{\mathbf{r}}'\right)\left(\vec{\mathbf{r}}\cdot\vec{\mathbf{r}}'\right)d\tau'\end{aligned} \qquad [3.94]$$

To express the integral appearing in equation [3.94] in terms of the components of the magnetic moment and the components of the electric quadrupole moment, let us use the identity $\left(\vec{\mathbf{r}}'\times\vec{\hat{\mathbf{J}}}'\right)\times\vec{\mathbf{r}} = \vec{\hat{\mathbf{J}}}'\left(\vec{\mathbf{r}}'\cdot\vec{\mathbf{r}}\right) - \vec{\mathbf{r}}'\left(\vec{\hat{\mathbf{J}}}'\cdot\vec{\mathbf{r}}\right)$ to rewrite:

$$\begin{aligned}\int_{\tau'}\left(\vec{\mathbf{r}}\cdot\vec{\mathbf{r}}'\right)\vec{\hat{\mathbf{J}}}'d\tau' = \frac{1}{2}\int_{\tau'}&\left\{\left(\vec{\mathbf{r}}'\cdot\vec{\mathbf{r}}\right)\vec{\hat{\mathbf{J}}}' + \left(\vec{\mathbf{r}}\cdot\vec{\hat{\mathbf{J}}}'\right)\vec{\mathbf{r}}'\right.\\ &\left.+\left(\vec{\mathbf{r}}'\times\vec{\hat{\mathbf{J}}}'\right)\times\vec{\mathbf{r}}\right\}d\tau'\end{aligned} \qquad [3.95]$$

The last term is equal to $\vec{\mathbf{m}}\times\vec{\mathbf{r}}$, where $\vec{\mathbf{m}}$ is the *magnetic moment*, associated with the current distribution, defined by:

$$\vec{\mathbf{m}} = \frac{1}{2}\int_{\tau'}\vec{\mathbf{r}}'\times\vec{\hat{\mathbf{J}}}'d\tau' \qquad [3.96]$$

Consequently, equation [3.94] takes the form:

$$\vec{\hat{\mathbf{A}}}_1\left(\vec{\mathbf{r}}\right) = -\frac{\mathrm{i}\mu_0 \mathrm{k} e^{\mathrm{i}kr}}{4\pi r^2}\left(1-\frac{1}{\mathrm{i}kr}\right)\left[\vec{\hat{\mathbf{m}}}\times\vec{\mathbf{r}}+\frac{1}{2}\int_{\tau'}\left\{\left(\vec{\mathbf{r}}'\cdot\vec{\mathbf{r}}\right)\vec{\hat{\mathbf{J}}}' + \left(\vec{\mathbf{r}}\cdot\vec{\hat{\mathbf{J}}}'\right)\vec{\mathbf{r}}'\right\}d\tau'\right] \quad [3.97]$$

Let us now focus on the contribution of the magnetic dipole moment (the first term in equation [3.97]) to the vector potential:

$$\vec{\hat{\mathbf{A}}}_M\left(\vec{\mathbf{r}}\right) = \frac{\mathrm{i}\mu_0 \mathrm{k} e^{\mathrm{i}kr}}{4\pi r}\left(1-\frac{1}{\mathrm{i}kr}\right)\left(\vec{\mathbf{e}}_r\times\vec{\hat{\mathbf{m}}}\right) \quad [3.98]$$

Compared with those for electric dipole fields – equations [3.73] and [3.78] – we observe that $\vec{\hat{\mathbf{A}}}_M$ has the same form as $\vec{\hat{\mathbf{B}}}$ $\left(\vec{\hat{\mathbf{B}}} \Longleftrightarrow \vec{\hat{\mathbf{A}}}_M\right)$. To go from equation [3.73] to [3.98], it is sufficient to replace $\vec{\hat{\mathbf{p}}}_0$ with $\frac{\mathrm{i}\vec{\hat{\mathbf{m}}}}{\omega}$ (i.e. use the correspondence $\vec{\hat{\mathbf{p}}}_0 \Longleftrightarrow \frac{\mathrm{i}\vec{\hat{\mathbf{m}}}}{\omega}$). Consequently, the field $\vec{\hat{\mathbf{B}}}_M = \vec{\nabla}\times\vec{\hat{\mathbf{A}}}_M$ will have the same form as the electric field of the electric dipole $\vec{\hat{\mathbf{E}}} = \frac{\mathrm{i}c^2}{\omega}\vec{\nabla}\times\vec{\hat{\mathbf{B}}}$, on the condition that the correspondence $\frac{\omega}{\mathrm{i}c^2}\vec{\hat{\mathbf{E}}} \Longleftrightarrow \vec{\hat{\mathbf{B}}}_M$ is used:

$$\begin{aligned}\vec{\hat{\mathbf{B}}}_M &= \frac{\omega}{\mathrm{i}c^2}\left\{\frac{c\omega\mathrm{k}\mu_0}{4\pi}\frac{e^{ikr}}{r}\vec{\mathbf{e}}_r\times\left(\frac{\mathrm{i}\vec{\hat{\mathbf{m}}}}{\omega}\times\vec{\mathbf{e}}_r\right)\right.\\ &\quad\left.+\frac{(\mathrm{i}kr-1)\,e^{ikr}}{4\pi\varepsilon_0 r^3}\left[\frac{\mathrm{i}\vec{\hat{\mathbf{m}}}}{\omega}-3\left(\vec{\mathbf{e}}_r\cdot\frac{\mathrm{i}\vec{\hat{\mathbf{m}}}}{\omega}\right)\vec{\mathbf{e}}_r\right]\right\}\\ &= \frac{\mathrm{k}^2\mu_0}{4\pi}\frac{e^{ikr}}{r}\vec{\mathbf{e}}_r\times\left(\vec{\hat{\mathbf{m}}}\times\vec{\mathbf{e}}_r\right)\\ &\quad+\frac{\mu_0}{4\pi}\frac{(\mathrm{i}kr-1)\,e^{ikr}}{r^3}\left[\vec{\hat{\mathbf{m}}}-3\left(\vec{\mathbf{e}}_r\cdot\vec{\hat{\mathbf{m}}}\right)\vec{\mathbf{e}}_r\right]\end{aligned} \quad [3.99]$$

To find the expression for $\vec{\hat{\mathbf{E}}}_M$, let us start with the equation $\vec{\nabla}\times\vec{\hat{\mathbf{E}}}_M = \mathrm{i}\omega\vec{\hat{\mathbf{B}}}_M \Longleftrightarrow \vec{\nabla}\times\vec{\hat{\mathbf{E}}}_M = \frac{\omega^2}{c^2}\vec{\hat{\mathbf{E}}} = \mathrm{i}\omega\vec{\nabla}\times\vec{\hat{\mathbf{B}}}$. This implies that $\vec{\hat{\mathbf{E}}}_M \Longleftrightarrow \mathrm{i}\omega\vec{\hat{\mathbf{B}}}$ and consequently:

$$\vec{\hat{\mathbf{E}}}_M = \mathrm{i}\omega\frac{\omega\mathrm{k}\mu_0}{4\pi}\left(1-\frac{1}{\mathrm{i}kr}\right)\frac{e^{ikr}}{r}\vec{\mathbf{e}}_r\times\frac{\mathrm{i}\vec{\hat{\mathbf{m}}}}{\omega} = -\frac{\omega\mathrm{k}\mu_0}{4\pi}\left(1-\frac{1}{\mathrm{i}kr}\right)\frac{e^{ikr}}{r}\vec{\mathbf{e}}_r\times\vec{\hat{\mathbf{m}}} \quad [3.100]$$

We see that the electric field of the radiation from a magnetic dipole is purely transverse to the direction of observation.

Let us now return to the remaining term of the vector potential [3.97], which we denote here by $\vec{\hat{\mathbf{A}}}_Q$:

$$\vec{\hat{\mathbf{A}}}_Q\left(\vec{\mathbf{r}}\right) = -\frac{\mathrm{i}\mu_0 \mathrm{k} e^{\mathrm{i}kr}}{8\pi r^2}\left(1 - \frac{1}{\mathrm{i}kr}\right)\int_{\tau'}\left[\left(\vec{\mathbf{r}}'\cdot\vec{\mathbf{r}}\right)\vec{\hat{\mathbf{J}}}' + \left(\vec{\mathbf{r}}\cdot\vec{\hat{\mathbf{J}}}'\right)\vec{\mathbf{r}}'\right]d\tau' \tag{3.101}$$

We will see in what follows that this component arises from oscillations of the electric quadrupole moment. To make the components of the electric quadrupole appear in the integral [3.101], let us use the following identities:

$$\begin{aligned}
&\vec{\nabla}'\cdot\left[\left(\vec{\mathbf{r}}'\cdot\vec{\mathbf{r}}\right)x'^{\alpha}\vec{\hat{\mathbf{J}}}'\right] = \left(\vec{\mathbf{r}}'\cdot\vec{\mathbf{r}}\right)x'^{\alpha}\left(\vec{\nabla}'\cdot\vec{\hat{\mathbf{J}}}'\right) + \vec{\hat{\mathbf{J}}}'\cdot\vec{\nabla}'\left[\left(\vec{\mathbf{r}}'\cdot\vec{\mathbf{r}}\right)x'^{\alpha}\right] \\
&= x_\beta x'^{\beta}x'^{\alpha}\left(\vec{\nabla}'\cdot\vec{\hat{\mathbf{J}}}'\right) + \hat{J}'^{\gamma}\partial'_{\gamma}\left(x_\beta x'^{\beta}x'^{\alpha}\right) = x_\beta x'^{\beta}x'^{\alpha}\left(\vec{\nabla}'\cdot\vec{\hat{\mathbf{J}}}'\right) + \\
&\hat{J}'^{\gamma}x_\beta\left(\partial'_{\gamma}x'^{\beta}x'^{\alpha} + x'^{\beta}\partial'_{\gamma}x'^{\alpha}\right) = x_\beta x'^{\beta}x'^{\alpha}\left(\vec{\nabla}'\cdot\vec{\hat{\mathbf{J}}}'\right) + x_\beta x'^{\alpha}\hat{J}'^{\beta} + x_\beta x'^{\beta}\hat{J}'^{\alpha} \\
&= x_\beta x'^{\beta}x'^{\alpha}\left(\vec{\nabla}'\cdot\vec{\hat{\mathbf{J}}}'\right) + \left(\vec{\mathbf{r}}\cdot\vec{\hat{\mathbf{J}}}'\right)x'^{\alpha} + \left(\vec{\mathbf{r}}\cdot\vec{\mathbf{r}}'\right)\hat{J}'^{\alpha}
\end{aligned} \tag{3.102}$$

The last two terms are simply the component α of the vector under the integral sign in [3.101]. It therefore follows that:

$$\begin{aligned}
&\int_{\tau'}\left[\left(\vec{\mathbf{r}}\cdot\vec{\hat{\mathbf{J}}}'\right)x'^{\alpha} + \left(\vec{\mathbf{r}}'\cdot\vec{\mathbf{r}}\right)\hat{J}'^{\alpha}\right]d\tau' = \int_{\tau'}\vec{\nabla}'\cdot\left[\left(\vec{\mathbf{r}}'\cdot\vec{\mathbf{r}}\right)x'^{\alpha}\vec{\hat{\mathbf{J}}}'\right]d\tau' - \\
&\int_{\tau'}x_\beta x'^{\beta}x'^{\alpha}\left(\vec{\nabla}'\cdot\vec{\hat{\mathbf{J}}}'\right)d\tau' = \oint_{S'}\left(\vec{\mathbf{r}}'\cdot\vec{\mathbf{r}}\right)x'^{\alpha}\vec{\hat{\mathbf{J}}}'\cdot d\vec{\mathbf{S}}' - \\
&\int_{\tau'}x_\beta x'^{\beta}x'^{\alpha}\left(\mathrm{i}\omega\hat{\rho}'\right)d\tau' = -\mathrm{i}\omega\int_{\tau'}x_\beta x'^{\beta}x'^{\alpha}\hat{\rho}\left(\vec{\mathbf{r}}'\right)d\tau'
\end{aligned} \tag{3.103}$$

The surface integral is, of course, zero. Repeating the same process for the other components, we find that:

$$\int_{\tau'}\left[\left(\vec{\mathbf{r}}\cdot\vec{\hat{\mathbf{J}}}'\right)\vec{\mathbf{r}}' + \left(\vec{\mathbf{r}}'\cdot\vec{\mathbf{r}}\right)\vec{\hat{\mathbf{J}}}'\right]d\tau' = -\mathrm{i}\omega\int_{\tau'}\left(\vec{\mathbf{r}}\cdot\vec{\mathbf{r}}'\right)\hat{\rho}\left(\vec{\mathbf{r}}'\right)\vec{\mathbf{r}}'d\tau' \tag{3.104}$$

Let us substitute the result obtained into equation [3.101]:

$$\vec{\hat{\mathbf{A}}}_Q(\vec{\mathbf{r}}) = -\frac{\mu_0 \omega \mathrm{k} e^{\mathrm{i}\mathrm{k}r}}{8\pi r^2}\left(1-\frac{1}{\mathrm{i}\mathrm{k}r}\right)\int_{\tau'}(\vec{\mathbf{r}}\cdot\vec{\mathbf{r}}')\,\hat{\rho}(\vec{\mathbf{r}}')\,\vec{\mathbf{r}}'d\tau' \qquad [3.105]$$

To relate $\vec{\hat{\mathbf{A}}}_Q$ with the quadrupole moment of the distribution $Q^{\alpha\beta} = \int_{\tau'}\left[3x'^{\alpha}x'^{\beta} - \delta^{\alpha\beta}r'^2\right]d\tau'$, we proceed as follows:

$$\begin{aligned}
\left(\vec{\hat{\mathbf{A}}}_Q\right)^{\alpha} &= -\frac{\mu_0 \omega \mathrm{k} e^{\mathrm{i}\mathrm{k}r}}{8\pi r^2}\left(1-\frac{1}{\mathrm{i}\mathrm{k}r}\right)\frac{1}{3}x_\beta\int_{\tau'}3x'^{\alpha}x'^{\beta}\hat{\rho}'d\tau' \\
&= -\frac{\mu_0 \omega \mathrm{k} e^{\mathrm{i}\mathrm{k}r}}{24\pi r^2}\left(1-\frac{1}{\mathrm{i}\mathrm{k}r}\right)\left[x_\beta Q^{\alpha\beta} + x_\beta\int_{\tau'}\delta^{\alpha\beta}r'^2\hat{\rho}'\right]d\tau' \\
&= -\frac{\mu_0 \omega \mathrm{k} e^{\mathrm{i}\mathrm{k}r}}{24\pi r^2}\left(1-\frac{1}{\mathrm{i}\mathrm{k}r}\right)\left[x_\beta Q^{\alpha\beta} + x^{\alpha}\int_{\tau'}r'^2\hat{\rho}'\right]d\tau' \\
&= f(r)\,x_\beta Q^{\alpha\beta} + g(r)\,x^{\alpha}
\end{aligned} \qquad [3.106]$$

where $f(r) = -\frac{\mu_0\omega \mathrm{k} e^{\mathrm{i}\mathrm{k}r}}{24\pi r^2}\left(1-\frac{1}{\mathrm{i}\mathrm{k}r}\right)$ and $g(r) = -\frac{\mu_0\omega \mathrm{k} e^{\mathrm{i}\mathrm{k}r}}{24\pi r^2}\left(1-\frac{1}{\mathrm{i}\mathrm{k}r}\right)\int_{\tau'}r'^2\hat{\rho}'d\tau'$. Let us first compute the field $\vec{\hat{\mathbf{B}}}$:

$$\begin{aligned}
\left(\vec{\hat{\mathbf{B}}}_Q\right)^{\alpha} &= \left(\vec{\nabla}\times\vec{\hat{\mathbf{A}}}_Q\right)^{\alpha} = \epsilon^{\alpha\beta\gamma}\partial_\beta\left(\vec{\hat{\mathbf{A}}}_Q\right)_\gamma = \epsilon^{\alpha\beta\gamma}\left(\frac{f'}{r}x_\beta x^\eta Q_{\gamma\eta} + f\delta^\eta_\beta Q_{\gamma\eta} + \right. \\
&\left.\frac{g'}{r}x_\beta x_\gamma + g\delta_{\beta\gamma}\right) = \epsilon^{\alpha\beta\gamma}\frac{f'}{r}x_\beta x^\eta Q_{\gamma\eta}
\end{aligned} \qquad [3.107]$$

because the last three terms are zero. Thus, we finally find:

$$\begin{aligned}
\vec{\hat{\mathbf{A}}}_Q &= -\frac{\mu_0 \omega \mathrm{k} e^{\mathrm{i}\mathrm{k}r}}{24\pi r}\left(1-\frac{1}{\mathrm{i}\mathrm{k}r}\right)\left(\overleftrightarrow{\mathbf{Q}}\dot{\otimes}\vec{\mathbf{e}}_r\right) \\
&\quad -\frac{\mu_0 \omega \mathrm{k} e^{\mathrm{i}\mathrm{k}r}}{24\pi r}\left(1-\frac{1}{\mathrm{i}\mathrm{k}r}\right)\left(\int_{\tau'}r'^2\hat{\rho}'d\tau'\right)\vec{\mathbf{e}}_r \qquad \text{(i)} \\
\vec{\hat{\mathbf{B}}}_Q &= -\frac{\mu_0 \omega \mathrm{i}\mathrm{k}^2 e^{\mathrm{i}\mathrm{k}r}}{24\pi r}\left(1-\frac{3}{\mathrm{i}\mathrm{k}r}+\frac{3}{(\mathrm{i}\mathrm{k}r)^2}\right)\vec{\mathbf{e}}_r\times\left(\overleftrightarrow{\mathbf{Q}}\dot{\otimes}\vec{\mathbf{e}}_r\right) \qquad \text{(ii)}
\end{aligned} \qquad [3.108]$$

In the radiation zone, these equations reduce to:

$$\begin{aligned}
\vec{\hat{\mathbf{A}}}_Q &= -\frac{\mu_0 \omega \mathrm{k} e^{\mathrm{i}\mathrm{k}r}}{24\pi r}\left(\overleftrightarrow{\mathbf{Q}}\dot{\otimes}\vec{\mathbf{e}}_r\right) - \frac{\mu_0 \omega \mathrm{k} e^{\mathrm{i}\mathrm{k}r}}{24\pi r}\left(\int_{\tau'}r'^2\hat{\rho}'d\tau'\right)\vec{\mathbf{e}}_r \qquad \text{(i)} \\
\vec{\hat{\mathbf{B}}}_Q &= -\frac{\mu_0 \omega \mathrm{i}\mathrm{k}^2 e^{\mathrm{i}\mathrm{k}r}}{24\pi r}\vec{\mathbf{e}}_r\times\left(\overleftrightarrow{\mathbf{Q}}\dot{\otimes}\vec{\mathbf{e}}_r\right) \equiv \mathrm{i}\mathrm{k}\vec{\mathbf{e}}_r\times\vec{\hat{\mathbf{A}}}_Q \qquad \text{(ii)}
\end{aligned} \qquad [3.109]$$

and consequently, the electric field will have the value:

$$\vec{\hat{\mathbf{E}}}_Q = c\vec{\hat{\mathbf{B}}}_Q \times \vec{\mathbf{e}}_r = \mathrm{i}\mathrm{k}c\left(\vec{\mathbf{e}}_r \times \vec{\hat{\mathbf{A}}}_Q\right) \times \vec{\mathbf{e}}_r \qquad [3.110]$$

It is clear that the second term of $\vec{\hat{\mathbf{A}}}_Q$ does not contribute to the expressions of the fields $\vec{\hat{\mathbf{E}}}_Q$ and $\vec{\hat{\mathbf{B}}}_Q$. Therefore, it can be omitted from the expression of $\vec{\hat{\mathbf{A}}}_Q$ without affecting the results. We then retain only:

$$\vec{\hat{\mathbf{A}}}_Q = -\frac{\mu_0\omega\mathrm{k}e^{\mathrm{i}\mathrm{k}r}}{24\pi r}\left(\overleftrightarrow{\mathbf{Q}}\dot{\otimes}\vec{\mathbf{e}}_r\right) = -\frac{\mu_0\omega\mathrm{k}e^{\mathrm{i}\mathrm{k}r}}{24\pi r^2}\left(\overleftrightarrow{\mathbf{Q}}\dot{\otimes}\vec{\mathbf{r}}\right) \qquad [3.111]$$

For simplification, let us set $\vec{\mathbf{X}} = \overleftrightarrow{\mathbf{Q}}\dot{\otimes}\vec{\mathbf{r}}$. The time-averaged Poynting vector is equal to:

$$\begin{aligned}\langle\vec{\mathbf{\Pi}}\rangle &= \frac{\vec{\hat{\mathbf{E}}}_Q \times \vec{\hat{\mathbf{A}}}^{\star}_Q}{2\mu_0} = \frac{\mu_0\omega^2\mathrm{k}^4c}{1152\pi^2r^4}\left[\left(\vec{\mathbf{e}}_r \times \vec{\mathbf{X}}\right) \times \vec{\mathbf{e}}_r\right] \times \left[\vec{\mathbf{e}}_r \times \vec{\mathbf{X}}\right] \\ &= \frac{\mu_0\omega^6}{1152\pi^2r^4c^3}\left[\vec{\mathbf{X}}^2 - \left(\vec{\mathbf{X}}\cdot\vec{\mathbf{e}}_r\right)^2\right]\vec{\mathbf{e}}_r \end{aligned} \qquad [3.112]$$

Consequently, the power radiated per unit of solid angle is:

$$\frac{dP}{d\Omega} = r^2\langle\vec{\mathbf{\Pi}}\rangle\cdot\vec{\mathbf{e}}_r = \frac{\mu_0\omega^6}{1152\pi^2r^2c^3}\left[x_\beta Q^{\alpha\beta}Q_{\alpha\gamma}x^\gamma - \frac{x_\beta Q^{\alpha\beta}x_\alpha x^\gamma Q_{\gamma\delta}x^\delta}{r^2}\right] \qquad [3.113]$$

To determine the total power radiated throughout space, one must integrate over the entire solid angle:

$$P = \int_0^{4\pi}\frac{\mu_0\omega^6}{1152\pi^2r^2c^3}\left[x_\beta Q^{\alpha\beta}Q_{\alpha\gamma}x^\gamma - \frac{x_\beta Q^{\alpha\beta}x_\alpha x^\gamma Q_{\gamma\delta}x^\delta}{r^2}\right]d\Omega \qquad [3.114]$$

Using the identities:

$$\begin{aligned}&\int_0^{4\pi} x_\alpha x^\beta d\Omega = \frac{4}{3}\pi r^2\delta^\beta_\alpha && \text{(i)} \\ &\int_0^{4\pi} x_\alpha x_\beta x^\gamma x^\delta d\Omega = \frac{4}{15}\pi r^4\left(\delta_{\alpha\beta}\delta^{\gamma\delta} + \delta^\gamma_\alpha\delta^\delta_\beta + \delta^\delta_\alpha\delta^\gamma_\beta\right) && \text{(ii)}\end{aligned} \qquad [3.115]$$

(these identities are easy to verify by expressing the coordinates $x^1 = r\sin\theta\cos\varphi$, $x^2 = r\sin\theta\sin\varphi$, and $x^3 = r\cos\theta$ in terms of spherical harmonics), we obtain:

$$\int_0^{4\pi}\left[x_\beta Q^{\alpha\beta}Q_{\alpha\gamma}x^\gamma - \frac{x_\beta Q^{\alpha\beta}x_\alpha x^\gamma Q_{\gamma\delta}x^\delta}{r^2}\right]d\Omega = \frac{4}{3}\pi r^2\left[Q_{\alpha\beta}Q^{\alpha\beta}-\right.$$

$$\left.\frac{1}{5}\left(Q^{\alpha\beta}\delta_{\alpha\beta}\delta_{\gamma\delta}Q^{\gamma\delta}+Q_{\alpha\gamma}Q^{\alpha\gamma}+Q_{\alpha\delta}Q^{\alpha\delta}\right)\right] = \frac{4}{3}\pi r^2\left[Q_{\alpha\beta}Q^{\alpha\beta}\right.$$

$$\left.-\frac{2}{5}Q_{\alpha\gamma}Q^{\alpha\gamma}\right] = \frac{4}{5}\pi r^2 Q_{\alpha\beta}Q^{\alpha\beta} \qquad [3.116]$$

since $Q^{\alpha\beta}\delta_{\alpha\beta} = \sum_i Q_{ii} = 0$ (zero trace tensor). Finally, substituting equation [3.116] into equation [3.114], we get:

$$P = \frac{\mu_0\omega^6}{1440\pi c^3}Q_{\alpha\beta}Q^{\alpha\beta} \qquad [3.117]$$

The power radiated by an electric quadrupole moment is proportional to ω^6. This means that it can dominate over that generated by an electric dipole moment at very high radiation frequencies.

EXAMPLE 3.2.– Determine the total power radiated by the quadrupole moment with linear oscillation, as shown in Figure 3.3(a). The charges in the quadrupole moment $-q_0$, $2q_0$ (the central charge) and $-q_0$ are situated, respectively, at the points $(0,0,a)$, $(0,0,0)$ and $(0,0,-a)$. The nonzero components of the quadrupole moment are:

$$Q_{xx} = \int q_0\left[-\delta^3\left(\vec{\mathbf{r}} - a\vec{\mathbf{k}}\right) + 2\delta^3\left(\vec{\mathbf{r}}\right) - \delta^3\left(\vec{\mathbf{r}} + a\vec{\mathbf{k}}\right)\right] \times \left(3x^2 - r^2\right)d\tau = 2q_0a^2 \qquad \text{(i)}$$

$$Q_{yy} = \int q_0\left[-\delta^3\left(\vec{\mathbf{r}} - a\vec{\mathbf{k}}\right) + 2\delta^3\left(\vec{\mathbf{r}}\right) - \delta^3\left(\vec{\mathbf{r}} + a\vec{\mathbf{k}}\right)\right] \times \left(3y^2 - r^2\right)d\tau = 2q_0a^2 \qquad \text{(ii)}$$

$$Q_{zz} = \int q_0\left[-\delta^3\left(\vec{\mathbf{r}} - a\vec{\mathbf{k}}\right) + 2\delta^3\left(\vec{\mathbf{r}}\right) - \delta^3\left(\vec{\mathbf{r}} + a\vec{\mathbf{k}}\right)\right] \times \left(3z^2 - r^2\right)d\tau = -4q_0a^2 \qquad \text{(iii)}$$

[3.118]

Consequently, the first and second terms of the expression between the square brackets in equation [3.113] are written as:

$$\begin{aligned}
Q_{xx}^2 x^2 + Q_{yy}^2 y^2 + Q_{zz}^2 z^2 &= 4q_0^2 a^4 \left(x^2 + y^2 + 4z^2\right) \\
&= 4q_0^2 a^4 r^2 \left(1 + 3\cos^2\theta\right) \\
\frac{1}{r^2}\left(Q_{xx}^2 x^4 + Q_{yy}^2 y^4 + Q_{zz}^2 z^4 + 2Q_{xx}Q_{yy}x^2y^2 + 2Q_{xx}Q_{zz}x^2z^2\right. & \\
\left.+2Q_{yy}Q_{zz}y^2z^2\right) &= \frac{4q_0^2 a^4}{r^2}\left(x^4 + y^4 + 4z^4 + 2x^2y^2 - 4x^2z^2 - 4y^2z^2\right) \\
&= \frac{4q_0^2 a^4}{r^2}\left(x^2 + y^2 - 2z^2\right)^2 = 4q_0^2 a^4 r^2 \left(1 - 3\cos^2\theta\right)^2
\end{aligned} \quad [3.119]$$

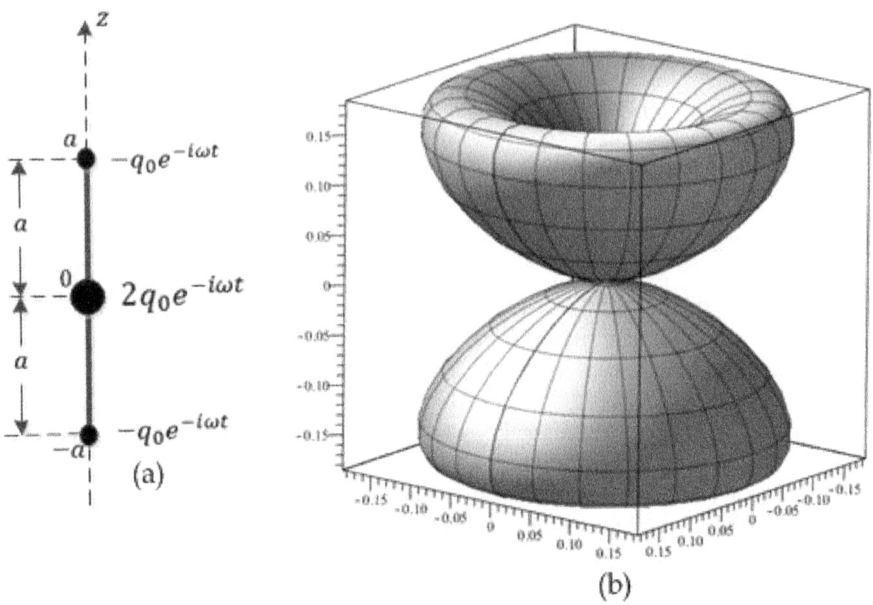

Figure 3.3. *(a) Example of the quadrupole moment with linear oscillation. (b) Spatial distribution of the radiated power (per unit of the solid angle). For a color version of this figure, see www.iste.co.uk/abdesselam/electrodynamics3.zip*

Taking into account the earlier results, we obtain:

$$\begin{aligned}
\frac{dP}{d\Omega} &= \frac{\mu_0 \omega^6}{1152\pi^2 c^3} 4a^4 q_0^2 \left(1 + 3\cos^2\theta - 1 + 6\cos^2\theta - 9\cos^4\theta\right) \\
&= \frac{\mu_0 \omega^6}{32\pi^2 c^3} a^4 q_0^2 \cos^2\theta \sin^2\theta
\end{aligned} \quad [3.120]$$

It is clear that the distribution (power) is independent of the angle φ, since the system is rotationally invariant around the Oz axis. The total average radiated power from the quadrupole moment can be obtained directly from equation [3.117]:

$$P = \frac{\mu_0 \omega^6}{1440\pi c^3}\left(4a^4 q_0^2 + 4a^4 q_0^2 + 16a^4 q_0^2\right) = \frac{\mu_0 \omega^6}{60\pi c^3} a^4 q_0^2 \quad [3.121]$$

This result can also be derived by directly integrating $\frac{dP}{d\Omega}$ in all space.

3.2.3. Multipole radiation of order $l \geq 2$

The evaluation of multipole terms for $l \geq 2$ using the previous methods becomes increasingly difficult, particularly when trying to separate the contributions of magnetic and electric moments. To compute these terms, we will introduce, in what follows, a more general approach based on the results of section 3.6 of Volume 2. Let us first summarize the key points of that section:

1) The 3D spherical solutions of vector wave equations (i.e. for $\vec{\hat{\mathbf{E}}}$ and $\vec{\hat{\mathbf{B}}}$) in vacuum can be derived from the Debye potential:

$$\psi_l^m\left(\vec{\mathbf{r}}\right) = \left(a_l^m j_l\left(\mathrm{k}r\right) + b_l^m n_l\left(\mathrm{k}r\right)\right) Y_l^m\left(\theta, \varphi\right) \quad [3.122]$$

2) There are two propagation modes (TE and TM modes). The TE fields are obtained from the expressions:

$$\begin{aligned} \vec{\hat{\mathbf{E}}}_{\mathrm{TE}} &= \mathrm{ik}\vec{\nabla} \times \vec{\mathbf{r}}\psi & \text{(i)} \\ \vec{\hat{\mathbf{B}}}_{\mathrm{TE}} &= \frac{1}{c}\vec{\nabla} \times \left(\vec{\nabla} \times \vec{\mathbf{r}}\psi\right) & \text{(ii)} \end{aligned} \quad [3.123]$$

while the TM fields are given by:

$$\begin{aligned} \vec{\hat{\mathbf{E}}}_{\mathrm{TM}} &= \vec{\nabla} \times \left(\vec{\nabla} \times \vec{\mathbf{r}}\psi\right) & \text{(i)} \\ \vec{\hat{\mathbf{B}}}_{\mathrm{TM}} &= -\frac{\mathrm{ik}}{c}\vec{\nabla} \times \vec{\mathbf{r}}\psi & \text{(ii)} \end{aligned} \quad [3.124]$$

To obtain outgoing waves of the form $\frac{e^{ikr}}{r}$, we take $b_l^m = \mathrm{i}a_l^m$ (i.e. the combination of $j_l\left(\mathrm{k}r\right)$ and $n_l\left(\mathrm{k}r\right)$ must produce $h_l^{(1)}\left(\mathrm{k}r\right)$).

3) In the TE mode (respectively, TM mode), all components of the fields $\vec{\hat{\mathbf{E}}}_{\mathrm{TE}}$ and $\vec{\hat{\mathbf{B}}}_{\mathrm{TE}}$ (respectively, $\vec{\hat{\mathbf{E}}}_{\mathrm{TM}}$ and $\vec{\hat{\mathbf{B}}}_{\mathrm{TM}}$) are determined from the radial component B_r^{TE} (respectively, E_r^{TM}). These radial components are written as:

$$\begin{aligned} B_r^{\mathrm{TE}} &= \frac{l\left(l+1\right)}{cr}\psi_l^m & \text{(i)} \\ E_r^{\mathrm{TM}} &= \frac{l\left(l+1\right)}{r}\psi_l^m & \text{(ii)} \end{aligned} \quad [3.125]$$

We therefore conclude that if E_r and B_r (or equivalently, H_r) are known, all other fields are completely determined.

In what follows, we will attempt to calculate the radiated components E_r and B_r from a harmonically oscillating current source (i.e. to find the relationship between ψ_l^m and the current source), thus obtaining the complete solution. If the source is described as an oscillating charge distribution, the continuity equation will be used to relate the current to the charge density. We will now establish the inhomogeneous wave equations for E_r and H_r. For harmonic fields, $\vec{\hat{\mathbf{H}}}$ is directly related to the current by:

$$\vec{\nabla} \times \vec{\hat{\mathbf{H}}} = \vec{\hat{\mathbf{J}}} - i\omega\varepsilon_0 \vec{\hat{\mathbf{E}}} \quad [3.126]$$

whereas $\vec{\hat{\mathbf{E}}}$ is given by:

$$\vec{\nabla} \times \vec{\hat{\mathbf{E}}} = i\omega\mu_0 \vec{\hat{\mathbf{H}}} \quad [3.127]$$

To derive an inhomogeneous wave equation for H_r, we proceed as follows:

$$\begin{aligned}
\Delta\left(\vec{\mathbf{r}} \cdot \vec{\hat{\mathbf{H}}}\right) &= 2\vec{\nabla} \cdot \vec{\hat{\mathbf{H}}} + \vec{\mathbf{r}} \cdot \Delta\vec{\hat{\mathbf{H}}} = \vec{\mathbf{r}} \cdot \Delta\vec{\hat{\mathbf{H}}} \\
&= \vec{\mathbf{r}} \cdot \left[-\vec{\nabla} \times \left(\vec{\nabla} \times \vec{\hat{\mathbf{H}}}\right) + \vec{\nabla}\left(\vec{\nabla} \cdot \vec{\hat{\mathbf{H}}}\right)\right] = -\vec{\mathbf{r}} \cdot \left[\vec{\nabla} \times \left(\vec{\nabla} \times \vec{\hat{\mathbf{H}}}\right)\right] \\
&= -\vec{\mathbf{r}} \cdot \left[\vec{\nabla} \times \left(\vec{\hat{\mathbf{J}}} - i\omega\varepsilon_0 \vec{\hat{\mathbf{E}}}\right)\right] = -\vec{\mathbf{r}} \cdot \left(\vec{\nabla} \times \vec{\hat{\mathbf{J}}}\right) + i\omega\varepsilon_0 \vec{\mathbf{r}} \cdot \left(\vec{\nabla} \times \vec{\hat{\mathbf{E}}}\right) \\
&= -\vec{\mathbf{r}} \cdot \left(\vec{\nabla} \times \vec{\hat{\mathbf{J}}}\right) - \omega^2\varepsilon_0\mu_0 \left(\vec{\mathbf{r}} \cdot \vec{\hat{\mathbf{H}}}\right)
\end{aligned} \quad [3.128]$$

which yields:

$$\left(\Delta + \frac{\omega^2}{c^2}\right)\left(\vec{\mathbf{r}} \cdot \vec{\hat{\mathbf{H}}}\right) = -\vec{\mathbf{r}} \cdot \left(\vec{\nabla} \times \vec{\hat{\mathbf{J}}}\right) \quad [3.129]$$

This latter equation admits as a solution:

$$\left(\vec{\mathbf{r}} \cdot \vec{\hat{\mathbf{H}}}\right) = \frac{1}{4\pi} \int_{\tau'} \frac{\vec{\mathbf{r}}\,' \cdot \left(\vec{\nabla}' \times \vec{\hat{\mathbf{J}}}\,'\right) e^{ik\left|\vec{\mathbf{r}} - \vec{\mathbf{r}}\,'\right|}}{\left|\vec{\mathbf{r}} - \vec{\mathbf{r}}\,'\right|} d\tau' \quad [3.130]$$

The same procedure can be repeated for $\vec{\mathbf{r}} \cdot \vec{\hat{\mathbf{E}}}$, except that now $\vec{\nabla} \cdot \vec{\hat{\mathbf{E}}}$ will not vanish. To avoid this problem, we use $\vec{\hat{\mathbf{E}}} - \frac{\vec{\hat{\mathbf{J}}}}{i\omega\varepsilon_0}$ instead of $\vec{\hat{\mathbf{E}}}$, since

$\vec{\nabla}\cdot\left(\vec{\hat{\mathbf{E}}}-\frac{\vec{\hat{\mathbf{J}}}}{\mathrm{i}\omega\varepsilon_0}\right)=0$. Repeating the same steps, we find the following:

$$
\begin{aligned}
&\Delta\left[\vec{\mathbf{r}}\cdot\left(\vec{\hat{\mathbf{E}}}-\frac{\vec{\hat{\mathbf{J}}}}{\mathrm{i}\omega\varepsilon_0}\right)\right]=2\vec{\nabla}\cdot\left(\vec{\hat{\mathbf{E}}}-\frac{\vec{\hat{\mathbf{J}}}}{\mathrm{i}\omega\varepsilon_0}\right)+\vec{\mathbf{r}}\cdot\Delta\left(\vec{\hat{\mathbf{E}}}-\frac{\vec{\hat{\mathbf{J}}}}{\mathrm{i}\omega\varepsilon_0}\right)\\
&=\vec{\mathbf{r}}\cdot\Delta\left(\vec{\hat{\mathbf{E}}}-\frac{\vec{\hat{\mathbf{J}}}}{\mathrm{i}\omega\varepsilon_0}\right)=\vec{\mathbf{r}}\cdot\left\{-\vec{\nabla}\times\left[\vec{\nabla}\times\left(\vec{\hat{\mathbf{E}}}-\frac{\vec{\hat{\mathbf{J}}}}{\mathrm{i}\omega\varepsilon_0}\right)\right]\right.\\
&\left.+\vec{\nabla}\left[\vec{\nabla}\cdot\left(\vec{\hat{\mathbf{E}}}-\frac{\vec{\hat{\mathbf{J}}}}{\mathrm{i}\omega\varepsilon_0}\right)\right]\right\}=-\vec{\mathbf{r}}\cdot\left\{\vec{\nabla}\times\left[\vec{\nabla}\times\left(\vec{\hat{\mathbf{E}}}-\frac{\vec{\hat{\mathbf{J}}}}{\mathrm{i}\omega\varepsilon_0}\right)\right]\right\}\\
&=-\vec{\mathbf{r}}\cdot\left[\vec{\nabla}\times\left(\vec{\nabla}\times\vec{\hat{\mathbf{E}}}\right)-\vec{\nabla}\times\left(\vec{\nabla}\times\frac{\vec{\hat{\mathbf{J}}}}{\mathrm{i}\omega\varepsilon_0}\right)\right]\\
&=-\vec{\mathbf{r}}\cdot\left[\vec{\nabla}\times\left(\mathrm{i}\omega\mu_0\vec{\hat{\mathbf{H}}}\right)-\vec{\nabla}\times\left(\vec{\nabla}\times\frac{\vec{\hat{\mathbf{J}}}}{\mathrm{i}\omega\varepsilon_0}\right)\right]\\
&=-\vec{\mathbf{r}}\cdot\left[\mathrm{i}\omega\mu_0\left(\vec{\hat{\mathbf{J}}}-\mathrm{i}\omega\varepsilon_0\vec{\hat{\mathbf{E}}}\right)-\vec{\nabla}\times\left(\vec{\nabla}\times\frac{\vec{\hat{\mathbf{J}}}}{\mathrm{i}\omega\varepsilon_0}\right)\right]\\
&=\omega^2\varepsilon_0\mu_0\,\vec{\mathbf{r}}\cdot\left(\vec{\hat{\mathbf{E}}}-\frac{\vec{\hat{\mathbf{J}}}}{\mathrm{i}\omega\varepsilon_0}\right)+\vec{\mathbf{r}}\cdot\left[\vec{\nabla}\times\left(\vec{\nabla}\times\frac{\vec{\hat{\mathbf{J}}}}{\mathrm{i}\omega\varepsilon_0}\right)\right]
\end{aligned}
\quad [3.131]
$$

We thus obtain the inhomogeneous wave equation:

$$\left(\Delta+\frac{\omega^2}{c^2}\right)\left[\vec{\mathbf{r}}\cdot\left(\vec{\hat{\mathbf{E}}}-\frac{\vec{\hat{\mathbf{J}}}}{\mathrm{i}\omega\varepsilon_0}\right)\right]=-\frac{\mathrm{i}}{\omega\varepsilon_0}\vec{\mathbf{r}}\cdot\left[\vec{\nabla}\times\left(\vec{\nabla}\times\vec{\hat{\mathbf{J}}}\right)\right] \quad [3.132]$$

whose solution is written as:

$$\vec{\mathbf{r}}\cdot\vec{\hat{\mathbf{E}}}=\frac{\vec{\mathbf{r}}\cdot\vec{\hat{\mathbf{J}}}}{\mathrm{i}\omega\varepsilon_0}+\frac{\mathrm{i}}{4\pi\omega\varepsilon_0}\int_{\tau'}\frac{\vec{\mathbf{r}}'\cdot\left[\vec{\nabla}'\times\left(\vec{\nabla}'\times\vec{\hat{\mathbf{J}}}'\right)\right]e^{\mathrm{ik}\left|\vec{\mathbf{r}}-\vec{\mathbf{r}}'\right|}}{\left|\vec{\mathbf{r}}-\vec{\mathbf{r}}'\right|}d\tau' \quad [3.133]$$

The integral in expression [3.133] is recognized as containing the Laplacian of the solenoidal current $\Delta\vec{\hat{\mathbf{J}}}_s=-\vec{\nabla}\times\left(\vec{\nabla}\times\vec{\hat{\mathbf{J}}}\right)$. Thus, formula [3.133] becomes:

$$\vec{\mathbf{r}} \cdot \hat{\vec{\mathbf{E}}} = \frac{1}{\mathrm{i}\omega\varepsilon_0}\left[\vec{\mathbf{r}} \cdot \hat{\vec{\mathbf{J}}} + \frac{1}{4\pi}\int_{\tau'} \frac{\left(\vec{\mathbf{r}}' \cdot \Delta' \hat{\vec{\mathbf{J}}}'_s\right) e^{\mathrm{ik}\left|\vec{\mathbf{r}} - \vec{\mathbf{r}}'\right|}}{\left|\vec{\mathbf{r}} - \vec{\mathbf{r}}'\right|} d\tau'\right] \qquad [3.134]$$

Now, substituting $\hat{\vec{\mathbf{J}}}$ with $\hat{\vec{\mathbf{J}}}_l + \hat{\vec{\mathbf{J}}}_s$ in equation [3.130] and considering that $\vec{\nabla} \times \hat{\vec{\mathbf{J}}}_l = 0$, we also obtain:

$$\vec{\mathbf{r}} \cdot \hat{\vec{\mathbf{H}}} = \frac{1}{4\pi}\int_{\tau'} \frac{\vec{\mathbf{r}}' \cdot \left(\vec{\nabla}' \times \hat{\vec{\mathbf{J}}}'_s\right) e^{\mathrm{ik}\left|\vec{\mathbf{r}} - \vec{\mathbf{r}}'\right|}}{\left|\vec{\mathbf{r}} - \vec{\mathbf{r}}'\right|} d\tau' \qquad [3.135]$$

We therefore conclude that the radiation field depends only on the solenoidal current.

Now using expansion [3.65] and assuming that $r' < r$, equation [3.130] becomes:

$$\vec{\mathbf{r}} \cdot \hat{\vec{\mathbf{H}}} = \mathrm{ik}\sum_{l,m} h_l^{(1)}(\mathrm{k}r)\, Y_l^m(\theta,\varphi) \int_{\tau'} \vec{\mathbf{r}}' \cdot \left[\vec{\nabla}' \times \hat{\vec{\mathbf{J}}}'\right] j_l(\mathrm{k}r')\, Y_l^{m\star}(\theta',\varphi')\, d\tau'$$

[3.136]

Comparing this result with the Debye potential [3.125]:

$$\begin{aligned}\vec{\mathbf{r}} \cdot \hat{\vec{\mathbf{H}}} &= \frac{l(l+1)}{c\mu_0}\sum_{l,m} a_l^m|_{\mathrm{TE}}\, h_l^{(1)}(\mathrm{k}r)\, Y_l^m(\theta,\varphi) \\ &= l(l+1)\sqrt{\frac{\varepsilon_0}{\mu_0}}\sum_{l,m} a_l^m|_{\mathrm{TE}}\, h_l^{(1)}(\mathrm{k}r)\, Y_l^m(\theta,\varphi)\end{aligned} \qquad [3.137]$$

we obtain the coefficient $a_l^m|_{\mathrm{TE}}$ for the TE wave (generated by the current $\hat{\vec{\mathbf{J}}}$):

$$a_l^m|_{\mathrm{TE}} = \frac{\mathrm{ik}}{l(l+1)}\sqrt{\frac{\mu_0}{\varepsilon_0}}\int_{\tau'} \vec{\mathbf{r}}' \cdot \left[\vec{\nabla}' \times \hat{\vec{\mathbf{J}}}'\right] j_l(\mathrm{k}r')\, Y_l^{m\star}(\theta',\varphi')\, d\tau'$$

[3.138]

The integral appearing in expression [3.138] can be simplified using the identity $\xi\vec{\mathbf{r}}' \cdot \left(\vec{\nabla}' \times \hat{\vec{\mathbf{J}}}'\right) = \hat{\vec{\mathbf{J}}}' \cdot \left(\vec{\nabla}' \times \vec{\mathbf{r}}'\xi\right) + \vec{\nabla}' \cdot \left(\hat{\vec{\mathbf{J}}}' \times \vec{\mathbf{r}}'\xi\right)$. The term $\vec{\nabla}' \cdot \left(\hat{\vec{\mathbf{J}}}' \times \vec{\mathbf{r}}'\xi\right)$ in the volume integral can be converted to a surface integral surrounding the source, and since $\vec{\hat{\mathbf{J}}}$ vanishes outside the source, this term disappears. Consequently, the coefficient $a_l^m|_{\text{TE}}$ takes its final form:

$$a_l^m|_{\text{TE}} = \frac{\mathrm{ik}}{l(l+1)}\sqrt{\frac{\mu_0}{\varepsilon_0}}\int_{\tau'} \vec{\hat{\mathbf{J}}}'\cdot\left[\vec{\nabla}'\times\vec{\mathbf{r}}' j_l(\mathrm{k}r')\, Y_l^{m\star}(\theta',\varphi')\right] d\tau' \quad [3.139]$$

In the same way, according to equation [3.133], outside the source we have:

$$\vec{\mathbf{r}}\cdot\vec{\hat{\mathbf{E}}} = -\frac{\mathrm{k}}{\omega\varepsilon_0}\sum_{l,m} h_l^{(1)}(\mathrm{k}r)\, Y_l^m(\theta,\varphi)\int_{\tau'}\vec{\mathbf{r}}'\cdot\left[\vec{\nabla}'\times\left(\vec{\nabla}'\times\vec{\hat{\mathbf{J}}}'\right)\right]$$

$$j_l(\mathrm{k}r')\, Y_l^{m\star}(\theta',\varphi')\, d\tau' = -\sqrt{\frac{\mu_0}{\varepsilon_0}}\sum_{l,m} h_l^{(1)}(\mathrm{k}r)\, Y_l^m(\theta,\varphi)$$

$$\int_{\tau'}\vec{\hat{\mathbf{J}}}'\cdot\left[\vec{\nabla}'\times\left(\vec{\nabla}'\times\vec{\mathbf{r}}' j_l(\mathrm{k}r')\, Y_l^{m\star}(\theta',\varphi')\right)\right] d\tau'$$

$$= \sum_{l,m} a_l^m|_{\text{TM}}\, l(l+1)\, h_l^{(1)}(\mathrm{k}r)\, Y_l^m(\theta,\varphi) \quad [3.140]$$

This corresponds to the TM component of the electric field given by equation [3.125], which allows us to determine the coefficient $a_l^m|_{\text{TM}}$ for the TM wave:

$$a_l^m|_{\text{TM}} = -\frac{1}{l(l+1)}\sqrt{\frac{\mu_0}{\varepsilon_0}}\int_{\tau'}\vec{\hat{\mathbf{J}}}'\cdot\left[\vec{\nabla}'\times\left(\vec{\nabla}'\times\vec{\mathbf{r}}' j_l(\mathrm{k}r')\, Y_l^{m\star}(\theta',\varphi')\right)\right] d\tau' \quad [3.141]$$

When $\mathrm{k}r'$ is very small, we can always replace $j_l(\mathrm{k}r')$ with its first-order approximation $\frac{(\mathrm{k}r')^l}{(2l+1)!!}$. Radiation fields from an arbitrary source can now be expressed as follows:

$$\vec{\hat{\mathbf{E}}} = \sum_{l,m} a_l^m|_{\text{TE}}\, \vec{\hat{\mathbf{E}}}_{lm}^{\text{TE}} + a_l^m|_{\text{TM}}\, \vec{\hat{\mathbf{E}}}_{lm}^{\text{TM}}$$

$$\vec{\hat{\mathbf{H}}} = \sum_{l,m} a_l^m|_{\text{TE}}\, \vec{\hat{\mathbf{H}}}_{lm}^{\text{TE}} + a_l^m|_{\text{TM}}\, \vec{\hat{\mathbf{H}}}_{lm}^{\text{TM}} \quad [3.142]$$

The expressions for the fields $\vec{\hat{\mathbf{E}}}_{lm}^{\text{TE}}$, $\vec{\hat{\mathbf{H}}}_{lm}^{\text{TE}}$, $\vec{\hat{\mathbf{E}}}_{lm}^{\text{TM}}$, and $\vec{\hat{\mathbf{H}}}_{lm}^{\text{TM}}$ are obtained by substituting the Debye potential [3.122] (with $a_l^m = 1$ and $b_l^m = \mathrm{i}$) into equations [3.124] and [3.125]:

$$
\begin{aligned}
\vec{\hat{\mathbf{E}}}_{lm}^{\mathrm{TE}} &= \mathrm{i}k\vec{\nabla} \times \left[\vec{\mathbf{r}} h_l^{(1)}(\mathrm{k}r) Y_l^m(\theta,\varphi)\right] = \mathrm{i}k h_l^{(1)}(\mathrm{k}r)\left[\frac{1}{\sin\theta}\frac{\partial Y_l^m(\theta,\varphi)}{\partial\varphi}\vec{\mathbf{e}}_\theta \right.\\
&\quad \left. -\frac{\partial Y_l^m(\theta,\varphi)}{\partial\theta}\vec{\mathbf{e}}_\varphi\right] = -\mathrm{i}k h_l^{(1)}(\mathrm{k}r)\,\vec{\mathbf{e}}_r \times \left[\frac{\partial Y_l^m(\theta,\varphi)}{\partial\theta}\vec{\mathbf{e}}_\theta \right.\\
&\quad \left. +\frac{1}{\sin\theta}\frac{\partial Y_l^m(\theta,\varphi)}{\partial\varphi}\vec{\mathbf{e}}_\varphi\right] = -\mathrm{i}k h_l^{(1)}(\mathrm{k}r)\,\vec{\mathbf{r}} \times \vec{\nabla} Y_l^m(\theta,\varphi)\\
\vec{\hat{\mathbf{H}}}_{lm}^{\mathrm{TE}} &= \sqrt{\frac{\varepsilon_0}{\mu_0}}\vec{\nabla} \times \left\{\vec{\nabla} \times \left[\vec{\mathbf{r}} h_l^{(1)}(\mathrm{k}r) Y_l^m(\theta,\varphi)\right]\right\}\\
&= \sqrt{\frac{\varepsilon_0}{\mu_0}}\vec{\nabla} \times \left\{h_l^{(1)}(\mathrm{k}r)\left[\frac{1}{\sin\theta}\frac{\partial Y_l^m(\theta,\varphi)}{\partial\varphi}\vec{\mathbf{e}}_\theta - \frac{\partial Y_l^m(\theta,\varphi)}{\partial\theta}\vec{\mathbf{e}}_\varphi\right]\right\}\\
&= \sqrt{\frac{\varepsilon_0}{\mu_0}}\left\{h_l^{(1)}(\mathrm{k}r)\left[-\frac{\partial}{\partial\theta}\left(\sin\theta\frac{\partial Y_l^m(\theta,\varphi)}{\partial\theta}\right) - \frac{1}{\sin\theta}\frac{\partial^2 Y_l^m(\theta,\varphi)}{\partial\varphi^2}\right]\frac{\vec{\mathbf{e}}_r}{r\sin\theta}\right.\\
&\quad \left. +\frac{\partial}{\partial r}\left[r h_l^{(1)}(\mathrm{k}r)\frac{\partial Y_l^m(\theta,\varphi)}{\partial\theta}\right]\frac{\vec{\mathbf{e}}_\theta}{r} + \frac{\partial}{\partial r}\left[\frac{r h_l^{(1)}(\mathrm{k}r)}{\sin\theta}\frac{\partial Y_l^m(\theta,\varphi)}{\partial\varphi}\right]\frac{\vec{\mathbf{e}}_\varphi}{r}\right\}\\
&= \sqrt{\frac{\varepsilon_0}{\mu_0}}\left[\frac{l(l+1)h_l^{(1)}(\mathrm{k}r)Y_l^m(\theta,\varphi)}{r}\vec{\mathbf{e}}_r + \frac{d\left(r h_l^{(1)}(\mathrm{k}r)\right)}{dr}\vec{\nabla} Y_l^m(\theta,\varphi)\right]
\end{aligned}
\tag{3.143}
$$

and:

$$
\begin{aligned}
\vec{\hat{\mathbf{E}}}_{lm}^{\mathrm{TM}} &= \frac{l(l+1)h_l^{(1)}(\mathrm{k}r)Y_l^m(\theta,\varphi)}{r}\vec{\mathbf{e}}_r + \frac{d\left(r h_l^{(1)}(\mathrm{k}r)\right)}{dr}\vec{\nabla} Y_l^m(\theta,\varphi)\\
\vec{\hat{\mathbf{H}}}_{lm}^{\mathrm{TM}} &= \mathrm{i}k\sqrt{\frac{\varepsilon_0}{\mu_0}}h_l^{(1)}(\mathrm{k}r)\,\vec{\mathbf{r}} \times \vec{\nabla} Y_l^m(\theta,\varphi)
\end{aligned}
\tag{3.144}
$$

3.2.4. *Energy and angular momentum of multipolar fields*

To calculate the energy radiated by an oscillating source, we simply use the large-scale asymptotic form (i.e. for $\mathrm{k}r \gg 1$) of $h_l^{(1)}(\mathrm{k}r) = (-\mathrm{i})^{l+1}\frac{e^{\mathrm{i}kr}}{\mathrm{k}r}$.

Using this simplification and keeping only the first-order terms, the fields $\vec{\hat{\mathbf{E}}}_{lm}^{\mathrm{TE}}$, $\vec{\hat{\mathbf{H}}}_{lm}^{\mathrm{TE}}$, $\vec{\hat{\mathbf{E}}}_{lm}^{\mathrm{TM}}$ and $\vec{\hat{\mathbf{H}}}_{lm}^{\mathrm{TM}}$ are expressed as:

$$
\begin{aligned}
\vec{\hat{\mathbf{E}}}_{lm}^{\mathrm{TE}} &= -\mathrm{i}^{-l}\frac{e^{\mathrm{i}kr}}{r}\vec{\mathbf{r}} \times \vec{\nabla} Y_l^m(\theta,\varphi) \quad \text{(i)}\\
\vec{\hat{\mathbf{H}}}_{lm}^{\mathrm{TE}} &= \mathrm{i}^{-l}\sqrt{\frac{\varepsilon_0}{\mu_0}}e^{\mathrm{i}kr}\vec{\nabla} Y_l^m(\theta,\varphi) \quad \text{(ii)}
\end{aligned}
\qquad [3.145]
$$

and:

$$
\begin{aligned}
\vec{\hat{\mathbf{E}}}_{lm}^{\mathrm{TM}} &= \mathrm{i}^{-l}e^{\mathrm{i}kr}\vec{\nabla} Y_l^m(\theta,\varphi) \quad \text{(i)}\\
\vec{\hat{\mathbf{H}}}_{lm}^{\mathrm{TM}} &= \mathrm{i}^{-l}\sqrt{\frac{\varepsilon_0}{\mu_0}}\frac{e^{\mathrm{i}kr}}{r}\vec{\mathbf{r}} \times \vec{\nabla} Y_l^m(\theta,\varphi) \quad \text{(ii)}
\end{aligned}
\qquad [3.146]
$$

The time-averaged Poynting vector is identical for both modes:

$$
\begin{aligned}
\langle\vec{\mathbf{\Pi}}\rangle &= \frac{1}{2}\Re\left[\vec{\hat{\mathbf{E}}}_{lm} \times \vec{\hat{\mathbf{H}}}_{lm}^{\star}\right] = \frac{1}{2r}\sqrt{\frac{\varepsilon_0}{\mu_0}}\left[\vec{\mathbf{r}} \times \vec{\nabla} Y_l^m(\theta,\varphi)\right] \times \vec{\nabla} Y_l^{m\star}(\theta,\varphi)\\
&= \frac{1}{2}\sqrt{\frac{\varepsilon_0}{\mu_0}}\left|\vec{\nabla} Y_l^m(\theta,\varphi)\right|^2\vec{\mathbf{e}}_r
\end{aligned}
\qquad [3.147]
$$

The total radiated power P from the source can be calculated by integrating equation [3.147] over the surface of a large sphere of radius R centered on the source:

$$
\begin{aligned}
P &= \oint\langle\vec{\mathbf{\Pi}}\rangle\cdot d\vec{\mathbf{S}} = \frac{1}{2}\sqrt{\frac{\varepsilon_0}{\mu_0}}\oint\left|\vec{\nabla} Y_l^m(\theta,\varphi)\right|^2\vec{\mathbf{e}}_r\cdot d\vec{\mathbf{S}}\\
&= \frac{1}{2}\sqrt{\frac{\varepsilon_0}{\mu_0}}\oint\left[\frac{1}{R^2}\frac{\partial Y_l^m}{\partial\theta}\frac{\partial Y_l^{m\star}}{\partial\theta} + \frac{1}{R^2\sin^2\theta}\frac{\partial Y_l^m}{\partial\varphi}\frac{\partial Y_l^{m\star}}{\partial\varphi}\right]R^2 d\Omega\\
&= \frac{1}{2}\sqrt{\frac{\varepsilon_0}{\mu_0}}\oint\left[\frac{\partial Y_l^m}{\partial\theta}\frac{\partial Y_l^{m\star}}{\partial\theta} + \frac{\cos^2\theta}{\sin^2\theta}\frac{\partial Y_l^m}{\partial\varphi}\frac{\partial Y_l^{m\star}}{\partial\varphi} + \frac{\partial Y_l^m}{\partial\varphi}\frac{\partial Y_l^{m\star}}{\partial\varphi}\right]d\Omega\\
&= \frac{1}{2}\sqrt{\frac{\varepsilon_0}{\mu_0}}\oint\left[\frac{1}{4}\left(e^{-\mathrm{i}\varphi}L_+Y_l^m + e^{\mathrm{i}\varphi}L_-Y_l^m\right)\left(e^{\mathrm{i}\varphi}L_+Y_l^{m\star} + e^{-\mathrm{i}\varphi}L_-Y_l^{m\star}\right) + \right.\\
&\left.\frac{1}{4}\left(e^{-\mathrm{i}\varphi}L_+Y_l^m - e^{\mathrm{i}\varphi}L_-Y_l^m\right)\left(e^{\mathrm{i}\varphi}L_+Y_l^{m\star} - e^{-\mathrm{i}\varphi}L_-Y_l^{m\star}\right) + m^2Y_l^mY_l^{m\star}\right]d\Omega\\
&= \frac{1}{2}\sqrt{\frac{\varepsilon_0}{\mu_0}}\oint\left[\frac{1}{2}\left(|L_+Y_l^m|^2 + |L_-Y_l^m|^2\right) + m^2\left|Y_l^{m+1}\right|^2\right]d\Omega\\
&= \frac{1}{2}\sqrt{\frac{\varepsilon_0}{\mu_0}}\oint\left[\frac{1}{2}(l-m)(l+m+1)\left|Y_l^{m+1}\right|^2 + (l+m)(l-m+1)\left|Y_l^{m-1}\right|^2 + \right.\\
&\left. m^2\left|Y_l^m\right|^2\right]d\Omega = \frac{1}{2}\sqrt{\frac{\varepsilon_0}{\mu_0}}\left[\frac{1}{2}(l-m)(l+m+1) + (l+m)(l-m+1) + m^2\right]\\
&= \frac{1}{2}\sqrt{\frac{\varepsilon_0}{\mu_0}}l(l+1)
\end{aligned}
\qquad [3.148]
$$

where the identities $L_z = -\mathrm{i}\frac{\partial}{\partial\varphi}$, $L_\pm = \pm e^{\pm \mathrm{i}\varphi}\left(\frac{\partial}{\partial\theta} \pm \mathrm{i}\cot\theta\frac{\partial}{\partial\varphi}\right)$, $L_z Y_l^m(\theta,\varphi) = mY_l^m(\theta,\varphi)$ and $L_\pm Y_l^m(\theta,\varphi) = \sqrt{(l \mp m)(l \pm m + 1)}Y_l^{m\pm 1}(\theta,\varphi)$ have been employed.

The angular momentum radiated through a spherical surface of radius r during the time interval dt is precisely the angular momentum contained within a shell of thickness dt:

$$d\vec{\mathbf{L}} = \oint_{\text{sphere}} \left(\vec{\mathbf{r}} \times \vec{\mathcal{P}}\right) r^2 c dt d\Omega = c\oint_{\text{sphere}} \vec{\mathbf{r}} \times \left(\frac{\vec{\mathbf{E}} \times \vec{\mathbf{H}}}{c^2}\right) r^2 c d\Omega dt$$
$$= \frac{r^2}{c}\oint_{\text{sphere}} \left[\vec{\mathbf{E}}\left(\vec{\mathbf{r}} \cdot \vec{\mathbf{H}}\right) - \vec{\mathbf{H}}\left(\vec{\mathbf{r}} \cdot \vec{\mathbf{E}}\right)\right] d\Omega dt \qquad [3.149]$$

Thus, the angular momentum radiated per unit time will be:

$$\frac{d\vec{\mathbf{L}}}{dt} = \frac{r^2}{c}\oint_{\text{sphere}} \left[\vec{\mathbf{E}}\left(\vec{\mathbf{r}} \cdot \vec{\mathbf{H}}\right) - \vec{\mathbf{H}}\left(\vec{\mathbf{r}} \cdot \vec{\mathbf{E}}\right)\right] d\Omega \qquad [3.150]$$

Averaging over one period, the final result is:

$$\langle\frac{d\vec{\mathbf{L}}}{dt}\rangle = \frac{r^2}{2c}\oint_{\text{sphere}} \Re\left[\vec{\hat{\mathbf{E}}}\left(\vec{\mathbf{r}} \cdot \vec{\hat{\mathbf{H}}}^\star\right) - \vec{\hat{\mathbf{H}}}^\star\left(\vec{\mathbf{r}} \cdot \vec{\hat{\mathbf{E}}}\right)\right] d\Omega \qquad [3.151]$$

The results are identical for both the TE and TM modes. To evaluate the integral [3.151], we select the TM mode. For this radiation, we have:

$$\begin{aligned}
&\vec{\hat{\mathbf{H}}}_{lm}^{\mathrm{TM}} \cdot \vec{\mathbf{r}} = 0\\
&\vec{\hat{\mathbf{E}}}_{lm}^{\mathrm{TM}} \cdot \vec{\mathbf{r}} = l(l+1)\,\mathrm{i}^{-l-1}\frac{e^{\mathrm{i}kr}}{kr}Y_l^m(\theta,\varphi)
\end{aligned} \qquad [3.152]$$

Substituting expressions [3.144] and [3.152] into [3.151], we obtain:

$$\langle\frac{d\vec{\mathbf{L}}}{dt}\rangle = \frac{l(l+1)}{2kc}\sqrt{\frac{\varepsilon_0}{\mu_0}}\oint_{\text{sphere}} \Re\left[\mathrm{i}\left(\vec{\mathbf{r}} \times \vec{\nabla}Y_l^{m\star}(\theta,\varphi)\right)Y_l^m(\theta,\varphi)\right] d\Omega$$
$$= \frac{l(l+1)}{2kc}\sqrt{\frac{\varepsilon_0}{\mu_0}}\oint_{\text{sphere}} \Re\left[\mathrm{i}Y_l^m\frac{\partial Y_l^{m\star}}{\partial\theta}\vec{\mathbf{e}}_\varphi - \frac{\mathrm{i}}{\sin\theta}Y_l^m\frac{\partial Y_l^{m\star}}{\partial\varphi}\vec{\mathbf{e}}_\theta\right] d\Omega$$
$$= \frac{l(l+1)}{2kc}\sqrt{\frac{\varepsilon_0}{\mu_0}}\oint_{\text{sphere}} \Re\left[-\frac{\mathrm{i}\cos\theta}{\sin\theta}\cos\varphi Y_l^m\frac{\partial Y_l^{m\star}}{\partial\varphi}\vec{\mathbf{i}}\right.$$
$$\left.-\frac{\mathrm{i}\cos\theta}{\sin\theta}\sin\varphi Y_l^m\frac{\partial Y_l^{m\star}}{\partial\varphi}\vec{\mathbf{j}} + \mathrm{i}Y_l^m\frac{\partial Y_l^{m\star}}{\partial\varphi}\vec{\mathbf{k}}\right] d\Omega$$
$$= \frac{l(l+1)}{2kc}\sqrt{\frac{\varepsilon_0}{\mu_0}}m\vec{\mathbf{k}}\oint_{\text{sphere}} |Y_l^m|^2 d\Omega = \frac{l(l+1)}{2kc}\sqrt{\frac{\varepsilon_0}{\mu_0}}m\vec{\mathbf{k}} \qquad [3.153]$$

We conclude that the radiated angular momentum is oriented along the Oz axis:

$$\langle\frac{dL_z}{dt}\rangle = \frac{m}{2\mathrm{k}c}\sqrt{\frac{\varepsilon_0}{\mu_0}}l\,(l+1) \qquad [3.154]$$

The ratio between the radiated angular momentum and the radiated power $P = \frac{dW}{dt}$ is given by:

$$\frac{\langle\frac{dL_z}{dt}\rangle}{\langle\frac{dW}{dt}\rangle} = \frac{m}{\mathrm{k}c} = \frac{m}{\omega} \qquad [3.155]$$

This result is consistent with quantum mechanics. According to quantum theory, atoms emit photons with energy $\hbar\omega$ and angular momentum proportional to $\hbar$.

3.3. Radiation from an extended source

When the dimensions of the system are not small compared to the wavelength of radiation, the multipole expansion of the vector potential is no longer valid, and the integral [3.48] must be evaluated directly. To this end, let us consider an infinitely thin conducting rod of length d. It is assumed that the current flowing through the rod is sinusoidal in time and has an amplitude that varies with z. The current density is therefore:

$$\overrightarrow{\mathbf{J}}(\overrightarrow{\mathbf{r}}) = \delta(x)\delta(y)I(z)e^{-\mathrm{i}\omega t}\overrightarrow{\mathbf{k}} \Longrightarrow \hat{\overrightarrow{\mathbf{J}}}(\overrightarrow{\mathbf{r}}) = \delta(x)\delta(y)I(z)\overrightarrow{\mathbf{k}} \qquad [3.156]$$

This system is called an *antenna*. The corresponding vector potential in the radiation zone is:

$$\hat{\overrightarrow{\mathbf{A}}}(\overrightarrow{\mathbf{r}}) = \frac{\mu_0}{4\pi}\frac{e^{\mathrm{i}\mathrm{k}r}}{r}\hat{\overrightarrow{\mathbf{N}}}\left(\overrightarrow{\mathrm{k}}\right) \qquad [3.157]$$

where:

$$\hat{\overrightarrow{\mathbf{N}}}\left(\overrightarrow{\mathrm{k}}\right) = \int_{\tau'} d\tau'\,\hat{\overrightarrow{\mathbf{J}}}\left(\overrightarrow{\mathbf{r}}'\right)e^{-\mathrm{i}\overrightarrow{\mathrm{k}}\cdot\overrightarrow{\mathbf{r}}'} = \overrightarrow{\mathbf{k}}\int dz' I\left(z'\right)e^{-\mathrm{i}\mathrm{k}z'\cos\theta} \qquad [3.158]$$

θ is the angle measured between the observation direction and the Oz axis. It can be assumed that the current in the rod results from a wave propagating in a waveguide and that its spatial dependence is sinusoidal in z, with a wavenumber $\mathrm{k} = \frac{\omega}{c}$. If the rod is fed with current at its center by a coaxial cable, as illustrated in Figure 3.4, the current must be zero at the ends and symmetric with respect to $z = 0$, since the two conductors of the coaxial feed cable must carry equal and opposite currents at every instant. We thus take:

$$I(z) = I_0 \sin \mathrm{k} \left(\frac{1}{2} d - |z| \right) \qquad [3.159]$$

By substituting [3.159] and [3.158], we obtain:

$$\begin{aligned}
\vec{\hat{\mathbf{N}}}\left(\vec{\mathrm{k}}\right) &= I_0 \vec{\mathbf{k}} \int_{-\frac{d}{2}}^{\frac{d}{2}} \sin\left(\frac{1}{2}\mathrm{k}d - \mathrm{k}|z|\right) e^{-\mathrm{i}\mathrm{k}z'\cos\theta} dz' \\
&= 2I_0 \vec{\mathbf{k}} \int_0^{\frac{d}{2}} \sin\left(\frac{1}{2}\mathrm{k}d - \mathrm{k}|z|\right) \cos\left(\mathrm{k}z'\cos\theta\right) dz' \\
&= I_0 \vec{\mathbf{k}} \int_0^{\frac{d}{2}} dz' \left\{ \sin\left[\frac{1}{2}\mathrm{k}d - \mathrm{k}z\left(1-\cos\theta\right)\right] + \sin\left[\frac{1}{2}\mathrm{k}d - \mathrm{k}z\left(1+\cos\theta\right)\right] \right\} \\
&= I_0 \vec{\mathbf{k}} \left\{ \frac{\cos\left[\frac{1}{2}\mathrm{k}d - \mathrm{k}z\left(1-\cos\theta\right)\right]}{\mathrm{k}\left(1-\cos\theta\right)} + \frac{\cos\left[\frac{1}{2}\mathrm{k}d - \mathrm{k}z\left(1+\cos\theta\right)\right]}{\mathrm{k}\left(1+\cos\theta\right)} \right\} \Bigg|_0^{\frac{d}{2}} \\
&= I_0 \vec{\mathbf{k}} \left[\frac{\cos\left(\frac{1}{2}\mathrm{k}d\cos\theta\right) - \cos\left(\frac{1}{2}\mathrm{k}d\right)}{\mathrm{k}\left(1-\cos\theta\right)} + \frac{\cos\left(\frac{1}{2}\mathrm{k}d\cos\theta\right) - \cos\left(\frac{1}{2}\mathrm{k}d\right)}{\mathrm{k}\left(1+\cos\theta\right)} \right] \\
&= \frac{2I_0}{\mathrm{k}} \left[\frac{\cos\left(\frac{1}{2}\mathrm{k}d\cos\theta\right) - \cos\left(\frac{1}{2}\mathrm{k}d\right)}{\sin^2\theta} \right] \vec{\mathbf{k}}
\end{aligned} \qquad [3.160]$$

and consequently:

$$\vec{\hat{\mathbf{A}}}\left(\vec{\mathbf{r}}\right) = \frac{\mu_0}{2\pi} \frac{e^{\mathrm{i}\mathrm{k}r}}{\mathrm{k}r} I_0 \left[\frac{\cos\left(\frac{1}{2}\mathrm{k}d\cos\theta\right) - \cos\left(\frac{1}{2}\mathrm{k}d\right)}{\sin^2\theta} \right] \vec{\mathbf{k}} \qquad [3.161]$$

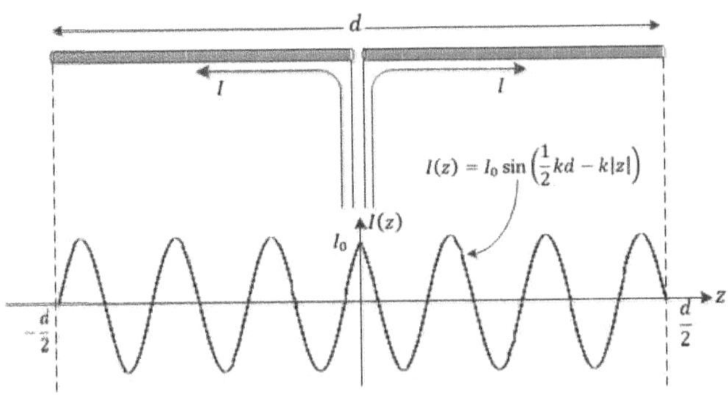

Figure 3.4. *Radiation from an extended source. For a color version of this figure, see www.iste.co.uk/abdesselam/electrodynamics3.zip*

In the radiation zone, we have:

$$\vec{\hat{\mathbf{H}}} = \frac{\mathrm{ik}}{4\pi}\frac{e^{\mathrm{i}kr}}{r}\vec{\mathbf{e}}_r \times \vec{\hat{\mathbf{N}}}\left(\vec{\mathrm{k}}\right) = -\frac{\mathrm{i}}{2\pi}\frac{e^{\mathrm{i}kr}}{r}I_0\left[\frac{\cos\left(\frac{1}{2}\mathrm{k}d\cos\theta\right) - \cos\left(\frac{1}{2}\mathrm{k}d\right)}{\sin\theta}\right]\vec{\mathbf{e}}_\varphi$$

$$\vec{\hat{\mathbf{E}}} = \sqrt{\frac{\mu_0}{\varepsilon_0}}\vec{\hat{\mathbf{H}}} \times \vec{\mathbf{e}}_r = -\sqrt{\frac{\mu_0}{\varepsilon_0}}\frac{\mathrm{i}}{2\pi}\frac{e^{\mathrm{i}kr}}{r}I_0\left[\frac{\cos\left(\frac{1}{2}\mathrm{k}d\cos\theta\right) - \cos\left(\frac{1}{2}\mathrm{k}d\right)}{\sin\theta}\right]\vec{\mathbf{e}}_\theta$$

[3.162]

Similarly, the radiated power, per unit of solid angle, is given as:

$$\frac{dP}{d\Omega} = \sqrt{\frac{\mu_0}{\varepsilon_0}}\frac{\mathrm{k}^2}{32\pi^2}\left|\vec{\mathbf{e}}_r \times \vec{\hat{\mathbf{N}}}\right|^2 = \sqrt{\frac{\mu_0}{\varepsilon_0}}\frac{I_0^2}{8\pi^2}\left[\frac{\cos\left(\frac{1}{2}\mathrm{k}d\cos\theta\right) - \cos\left(\frac{1}{2}\mathrm{k}d\right)}{\sin\theta}\right]^2$$

[3.163]

The angular distribution obviously depends on kd, as illustrated in Figure 3.5.

Here are a few examples:

1) Short antenna ($\mathrm{k}d \ll 1$): since $\cos\left(\frac{1}{2}\mathrm{k}d\cos\theta\right) - \cos\left(\frac{1}{2}\mathrm{k}d\right) \simeq \frac{1}{8}\left(\mathrm{k}d\right)^2 \left(1 - \cos^2\theta\right) = \frac{1}{8}\left(\mathrm{k}d\right)^2 \sin^2\theta$, it follows that:

$$\frac{dP}{d\Omega} = \sqrt{\frac{\mu_0}{\varepsilon_0}}\frac{I_0^2}{512\pi^2}\mathrm{k}^4 d^4 \sin^2\theta \qquad [3.164]$$

This corresponds to the radiation pattern of an electric dipole with a magnitude $p_0 = \frac{I_0 d^2}{4c}$.

2) If $\mathrm{k}d = \pi$, that is, $d = \frac{\lambda}{2}$ (half-wave antenna), therefore:

$$\frac{dP}{d\Omega} = \sqrt{\frac{\mu_0}{\varepsilon_0}}\frac{I_0^2}{8\pi^2}\frac{\cos^2\left(\frac{\pi}{2}\cos\theta\right)}{\sin^2\theta} \qquad [3.165]$$

3) If $\mathrm{k}d = 2\pi$, that is, $d = \lambda$, equation [3.163] reduces to:

$$\frac{dP}{d\Omega} = \sqrt{\frac{\mu_0}{\varepsilon_0}}\frac{I_0^2}{8\pi^2}\left[\frac{\cos\left(\pi\cos\theta\right) + 1}{\sin\theta}\right]^2 = \sqrt{\frac{\mu_0}{\varepsilon_0}}\frac{I_0^2}{2\pi^2}\frac{\cos^4\left(\frac{\pi}{2}\cos\theta\right)}{\sin^2\theta} \qquad [3.166]$$

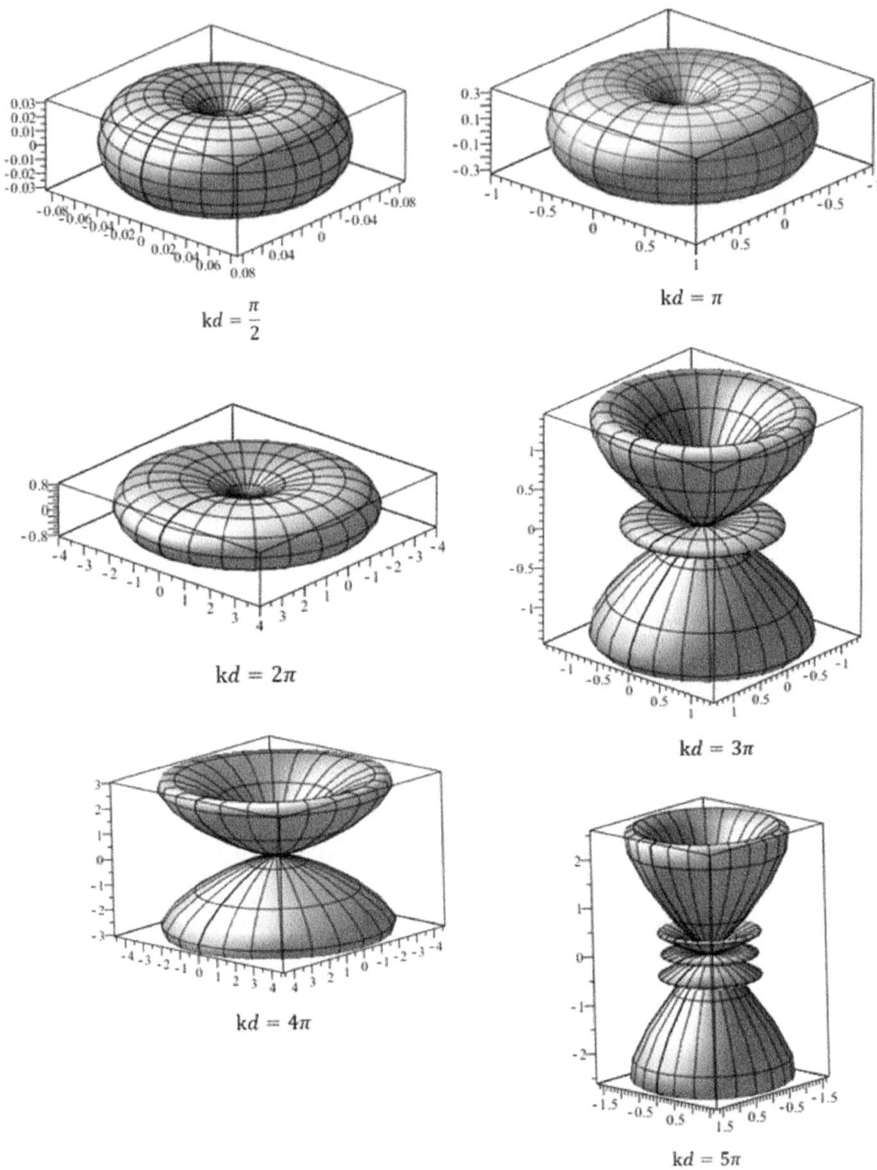

Figure 3.5. *Angular dependence of the radiation for* $\mathrm{kd} = \frac{\pi}{2}$, π, 3π, 4π, 4π. *For a color version of this figure, see www.iste.co.uk/abdesselam/electrodynamics3.zip*

The total power radiated in all space is equal to:

$$P = \sqrt{\frac{\mu_0}{\varepsilon_0}}\frac{I_0^2}{8\pi^2}\int_0^{\pi}\int_0^{2\pi}\left[\frac{\cos\left(\frac{1}{2}\mathsf{k}d\cos\theta\right)-\cos\left(\frac{1}{2}\mathsf{k}d\right)}{\sin\theta}\right]^2\sin\theta d\theta d\varphi$$

$$= \frac{I_0^2}{2}\frac{1}{4\pi}\sqrt{\frac{\mu_0}{\varepsilon_0}}\times\left[2\int_0^{\pi}\left[\frac{\cos\left(\frac{1}{2}\mathsf{k}d\cos\theta\right)-\cos\left(\frac{1}{2}\mathsf{k}d\right)}{\sin\theta}\right]^2\sin\theta d\theta\right]$$

$$\simeq \frac{I_0^2}{2}\frac{1}{4\pi}\sqrt{\frac{\mu_0}{\varepsilon_0}}\begin{cases}0.224 & \mathsf{k}d=\frac{\pi}{2}\\ 2.438 & \mathsf{k}d=\pi\\ 6.636 & \mathsf{k}d=2\pi\\ 3.516 & \mathsf{k}d=3\pi\end{cases} \qquad [3.167]$$

The integral in brackets is evaluated numerically. It can be seen that P varies as the square of the current. The *radiation resistance* R is defined by analogy to the Ohm law for alternating currents:

$$P = \frac{1}{2}RI_0^2 \qquad [3.168]$$

The factor $\frac{1}{2}$ appears when averaging over time. In particular, we have:

$$R \simeq \frac{1}{4\pi}\sqrt{\frac{\mu_0}{\varepsilon_0}}\begin{cases}0.224 & \mathsf{k}d=\frac{\pi}{2}\\ 2.438 & \mathsf{k}d=\pi\\ 6.636 & \mathsf{k}d=2\pi\\ 3.516 & \mathsf{k}d=3\pi\end{cases} \qquad [3.169]$$

EXAMPLE 3.3. ANTENNA ARRAY. Let us now consider an array of N identical antennas located at positions $\vec{\mathbf{r}}_j$ $(j = 1, \ldots, N)$, each fed with currents of the same waveform but different amplitudes $I_j = \gamma_j I_0$, where γ_j is a complex number. The antenna located at position $\vec{\mathbf{r}}_j$ and fed current $\gamma_j I_0$, resulting from current density $\gamma_j \vec{\hat{\mathbf{J}}}$, would produce the following radiation vector:

$$\vec{\hat{\mathbf{N}}}_j\left(\vec{\mathsf{k}}\right) = \gamma_j\int_{\tau'} d\tau' \vec{\hat{\mathbf{J}}}\left(\vec{\mathbf{r}}' - \vec{\mathbf{r}}_j\right)e^{-i\vec{\mathsf{k}}\cdot\vec{\mathbf{r}}'} = \gamma_j e^{-i\vec{\mathsf{k}}\cdot\vec{\mathbf{r}}_j}\vec{\hat{\mathbf{N}}}_0\left(\vec{\mathsf{k}}\right) \qquad [3.170]$$

where $\vec{\hat{\mathbf{N}}}_0\left(\vec{\mathsf{k}}\right) = \int_{\tau'} d\tau' \vec{\hat{\mathbf{J}}}\left(\vec{\mathbf{r}}'\right)e^{-i\vec{\mathsf{k}}\cdot\vec{\mathbf{r}}'}$. We assume that the antennas are arranged along the Ox axis, spaced by a distance a from each other, and oriented along the Oz axis, as illustrated in Figure 3.6. If, in addition, the antennas are fed in phase with currents of the same amplitude $(\gamma_j = 1)$, we obtain:

$$\vec{\hat{\mathbf{N}}}\left(\vec{\mathsf{k}}\right) = \vec{\hat{\mathbf{N}}}_0\left(\vec{\mathsf{k}}\right)\sum_{j=1}^{N} e^{-\mathrm{i}\vec{\mathbf{k}}\cdot\vec{\mathbf{r}}_j} = \vec{\hat{\mathbf{N}}}_0\left(\vec{\mathsf{k}}\right)\sum_{j=1}^{N} e^{-\mathrm{i}x_j\vec{\mathbf{k}}\cdot\vec{\mathbf{i}}}$$

$$= \begin{cases} \vec{\hat{\mathbf{N}}}_0\left(\vec{\mathsf{k}}\right)\displaystyle\sum_{j=-M}^{M} e^{-\mathrm{i}ja\mathsf{k}\sin\theta\cos\varphi} & \text{pour } N = 2M+1 \\ \vec{\hat{\mathbf{N}}}_0\left(\vec{\mathsf{k}}\right)\displaystyle\sum_{j=-M+1}^{M} e^{-\mathrm{i}\left(j-\frac{1}{2}\right)a\mathsf{k}\sin\theta\cos\varphi} & \text{pour } N = 2M \end{cases}$$

$$= \begin{cases} \vec{\hat{\mathbf{N}}}_0\left(\vec{\mathsf{k}}\right)\dfrac{\sin\left[\frac{1}{2}\left(2M+1\right)a\mathsf{k}\sin\theta\cos\varphi\right]}{\sin\left[\frac{1}{2}a\mathsf{k}\sin\theta\cos\varphi\right]} & \text{pour } N = 2M+1 \\ \vec{\hat{\mathbf{N}}}_0\left(\vec{\mathsf{k}}\right)\dfrac{\sin\left[Ma\mathsf{k}\sin\theta\cos\varphi\right]}{\sin\left[\frac{1}{2}a\mathsf{k}\sin\theta\cos\varphi\right]} & \text{pour } N = 2M \end{cases}$$

$$= \vec{\hat{\mathbf{N}}}_0\left(\vec{\mathsf{k}}\right)\frac{\sin\left[\frac{N}{2}a\mathsf{k}\sin\theta\cos\varphi\right]}{\sin\left[\frac{1}{2}a\mathsf{k}\sin\theta\cos\varphi\right]} \qquad [3.171]$$

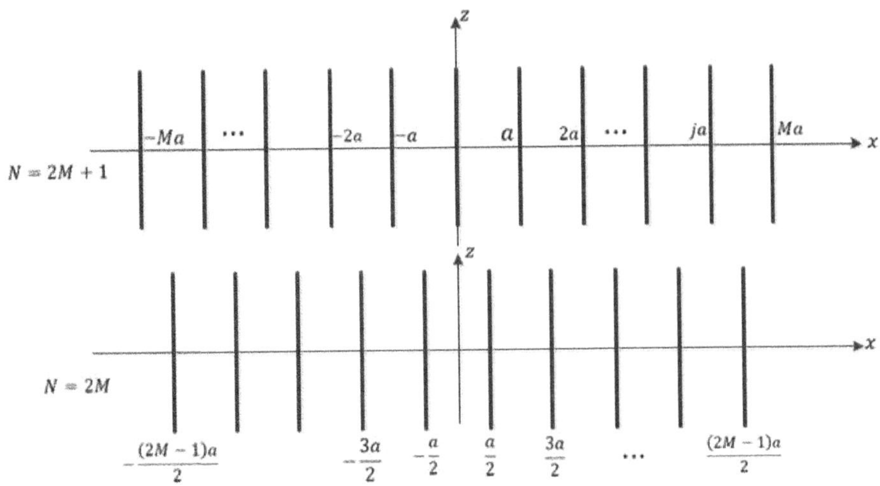

Figure 3.6. *Linear antenna array arranged along the Ox axis with spacing a between elements*

Consequently, the power radiated per unit of the solid angle will be:

$$\frac{dP}{d\Omega} = \sqrt{\frac{\mu_0}{\varepsilon_0}}\frac{\mathsf{k}^2}{32\pi^2}\left|\vec{\hat{\mathbf{N}}}_0\right|^2\sin^2\theta\frac{\sin^2\left[\frac{N}{2}a\mathsf{k}\sin\theta\cos\varphi\right]}{\sin^2\left[\frac{1}{2}a\mathsf{k}\sin\theta\cos\varphi\right]}$$

$$= \left.\frac{dP}{d\Omega}\right|_0 \frac{\sin^2\left[\frac{N}{2}a\mathrm{k}\sin\theta\cos\varphi\right]}{\sin^2\left[\frac{1}{2}a\mathrm{k}\sin\theta\cos\varphi\right]}$$

$$= \sqrt{\frac{\mu_0}{\varepsilon_0}}\frac{I_0^2}{8\pi^2}\left[\frac{\cos\left(\frac{1}{2}\mathrm{k}d\cos\theta\right)-\cos\left(\frac{1}{2}\mathrm{k}d\right)}{\sin\theta}\right]^2 \frac{\sin^2\left[\frac{N}{2}a\mathrm{k}\sin\theta\cos\varphi\right]}{\sin^2\left[\frac{1}{2}a\mathrm{k}\sin\theta\cos\varphi\right]} \quad [3.172]$$

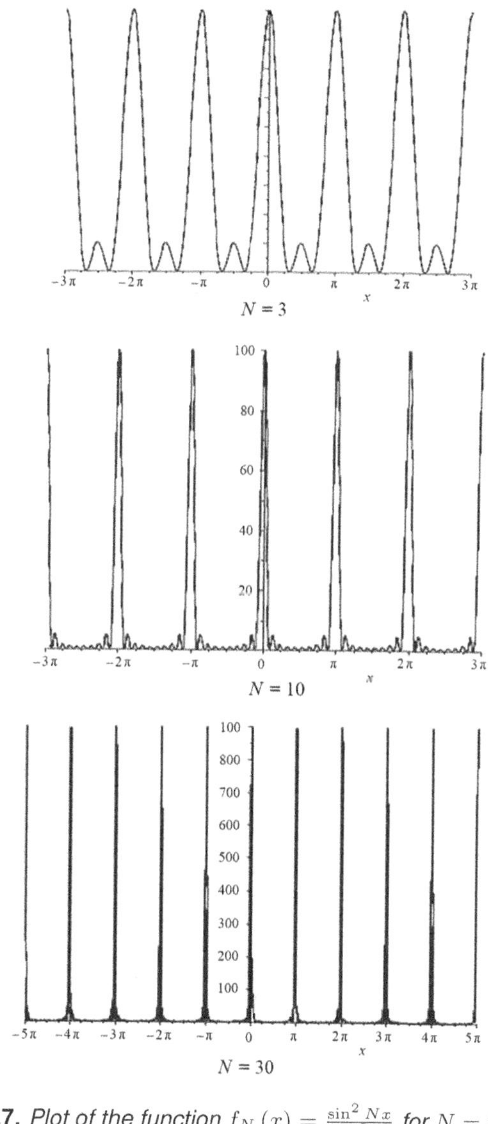

Figure 3.7. *Plot of the function $f_N(x) = \frac{\sin^2 Nx}{\sin^2 x}$ for $N = 3,\ 10,\ 30$. As $N \longrightarrow \infty$, $f_N(x)$ becomes proportional to the Dirac comb*

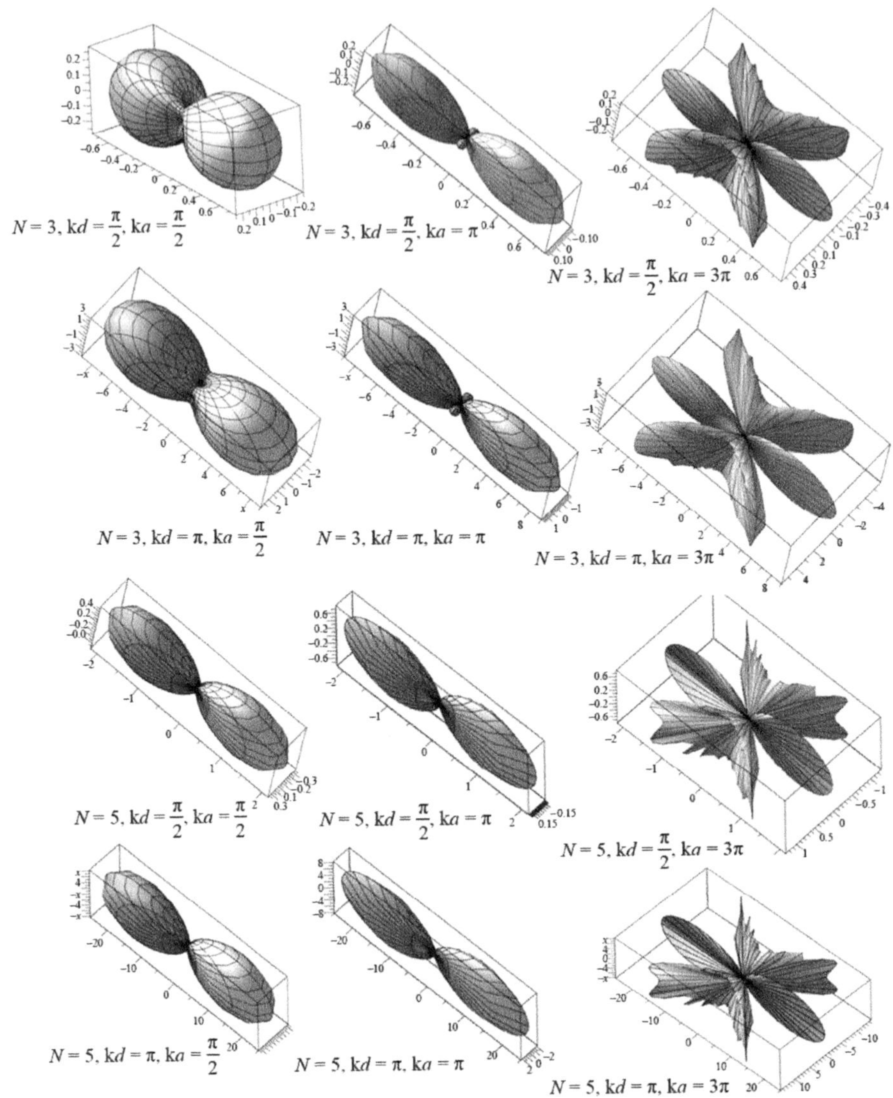

Figure 3.8A. *Angular distribution of the power radiated by* $N = 3,\ 5,\ 10$ *antennas of length* $d = \frac{\lambda}{4},\ \frac{\lambda}{2}$*, fed in phase and separated by a distance* $a = \frac{\lambda}{4},\ \frac{\lambda}{2},\ \frac{3\lambda}{2}$*. For a color version of this figure, see www.iste.co.uk/abdesselam/electrodynamics3.zip*

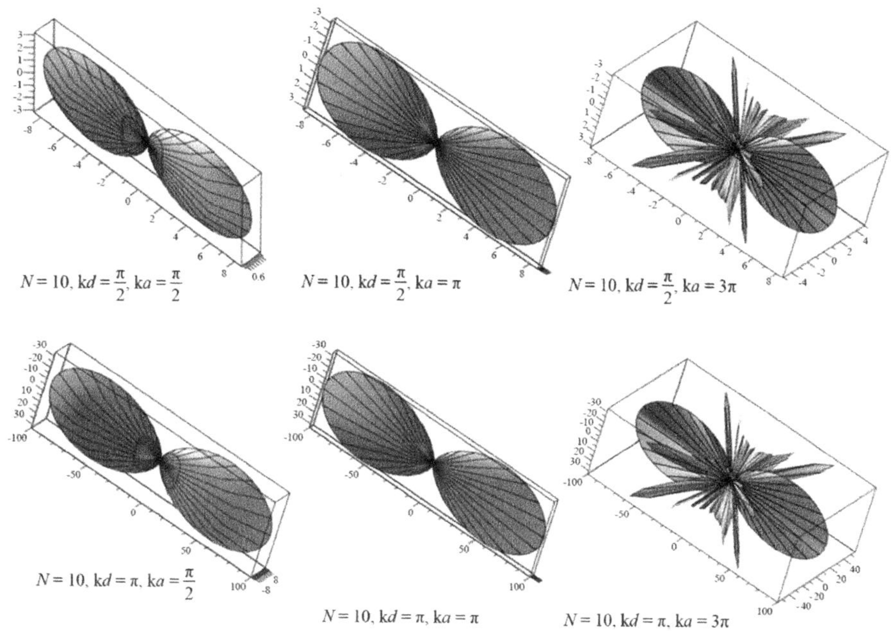

Figure 3.8B. *Angular distribution of the power radiated by $N = 3, 5, 10$ antennas of length $d = \frac{\lambda}{4}, \frac{\lambda}{2}$, fed in phase and separated by a distance $a = \frac{\lambda}{4}, \frac{\lambda}{2}, \frac{3\lambda}{2}$ (continued). For a color version of this figure, see www.iste.co.uk/abdesselam/electrodynamics3.zip*

We observe the following:

1) $\frac{dP}{d\Omega}$ appears here as the power radiated by a single antenna placed at the point $x = 0$, modulated by the function $f_N(x) = \frac{\sin^2 Nx}{\sin^2 x}$.

2) $\frac{dP}{d\Omega}$ now depends on θ and φ.

3) The variations of the function $f_N(x) = \frac{\sin^2 Nx}{\sin^2 x}$ are illustrated in Figure 3.7. It is a function with period π in x. The first zero of this function is at $x = \frac{\pi}{N}$. The larger N, the sharper the peaked functions become around the values $x = n\pi$ $(n \in \mathbb{Z})$. As $N \to \infty$, the function $f_N(x)$ resembles a Dirac comb. In fact, we have:

$$\lim_{N \longrightarrow \infty} \frac{1}{N} \frac{\sin^2 Nx}{\sin^2 x} = \pi \sum_{n \in \mathbb{Z}} \delta(x - n\pi) \qquad [3.173]$$

The angular distributions of $\frac{dP}{d\Omega}$ for different values of N, kd, and ka are presented in Figure 3.8. Radiation is directed along the Oy axis. The larger N is, the more directional the radiation becomes.

Chapter 4

Radiation from Point Charges

4.1. Lińard–Wiechert potentials

In what follows, we will focus on the radiation produced by an accelerated point charge. First, we calculate the electric and magnetic fields generated by this charge. Then, we will discuss specific cases: a particle in uniform motion, linear acceleration and circular acceleration.

Consider a point charge in an arbitrary motion, whose position as a function of time is $\vec{\mathbf{r}}_q(t')$. The charge density associated with the moving particle is given by $\rho\left(\vec{\mathbf{r}}', t'\right) = q\delta^3\left(\vec{\mathbf{r}}' - \vec{\mathbf{r}}_q(t')\right)$. Using the Green functions [3.32], the particle produces at the point $\left(\vec{\mathbf{r}}, t\right)$ the electric potential:

$$\begin{aligned} V\left(\vec{\mathbf{r}}, t\right) &= \frac{1}{4\pi\varepsilon_0}\int \frac{q\delta\left(\vec{\mathbf{r}}' - \vec{\mathbf{r}}_q(t')\right)\delta\left(t - t' - \frac{\left|\vec{\mathbf{r}} - \vec{\mathbf{r}}'\right|}{c}\right)}{\left|\vec{\mathbf{r}} - \vec{\mathbf{r}}'\right|} d\tau' dt' \\ &= \frac{q}{4\pi\varepsilon_0}\int \frac{\delta\left(t - t' - \frac{\left|\vec{\mathbf{r}} - \vec{\mathbf{r}}_q(t')\right|}{c}\right)}{\left|\vec{\mathbf{r}} - \vec{\mathbf{r}}_q(t')\right|} dt' \\ &= \frac{q}{4\pi\varepsilon_0}\int \frac{1}{R}\delta\left(t - t' - \frac{R}{c}\right) dt' \end{aligned} \qquad [4.1]$$

where $R = \left|\vec{\mathbf{r}} - \vec{\mathbf{r}}_q(t')\right|$. To evaluate the integral [4.1], we can proceed in two different (but equivalent) ways:

1) Using the identity $\delta\left(g\left(t'\right)\right) = \sum_i \frac{\delta\left(t' - t'_i\right)}{\left|g'\left(t'_i\right)\right|}$, where t'_i are the roots of the function $g\left(t'\right)$, it is possible to simplify the integral [4.1]. In our case, we have $g\left(t'\right) = t - \frac{\left|\vec{\mathbf{r}} - \vec{\mathbf{r}}_q(t')\right|}{c} - t'$. This gives:

$$\left|\frac{dg}{dt'}\right| = \left|-1 - \frac{d}{dt'}\frac{\left|\vec{\mathbf{r}} - \vec{\mathbf{r}}_q(t')\right|}{c}\right| = \left|-1 + \frac{\left(\vec{\mathbf{r}} - \vec{\mathbf{r}}_q(t')\right)}{c\left|\vec{\mathbf{r}} - \vec{\mathbf{r}}_q(t')\right|} \cdot \frac{d\vec{\mathbf{r}}_q}{dt'}\right|$$

$$= \left|-1 + \frac{\vec{\mathbf{v}}_q(t') \cdot \vec{\mathbf{n}}(t')}{c}\right| = 1 - \vec{\beta}(t') \cdot \vec{\mathbf{n}}(t') \qquad [4.2]$$

where we have defined $\vec{\mathbf{n}}(t') = \frac{\vec{\mathbf{r}} - \vec{\mathbf{r}}_q(t')}{\left|\vec{\mathbf{r}} - \vec{\mathbf{r}}_q(t')\right|}$ and $\vec{\beta}(t') = \frac{\vec{\mathbf{v}}_q(t')}{c}$. Note that $1 - \vec{\beta} \cdot \vec{\mathbf{n}}$ is always positive since $\left|\vec{\beta}\right| \ll 1$. Consequently, we obtain:

$$\delta\left(t - t' - \frac{\left|\vec{\mathbf{r}} - \vec{\mathbf{r}}_q(t')\right|}{c}\right) = \frac{\delta(t' - t_{\text{ret.}})}{1 - \vec{\beta}(t_{\text{ret.}}) \cdot \vec{\mathbf{n}}(t_{\text{ret.}})} \qquad [4.3]$$

where $t_{\text{ret.}} = t - \frac{\left|\vec{\mathbf{r}} - \vec{\mathbf{r}}_q(t_{\text{ret.}})\right|}{c}$ is the retarded time. Using formula [4.3], it is now easy to carry out the integration of [4.1] on t':

$$V(\vec{\mathbf{r}}, t) = \frac{q}{4\pi\varepsilon_0}\int \frac{\delta\left(t - t' - \frac{R}{c}\right)}{R} dt' = \frac{q}{4\pi\varepsilon_0}\left[\frac{1}{R\left(1 - \vec{\beta} \cdot \vec{\mathbf{n}}\right)}\right]_{\text{ret.}} \qquad [4.4]$$

where the notation $[\cdots]_{\text{ret.}}$ means that the argument is evaluated at the retarded time defined by $t_{\text{ret.}} = t - \frac{\left|\vec{\mathbf{r}} - \vec{\mathbf{r}}_q(t_{\text{ret.}})\right|}{c}$.

2) We can also perform the change of variable $u = t - t' - \frac{R}{c}$. It is easy to verify that:

$$\frac{du}{dt'} = -1 - \frac{1}{c}\frac{dR}{dt} = -1 + \frac{1}{c}\frac{\left(\vec{\mathbf{r}} - \vec{\mathbf{r}}_q\right)}{R} \cdot \vec{\mathbf{v}}_q = -1 + \vec{\mathbf{n}} \cdot \vec{\beta} \qquad [4.5]$$

which immediately gives:

$$V(\vec{\mathbf{r}}, t) = \frac{q}{4\pi\varepsilon_0}\int_{-\infty}^{\infty} \frac{\delta(u)}{R\left(1 - \vec{\mathbf{n}} \cdot \vec{\beta}\right)} du = \frac{q}{4\pi\varepsilon_0}\left[\frac{1}{R\left(1 - \vec{\beta} \cdot \vec{\mathbf{n}}\right)}\right]_{\text{ret.}} \qquad [4.6]$$

Similarly, the potential vector produced by the moving charge is given by:

$$\vec{\mathbf{A}} = \frac{\mu_0}{4\pi}\int \frac{q\vec{\mathbf{v}}_q(t')\,\delta\left(\vec{\mathbf{r}}' - \vec{\mathbf{r}}_q\right)\delta\left(t - t' - \frac{\left|\vec{\mathbf{r}} - \vec{\mathbf{r}}'\right|}{c}\right)}{\left|\vec{\mathbf{r}} - \vec{\mathbf{r}}'\right|} d\tau' dt'$$

$$
\begin{aligned}
&= \frac{q\mu_0}{4\pi}\int \frac{\vec{\mathbf{v}}_q(t')\,\delta\left(t-t'-\frac{|\vec{\mathbf{r}}-\vec{\mathbf{r}}_q(t')|}{c}\right)}{\left|\vec{\mathbf{r}}-\vec{\mathbf{r}}_q(t')\right|}dt' \\
&= \frac{q}{4\pi\varepsilon_0 c}\int \frac{\vec{\beta}(t')\,\delta\left(t-t'-\frac{R}{c}\right)}{R}dt' \\
&= \frac{q}{4\pi\varepsilon_0 c}\int \frac{\vec{\beta}\,\delta(u)}{R\left(1-\vec{\beta}\cdot\vec{\mathbf{n}}\right)}du = \frac{q}{4\pi\varepsilon_0 c}\left[\frac{\vec{\beta}}{R\left(1-\vec{\beta}\cdot\vec{\mathbf{n}}\right)}\right]_{\text{ret.}}
\end{aligned}
\qquad [4.7]
$$

The potentials [4.4] and [4.7] are called, respectively, the *scalar and vector Liénard–Wiechert potentials*. They describe, in a classical context, the electromagnetic effects produced by a moving point charge.

4.2. Fields of a moving point charge

To obtain the fields produced by the moving point charge, we apply the relations $\vec{\mathbf{E}} = -\vec{\nabla}V - \frac{\partial\vec{\mathbf{A}}}{\partial t}$ and $\vec{\mathbf{B}} = \vec{\nabla}\times\vec{\mathbf{A}}$. Before that, let us recall the following properties: $\vec{\nabla}f(R) = \frac{df}{dR}\vec{\mathbf{n}}$ and $\vec{\nabla}\times\left(f(R)\,\vec{\beta}\right) = \vec{\nabla}f\times\vec{\beta} = \frac{df}{dR}\vec{\mathbf{n}}\times\vec{\beta}$.

Let us begin with the electric field $\vec{\mathbf{E}}$:

$$
\begin{aligned}
-\vec{\nabla}V &= -\frac{q}{4\pi\varepsilon_0}\int_{-\infty}^{\infty} dt'\left[-\frac{1}{R^2}\delta\left(t-t'-\frac{R}{C}\right)-\frac{1}{cR}\delta'\left(t-t'-\frac{R}{c}\right)\right]\vec{\mathbf{n}} \\
-\frac{\partial\vec{\mathbf{A}}}{\partial t} &= -\frac{q}{4\pi\varepsilon_0 c}\int_{-\infty}^{\infty} dt'\frac{\vec{\beta}}{R}\delta'\left(t-t'-\frac{R}{c}\right)
\end{aligned}
\qquad [4.8]
$$

To calculate these two integrals, let us perform the change of variable $u = t-t'-\frac{R}{c}$:

$$
\begin{aligned}
\vec{\mathbf{E}} &= \frac{q}{4\pi\varepsilon_0}\int_{-\infty}^{\infty}\frac{du}{1-\vec{\mathbf{n}}\cdot\vec{\beta}}\left[\frac{1}{R^2}\delta(u)+\frac{1}{cR}\delta'(u)\right]\vec{\mathbf{n}} \\
&\quad -\frac{q}{4\pi\varepsilon_0 c}\int_{-\infty}^{\infty}\frac{du}{1-\vec{\mathbf{n}}\cdot\vec{\beta}}\frac{\vec{\beta}}{R}\delta'(u) \\
&= \frac{q}{4\pi\varepsilon_0}\int_{-\infty}^{\infty}\frac{du}{1-\vec{\mathbf{n}}\cdot\vec{\beta}}\left[\frac{\vec{\mathbf{n}}}{R^2}\delta(u)+\frac{\vec{\mathbf{n}}-\vec{\beta}}{cR}\delta'(u)\right]
\end{aligned}
\qquad [4.9]
$$

Using the integration properties of the Dirac delta function $\int_{-\infty}^{\infty} f(x)\,\delta(x-a)\,dx = f(a)$ and $\int_{-\infty}^{\infty} f(x)\,\delta'(x-a)\,dx = -f'(a)$, we find:

$$\begin{aligned}\vec{\mathbf{E}} &= \frac{q}{4\pi\varepsilon_0}\left[\frac{\vec{\mathbf{n}}}{R^2\left(1-\vec{\mathbf{n}}\cdot\vec{\beta}\right)} - \frac{d}{du}\left(\frac{\vec{\mathbf{n}}-\vec{\beta}}{cR\left(1-\vec{\mathbf{n}}\cdot\vec{\beta}\right)}\right)\right]_{\text{ret.}} \\ &= \frac{q}{4\pi\varepsilon_0}\left[\frac{\vec{\mathbf{n}}}{R^2\left(1-\vec{\mathbf{n}}\cdot\vec{\beta}\right)} + \frac{1}{1-\vec{\mathbf{n}}\cdot\vec{\beta}}\frac{d}{dt'}\left(\frac{\vec{\mathbf{n}}-\vec{\beta}}{cR\left(1-\vec{\mathbf{n}}\cdot\vec{\beta}\right)}\right)\right]_{\text{ret.}}\end{aligned} \quad [4.10]$$

To expand expression [4.10], we need the following derivatives:

$$\begin{aligned}&\frac{1}{c}\frac{dR}{dt'} = -\frac{\left(\vec{\mathbf{r}}-\vec{\mathbf{r}}_q\right)\cdot\vec{\mathbf{v}}_q}{cR} = -\vec{\mathbf{n}}\cdot\vec{\beta} \\ &\frac{1}{c}\frac{d}{dt'}\left(\frac{1}{R}\right) = -\frac{1}{R^2}\frac{1}{c}\frac{dR}{dt'} = \frac{\vec{\mathbf{n}}\cdot\vec{\beta}}{R^2} \\ &\frac{1}{c}\frac{d\vec{\mathbf{n}}}{dt'} = -\frac{\vec{\mathbf{v}}_q}{cR} + \frac{\vec{\mathbf{n}}\cdot\vec{\beta}}{R^2}\vec{\mathbf{R}} = \frac{-\vec{\beta}+\left(\vec{\mathbf{n}}\cdot\vec{\beta}\right)\vec{\mathbf{n}}}{R} = \frac{\vec{\mathbf{n}}\times\left(\vec{\mathbf{n}}\times\vec{\beta}\right)}{R} \\ &\frac{1}{c}\frac{d\vec{\beta}}{dt'} = \frac{\vec{\mathbf{a}}}{c^2} \\ &\frac{1}{c}\frac{d}{dt'}\frac{1}{1-\vec{\beta}\cdot\vec{\mathbf{n}}} = \frac{1}{\left(1-\vec{\beta}\cdot\vec{\mathbf{n}}\right)^2}\left[\vec{\beta}\cdot\frac{\vec{\mathbf{n}}}{cdt'} + \vec{\mathbf{n}}\cdot\frac{\vec{\beta}}{cdt'}\right] \\ &\qquad = \frac{1}{\left(1-\vec{\beta}\cdot\vec{\mathbf{n}}\right)^2}\left[\frac{\vec{\mathbf{n}}\cdot\vec{\mathbf{a}}}{c^2} - \frac{\beta^2}{R} + \frac{\left(\vec{\mathbf{n}}\cdot\vec{\beta}\right)^2}{R}\right]\end{aligned} \quad [4.11]$$

(here $\vec{\mathbf{a}}$ is the acceleration), which gives:

$$R^2c^2\left(1-\vec{\mathbf{n}}\cdot\vec{\beta}\right)^2\frac{1}{c}\frac{d}{dt'}\left[\frac{\vec{\mathbf{n}}-\vec{\beta}}{R\left(1-\vec{\mathbf{n}}\cdot\vec{\beta}\right)}\right] = Rc^2\left(1-\vec{\mathbf{n}}\cdot\vec{\beta}\right)$$

$$\times\left[\frac{-\vec{\beta}+\left(\vec{\mathbf{n}}\cdot\vec{\beta}\right)\vec{\mathbf{n}}}{R} - \frac{\vec{\mathbf{a}}}{c^2}\right] + c^2\left(\vec{\mathbf{n}}-\vec{\beta}\right)\left\{\vec{\beta}\cdot\vec{\mathbf{n}}\left(1-\vec{\mathbf{n}}\cdot\vec{\beta}\right)+\right.$$

$$\left.R\left[\frac{\vec{\mathbf{n}}\cdot\vec{\mathbf{a}}}{c^2} - \frac{\beta^2}{R} + \frac{\left(\vec{\mathbf{n}}\cdot\vec{\beta}\right)^2}{R}\right]\right\} = c^2\left\{\left(1-\vec{\mathbf{n}}\cdot\vec{\beta}\right)\left[-\vec{\beta}+\left(\vec{\mathbf{n}}\cdot\vec{\beta}\right)\vec{\mathbf{n}}\right]+\right.$$

$$\left(\vec{\mathbf{n}} - \vec{\beta}\right)\left[\vec{\beta}\cdot\vec{\mathbf{n}} - \beta^2\right]\Big\} + R\left[-\left(1 - \vec{\mathbf{n}}\cdot\vec{\beta}\right)\vec{\mathbf{a}} + \left(\vec{\mathbf{n}} - \vec{\beta}\right)\left(\vec{\mathbf{n}}\cdot\vec{\mathbf{a}}\right)\right]$$
$$= c^2\left[\left(\vec{\mathbf{n}}\cdot\vec{\beta}\right)\left(2 - \vec{\mathbf{n}}\cdot\vec{\beta}\right)\vec{\mathbf{n}} - \vec{\beta}\left(1-\beta^2\right) - \beta^2\vec{\mathbf{n}}\right]$$
$$+R\vec{\mathbf{n}}\times\left[\left(\vec{\mathbf{n}} - \vec{\beta}\right)\times\vec{\mathbf{n}}\right] \qquad [4.12]$$

and consequently:

$$\frac{1}{c}\frac{d}{dt'}\left[\frac{\vec{\mathbf{n}} - \vec{\beta}}{R\left(1 - \vec{\mathbf{n}}\cdot\vec{\beta}\right)}\right] = \frac{\left[\left(\vec{\mathbf{n}}\cdot\vec{\beta}\right)\left(2 - \vec{\mathbf{n}}\cdot\vec{\beta}\right)\vec{\mathbf{n}} - \vec{\beta}\left(1-\beta^2\right) - \beta^2\vec{\mathbf{n}}\right]}{R^2\left(1 - \vec{\mathbf{n}}\cdot\vec{\beta}\right)^2}$$
$$+\frac{\vec{\mathbf{n}}\times\left[\left(\vec{\mathbf{n}} - \vec{\beta}\right)\times\vec{\mathbf{n}}\right]}{Rc^2\left(1 - \vec{\mathbf{n}}\cdot\vec{\beta}\right)^2} \qquad [4.13]$$

Substituting the result obtained in [4.10], we get:

$$\vec{\mathbf{E}} = \frac{q}{4\pi\varepsilon_0}\left[\frac{\vec{\mathbf{n}}}{R^2\left(1 - \vec{\mathbf{n}}\cdot\vec{\beta}\right)} + \frac{\vec{\mathbf{n}}\times\left[\left(\vec{\mathbf{n}} - \vec{\beta}\right)\times\vec{\mathbf{a}}\right]}{Rc^2\left(1 - \vec{\mathbf{n}}\cdot\vec{\beta}\right)^3}\right.$$
$$\left.+\frac{\left[\left(\vec{\mathbf{n}}\cdot\vec{\beta}\right)\left(2 - \vec{\mathbf{n}}\cdot\vec{\beta}\right)\vec{\mathbf{n}} - \vec{\beta}\left(1-\beta^2\right) - \beta^2\vec{\mathbf{n}}\right]}{R^2\left(1 - \vec{\mathbf{n}}\cdot\vec{\beta}\right)^3}\right]_{\text{ret.}}$$
$$= \frac{q}{4\pi\varepsilon_0}\left[\frac{\left(\vec{\mathbf{n}} - \vec{\beta}\right)\left(1-\beta^2\right)}{R^2\left(1 - \vec{\mathbf{n}}\cdot\vec{\beta}\right)^3} + \frac{\vec{\mathbf{n}}\times\left[\left(\vec{\mathbf{n}} - \vec{\beta}\right)\times\vec{\mathbf{a}}\right]}{Rc^2\left(1 - \vec{\mathbf{n}}\cdot\vec{\beta}\right)^3}\right]_{\text{ret.}}$$
$$= \frac{q}{4\pi\varepsilon_0}\left[\frac{\vec{\mathbf{n}} - \vec{\beta}}{\gamma^2R^2\left(1 - \vec{\mathbf{n}}\cdot\vec{\beta}\right)^3} + \frac{\vec{\mathbf{n}}\times\left[\left(\vec{\mathbf{n}} - \vec{\beta}\right)\times\vec{\mathbf{a}}\right]}{Rc^2\left(1 - \vec{\mathbf{n}}\cdot\vec{\beta}\right)^3}\right]_{\text{ret.}} \qquad [4.14]$$

where $\gamma = \frac{1}{\sqrt{1-\beta^2}}$. There is another form for $\vec{\mathbf{E}}$ (which will be used in section 4.5):

$$\begin{aligned}\vec{\mathbf{E}} &= \frac{q}{4\pi\varepsilon_0}\left[\frac{\vec{\mathbf{n}}-\vec{\beta}}{\gamma^2R^2\left(1-\vec{\mathbf{n}}\cdot\vec{\beta}\right)^3}+\frac{\vec{\mathbf{n}}\times(\vec{\mathbf{n}}\times\vec{\mathbf{a}})+\vec{\mathbf{n}}\times\left(\vec{\mathbf{a}}\times\vec{\beta}\right)}{Rc^2\left(1-\vec{\mathbf{n}}\cdot\vec{\beta}\right)^3}\right]_{\text{ret.}}\\ &= \frac{q}{4\pi\varepsilon_0}\left[\frac{\vec{\mathbf{n}}-\vec{\beta}}{\gamma^2R^2\left(1-\vec{\mathbf{n}}\cdot\vec{\beta}\right)^3}+\frac{(\vec{\mathbf{n}}\cdot\vec{\mathbf{a}})\,\vec{\mathbf{n}}-\vec{\mathbf{a}}+\vec{\mathbf{n}}\times\left(\vec{\mathbf{a}}\times\vec{\beta}\right)}{Rc^2\left(1-\vec{\mathbf{n}}\cdot\vec{\beta}\right)^3}\right]_{\text{ret.}}\\ &= \frac{q}{4\pi\varepsilon_0}\left[\frac{\vec{\mathbf{n}}-\vec{\beta}}{\gamma^2R^2\left(1-\vec{\mathbf{n}}\cdot\vec{\beta}\right)^3}+\frac{-\vec{\mathbf{a}}_\perp+\vec{\mathbf{n}}\times\left(\vec{\mathbf{a}}\times\vec{\beta}\right)}{Rc^2\left(1-\vec{\mathbf{n}}\cdot\vec{\beta}\right)^3}\right]_{\text{ret.}}\end{aligned} \quad [4.15]$$

where $\vec{\mathbf{a}}_\perp$ is the acceleration component perpendicular to $\vec{\mathbf{n}}$.

The magnetic field is also calculated following the same steps:

$$\begin{aligned}\vec{\mathbf{B}} &= \vec{\nabla}\times\vec{\mathbf{A}} = \frac{q}{4\pi\varepsilon_0 c}\int\vec{\nabla}\times\left[\frac{\delta\left(t-t'-\frac{R}{c}\right)}{R}\vec{\beta}\right]dt'\\ &= \frac{q}{4\pi\varepsilon_0 c}\int\frac{d}{dR}\left[\frac{\delta\left(t-t'-\frac{R}{c}\right)}{R}\right]\vec{\mathbf{n}}\times\vec{\beta}\,dt'\\ &= \frac{q}{4\pi\varepsilon_0 c}\int\left[-\frac{1}{R^2}\delta\left(t-t'-\frac{R}{c}\right)-\frac{1}{Rc}\delta'\left(t-t'-\frac{R}{c}\right)\right]\vec{\mathbf{n}}\times\vec{\beta}\,dt'\end{aligned} \quad [4.16]$$

By making the same change of variable, we obtain:

$$\begin{aligned}\vec{\mathbf{B}} &= \frac{q}{4\pi\varepsilon_0 c}\int\left[-\frac{1}{R^2}\delta(u)-\frac{1}{Rc}\delta'(u)\right]\frac{\vec{\mathbf{n}}\times\vec{\beta}}{1-\vec{\mathbf{n}}\cdot\vec{\beta}}du\\ &= \frac{q}{4\pi\varepsilon_0 c}\left[-\frac{\vec{\mathbf{n}}\times\vec{\beta}}{R^2\left(1-\vec{\mathbf{n}}\cdot\vec{\beta}\right)}+\frac{d}{du}\left(\frac{\vec{\mathbf{n}}\times\vec{\beta}}{Rc\left(1-\vec{\mathbf{n}}\cdot\vec{\beta}\right)}\right)\right]_{\text{ret.}}\\ &= \frac{q}{4\pi\varepsilon_0 c}\left[\frac{\vec{\beta}\times\vec{\mathbf{n}}}{R^2\left(1-\vec{\mathbf{n}}\cdot\vec{\beta}\right)}+\frac{1}{1-\vec{\mathbf{n}}\cdot\vec{\beta}}\frac{d}{dt'}\left(\frac{\vec{\beta}\times\vec{\mathbf{n}}}{Rc\left(1-\vec{\mathbf{n}}\cdot\vec{\beta}\right)}\right)\right]_{\text{ret.}}\end{aligned} \quad [4.17]$$

Since:

$$\frac{d}{dt'}\frac{\vec{\beta}\times\vec{\mathbf{n}}}{cR\left(1-\vec{\beta}\cdot\vec{\mathbf{n}}\right)}=\frac{\vec{\beta}\times\vec{\mathbf{n}}}{\left(1-\vec{\beta}\cdot\vec{\mathbf{n}}\right)^{2}}\left[\frac{\vec{\mathbf{n}}\cdot\vec{\mathbf{a}}}{c^{2}R}-\frac{\beta^{2}}{R^{2}}+\frac{\vec{\beta}\cdot\vec{\mathbf{n}}}{R^{2}}\right]+$$

$$\frac{1}{R\left(1-\vec{\beta}\cdot\vec{\mathbf{n}}\right)}\left[\frac{\vec{\mathbf{a}}\times\vec{\mathbf{n}}}{c^{2}}+\frac{\left(\vec{\mathbf{n}}\cdot\vec{\beta}\right)\vec{\beta}\times\vec{\mathbf{n}}}{R}\right]$$

$$=\frac{\vec{\beta}\times\vec{\mathbf{n}}}{R^{2}\left(1-\vec{\beta}\cdot\vec{\mathbf{n}}\right)^{2}}\left[-\beta^{2}+2\left(\vec{\beta}\cdot\vec{\mathbf{n}}\right)-\left(\vec{\beta}\cdot\vec{\mathbf{n}}\right)^{2}\right]$$

$$\frac{1}{Rc^{2}\left(1-\vec{\beta}\cdot\vec{\mathbf{n}}\right)^{2}}\left[\vec{\mathbf{a}}\times\vec{\mathbf{n}}\left(1-\vec{\mathbf{n}}\cdot\vec{\beta}\right)+\left(\vec{\mathbf{a}}\cdot\vec{\mathbf{n}}\right)\left(\vec{\beta}\times\vec{\mathbf{n}}\right)\right] \qquad [4.18]$$

it follows that:

$$\vec{\mathbf{B}}=\frac{q}{4\pi\varepsilon_0 c}\left[\frac{\left(\vec{\beta}\times\vec{\mathbf{n}}\right)\left(1-\beta^{2}\right)}{R^{2}\left(1-\vec{\mathbf{n}}\cdot\vec{\beta}\right)^{3}}+\frac{\vec{\mathbf{a}}\times\vec{\mathbf{n}}\left(1-\vec{\mathbf{n}}\cdot\vec{\beta}\right)}{Rc^{2}\left(1-\vec{\beta}\cdot\vec{\mathbf{n}}\right)^{3}}+\frac{\left(\vec{\mathbf{a}}\cdot\vec{\mathbf{n}}\right)\left(\vec{\beta}\times\vec{\mathbf{n}}\right)}{Rc^{2}\left(1-\vec{\beta}\cdot\vec{\mathbf{n}}\right)^{3}}\right]_{\text{ret.}}=\frac{q}{4\pi\varepsilon_0 c}\left[\frac{\left(\vec{\beta}\times\vec{\mathbf{n}}\right)}{\gamma^{2}R^{2}\left(1-\vec{\mathbf{n}}\cdot\vec{\beta}\right)^{3}}+\frac{\vec{\mathbf{a}}\times\vec{\mathbf{n}}\left(1-\vec{\mathbf{n}}\cdot\vec{\beta}\right)+\left(\vec{\mathbf{a}}\cdot\vec{\mathbf{n}}\right)\left(\vec{\beta}\times\vec{\mathbf{n}}\right)}{Rc^{2}\left(1-\vec{\beta}\cdot\vec{\mathbf{n}}\right)^{3}}\right]_{\text{ret.}} \qquad [4.19]$$

To conclude, let us give the relation between the fields $\vec{\mathbf{E}}$ and $\vec{\mathbf{B}}$.

According to [4.14], we also write:

$$\vec{\mathbf{E}}=\frac{q}{4\pi\varepsilon_0}\left[\frac{\vec{\mathbf{n}}-\vec{\beta}}{\gamma^{2}R^{2}\left(1-\vec{\mathbf{n}}\cdot\vec{\beta}\right)^{3}}+\frac{\left(\vec{\mathbf{n}}\cdot\vec{\mathbf{a}}\right)\left(\vec{\mathbf{n}}-\vec{\beta}\right)-\left(1-\vec{\mathbf{n}}\cdot\vec{\beta}\right)\vec{\mathbf{a}}}{Rc^{2}\left(1-\vec{\mathbf{n}}\cdot\vec{\beta}\right)^{3}}\right]_{\text{ret.}} \qquad [4.20]$$

It follows that:

$$\vec{\mathbf{n}} \times \vec{\mathbf{E}} = \frac{q}{4\pi\varepsilon_0}\left[\frac{\vec{\beta} \times \vec{\mathbf{n}}}{\gamma^2 R^2 \left(1 - \vec{\mathbf{n}} \cdot \vec{\beta}\right)^3} + \frac{\left(\vec{\mathbf{n}} \cdot \vec{\mathbf{a}}\right)\left(\vec{\beta} \times \vec{\mathbf{n}}\right)}{Rc^2 \left(1 - \vec{\mathbf{n}} \cdot \vec{\beta}\right)^3} + \frac{\left(1 - \vec{\mathbf{n}} \cdot \vec{\beta}\right) \vec{\mathbf{a}} \times \vec{\mathbf{n}}}{Rc^2 \left(1 - \vec{\mathbf{n}} \cdot \vec{\beta}\right)^3}\right]_{\text{ret.}} = c\vec{\mathbf{B}} \qquad [4.21]$$

4.3. Radiation from a charge in uniform motion

When the motion of the point charge is uniform (i.e. $\vec{\mathbf{a}} = 0$), the electric [4.14] and magnetic [4.19] fields can be written as:

$$\vec{\mathbf{E}}\left(\vec{\mathbf{r}}, t\right) = \frac{q}{4\pi\varepsilon_0}\left[\frac{\vec{\mathbf{n}} - \vec{\beta}}{\gamma^2 R^2 \left(1 - \vec{\mathbf{n}} \cdot \vec{\beta}\right)^3}\right]_{\text{ret.}} = \frac{q}{4\pi\varepsilon_0}\left[\frac{\vec{\mathbf{R}}_0 - R_0\vec{\beta}}{\gamma^2 \left(R_0 - \vec{\mathbf{R}}_0 \cdot \vec{\beta}\right)^3}\right]$$
$$\vec{\mathbf{B}}\left(\vec{\mathbf{r}}, t\right) = \frac{q}{4\pi\varepsilon_0 c}\left[\frac{\left(\vec{\beta} \times \vec{\mathbf{n}}\right)}{\gamma^2 R^2 \left(1 - \vec{\mathbf{n}} \cdot \vec{\beta}\right)^3}\right]_{\text{ret.}} = \frac{q}{4\pi\varepsilon_0 c}\left[\frac{\left(\vec{\beta} \times \vec{\mathbf{R}}_0\right)}{\gamma^2 \left(R_0 - \vec{\mathbf{R}}_0 \cdot \vec{\beta}\right)^3}\right] \qquad [4.22]$$

where $\vec{\mathbf{R}}_0 = \vec{\mathbf{R}}_{\text{ret.}} = \vec{\mathbf{r}} - c\vec{\beta}\, t_{\text{ret.}} = \vec{\mathbf{r}} - c\vec{\beta}\, t + \vec{\beta}\, R_0$. Setting $\vec{\mathbf{R}} = \vec{\mathbf{r}} - c\vec{\beta}\, t$, we obtain the identities:

$$\vec{\mathbf{R}}_0 - R_0\vec{\beta} = \vec{\mathbf{R}}$$
$$\vec{\beta} \times \vec{\mathbf{R}}_0 = \vec{\beta} \times \vec{\mathbf{R}} \qquad [4.23]$$

To express the denominators of fields [4.22] in terms of $\vec{\mathbf{R}}$, let us square the first identity from [4.23]:

$$R^2 = R_0^2\left(1 + \beta^2\right) - 2R_0\left(\vec{\mathbf{R}}_0 \cdot \vec{\beta}\right) \qquad [4.24]$$

The second identity of [4.23] also allows us to write $\left(\vec{\beta} \times \vec{\mathbf{R}}_0\right)^2 = \left(\vec{\beta} \times \vec{\mathbf{R}}\right)^2$ $= \beta^2 R^2 \sin^2\theta$, where θ is the angle measured between $\vec{\beta}$ and $\vec{\mathbf{R}}$. This shows that:

$$\beta^2 R^2 \sin^2\theta = R_0^2\beta^2 - \left(\vec{\beta} \cdot \vec{\mathbf{R}}_0\right)^2 \qquad [4.25]$$

By subtracting equations [4.24] and [4.25] from each other, we find the following:

$$R^2\left(1-\beta^2\sin^2\theta\right)=R_0^2-2R_0\left(\overrightarrow{\mathbf{R}}_0\cdot\overrightarrow{\beta}\right)+\left(\overrightarrow{\beta}\cdot\overrightarrow{\mathbf{R}}_0\right)^2=\left(R_0-\overrightarrow{\mathbf{R}}_0\cdot\overrightarrow{\beta}\right)^2 \qquad [4.26]$$

Taking into account identities [4.23] and [4.26], the fields can be written as:

$$\begin{aligned}\overrightarrow{\mathbf{E}}\left(\overrightarrow{\mathbf{r}},t\right)&=\frac{q}{4\pi\varepsilon_0}\left[\frac{\overrightarrow{\mathbf{R}}}{\gamma^2R^3\left(1-\beta^2\sin^2\theta\right)^{3/2}}\right]\\ \overrightarrow{\mathbf{B}}\left(\overrightarrow{\mathbf{r}},t\right)&=\frac{q}{4\pi\varepsilon_0 c}\left[\frac{\left(\overrightarrow{\beta}\times\overrightarrow{\mathbf{R}}\right)}{\gamma^2R^3\left(1-\beta^2\sin^2\theta\right)^{3/2}}\right]\end{aligned} \qquad [4.27]$$

The electric field is directed along the instantaneous radial direction, not the retarded one.

4.4. Radiation from a slowly moving charge

When the charge velocity is low compared to c, the terms involving $\overrightarrow{\beta}$ and the retardation can be neglected. Thus, we obtain:

$$\begin{aligned}\overrightarrow{\mathbf{E}}&=\frac{q}{4\pi\varepsilon_0}\left[\frac{\overrightarrow{\mathbf{n}}}{R^2}+\frac{\overrightarrow{\mathbf{n}}\times\left(\overrightarrow{\mathbf{n}}\times\overrightarrow{\mathbf{a}}\right)}{Rc^2}\right]\\ \overrightarrow{\mathbf{B}}&=\frac{q}{4\pi\varepsilon_0}\frac{\overrightarrow{\mathbf{a}}\times\overrightarrow{\mathbf{n}}}{Rc^3}\end{aligned} \qquad [4.28]$$

The fields $\overrightarrow{\mathbf{E}}_R=\frac{q}{4\pi\varepsilon_0}\frac{\overrightarrow{\mathbf{n}}\times(\overrightarrow{\mathbf{n}}\times\overrightarrow{\mathbf{a}})}{Rc^2}$ and $\overrightarrow{\mathbf{B}}_R=\frac{q}{4\pi\varepsilon_0}\frac{\overrightarrow{\mathbf{a}}\times\overrightarrow{\mathbf{n}}}{Rc^3}$ are the radiation fields of a slowly moving particle – they decrease as $\frac{1}{R}$. Since $\overrightarrow{\mathbf{E}}_R$ is perpendicular to $\overrightarrow{\mathbf{n}}$ and $\overrightarrow{\mathbf{B}}_R=\overrightarrow{\mathbf{n}}\times\overrightarrow{\mathbf{E}}_R$, the radiation emitted is transverse. The Poynting vector (in the limit $R\longrightarrow\infty$) is given by:

$$\begin{aligned}\overrightarrow{\Pi}&=\frac{\overrightarrow{\mathbf{E}}_R\times\overrightarrow{\mathbf{B}}_R}{\mu_0}=\left(\frac{q}{4\pi\varepsilon_0}\right)^2\frac{\left[\overrightarrow{\mathbf{n}}\times\left(\overrightarrow{\mathbf{n}}\times\overrightarrow{\mathbf{a}}\right)\right]\times\left(\overrightarrow{\mathbf{a}}\times\overrightarrow{\mathbf{n}}\right)}{\mu_0R^2c^5}\\ &=\left(\frac{q}{4\pi\varepsilon_0}\right)^2\frac{\left(\overrightarrow{\mathbf{n}}\times\overrightarrow{\mathbf{a}}\right)^2\overrightarrow{\mathbf{n}}}{\mu_0R^2c^5}\end{aligned} \qquad [4.29]$$

where we have neglected the terms in $\frac{1}{R^3}$, etc. If θ is the angle between $\overrightarrow{\mathbf{n}}$ and $\overrightarrow{\mathbf{a}}$, then:

$$\overrightarrow{\Pi}=\left(\frac{q}{4\pi\varepsilon_0}\right)^2\frac{a^2\sin^2\theta}{\mu_0R^2c^5}\overrightarrow{\mathbf{n}} \qquad [4.30]$$

It follows that the radiated power, per unit solid angle, is given by:

$$\frac{dP}{d\Omega} = \frac{q^2 a^2 \sin^2\theta}{16\pi^2 \varepsilon_0 c^3} \qquad [4.31]$$

The total radiated power is obtained by integrating over the entire solid angle:

$$P = \frac{q^2 a^2}{16\pi^2 \varepsilon_0 c^3} \int_0^{\pi} d\theta \int_0^{2\pi} d\varphi \sin^2\theta = \frac{q^2 a^2}{8\pi \varepsilon_0 c^3} \int_0^{\pi} d\theta \sin^3\theta = \frac{q^2 a^2}{6\pi \varepsilon_0 c^3} \qquad [4.32]$$

This is the *Larmor formula*: the power of the radiation emitted by a non-relativistic accelerated particle is proportional to a^2. This radiation is of the same nature as that of an oscillating electric dipole in the direction $\vec{\mathbf{n}}$, except that here it is instantaneous radiation rather than monochromatic radiation.

4.5. Radiation from a relativistic charge

When the speed of the charged particle is not small compared to c, it is more convenient to write the radiation fields in the following form:

$$\begin{aligned} \vec{\mathbf{E}}_R &= \frac{q}{4\pi\varepsilon_0} \left[\frac{\vec{\mathbf{n}} \times \left[\left(\vec{\mathbf{n}} - \vec{\beta} \right) \times \vec{\mathbf{a}} \right]}{Rc^2 \left(1 - \vec{\mathbf{n}} \cdot \vec{\beta} \right)^3} \right]_{\text{ret.}} \\ \vec{\mathbf{B}}_R &= \left[\frac{\vec{\mathbf{n}} \times \vec{\mathbf{E}}_R}{c} \right]_{\text{ret.}} \end{aligned} \qquad [4.33]$$

The component of the Poynting vector along $\vec{\mathbf{n}}$ is given by:

$$\begin{aligned} \vec{\Pi} \cdot \vec{\mathbf{n}} &= \left(\frac{\vec{\mathbf{E}}_R \times \vec{\mathbf{B}}_R}{\mu_0} \right) \cdot \vec{\mathbf{n}} = \left(\frac{q}{4\pi\varepsilon_0} \right)^2 \left[\frac{\left| \vec{\mathbf{n}} \times \left[\left(\vec{\mathbf{n}} - \vec{\beta} \right) \times \vec{\mathbf{a}} \right] \right|^2}{\mu_0 R^2 c^5 \left(1 - \vec{\mathbf{n}} \cdot \vec{\beta} \right)^6} \right]_{\text{ret.}} \\ &= \frac{q^2}{16\pi^2 \varepsilon_0 c^3} \left[\frac{\left| \vec{\mathbf{n}} \times \left[\left(\vec{\mathbf{n}} - \vec{\beta} \right) \times \vec{\mathbf{a}} \right] \right|^2}{R^2 \left(1 - \vec{\mathbf{n}} \cdot \vec{\beta} \right)^6} \right]_{\text{ret.}} \end{aligned} \qquad [4.34]$$

It is useful to note that $\vec{\Pi} \cdot \vec{\mathbf{n}}$ represents the energy flux propagating in the direction $\vec{\mathbf{n}}$, as detected by an observer at $\vec{\mathbf{r}}$ (between observation times t and $t + \Delta t$). This same energy was emitted during a previous time interval $\Delta t'$ (i.e. between the emission times t' and $t' + \Delta t'$), which differs in duration from the interval Δt during which it is observed. This means that the rate of reception is different from the rate of emission. In other words, the power received per unit solid angle is given by:

$$\frac{dP_{\text{reç.}}}{d\Omega} = \lim_{\Delta t \longrightarrow 0} \frac{\Delta}{\Delta t}\left(\frac{dW}{d\Omega}\right) = R^2\left(\overrightarrow{\mathbf{\Pi}} \cdot \overrightarrow{\mathbf{n}}\right) \qquad [4.35]$$

while the power radiated at the point $\overrightarrow{\mathbf{r}}$ (between the emission times t' and $t' + \Delta t'$) by the particle, per unit of solid angle, will be:

$$\begin{aligned}\frac{dP_{\text{rad.}}}{d\Omega} &= \lim_{\Delta t' \longrightarrow 0} \frac{\Delta}{\Delta t'}\left(\frac{dW}{d\Omega}\right) = \lim_{\Delta t' \longrightarrow 0} \frac{\Delta}{\Delta t}\left(\frac{dW}{d\Omega}\right)\frac{\Delta t}{\Delta t'} \\ &= \frac{dP_{\text{reç.}}}{d\Omega}\frac{dt}{dt'} = R^2\left(\overrightarrow{\mathbf{\Pi}} \cdot \overrightarrow{\mathbf{n}}\right)\frac{dt}{dt'}\end{aligned} \qquad [4.36]$$

Since $\frac{dt}{dt'} = 1 + \frac{1}{c}\frac{dR}{dt'} = 1 - \overrightarrow{\mathbf{n}} \cdot \overrightarrow{\beta}$, we finally obtain:

$$\frac{dP_{\text{rad.}}}{d\Omega} = \frac{q^2}{16\pi^2\varepsilon_0 c^3}\left[\frac{\left|\overrightarrow{\mathbf{n}} \times \left[\left(\overrightarrow{\mathbf{n}} - \overrightarrow{\beta}\right) \times \overrightarrow{\mathbf{a}}\right]\right|^2}{\left(1 - \overrightarrow{\mathbf{n}} \cdot \overrightarrow{\beta}\right)^5}\right]_{\text{ret.}} \qquad [4.37]$$

This expression is rather complicated to compute. For this reason, we will examine the following cases separately: (1) the velocity is parallel to the acceleration (uniformly accelerated linear motion); (2) the velocity is perpendicular to the acceleration (circular motion).

4.5.1. *Velocity and acceleration parallel*

If $\overrightarrow{\mathbf{n}}$ makes an angle θ with $\overrightarrow{\beta}$ (and $\overrightarrow{\mathbf{a}}$), then $\overrightarrow{\mathbf{n}} \cdot \overrightarrow{\beta} = \beta\cos\theta$ and $\left(\overrightarrow{\mathbf{n}} - \overrightarrow{\beta}\right) \times \overrightarrow{\mathbf{a}} = \overrightarrow{\mathbf{n}} \times \overrightarrow{\mathbf{a}}$. This results in the fact that $\left|\overrightarrow{\mathbf{n}} \times \left(\overrightarrow{\mathbf{n}} \times \overrightarrow{\mathbf{a}}\right)\right| = a\sin\theta$, as illustrated in Figure 4.1(a).

By substituting these results into [4.37], we find:

$$\frac{dP_{\text{rad.}}}{d\Omega} = \frac{q^2 a^2 \sin^2\theta}{16\pi^2\varepsilon_0 c^3 \left(1 - \beta\cos\theta\right)^5} \qquad [4.38]$$

The angular distributions of $\frac{dP_{\text{rad.}}}{d\Omega}$ for $\beta \longrightarrow 0$, $\beta = 0.2$, $\beta = 0.5$ and $\beta = 0.95$ are illustrated in Figure 4.2. The following observations can be made:

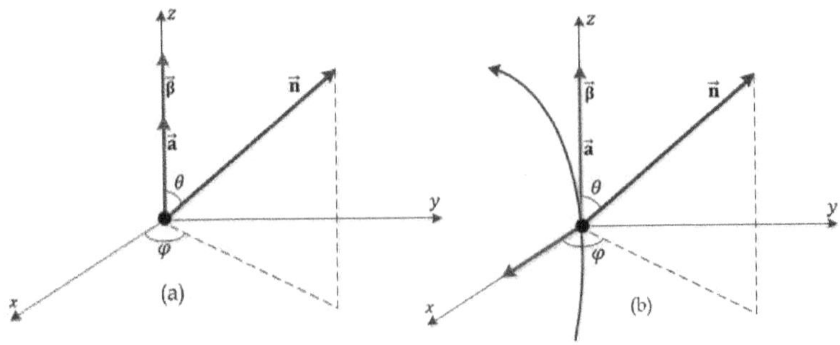

Figure 4.1. *Radiation from an accelerated charged particle: (a) velocity and acceleration parallel; (b) acceleration perpendicular to velocity. For a color version of this figure, see www.iste.co.uk/abdesselam/electrodynamics3.zip*

1) For $\beta \ll 1$, we recover the non-relativistic result [4.31].

2) When $\beta \longrightarrow 1$ (ultra-relativistic particle), there is an increase in radiation in the direction of motion (forward beaming). It can be shown that an ultra-relativistic particle emits a maximum amount of radiation at $\theta_{\max} = \frac{1}{2}\sqrt{1-\beta^2}$. The closer the particle's speed to c, the more radiation is concentrated in the forward direction. To verify this, let us set $u = \cos\theta$. The derivative of $\frac{dP_{\text{rad.}}}{d\Omega}$ with respect to u is equal to:

$$\frac{d}{du}\left(\frac{dP_{\text{rad.}}}{d\Omega}\right) = \frac{q^2 a^2}{16\pi^2\varepsilon_0 c^3}\,\frac{-2u\,(1-\beta u) + 5\beta\,(1-u^2)}{(1-\beta u)^6} \qquad [4.39]$$

This derivative vanishes in the direction where the radiation is maximal, that is, for the value:

$$u = \frac{1}{3\beta}\left(\sqrt{1+15\beta^3} - 1\right) \qquad [4.40]$$

which leads to:

$$\theta_{\max} = \cos^{-1}\left[\frac{1}{3\beta}\left(\sqrt{1+15\beta^3} - 1\right)\right] \qquad [4.41]$$

For an ultra-relativistic particle, β is very close to 1. We can therefore make the following approximation: $\beta = \sqrt{1-\frac{1}{\gamma^2}} \simeq 1 - \frac{1}{2\gamma^2}$, which indeed gives:

$$\begin{aligned}\theta_{\max} &= \cos^{-1}\left\{\frac{1}{3}\left(1+\frac{1}{2\gamma^2}\right)\left[4\left(1-\frac{15}{32\gamma^2}\right) - 1\right]\right\} \\ &\simeq \cos^{-1}\left(1-\frac{1}{8\gamma^2}\right) \simeq \frac{1}{2\gamma} = \frac{1}{2}\sqrt{1-\beta^2}\end{aligned} \qquad [4.42]$$

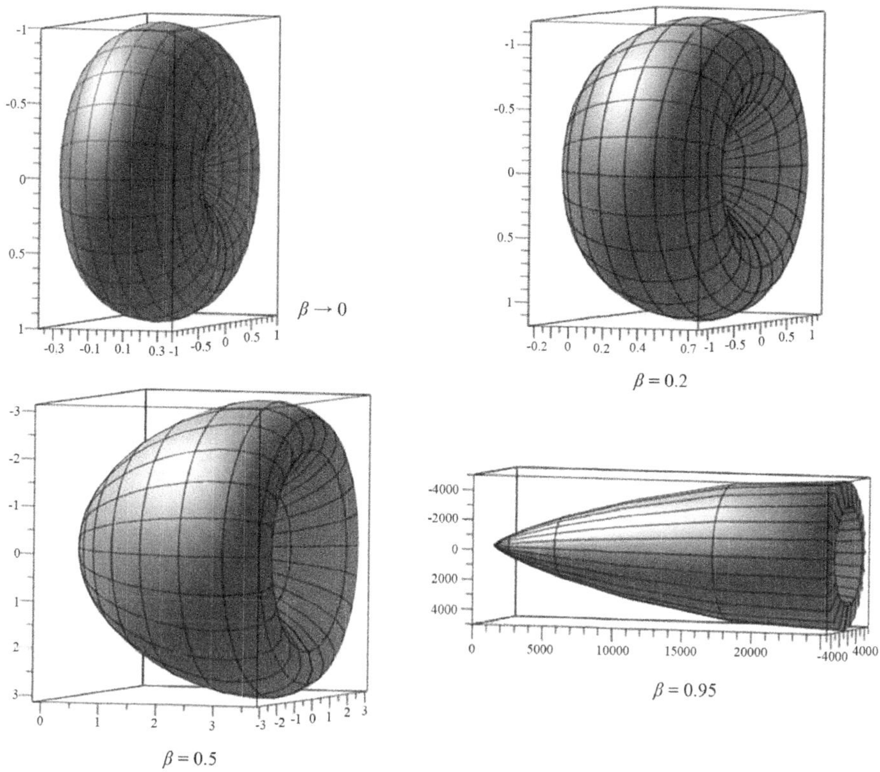

Figure 4.2. *Angular distributions of the radiation from a charged particle accelerated along the direction of its velocity for $\beta \longrightarrow 0$, $\beta = 0.2$, $\beta = 0.5$ and $\beta = 0.95$. For a color version of this figure, see www.iste.co.uk/abdesselam/electrodynamics3.zip*

3) Since the entire radiation from ultra-relativistic charges is concentrated in the forward direction, that is, in the region of small θ, we can therefore make the approximation:

$$\frac{dP_{\text{rad.}}}{d\Omega} \approx \frac{2q^2a^2\theta^2}{\pi^2\varepsilon_0 c^3\left[2\left(1-\beta\right)+\beta\theta^2\right]^5} \approx \frac{2q^2a^2\gamma^8}{\pi^2\varepsilon_0 c^3}\frac{\gamma^2\theta^2}{\left(1+\gamma^2\theta^2\right)^5} \qquad [4.43]$$

This distribution is represented in Figure 4.3: it reaches its peak at $\gamma\theta = \frac{1}{2}$ and drops to half its maximum at $\gamma\theta = 0.23$ and $\gamma\theta = 0.92$. In this ultra-relativistic limit, the root mean square of the radiation angle is equal to:

$$\langle\theta^2\rangle^{1/2} = \frac{1}{\gamma} = \frac{mc^2}{E} \qquad [4.44]$$

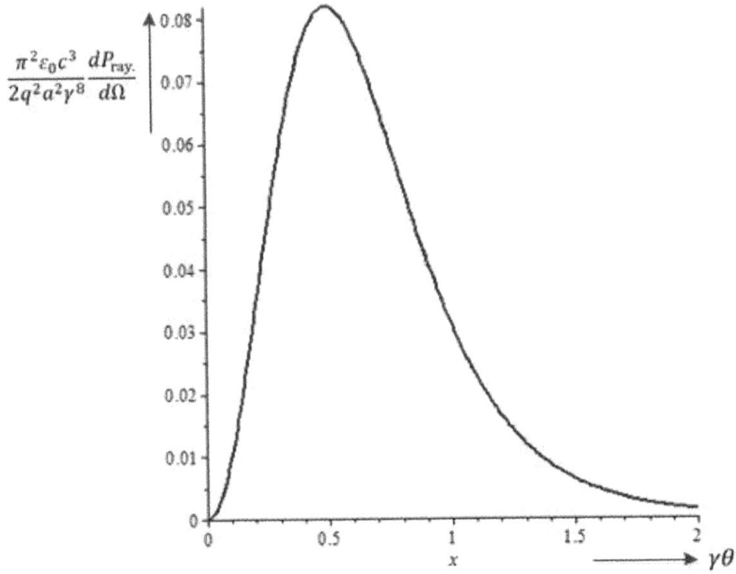

Figure 4.3. *Angular distribution of radiation for an ultra-relativistic charged particle*

To compute the total radiated power, we must integrate [4.38] over the entire space:

$$\begin{aligned}
P_{\text{rad.}} &= \frac{q^2a^2}{16\pi^2\varepsilon_0 c^3}\int_{\theta=0}^{\pi}\int_{\varphi=0}^{2\pi}\frac{\sin^2\theta}{(1-\beta\cos\theta)^5}\sin\theta d\theta d\varphi \\
&= \frac{q^2a^2}{8\pi\varepsilon_0 c^3}\int_0^{\pi}\frac{\sin^3\theta d\theta}{(1-\beta\cos\theta)^5} \\
&= \frac{q^2a^2}{8\pi\varepsilon_0 c^3}\int_{-1}^{1}\frac{(1-u^2)\,du}{(1-\beta u)^5} = \frac{q^2a^2}{8\pi\varepsilon_0 c^3}\int_{1-\beta}^{1+\beta}\frac{(\beta^2-x^2+2x-1)\,dx}{\beta^3x^5} \\
&= \frac{q^2a^2}{6\pi\varepsilon_0 c^3}\frac{1}{(1-\beta^2)^3} = \frac{q^2a^2\gamma^6}{6\pi\varepsilon_0 c^3}
\end{aligned} \quad [4.45]$$

In the case of a linearly accelerated particle, the total radiated power is proportional to γ^6. It can be seen that this power becomes infinite when the velocity of the charged particle reaches the speed of light c. Therefore, a charged particle cannot be accelerated beyond c – it would emit an infinite amount of radiation.

4.5.2. *Acceleration perpendicular to velocity*

Before continuing our discussion, let us recall some results from mechanics.

EXAMPLE 4.1.– Let us consider a point charge that undergoes arbitrary motion in a plane, here denoted as the plane Ozx (the position is identified by polar coordinates ρ and ϕ, where ϕ is the angle between the position vector and the Oz axis). The position, velocity and acceleration vectors are written, respectively, as follows:

$$\begin{aligned}
&\vec{\mathbf{r}} = \rho\left(\cos\phi\,\vec{\mathbf{k}} + \sin\phi\,\vec{\mathbf{i}}\right)\\
&\vec{\mathbf{v}} = \dot{\rho}\left(\cos\phi\,\vec{\mathbf{k}} + \sin\phi\,\vec{\mathbf{i}}\right) + \rho\dot{\phi}\left(-\sin\phi\,\vec{\mathbf{k}} + \cos\phi\,\vec{\mathbf{i}}\right)\\
&\vec{\mathbf{a}} = \left(\ddot{\rho} - \rho\dot{\phi}^2\right)\left(\cos\phi\,\vec{\mathbf{k}} + \sin\phi\,\vec{\mathbf{i}}\right) + \left(\rho\ddot{\phi} + 2\dot{\rho}\dot{\phi}\right)\left(-\sin\phi\,\vec{\mathbf{k}} + \cos\phi\,\vec{\mathbf{i}}\right)
\end{aligned} \qquad [4.46]$$

which gives:

$$\vec{\mathbf{v}}\cdot\vec{\mathbf{a}} = \rho^2\dot{\phi}\ddot{\phi} + \rho\dot{\rho}\dot{\phi}^2 + \dot{\rho}\ddot{\rho} = \frac{1}{2}\frac{d}{dt}\left(\rho^2\dot{\phi}^2 + \dot{\rho}^2\right) \qquad [4.47]$$

For $\vec{\mathbf{a}}$ to be perpendicular to $\vec{\mathbf{v}}$, we must have $\rho^2\dot{\phi}^2 + \dot{\rho}^2 = A^2$, where A is a constant. This means that:

$$\begin{aligned}
&\dot{\rho} = A\cos\psi\\
&\rho\dot{\phi} = A\sin\psi
\end{aligned} \qquad [4.48]$$

By differentiating equations [4.48], we obtain the following:

$$\begin{aligned}
&\ddot{\rho} = -A\sin\psi\dot{\psi}\\
&\rho\ddot{\phi} = A\cos\psi\dot{\psi} - \dot{\rho}\dot{\phi} = A\cos\psi\left(\dot{\psi} - \dot{\phi}\right)
\end{aligned} \qquad [4.49]$$

Substituting [4.48] and [4.49] into [4.46], we get:

$$\begin{aligned}
&\vec{\mathbf{v}} = A\left[\cos(\phi+\psi)\,\vec{\mathbf{k}} + \sin(\phi+\psi)\,\vec{\mathbf{i}}\right] \equiv A\vec{\mathbf{k}}'\\
&\vec{\mathbf{a}} = A\left(\dot{\phi}+\dot{\psi}\right)\left[-\sin(\phi+\psi)\,\vec{\mathbf{k}} + \cos(\phi+\psi)\,\vec{\mathbf{i}}\right] \equiv A\left(\dot{\phi}+\dot{\psi}\right)\vec{\mathbf{i}}'
\end{aligned} \qquad [4.50]$$

where $\vec{\mathbf{k}}' = \cos(\phi+\psi)\,\vec{\mathbf{k}} + \sin(\phi+\psi)\,\vec{\mathbf{i}}$ and $\vec{\mathbf{i}}' = -\sin(\phi+\psi)\,\vec{\mathbf{k}} + \cos(\phi+\psi)\,\vec{\mathbf{i}}$. ■

Let us choose a coordinate system (attached to the charged particle) in which, at time $t' = t - \frac{R}{C}$, $\vec{\beta}$ lies along the Oz axis and $\vec{\mathbf{a}}$ along the Ox axis. The observer, located at a distance R from the particle, has angular coordinates θ and φ, with $\vec{\beta} = (0, 0, \beta)$, $\vec{\mathbf{a}} = (a, 0, 0)$ and $\vec{\mathbf{n}} = (\sin\theta\cos\varphi, \sin\theta\sin\varphi, \cos\theta)$. Consequently, it follows that:

$$\vec{n} \times \left[\left(\vec{n} - \vec{\beta}\right) \times \vec{a}\right] = \left(\vec{n} \cdot \vec{a}\right)\left(\vec{n} - \vec{\beta}\right) - \left[\vec{n} \cdot \left(\vec{n} - \vec{\beta}\right)\right]\vec{a}$$
$$= a \sin\theta \cos\varphi \left(\vec{n} - \vec{\beta}\right) - \left(1 - \beta\cos\theta\right)\vec{a} \qquad [4.51]$$

Squaring equation [4.51]:

$$\left|\vec{n} \times \left[\left(\vec{n} - \vec{\beta}\right) \times \vec{a}\right]\right|^2 = a^2 \sin^2\theta \cos^2\varphi \left(1 - 2\vec{n} \cdot \vec{\beta} + \beta^2\right)$$
$$+ \left(1 - \beta\cos\theta\right)^2 a^2 - 2a \sin\theta \cos\varphi \left(1 - \beta\cos\theta\right)\left(\vec{n} - \vec{\beta}\right) \cdot \vec{a}$$
$$= a^2 \sin^2\theta \cos^2\varphi \left(1 - 2\beta\cos\theta + \beta^2\right) + \left(1 - \beta\cos\theta\right)^2 a^2$$
$$-2a^2 \sin^2\theta \cos^2\varphi \left(1 - \beta\cos\theta\right)$$
$$= a^2 \sin^2\theta \cos^2\varphi \left(\beta^2 - 1\right)\left(1 - \beta\cos\theta\right)^2 a^2$$
$$= a^2 \left(1 - \beta\cos\theta\right)^2 \left[1 - \frac{\sin^2\theta \cos^2\varphi}{\gamma^2 \left(1 - \beta\cos\theta\right)^2}\right] \qquad [4.52]$$

By writing [4.52] in [4.37], we get:

$$\frac{dP_{\text{rad.}}}{d\Omega} = \frac{q^2}{16\pi^2\varepsilon_0 c^3} \frac{a^2}{\left(1 - \beta\cos\theta\right)^3} \left[1 - \frac{\sin^2\theta \cos^2\varphi}{\gamma^2 \left(1 - \beta\cos\theta\right)^2}\right] \qquad [4.53]$$

The angular distributions of $\frac{dP_{\text{rad.}}}{d\Omega}$ for $\beta \longrightarrow 0$, $\beta = 0.2$, $\beta = 0.5$ and $\beta = 0.9$ are illustrated in Figure 4.4.

Here are some observations:

1) The radiation is mainly directed in the plane perpendicular to the acceleration (i.e. at $\varphi = \frac{\pi}{2}$ or $\varphi = \frac{3\pi}{2}$), and predominantly forward (in the direction of the particle's velocity), except when $\beta = 0$. The higher the velocity, the more the radiation is elongated in the direction of $\vec{\beta}$.

2) In the plane $\varphi = 0$ (the same results hold for $\varphi = \pi$), expression [4.53] simplifies as follows:

$$\left.\frac{dP_{\text{rad.}}}{d\Omega}\right|_{\varphi=0} = \frac{q^2 a^2}{16\pi^2\varepsilon_0 c^3} \frac{\left(\beta - \cos\theta\right)^2}{\left(1 - \beta\cos\theta\right)^5} \qquad [4.54]$$

- The radiation vanishes in the direction $\theta = \theta_c = \cos^{-1}\beta$. For an ultra-relativistic charge, this gives approximately $\theta_c \simeq \frac{1}{\gamma}$.

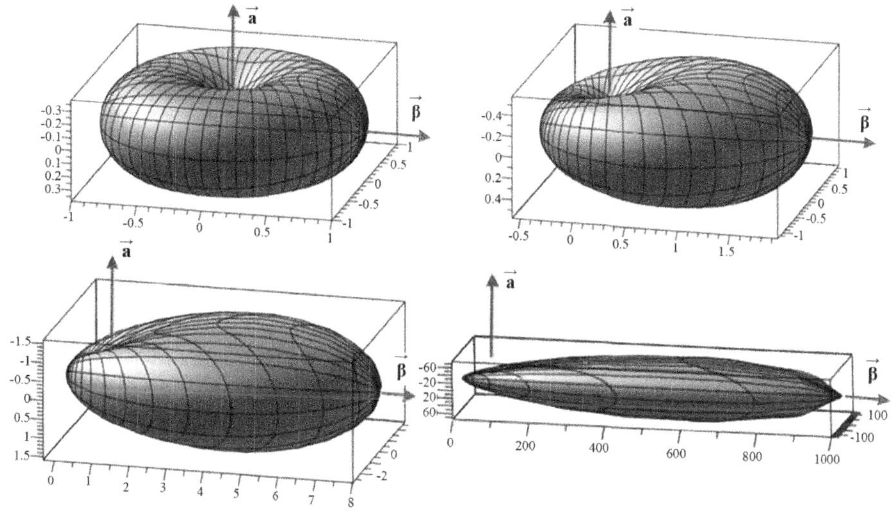

Figure 4.4. *3D view of the radiation from a charged particle accelerated perpendicular to its velocity, for $\beta \longrightarrow 0$, $\beta = 0.2$, $\beta = 0.5$ and $\beta = 0.9$. For a color version of this figure, see www.iste.co.uk/abdesselam/electrodynamics3.zip*

- Function [4.54] has extrema at $\theta = 0$, $\theta = \cos^{-1}\left(\frac{5\beta^2-2}{3\beta}\right)$, and $\theta = \pi$. To verify this, let us differentiate [4.54] with respect to θ:

$$\frac{d}{d\theta}\left.\frac{dP_{\text{rad.}}}{d\Omega}\right|_{\varphi=0} = \frac{q^2a^2}{16\pi^2\varepsilon_0 c^3}\frac{\sin\theta\,(\beta-\cos\theta)\left(2-5\beta^2+3\beta\cos\theta\right)}{(1-\beta\cos\theta)^6} \qquad [4.55]$$

which shows that $\left.\frac{dP_{\text{rad.}}}{d\Omega}\right|_{\varphi=0}$ reaches its extrema at the angles $\theta = 0$, $\theta = \cos^{-1}\left(\frac{5\beta^2-2}{3\beta}\right)$, and $\theta = \pi$.

Thus, we find:

$$\begin{aligned}
\frac{dP_{\text{rad.}}}{d\Omega}(0,0) &= \frac{q^2a^2}{16\pi^2\varepsilon_0 c^3}\frac{1}{(1-\beta)^3}\\
\frac{dP_{\text{rad.}}}{d\Omega}\left(\cos^{-1}\left[\frac{5\beta^2-2}{3\beta}\right],0\right) &= \frac{q^2a^2}{16\pi^2\varepsilon_0 c^3}\frac{108}{3125\beta^2(1-\beta)^3}\\
\frac{dP_{\text{rad.}}}{d\Omega}(\pi,0) &= \frac{q^2a^2}{16\pi^2\varepsilon_0 c^3}\frac{1}{(1+\beta)^3}
\end{aligned} \qquad [4.56]$$

It is zero (minimal) for $\theta = \cos^{-1}\beta$. In the ultra-relativistic limit, equations [4.56] reduce to:

$$\begin{aligned}
\frac{dP_{\text{rad.}}}{d\Omega}(0,0) &\simeq \frac{q^2a^2\gamma^6}{2\pi^2\varepsilon_0c^3}\\
\frac{dP_{\text{rad.}}}{d\Omega}\left(\sqrt{\frac{7}{3}}\frac{1}{\gamma},0\right) &\simeq \frac{54q^2a^2\gamma^6}{3125\pi^2\varepsilon_0c^3}\\
\frac{dP_{\text{rad.}}}{d\Omega}(\pi,0) &\simeq \frac{q^2a^2}{148\pi^2\varepsilon_0c^3}
\end{aligned} \qquad [4.57]$$

3) In the ultra-relativistic limit, we can restrict ourselves to small angles in the power distribution (since all the power is concentrated in this region):

$$\frac{dP_{\text{rad.}}}{d\Omega} = \frac{q^2}{2\pi^2\varepsilon_0c^3}\frac{a^2\gamma^6}{(1+\gamma^2\theta^2)^3}\left[1-\frac{4\gamma^2\theta^2\cos^2\varphi}{(1+\gamma^2\theta^2)^2}\right] \qquad [4.58]$$

To obtain the total radiated power over the entire space, let us integrate over the variables θ and φ:

$$\begin{aligned}
P_{\text{rad.}} &= \frac{q^2a^2}{16\pi^2\varepsilon_0c^3}\int_0^\pi \frac{\sin\theta d\theta}{(1-\beta\cos\theta)^3}\int_0^{2\pi} d\varphi\left[1-\frac{\sin^2\theta\cos^2\varphi}{\gamma^2(1-\beta\cos\theta)^2}\right]\\
&= \frac{q^2a^2}{16\pi^2\varepsilon_0c^3}\int_0^\pi \frac{\sin\theta d\theta}{(1-\beta\cos\theta)^3}\left[2\pi-\frac{\pi\sin^2\theta}{\gamma^2(1-\beta\cos\theta)^2}\right]\\
&= \frac{q^2a^2}{16\pi\varepsilon_0c^3}\int_0^\pi \frac{\sin\theta d\theta\left[(1+\beta^2)\left(1+\cos^2\theta\right)-4\beta\cos\theta\right]}{(1-\beta\cos\theta)^5}\\
&= \frac{q^2a^2}{16\pi\varepsilon_0c^3}\int_{-1}^1 \frac{du\left[(1+\beta^2)\left(1+u^2\right)-4\beta u\right]}{(1-\beta u)^5}\\
&= \frac{q^2a^2}{16\pi\varepsilon_0c^3}\int_{1-\beta}^{1+\beta} \frac{\beta^4+\beta^2x^2+2\beta^2x-2\beta^2+x^2-2x+1}{\beta^3x^5}dx\\
&= \frac{q^2a^2}{16\pi\varepsilon_0c^3}\frac{8}{3(1-\beta^2)^2} = \frac{q^2a^2\gamma^4}{6\pi\varepsilon_0c^3}
\end{aligned} \qquad [4.59]$$

The total power radiated by a particle in a circular orbit is proportional to γ^4. It follows that for a given velocity and acceleration, a particle in circular motion radiates less than a particle on a linear trajectory.

Let us study the case where the force is constant. The force exerted on a relativistic particle is:

$$\vec{\mathbf{f}} = \frac{d\vec{\mathbf{p}}}{dt} = \frac{d}{dt}\left(m\gamma\vec{\mathbf{v}}\right) = m\gamma\vec{\mathbf{a}} + m\gamma^3\left(\vec{\beta}\cdot\vec{\mathbf{a}}\right)\vec{\beta} \qquad [4.60]$$

and consequently:

$$\vec{\mathbf{f}}^2 = m^2\gamma^2 a^2 + m^2\gamma^6\left(\vec{\beta}\cdot\vec{\mathbf{a}}\right)^2\beta^2 + 2m^2\gamma^4\left(\vec{\beta}\cdot\vec{\mathbf{a}}\right)^2$$
$$= m^2\gamma^2 a_\perp^2 + m^2\gamma^2 a_\parallel^2 + m^2\gamma^6 a_\parallel^2\beta^4 + 2m^2\gamma^4 a_\parallel^2\beta^2 = m^2\gamma^2 a_\perp^2 + m^2\gamma^6 a_\parallel^2 \quad [4.61]$$

As a function of the force $\vec{\mathbf{f}}$ acting on the particle, the total radiated power is therefore:

$$P = \begin{cases} \dfrac{q^2 a_\parallel^2\gamma^6}{6\pi\varepsilon_0 c^3} = \dfrac{q^2\vec{\mathbf{f}}^2}{6\pi\varepsilon_0 m^2 c^3} & \vec{\mathbf{a}} \parallel \vec{\beta} \\ \dfrac{q^2 a_\perp^2\gamma^4}{6\pi\varepsilon_0 c^3} = \dfrac{q^2\vec{\mathbf{f}}^2\gamma^2}{6\pi\varepsilon_0 m^2 c^3} & \vec{\mathbf{a}} \perp \vec{\beta} \end{cases} \quad [4.62]$$

Under the same force, a charged particle radiates more energy in a circular orbit than along a linear trajectory. It is therefore more difficult to accelerate an ultra-relativistic charge ($\gamma \gg 1$) in circular motion than along a straight path.

Here are a few remarks:

1) In a linear accelerator, the motion is one-dimensional. The radiation emitted by a particle undergoing linear acceleration is so weak that it has never been observed. This is easy to understand: let us consider the example of a linearly accelerated electron. According to [4.62], the power radiated by the linearly accelerated electron is:

$$P = \frac{e^2\vec{\mathbf{f}}^2}{6\pi\varepsilon_0 m^2 c^3} = \frac{e^2}{6\pi\varepsilon_0 m^2 c^3}\left|\frac{d\vec{\mathbf{p}}}{dt}\right|^2 \quad [4.63]$$

Since the rate of change of momentum is equal to the change in the particle's energy per unit distance, we write:

$$P = \frac{e^2}{6\pi\varepsilon_0 m^2 c^3}\left(\frac{dE}{dx}\right)^2 \quad [4.64]$$

This formula shows that for linear motion, the radiated power depends only on the external forces that determine the rate of change of the particle's energy with respect to distance, and not on the actual energy or momentum of the particle. The ratio of the radiated power to the power supplied to the particle by external sources is given by:

$$\frac{P}{\frac{dE}{dt}} = \frac{e^2}{6\pi\varepsilon_0 m^2 c^3}\left(\frac{dE}{dx}\right)^2\frac{1}{\frac{dE}{dx}\frac{dx}{dt}} = \frac{e^2}{6\pi\varepsilon_0 m^2 c^3 v}\frac{dE}{dx} \quad [4.65]$$

In the case of an ultra-relativistic electron, the previous equation reduces to:

$$\frac{P}{\frac{dE}{dt}} = \frac{e^2}{6\pi\varepsilon_0 m^2 c^4}\frac{dE}{dx} = \frac{2}{3}\frac{r_e}{mc^2}\frac{dE}{dx} \quad [4.66]$$

where $r_e = \frac{e^2}{4\pi\varepsilon_0 mc^2}$ is the classical radius of the electron. For the electron's radiation to be significant, the radiated power would need to be of the same order as the power supplied to the electron, that is:

$$\frac{dE}{dx} \sim \frac{mc^2}{r_e} \quad [4.67]$$

In other words, the energy supplied to the electron over a distance on the order of its radius would need to be comparable to its rest energy. We therefore conclude that the radiated energy is entirely negligible compared to the energy supplied to the particle.

2) The situation changes radically for circular accelerators. Let us consider an electron moving at constant speed along a circle of radius ρ. The acceleration is entirely directed toward the center of the circle and has a magnitude of $a = \frac{c^2\beta^2}{\rho}$. According to [4.62], this electron radiates the following power:

$$P = \frac{e^2\gamma^4 a^2}{6\pi\varepsilon_0 c^3} = \frac{e^2 c\gamma^4\beta^4}{6\pi\varepsilon_0\rho^2} \quad [4.68]$$

As a function of the electron's energy, the previous formula can also be written as:

$$P = \frac{e^2 c\beta^4}{6\pi\varepsilon_0\rho^2}\left(\frac{E}{mc^2}\right)^4 \quad [4.69]$$

The energy emitted per cycle of the motion is $\Delta E = PT$, where T is the period of the motion, $T = \frac{2\pi\rho}{c\beta}$. Thus, we obtain:

$$\Delta E = \frac{e^2\beta^3}{3\varepsilon_0\rho}\left(\frac{E}{mc^2}\right)^4 \quad [4.70]$$

which is a non-negligible quantity in practical terms.

3) For a charged particle undergoing arbitrary ultra-relativistic motion, the radiation emitted at any given instant can be considered as a coherent superposition of contributions from the components of the acceleration parallel and perpendicular to the velocity. However, as we have just seen, when the parallel and perpendicular forces are of comparable magnitude, the radiation from the parallel component is negligible (of the order of $\frac{1}{\gamma^2}$) compared to that from the perpendicular component.

Consequently, the parallel component of the acceleration can be neglected, and the radiation intensity can be approximated by that due solely to the perpendicular component. In other words, the radiation emitted by a charged particle in arbitrary ultra-relativistic motion is approximately the same as that emitted by a particle moving instantaneously along the arc of a circular trajectory whose radius of curvature ρ is given by:

$$\rho = \frac{c^2\beta^2}{a_\perp} \simeq \frac{c^2}{a_\perp} \quad [4.71]$$

where $a_\perp$ is the perpendicular component of the acceleration. The angular distribution of the radiation is given by [4.53] or [4.58]. It corresponds to a narrow cone or a spotlight-like radiation beam directed along the instantaneous velocity vector of the charge. The confinement of the radiation around the direction of the charged particle's velocity has important consequences. The radiation will only be visible if the particle's velocity is directed toward the observer. For a particle in arbitrary motion, the observer will detect a very short radiation pulse (or a series of pulses if the motion is periodic), as illustrated in Figure 4.5. This is referred to as the headlight effect. Since the angular width of the beam is on the order of γ^{-1}, the particle will illuminate the observer only during a time interval:

$$\Delta t' = \frac{\rho \times \frac{1}{\gamma}}{v} \sim \frac{\rho}{\gamma c} \quad [4.72]$$

in terms of its own time, where ρ is the radius of curvature [4.71]. However, the observer sees it at time interval $\Delta t \sim \langle \frac{dt}{dt'} \rangle \Delta t'$, where $\langle \frac{dt}{dt'} \rangle = \langle 1 - \vec{\mathbf{n}} \cdot \vec{\beta} \rangle \sim \frac{1}{\gamma^2}$.

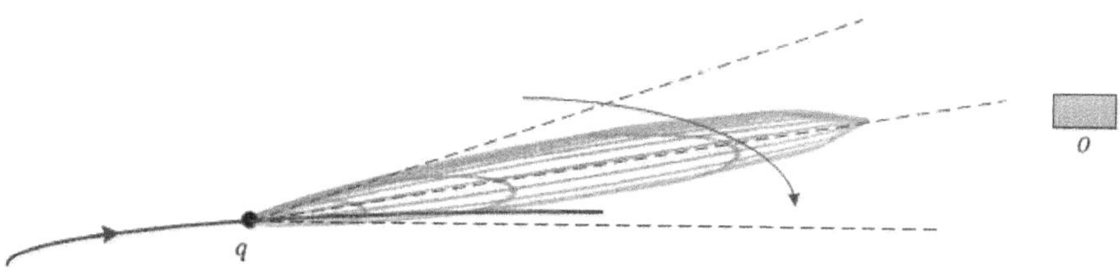

Figure 4.5. *Headlight effect: the radiating particle illuminates the observer located at point O only during the time interval Δt. For a color version of this figure, see www.iste.co.uk/abdesselam/electrodynamics3.zip*

Consequently, the radiation will illuminate the observer only during the time interval:

$$\Delta t \sim \frac{\rho}{\gamma^3 c} \quad [4.73]$$

4.5.3. *General case*

For any arbitrary velocity and acceleration, we have:

$$P_{\text{rad.}} = \frac{q^2}{16\pi^2\varepsilon_0 c^3}\left[\int \frac{d\Omega}{\left(1-\vec{\mathbf{n}}\cdot\vec{\beta}\right)^5}\left|\vec{\mathbf{n}}\times\left[\left(\vec{\mathbf{n}}-\vec{\beta}\right)\times\vec{\mathbf{a}}\right]\right|^2\right]_{\text{ret.}} \quad [4.74]$$

Expressing $\vec{\mathbf{n}} = (\sin\theta\cos\varphi, \sin\theta\sin\varphi, \cos\theta)$, $\vec{\beta} = (0,0,\beta)$ and $\vec{\mathbf{a}} = \vec{\mathbf{a}}_{\|} + \vec{\mathbf{a}}_{\perp}$, with $\vec{\mathbf{a}}_{\|} = (0,0,a_{\|})$, and $\vec{\mathbf{a}}_{\perp} = (a_{\perp},0,0)$, we get:

$$\vec{\mathbf{n}}\times\left[\left(\vec{\mathbf{n}}-\vec{\beta}\right)\times\vec{\mathbf{a}}\right] = \left(\vec{\mathbf{n}}\cdot\vec{\mathbf{a}}\right)\left(\vec{\mathbf{n}}-\vec{\beta}\right) - \left(1-\vec{\mathbf{n}}\cdot\vec{\beta}\right)\vec{\mathbf{a}} \quad [4.75]$$

$$= \left(\sin\theta\cos\varphi a_{\perp} + \cos\theta a_{\|}\right)\left(\vec{\mathbf{n}}-\vec{\beta}\right) - (1-\beta\cos\theta)\,\vec{\mathbf{a}}$$

which yields:

$$\left|\vec{\mathbf{n}}\times\left[\left(\vec{\mathbf{n}}-\vec{\beta}\right)\times\vec{\mathbf{a}}\right]\right|^2 = \left(\sin\theta\cos\varphi a_{\perp} + \cos\theta a_{\|}\right)^2\left(1+\beta^2-2\beta\cos\theta\right) +$$

$$(1-\beta\cos\theta)^2\left(a_{\|}^2+a_{\perp}^2\right) - 2\left(\sin\theta\cos\varphi a_{\perp}+\cos\theta a_{\|}\right)\left(\sin\theta\cos\varphi a_{\perp}+\right.$$

$$\left.\cos\theta a_{\|} - \beta a_{\|}\right)(1-\beta\cos\theta) = -\left(1-\beta^2\right)\left(\sin\theta\cos\varphi a_{\perp}+\cos\theta a_{\|}\right)^2 +$$

$$(1-\beta\cos\theta)^2\left(a_{\|}^2+a_{\perp}^2\right) + 2\beta a_{\|}\left(\sin\theta\cos\varphi a_{\perp}+\cos\theta a_{\|}\right)(1-\beta\cos\theta) \quad [4.76]$$

Let us insert [4.76] into [4.74] and then integrate over the variable φ:

$$P_{\text{rad.}} = \frac{q^2}{8\pi\varepsilon_0 c^3}\int_0^{\pi}\frac{\sin\theta d\theta}{(1-\beta\cos\theta)^5}\left[-\left(1-\beta^2\right)\left(\frac{1}{2}\sin^2\theta a_{\perp}^2+\cos^2\theta a_{\|}^2\right)+\right.$$

$$\left.(1-\beta\cos\theta)^2\left(a_{\|}^2+a_{\perp}^2\right)+2\beta\cos\theta a_{\|}^2(1-\beta\cos\theta)\right]$$

$$= \frac{q^2}{8\pi\varepsilon_0 c^3}\int_0^{\pi}\frac{\sin\theta d\theta}{(1-\beta\cos\theta)^5}\left\{\left[\frac{1}{2}\left(1+\beta^2\right)\left(\cos^2\theta+1\right)\right.\right.$$

$$\left.\left.-2\beta\cos\theta\right]a_{\perp}^2+\sin^2\theta a_{\|}^2\right\} = \frac{q^2}{8\pi\varepsilon_0 c^3}\int_{-1}^{1}\frac{du}{(1-\beta u)^5}\times$$

$$\left\{\left[\frac{1}{2}\left(1+\beta^2\right)\left(u^2+1\right)-2\beta u\right]a_{\perp}^2+\left(1-u^2\right)a_{\|}^2\right\}$$

$$= \frac{q^2}{8\pi\varepsilon_0 c^3}\int_{1-\beta}^{1+\beta} dx\left[\left(\frac{\beta^4+\beta^2x^2+2\beta^2x-2\beta^2+x^2-2x+1}{2\beta^3x^5}\right)a_\perp^2 + \left(\frac{\beta^2-x^2+2x-1}{\beta^3x^5}\right)a_\parallel^2\right] = \frac{q^2}{6\pi\varepsilon_0 c^3}\left(\gamma^4 a_\perp^2+\gamma^6 a_\parallel^2\right)$$

$$= \frac{q^2\gamma^6}{6\pi\varepsilon_0 c^3}\left[\left(1-\beta^2\right)a_\perp^2 + a_\parallel^2\right] \qquad [4.77]$$

This is the *relativistic Larmor formula*. It is easy to verify that the cases $\vec{\mathbf{a}} \perp \vec{\beta}$ and $\vec{\mathbf{a}} \parallel \vec{\beta}$ are recovered from this formula. If θ_0 denotes the angle between $\vec{\mathbf{a}}$ and $\vec{\beta}$, then $a_\perp = a\sin\theta_0$ and $a_\parallel = a\cos\theta_0$. The earlier equation is then written as:

$$P_{\text{rad.}} = \frac{q^2\gamma^6}{6\pi\varepsilon_0 c^3}\left[a^2-\beta^2a^2\sin^2\theta_0\right] = \frac{q^2\gamma^6}{6\pi\varepsilon_0 c^3}\left[\vec{\mathbf{a}}^2-\left(\vec{\mathbf{a}}\times\vec{\beta}\right)^2\right] \qquad [4.78]$$

EXAMPLE 4.2.– A non-relativistic particle of charge $Z_m e$, mass m and kinetic energy $\frac{1}{2}mv_0^2$ undergoes a head-on collision with a fixed central force field of finite range. The interaction is repulsive and is described by a potential $V(r)$, which exceeds $\frac{1}{2}mv_0^2$ as the particle approaches the target. We aim to calculate the radiated energy. In the non-relativistic limit, we can use the Larmor formula [4.62] (non-relativistic):

$$P(t) = \frac{(Z_m e)^2}{6\pi\varepsilon_0 m^2c^3}\left|\frac{d\vec{\mathbf{p}}}{dt}\right|^2 = \frac{(Z_m e)^2}{6\pi\varepsilon_0 m^2c^3}\left(\frac{dV(r)}{dr}\right)^2 \qquad [4.79]$$

where $\frac{d\vec{\mathbf{p}}}{dt} = -\frac{dV(r)}{dr}\vec{\mathbf{e}}_r$ (according to Newton's second law). The radiated energy is given by integrating the power over time:

$$\Delta W = \int_{-\infty}^{\infty} P(t)\,dt \qquad [4.80]$$

This last equation can be converted into an integral over the trajectory of the particle:

$$\Delta W = 2\int_{r_{\min}}^{\infty}\frac{P}{dr/dt}dr \qquad [4.81]$$

The factor 2 appears here due to the symmetry of the trajectory. The velocity $\frac{dr}{dt}$ can be obtained from the law of energy conservation, $E = \frac{1}{2}m\left(\frac{dr}{dt}\right)^2 + V(r)$, which leads to $\frac{dr}{dt} = \sqrt{\frac{2[E-V(r)]}{m}}$. By substituting the expressions for $P(t)$ and $\frac{dr}{dt}$ into [4.81], we obtain:

$$\Delta W = 2\frac{(Z_m e)^2}{6\pi\varepsilon_0 m^2c^3}\sqrt{\frac{m}{2}}\int_{r_{\min}}^{\infty}\left(\frac{dV}{dr}\right)^2\frac{dr}{\sqrt{E-V(r)}} \qquad [4.82]$$

Since the velocity (and therefore the kinetic energy) vanishes as the particle approaches the target, the total energy equals the potential energy at the point of closest approach, that is, $E = V(r_{\min})$. Using this, we finally have:

$$\Delta W = \frac{(Z_m e)^2}{3\pi\varepsilon_0 m^2 c^3}\sqrt{\frac{m}{2}}\int_{r_{\min}}^{\infty}\left(\frac{dV}{dr}\right)^2 \frac{dr}{\sqrt{V(r_{\min}) - V(r)}} \qquad [4.83]$$

If the interaction potential is of the Coulomb type, that is, of the form $V(r) = \frac{Z_m Z_f e^2}{r}$, then $\frac{dV}{dr} = -\frac{Z_m Z_f e^2}{r^2}$. Substituting this expression into [4.83], we get:

$$\begin{aligned}
\Delta W &= \frac{Z_m^3 Z_f e^5}{3\pi\varepsilon_0 m^2 c^3}\sqrt{\frac{Z_m Z_f m r_{min}}{2}}\int_{r_{\min}}^{\infty}\frac{1}{r^{7/2}}\frac{dr}{\sqrt{r - r_{\min}}} \\
&= \frac{Z_m^3 Z_f e^5}{3\pi\varepsilon_0 m^2 c^3 r_{\min}^3}\sqrt{\frac{Z_m Z_f m r_{min}}{2}}\int_{1}^{\infty}\frac{1}{r^{7/2}}\frac{dr}{\sqrt{r - 1}} \\
&= \frac{Z_m^3 Z_f e^5}{3\pi\varepsilon_0 m^2 c^3 r_{\min}^3}\sqrt{\frac{Z_m Z_f m r_{min}}{2}}\int_{0}^{\infty}\frac{1}{\cosh^7 x}\frac{2\cosh x \sinh x dx}{\sqrt{\cosh^2 x - 1}} \\
&= \frac{Z_m^3 Z_f e^5}{3\pi\varepsilon_0 m^2 c^3 r_{\min}^3}\sqrt{\frac{Z_m Z_f m r_{min}}{2}}\left[2\int_{0}^{\infty}\frac{dx}{\cosh^6 x}\right] \\
&= \frac{Z_m^3 Z_f e^5}{3\pi\varepsilon_0 m^2 c^3 r_{\min}^3}\sqrt{\frac{Z_m Z_f m r_{min}}{2}}\frac{16}{15} = \frac{8\sqrt{2} Z_m^{7/2} Z_f^{3/2} e^5}{45\pi\varepsilon_0 m^{3/2} c^3 r_{\min}^{5/2}}
\end{aligned} \qquad [4.84]$$

where we have used the identity, $\int \frac{dx}{\cosh^6 x} = \frac{\sinh x}{5\cosh^5 x} + \frac{4\sinh x}{15\cosh^3 x} + \frac{8}{15}\tanh x$. The quantity $r_{\min}$ can be related to the velocity v_0 at infinity using the energy conservation law $\frac{Z_m Z_f e^2}{r_{\min}} = \frac{1}{2}mv_0^2$, which gives:

$$r_{\min} = \frac{2Z_m Z_f e^2}{mv_0^2} \qquad [4.85]$$

By substituting this result into the expression for the radiated energy [4.84], we finally obtain the following.

$$\Delta W = \frac{2Z_m m v_0^5}{45\pi\varepsilon_0 c^3 Z_f} \qquad [4.86]$$

EXAMPLE 4.3.– Let us generalize the results of Example 4.2 to the case of non-zero angular momentum. Returning to equation [4.82], the radial velocity $\frac{dr}{dt}$ is obtained from the energy conservation equation $E = \frac{1}{2}m\left(\frac{dr}{dt}\right)^2 + \frac{L^2}{2mr^2} + V(r)$, which shows that $\frac{dr}{dt} = \sqrt{\frac{2}{m}}\left(E - V(r) - \frac{L^2}{2mr^2}\right)^{1/2}$. Substituting this expression into [4.82], we deduce that:

$$\Delta W = \frac{(Z_m e)^2}{3\pi\varepsilon_0 m^2 c^3}\sqrt{\frac{m}{2}}\int_{r_{\min}}^{\infty}\left(\frac{dV}{dr}\right)^2 \frac{dr}{\sqrt{E - V(r) - \frac{L^2}{2mr^2}}} \qquad [4.87]$$

By setting $V(r) = \frac{Z_m Z_f e^2}{r}$ (repulsive Coulomb potential), $L = mbv_0$ and $E = \frac{1}{2}mv_0^2$, equation [4.87] becomes:

$$\begin{aligned}\Delta W &= \frac{Z_m^4 Z_f^2 e^6}{3\pi\varepsilon_0 m^2 c^3 v_0}\int_{r_{\min}}^{\infty} r^{-4}\left[1 - \frac{2Z_m Z_f e^2}{mv_0^2 r} - \frac{b^2}{r^2}\right]^{-1/2} dr \\ &= \frac{Z_m m v_0^5}{3\pi\varepsilon_0 Z_f c^3 \xi^3}\int_{r_{\min}}^{\infty}\left(\frac{b}{r}\right)^2\left[1 - \frac{2b}{\xi r} - \frac{b^2}{r^2}\right]^{-1/2}\frac{bdr}{r^2} \\ &= \frac{Z_m m v_0^5}{3\pi\varepsilon_0 Z_f c^3 \xi^3}\int_0^{x_{\max}}\frac{x^2}{\sqrt{1 - \frac{2x}{\xi} - x^2}}dx \\ &= \frac{Z_m m v_0^5}{3\pi\varepsilon_0 Z_f c^3 \xi^3}\int_0^{x_+}\frac{x^2}{\sqrt{(x - x_-)(x_+ - x)}}dx \end{aligned} \qquad [4.88]$$

where we have used $\xi = \frac{mbv_0^2}{Z_m Z_f e^2}$ and the change of variable $x = \frac{b}{r}$. In the last line, x_+ and x_- are the roots of equation $1 - \frac{2x}{\xi} - x^2 = 0$, that is, $x_\pm = -\frac{1}{\xi} \pm \sqrt{\frac{1}{\xi^2} + 1}$. To compute the integral [4.88], we use (established by Euler's method): $\int \frac{x^2 dx}{\sqrt{1+(x_+ + x_-)x - x^2}} = \frac{-2x-3(x_+ + x_-)}{4}\sqrt{1+(x_+ + x_-)x - x^2} + \frac{3(x_+ + x_-)^2+4}{8}\sin^{-1}\left(\frac{2x-(x_+ + x_-)}{\sqrt{(x_+ + x_-)^2+4}}\right)$. Thus, we find:

$$\begin{aligned}\int_0^{x_+}\frac{x^2 dx}{\sqrt{1 + (x_+ + x_-)x - x^2}} &= \frac{3(x_+ + x_-)}{4} + \frac{3(x_+ + x_-)^2 + 4}{8}\times \\ &\left[\sin^{-1}\left(\frac{x_+ - x_-}{\sqrt{(x_+ + x_-)^2 + 4}}\right) + \sin^{-1}\left(\frac{x_+ + x_-}{\sqrt{(x_+ + x_-)^2 + 4}}\right)\right] \\ &= -\frac{3}{2\xi} + \frac{1}{2}\left(\frac{3}{\xi^2} + 1\right)\left[\frac{\pi}{2} + \sin^{-1}\left(-\frac{1}{\sqrt{1+\xi^2}}\right)\right] \\ &= -\frac{3}{2\xi} + \frac{1}{2}\left(\frac{3}{\xi^2} + 1\right)\tan^{-1}\xi \end{aligned} \qquad [4.89]$$

since $\sin^{-1} x + \sin^{-1} y = \tan^{-1} \frac{x\sqrt{1-y^2}+y\sqrt{1-x^2}}{\sqrt{1-x^2}\sqrt{1-y^2}-xy}$ when $xy < 0$. By inserting this result into [4.88], we finally obtain the following:

$$\Delta W = -\frac{Z_m m v_0^5}{2\pi\varepsilon_0 Z_f c^3}\left[-\frac{1}{\xi^4} + \frac{1}{\xi^5}\left(1 + \frac{\xi^2}{3}\right)\tan^{-1}\xi\right] \qquad [4.90]$$

If $\xi \longrightarrow 0$, we naturally recover formula [4.86].

EXAMPLE 4.4. ELECTRON INSIDE A CAPACITOR.– An electron of charge $-e$ and mass m, initially at rest, is accelerated through a parallel plate capacitor maintained at a potential difference V. The distance between the two plates is d. Assuming non-relativistic motion, let us determine the total energy radiated by the electron during the process. The field inside a parallel plate capacitor is constant, which means that the acceleration and radiation of the particle are also constant:

$$P = \frac{e^2 a^2}{6\pi\varepsilon_0 c^3} \qquad [4.91]$$

with:

$$a = \frac{F}{m} = -\frac{eE}{m} = -\frac{eV}{md} \qquad [4.92]$$

Consequently, the energy radiated during its passage is given by:

$$W = \int_0^{\sqrt{2d/a}} P dt = \frac{\sqrt{2} e^{7/2} V^{3/2}}{6\pi\varepsilon_0 c^3 m^{3/2} d} \qquad [4.93]$$

where $\sqrt{2d/a}$ represents the duration of the traversal.

4.6. Frequency distribution of the radiated energy

The general form of the radiated power per unit of solid angle is:

$$\frac{dP_{\text{rad.}}(t)}{d\Omega} = \left[\frac{R^2\left|\vec{\mathbf{E}}(t)\right|^2}{\mu_0 c}\right]_{\text{ret.}} = \left|\vec{\mathsf{A}}(t)\right|^2 \qquad [4.94]$$

where:

$$\vec{\mathsf{A}}(t) = \frac{1}{\sqrt{\mu_0 c}}\left[R\vec{\mathbf{E}}(t)\right]_{\text{ret.}} \qquad [4.95]$$

$\overrightarrow{\mathbf{E}}_R$ here is the electric field given in [4.33]. In [4.94], instantaneous power is expressed in terms of observer time. Consequently, the resulting frequency spectrum will be in terms of the observer frequencies.

The total energy radiated per unit solid angle is the time integral of [4.94]:

$$\frac{dW}{d\Omega} = \int_{-\infty}^{\infty} \left|\overrightarrow{\mathsf{A}}(t)\right|^2 dt \qquad [4.96]$$

It can alternatively be expressed as an integral over a frequency spectrum using Fourier transforms. We introduce the Fourier transform $\overrightarrow{\hat{\mathsf{A}}}(\omega)$ of $\overrightarrow{\mathsf{A}}(t)$ as follows:

$$\overrightarrow{\hat{\mathsf{A}}}(\omega) = \frac{1}{\sqrt{2\pi}} \int_{-\infty}^{\infty} \overrightarrow{\mathsf{A}}(t)\, e^{i\omega t} dt \qquad [4.97]$$

and, conversely:

$$\overrightarrow{\mathsf{A}}(t) = \frac{1}{\sqrt{2\pi}} \int_{-\infty}^{\infty} \overrightarrow{\hat{\mathsf{A}}}(\omega)\, e^{-i\omega t} d\omega \qquad [4.98]$$

Equation [4.96] can be written as:

$$\frac{dW}{d\Omega} = \frac{1}{2\pi} \int_{-\infty}^{\infty} dt \int_{-\infty}^{\infty} d\omega \int_{-\infty}^{\infty} d\omega' \,\overrightarrow{\hat{\mathsf{A}}}^{\star}(\omega) \cdot \overrightarrow{\hat{\mathsf{A}}}(\omega)\, e^{i(\omega' - \omega)t} \qquad [4.99]$$

By changing the order of time and frequency integration, we see that the time integral is simply a Fourier representation of the function $\delta(\omega - \omega')$. Consequently, the energy per unit solid angle becomes:

$$\frac{dW}{d\Omega} = \int_{-\infty}^{\infty} \left|\overrightarrow{\hat{\mathsf{A}}}(\omega)\right|^2 d\omega \qquad [4.100]$$

It is customary to integrate only on positive frequencies, since the sign of the frequency has no physical meaning. Then, the relation becomes:

$$\frac{dW}{d\Omega} = \int_{0}^{\infty} \frac{dI(\omega)}{d\Omega} d\omega \qquad [4.101]$$

which defines a quantity $\frac{dI(\omega)}{d\Omega}$, which is the energy radiated per unit of solid angle and per unit frequency interval:

$$\frac{dI(\omega)}{d\Omega} = \left|\overrightarrow{\hat{\mathsf{A}}}(\omega)\right|^2 + \left|\overrightarrow{\hat{\mathsf{A}}}(-\omega)\right|^2 \qquad [4.102]$$

If $\overrightarrow{\mathsf{A}}(t)$ is real, it is clear that $\overrightarrow{\hat{\mathsf{A}}}(-\omega) = \overrightarrow{\hat{\mathsf{A}}}^{\star}(\omega)$ (from [4.97]). It therefore follows that:

$$\frac{dI(\omega)}{d\Omega} = 2\left|\overrightarrow{\hat{\mathsf{A}}}(\omega)\right|^2 \qquad [4.103]$$

This result quantitatively relates the time-dependent behavior of the radiated power to the spectrum of the radiated energy.

Using expression [4.33] for the electric field of an accelerated charge, we can obtain a general formula for the energy radiated per unit solid angle and per unit frequency interval in terms of an integral over the trajectory of the particle. To do this, we must first compute the Fourier transform [4.97] of $\overrightarrow{\mathsf{A}}$ given by [4.95]:

$$\overrightarrow{\hat{\mathsf{A}}}(\omega) = \sqrt{\frac{q^2\mu_0}{32\pi^3 c}} \int_{-\infty}^{\infty} \left[\frac{\vec{\mathbf{n}} \times \left[\left(\vec{\mathbf{n}} - \vec{\beta}\right) \times \vec{\mathbf{a}}\right]}{\left(1 - \vec{\mathbf{n}} \cdot \vec{\beta}\right)^3} e^{i\omega t}\right]_{\text{ret.}} dt \qquad [4.104]$$

where *ret.* means that the integral is evaluated at the retarded time $t' = t - \frac{R(t')}{c}$. By changing the integration variable from t to t', we obtain:

$$\overrightarrow{\hat{\mathsf{A}}}(\omega) = \sqrt{\frac{q^2\mu_0}{32\pi^3 c}} \int_{-\infty}^{\infty} \frac{\vec{\mathbf{n}} \times \left[\left(\vec{\mathbf{n}} - \vec{\beta}\right) \times \vec{\mathbf{a}}\right]}{\left(1 - \vec{\mathbf{n}} \cdot \vec{\beta}\right)^2} e^{i\omega\left(t' + \frac{R(t')}{c}\right)} dt' \qquad [4.105]$$

In the far zone, we can use the approximation $R = \left|\vec{\mathbf{r}} - \vec{\mathbf{r}}_q(t')\right| \approx r - \vec{\mathbf{n}} \cdot \vec{\mathbf{r}}_q(t')$, where r is the distance between an origin O and the observation point P. Here, $\vec{\mathbf{r}}_q(t')$ is the position of the particle relative to O, as shown in Figure 4.6. Equation [4.105] then becomes:

$$\overrightarrow{\mathsf{A}}(\omega) = \sqrt{\frac{q^2\mu_0}{32\pi^3 c}} e^{i\frac{\omega r}{c}} \int_{-\infty}^{\infty} \frac{\vec{\mathbf{n}} \times \left[\left(\vec{\mathbf{n}} - \vec{\beta}\right) \times \vec{\mathbf{a}}\right]}{\left(1 - \vec{\mathbf{n}} \cdot \vec{\beta}\right)^2} e^{i\omega\left(t' - \frac{\vec{\mathbf{n}} \cdot \vec{\mathbf{r}}_q(t')}{c}\right)} dt' \qquad [4.106]$$

Consequently, the energy radiated per unit of the solid angle and per unit of the frequency interval is:

$$\frac{dI(\omega)}{d\Omega} = \frac{q^2\mu_0}{16\pi^3 c}\left|\int_{-\infty}^{\infty} \frac{\vec{\mathbf{n}} \times \left[\left(\vec{\mathbf{n}} - \vec{\beta}\right) \times \vec{\mathbf{a}}\right]}{\left(1 - \vec{\mathbf{n}} \cdot \vec{\beta}\right)^2} e^{i\omega\left(t' - \frac{\vec{\mathbf{n}} \cdot \vec{\mathbf{r}}_q(t')}{c}\right)} dt'\right|^2 \qquad [4.107]$$

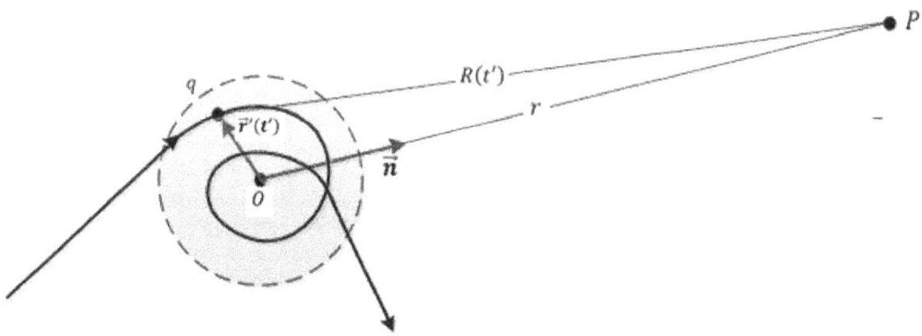

Figure 4.6. *Energy radiated by an accelerated charge per unit solid angle and per unit frequency interval. For a color version of this figure, see www.iste.co.uk/abdesselam/electrodynamics3.zip*

If the motion of the particle is specified, that is, if $\vec{\mathrm{r}}_q(t')$ is known, then $\vec{\beta}$ and $\vec{\mathrm{a}}$ can be calculated, and the integral can be evaluated as a function of ω and the direction $\vec{\mathrm{n}}$. Note that the time integration interval is limited to the times during which the acceleration is non-zero. When multiple charges are involved, we must sum all of them within the amplitude. To simplify this integral, let us first observe that:

$$\frac{d}{dt'}\left[\frac{\vec{\mathrm{n}}\times\left(\vec{\mathrm{n}}\times\vec{\beta}\right)}{\left(1-\vec{\beta}\cdot\vec{\mathrm{n}}\right)}\right] = -\frac{\vec{\mathrm{n}}\times\left(\vec{\mathrm{n}}\times\vec{\beta}\right)}{c\left(1-\vec{\beta}\cdot\vec{\mathrm{n}}\right)^2}\left(-\vec{\mathrm{n}}\cdot\vec{\mathrm{a}}\right) + \frac{\vec{\mathrm{n}}\times\left(\vec{\mathrm{n}}\times\vec{\mathrm{a}}\right)}{c\left(1-\vec{\beta}\cdot\vec{\mathrm{n}}\right)^2}\left(1-\vec{\mathrm{n}}\cdot\vec{\beta}\right) \qquad [4.108]$$

where the terms in $\frac{1}{R}$ that arise from the derivative of $-\vec{\mathrm{n}}$ have been neglected. The terms in [4.108] can also be regrouped as follows:

$$\frac{d}{dt'}\left[\frac{\vec{\mathrm{n}}\times\left(\vec{\mathrm{n}}\times\vec{\beta}\right)}{\left(1-\vec{\beta}\cdot\vec{\mathrm{n}}\right)}\right] = \frac{\vec{\mathrm{n}}\times\left(\vec{\mathrm{n}}\times\vec{\mathrm{a}}\right)}{c\left(1-\vec{\beta}\cdot\vec{\mathrm{n}}\right)^2}$$

$$+\frac{\vec{\mathrm{n}}\times\left[\vec{\mathrm{n}}\times\left\{\left(\vec{\mathrm{n}}\cdot\vec{\mathrm{a}}\right)\vec{\beta}-\left(\vec{\beta}\cdot\vec{\mathrm{n}}\right)\vec{\mathrm{a}}\right\}\right]}{c\left(1-\vec{\beta}\cdot\vec{\mathrm{n}}\right)^2}$$

$$= \frac{\vec{\mathrm{n}}\times\left(\vec{\mathrm{n}}\times\vec{\mathrm{a}}\right)}{c\left(1-\vec{\beta}\cdot\vec{\mathrm{n}}\right)^2} + \frac{\vec{\mathrm{n}}\times\left[\vec{\mathrm{n}}\times\left\{\vec{\mathrm{n}}\times\left(\vec{\beta}\times\vec{\mathrm{a}}\right)\right\}\right]}{c\left(1-\vec{\beta}\cdot\vec{\mathrm{n}}\right)^2}$$

$$= \frac{\vec{n} \times (\vec{n} \times \vec{a})}{c\left(1 - \vec{\beta} \cdot \vec{n}\right)^2} + \frac{\vec{n} \times \left[-\left(\vec{\beta} \times \vec{a}\right)\right]}{c\left(1 - \vec{\beta} \cdot \vec{n}\right)^2} = \frac{\vec{n} \times \left[\left(\vec{n} - \vec{\beta}\right) \times \vec{a}\right]}{c\left(1 - \vec{\beta} \cdot \vec{n}\right)^2} \tag{4.109}$$

Let us substitute this identity into [4.107] and then perform an integration by parts:

$$\frac{dI(\omega)}{d\Omega} = \frac{q^2 \mu_0 c}{16\pi^3} \left| \int_{-\infty}^{\infty} \frac{d}{dt'} \left[\frac{\vec{n} \times \left(\vec{n} \times \vec{\beta}\right)}{\left(1 - \vec{\beta} \cdot \vec{n}\right)} \right] e^{i\omega\left(t' - \frac{\vec{n} \cdot \vec{r}_q(t')}{c}\right)} dt' \right|^2$$

$$= \frac{q^2 \mu_0 c}{16\pi^3} \left| - \int_{-\infty}^{\infty} i\omega \left(1 - \vec{\beta} \cdot \vec{n}\right) \frac{\vec{n} \times \left(\vec{n} \times \vec{\beta}\right)}{\left(1 - \vec{\beta} \cdot \vec{n}\right)} e^{i\omega\left(t' - \frac{\vec{n} \cdot \vec{r}_q(t')}{c}\right)} dt' \right|^2$$

$$= \frac{q^2 \omega^2 \mu_0 c}{16\pi^3} \left| \int_{-\infty}^{\infty} \vec{n} \times \left(\vec{n} \times \vec{\beta}\right) e^{i\omega\left(t' - \frac{\vec{n} \cdot \vec{r}_q(t')}{c}\right)} dt' \right|^2 \tag{4.110}$$

When there are multiple charges instead of just one, the following replacement must be made in [4.110]:

$$q\vec{\beta} e^{-i(\omega/c)\vec{n} \cdot \vec{r}_q} \longrightarrow \sum_{j=1}^{N} q_j \vec{\beta}_j e^{-i(\omega/c)\vec{n} \cdot \vec{r}_j} \tag{4.111}$$

If it is a continuous distribution of (moving) charges, the sum over j becomes an integral over the current density:

$$q\vec{\beta} e^{-i(\omega/c)\vec{n} \cdot \vec{r}_q} \longrightarrow \frac{1}{c} \int d\tau' \vec{J}\left(\vec{r}', t'\right) e^{-i(\omega/c)\vec{n} \cdot \vec{r}'} \tag{4.112}$$

The intensity distribution then becomes:

$$\frac{dI(\omega)}{d\Omega} = \frac{\omega^2 \mu_0}{16\pi^3 c} \left| \int_{-\infty}^{\infty} dt' \int d\tau' \vec{n} \times \left[\vec{n} \times \vec{J}\left(\vec{r}', t'\right)\right] e^{i\omega\left(t' - \frac{\vec{n} \cdot \vec{r}'}{c}\right)} \right|^2 \tag{4.113}$$

EXAMPLE 4.5. RADIATION FROM A MAGNETIC MOMENT.– The radiation associated with a moving magnetic moment is of particular interest. It can be easily expressed by recalling that a magnetization density $\vec{\mathbf{M}}$ is equivalent to a current $\vec{\mathbf{J}}_M = \vec{\nabla} \times \vec{\mathbf{M}}$. By substituting $\vec{\mathbf{J}}_M$ into [4.113], we deduce that:

$$\begin{aligned}\frac{dI(\omega)}{d\Omega} &= \frac{\omega^2\mu_0}{16\pi^3 c}\left|\int_{-\infty}^{\infty} dt' \int d\tau' \vec{\mathbf{n}} \times \left[\vec{\mathbf{n}} \times \left(\vec{\nabla} \times \vec{\mathbf{M}}'\right)\right] e^{i\omega\left(t' - \frac{\vec{\mathbf{n}} \cdot \vec{\mathbf{r}}'}{c}\right)}\right|^2 \\ &= \frac{\omega^2\mu_0}{16\pi^3 c}\left|\int_{-\infty}^{\infty} dt' \int d\tau' \vec{\mathbf{n}} \times \left[\vec{\mathbf{n}} \times \left(\vec{\nabla} e^{i\omega\left(t' - \frac{\vec{\mathbf{n}} \cdot \vec{\mathbf{r}}'}{c}\right)} \times \vec{\mathbf{M}}'\right)\right]\right|^2 \\ &= \frac{\omega^2\mu_0}{16\pi^3 c}\left|-\frac{i\omega}{c}\int_{-\infty}^{\infty} dt' \int d\tau' \vec{\mathbf{n}} \times \left[\vec{\mathbf{n}} \times \left(\vec{\mathbf{n}} \times \vec{\mathbf{M}}'\right)\right] e^{i\omega\left(t' - \frac{\vec{\mathbf{n}} \cdot \vec{\mathbf{r}}'}{c}\right)}\right|^2 \\ &= \frac{\omega^4\mu_0}{16\pi^3 c^3}\left|\int_{-\infty}^{\infty} dt' \int d\tau' \vec{\mathbf{n}} \times \vec{\mathbf{M}}' e^{i\omega\left(t' - \frac{\vec{\mathbf{n}} \cdot \vec{\mathbf{r}}'}{c}\right)}\right|^2 \end{aligned} \quad [4.114]$$

If the magnetization corresponds to a point magnetic moment $\vec{\mathbf{m}}$ at the point $\vec{\mathbf{r}}_m$, then:

$$\vec{\mathbf{M}}\left(\vec{\mathbf{r}}', t'\right) = \vec{\mathbf{m}}\left(t'\right)\delta^3\left(\vec{\mathbf{r}}' - \vec{\mathbf{r}}_m\right) \quad [4.115]$$

Consequently, the energy radiated per unit solid angle and per unit frequency interval will be:

$$\frac{dI_m(\omega)}{d\Omega} = \frac{\omega^4\mu_0}{16\pi^3 c^3}\left|\int_{-\infty}^{\infty} dt' \vec{\mathbf{n}} \times \vec{\mathbf{m}}'\left(t'\right) e^{i\omega\left(t' - \frac{\vec{\mathbf{n}} \cdot \vec{\mathbf{r}}_m}{c}\right)}\right|^2 \quad [4.116]$$

We observe that the energy radiated by a magnetic moment is proportional to ω^4 (apart from the fact that the integrals depend on frequency), while the energy radiated by an accelerated charge is of the order of ω^2. ■

When the particle undergoes cyclic motion (with frequency ω_0), a Fourier series (instead of the Fourier transform) must be used to express the frequency distribution of the radiation. Indeed, the radiation fields produced during this cyclical motion are also periodic. Consequently, the function $\vec{\mathsf{A}}(t)$ must be replaced by a Fourier series, that is, by a sum over frequencies that are integer multiples of the rotation frequency of the particle:

$$\vec{\mathsf{A}}(t) = \frac{1}{\sqrt{\mu_0 c}}\left[R\vec{\mathbf{E}}(t)\right]_{\text{ret.}} = \sum_{n=-\infty}^{\infty} \vec{\mathsf{A}}_n e^{-in\omega_0 t} \quad [4.117]$$

with:

$$\vec{\mathsf{A}}_n = \frac{\omega_0}{2\pi}\int_0^{2\pi/\omega_0} \vec{\mathbf{A}}(t)\, e^{in\omega_0 t} dt \quad [4.118]$$

The energy received during one cycle, per unit solid angle, at a point $\vec{\mathbf{r}}$ is the integral of $\frac{dP}{d\Omega}$ over one period:

$$\begin{aligned}\frac{dW}{d\Omega} &= \int_0^{2\pi/\omega_0} \frac{dP(t)}{d\Omega} dt = \int_0^{2\pi/\omega_0} \left|\vec{\mathsf{A}}(t)\right|^2 dt \\ &= \sum_{m,n} \int_0^{2\pi/\omega_0} \left(\vec{\mathsf{A}}_n \cdot \vec{\mathsf{A}}^\star_m\right) e^{-i(n-m)\omega_0 t} dt \\ &= \frac{2\pi}{\omega_0} \sum_{n=-\infty}^{\infty} \left|\vec{\mathsf{A}}_n\right|^2 = \frac{4\pi}{\omega_0} \sum_{n=1}^{\infty} \left|\vec{\mathsf{A}}_n\right|^2 \end{aligned} \qquad [4.119]$$

The $n = 0$ term has been omitted from the final expression, since there is no radiation at zero frequency. The energy radiated at the frequency $n\omega_0$ is determined as follows:

$$\begin{aligned}\vec{\mathsf{A}}_n &= \frac{\omega_0}{2\pi}\sqrt{\frac{q^2\mu_0}{16\pi^2 c}} \int_0^{2\pi/\omega_0} \frac{\vec{\mathbf{n}} \times \left[\left(\vec{\mathbf{n}} - \vec{\beta}\right) \times \vec{\mathbf{a}}\right]}{\left(1 - \vec{\mathbf{n}} \cdot \vec{\beta}\right)^3} e^{in\omega_0 t} dt \\ &= \frac{\omega_0}{2\pi}\sqrt{\frac{q^2\mu_0}{16\pi^2 c}} \int_0^{2\pi/\omega_0} \frac{\vec{\mathbf{n}} \times \left[\left(\vec{\mathbf{n}} - \vec{\beta}\right) \times \vec{\mathbf{a}}\right]}{\left(1 - \vec{\mathbf{n}} \cdot \vec{\beta}\right)^2} e^{in\omega_0 \left(t' + \frac{R(t')}{c}\right)} dt' \end{aligned} \qquad [4.120]$$

We can now perform the same integration by parts (to obtain equation [4.99]):

$$\begin{aligned}\vec{\mathsf{A}}_n &= \frac{\omega_0}{2\pi}\sqrt{\frac{q^2\mu_0 c}{16\pi^2}} \int_0^{2\pi/\omega_0} \frac{d}{dt}\left(\frac{\vec{\mathbf{n}} \times \left(\vec{\mathbf{n}} \times \vec{\beta}\right)}{1 - \vec{\mathbf{n}} \cdot \vec{\beta}}\right) e^{in\omega_0 \left(t' + \frac{R(t')}{c}\right)} dt' \\ &= \frac{\omega_0}{2\pi}\sqrt{\frac{q^2\mu_0 c}{16\pi^2}} \frac{\vec{\mathbf{n}} \times \left(\vec{\mathbf{n}} \times \vec{\beta}\right)}{1 - \vec{\mathbf{n}} \cdot \vec{\beta}} e^{in\omega_0 \left(t' + \frac{R(t')}{c}\right)} \Bigg|_0^{2\pi/\omega_0} \\ &\quad - \frac{\omega_0}{2\pi}\sqrt{\frac{q^2\mu_0 c}{16\pi^2}} \int_0^{2\pi/\omega_0} in\omega_0 \, \vec{\mathbf{n}} \times \left(\vec{\mathbf{n}} \times \vec{\beta}\right) e^{in\omega_0 \left(t' + \frac{R(t')}{c}\right)} dt' \\ &= -\frac{in\omega_0^2}{2\pi}\sqrt{\frac{q^2\mu_0 c}{16\pi^2}} \int_0^{2\pi/\omega_0} \vec{\mathbf{n}} \times \left(\vec{\mathbf{n}} \times \vec{\beta}\right) e^{in\omega_0 \left(t' + \frac{R(t')}{c}\right)} dt' \end{aligned} \qquad [4.121]$$

Note that the term $\frac{\omega_0}{2\pi}\sqrt{\frac{q^2\mu_0 c}{16\pi^2}}\frac{\vec{n}\times\left(\vec{n}\times\vec{\beta}\right)}{1-\vec{n}\cdot\vec{\beta}}e^{in\omega_0\left(t'+\frac{R(t')}{c}\right)}\Bigg|_0^{2\pi/\omega_0}$ is zero, since the ends of the integration interval correspond to the same point on the periodic orbit of the particle. From equation [4.121], we can easily compute $\left|\vec{A}_n\right|^2$, and consequently $\frac{dW}{d\Omega}$:

$$\left|\vec{A}_n\right|^2 = \frac{n^2\omega_0^4 q^2\mu_0 c}{64\pi^4}\left|\int_0^{2\pi/\omega_0}\vec{n}\times\left(\vec{n}\times\vec{\beta}\right)e^{in\omega_0\left(t'+\frac{R(t')}{c}\right)}dt'\right|^2 \qquad [4.122]$$

$$\frac{dW}{d\Omega} = \sum_{n=1}^{\infty}\frac{n^2\omega_0^3 q^2\mu_0 c}{16\pi^3}\left|\int_0^{2\pi/\omega_0}\vec{n}\times\left(\vec{n}\times\vec{\beta}\right)e^{in\omega_0\left(t'+\frac{R(t')}{c}\right)}dt'\right|^2 \qquad [4.123]$$

Note also that the distribution of the time-averaged power is simply:

$$\frac{dP}{d\Omega} = \frac{\omega_0}{2\pi}\frac{dW}{d\Omega} = \sum_{n=1}^{\infty}\frac{n^2\omega_0^4 q^2\mu_0 c}{32\pi^4}\left|\int_0^{2\pi/\omega_0}\vec{n}\times\left(\vec{n}\times\vec{\beta}\right)e^{in\omega_0\left(t'+\frac{R(t')}{c}\right)}dt'\right|^2$$

[4.124]

4.7. Charge in instantaneous circular motion

In section 4.5.2, we saw that the radiation emitted by an ultra-relativistic particle undergoing arbitrary accelerations is equivalent to that emitted by a particle moving instantaneously at constant speed along an appropriate circular trajectory. The radiation is emitted in a narrow cone in the direction of the velocity vector. Furthermore, this radiation is seen by the observer as a short pulse when the spotlight beam sweeps across the observation point. To find the energy distribution as a function of frequency and angle, it is necessary to compute the integral in [4.110]. Since the duration of the pulse, $\Delta t \sim \frac{\rho}{\gamma c}$, is very short, we only need to know the velocity $\vec{\beta}$ and position $\vec{r}\,(t')$ over a small arc of the trajectory whose tangent approximately points in the direction of the observation point. Figure 4.7 shows a suitable coordinate system. At each instant, an arbitrary motion can be considered as a combination of a translation and a uniform circular rotation around an appropriate circle of radius ρ. The trajectory segment lies in the Oxy plane, with an instantaneous curvature radius ρ. Since the integral is evaluated on the trajectory, the unit vector $\vec{n}$ can be chosen (for simplicity and without loss of generality) to lie in the Oxz plane, forming an angle θ (the colatitude) with the Ox axis. Recall that there is a significant radiation intensity only for small angles θ. It is assumed that the particle is located at the origin O at time $t = 0$.

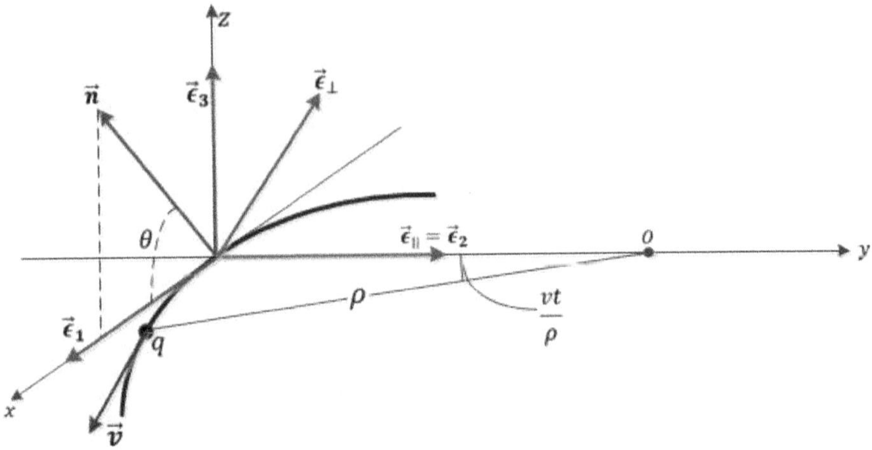

Figure 4.7. *The radiation produced during translation is negligible compared to that of instantaneous rotation*

Let $\vec{\epsilon}_1$, $\vec{\epsilon}_2$ and $\vec{\epsilon}_3$ be the unit vectors along the axes Ox, Oy and Oz, respectively. The vectors $\vec{\mathbf{n}}$, $\vec{\mathbf{r}}(t')$ and $\vec{\beta}$ are expressed as follows: $\vec{\mathbf{n}} = \cos\theta\,\vec{\epsilon}_1 + \sin\theta\,\vec{\epsilon}_3$, $\vec{\mathbf{r}}(t') = \rho\sin\left(\frac{vt'}{\rho}\right)\vec{\epsilon}_1 + \rho\left[1 - \cos\left(\frac{vt'}{\rho}\right)\right]\vec{\epsilon}_2$ and $\vec{\beta} = \frac{v}{c}\cos\left(\frac{vt'}{\rho}\right)\vec{\epsilon}_1 + \frac{v}{c}\sin\left(\frac{vt'}{\rho}\right)\vec{\epsilon}_2$. We also set $\vec{\epsilon}_\parallel = \vec{\epsilon}_2$ ($\vec{\epsilon}_\parallel$ corresponds to the polarization in the plane of the orbit) and $\vec{\epsilon}_\perp = \vec{\mathbf{n}} \times \vec{\epsilon}_2$ ($\vec{\epsilon}_\perp$ is the orthogonal polarization vector corresponding, approximately, to the polarization perpendicular to the plane of the orbit, for small values of θ). The vector part of the integral [4.110] then takes the form:

$$\vec{\mathbf{n}} \times \left(\vec{\mathbf{n}} \times \vec{\beta}\right) = \vec{\mathbf{n}} \times \left[\frac{v}{c}\cos\left(\frac{vt'}{\rho}\right)\overrightarrow{\mathbf{n} \times \epsilon_1} + \frac{v}{c}\rho\sin\left(\frac{vt'}{\rho}\right)\vec{\mathbf{n}} \times \vec{\epsilon}_2\right]$$

$$= \vec{\mathbf{n}} \times \left[\frac{v}{c}\cos\left(\frac{vt'}{\rho}\right)\sin\theta\,\vec{\epsilon}_\parallel + \frac{v}{c}\sin\left(\frac{vt'}{\rho}\right)\vec{\epsilon}_\perp\right]$$

$$= \frac{v}{c}\left[\cos\left(\frac{vt'}{\rho}\right)\sin\theta\,\vec{\mathbf{n}} \times \vec{\epsilon}_\parallel + \right.$$

$$\left.\sin\left(\frac{vt'}{\rho}\right)\vec{\mathbf{n}} \times \vec{\epsilon}_\perp\right] = \frac{v}{c}\left[\cos\left(\frac{vt'}{\rho}\right)\sin\theta\,\vec{\epsilon}_\perp - \sin\left(\frac{vt'}{\rho}\right)\vec{\epsilon}_\parallel\right] \quad [4.125]$$

The argument of the exponential in [4.110] also becomes:

$$\omega\left(t'-\frac{\vec{\mathbf{r}}\,(t')\cdot\vec{\mathbf{n}}}{c}\right)=\omega\left[t'-\frac{\rho}{c}\sin\left(\frac{vt'}{\rho}\right)\cos\theta\right] \qquad [4.126]$$

Since we must restrict ourselves to small angles θ and relatively short times around $t=0$, we can expand the two trigonometric functions in [4.126] to obtain:

$$\begin{aligned}\omega\left(t'-\frac{\vec{\mathbf{r}}\,(t')\cdot\vec{\mathbf{n}}}{c}\right)&\simeq\omega\left[\left(1-\beta+\frac{\theta^2}{2}\right)t'+\frac{c^2}{6\rho^2}t'^3\right]\\&\simeq\frac{\omega}{2}\left[\left(\frac{1}{\gamma^2}+\theta^2\right)t'+\frac{c^2}{3\rho^2}t'^3\right]\end{aligned} \qquad [4.127]$$

where β has been replaced by $1-\frac{1}{2}\gamma^{-2}$. Using the fact that $t\sim\frac{\rho}{c}\gamma^{-1}$ and $\theta\sim\gamma^{-1}$, it is easy to verify that only terms of order γ^{-3} have been retained. The other (higher order) terms have been neglected. Applying the same approximations as in [4.125], we obtain $\vec{\mathbf{n}}\times\left(\vec{\mathbf{n}}\times\vec{\beta}\right)\simeq\left[\theta\vec{\epsilon}_{\perp}-\frac{c}{\rho}t'\vec{\epsilon}_{\parallel}\right]$. Consequently, the distribution of the radiated energy [4.110] becomes:

$$\frac{dI\,(\omega)}{d\Omega}=\frac{q^2\omega^2\mu_0 c}{16\pi^3\varepsilon_0}\left|-\vec{\epsilon}_{\parallel}\mathsf{A}_{\parallel}\,(\omega)+\vec{\epsilon}_{\perp}\mathsf{A}_{\perp}\,(\omega)\right|^2 \qquad [4.128]$$

where:

$$\begin{aligned}\mathsf{A}_{\parallel}\,(\omega)&\simeq\frac{c}{\rho}\int_{-\infty}^{\infty}t'\exp\left\{\mathrm{i}\frac{\omega}{2}\left[\left(\frac{1}{\gamma^2}+\theta^2\right)t'+\frac{c^2}{3\rho^2}t'^3\right]\right\}dt'\\\mathsf{A}_{\perp}\,(\omega)&\simeq\theta\int_{-\infty}^{\infty}\exp\left\{\mathrm{i}\frac{\omega}{2}\left[\left(\frac{1}{\gamma^2}+\theta^2\right)t'+\frac{c^2}{3\rho^2}t'^3\right]\right\}dt'\end{aligned} \qquad [4.129]$$

The fact that the integration limits in [4.129] are $t'=\pm\infty$ seems to contradict the approximations that were previously used. The explanation is that beyond the time interval $\Delta t\sim\frac{\rho}{c}\gamma^{-1}$, the phase within the integrals [4.129] oscillates very rapidly for most frequencies, which means that the integrals have no significant contribution beyond $\frac{\rho}{c}\gamma^{-1}$. Therefore, the upper and lower limits of the integrals can be considered infinite without introducing any error.

By changing the variable $x=\frac{ct'}{\rho\left(\frac{1}{\gamma^2}+\theta^2\right)^{1/2}}$ and setting $\xi=\frac{\omega\rho}{3c}\left(\frac{1}{\gamma^2}+\theta^2\right)^{3/2}$, the integrals [4.129] become:

$$\begin{aligned}\mathsf{A}_{\parallel}\,(\omega)&\simeq\frac{\rho}{c}\left(\frac{1}{\gamma^2}+\theta^2\right)\int_{-\infty}^{\infty}x\exp\left[\mathrm{i}\frac{3}{2}\xi\left(x+\frac{1}{3}x^3\right)\right]dx\\\mathsf{A}_{\perp}\,(\omega)&\simeq\frac{\rho}{c}\theta\left(\frac{1}{\gamma^2}+\theta^2\right)^{1/2}\int_{-\infty}^{\infty}\exp\left[\mathrm{i}\frac{3}{2}\xi\left(x+\frac{1}{3}x^3\right)\right]dx\end{aligned} \qquad [4.130]$$

The integrals in [4.130] can be identified as *Airy integrals*, or alternatively, as modified Bessel functions:

$$\begin{aligned}\int_0^\infty x\sin\left[\frac{3}{2}\xi\left(x+\frac{1}{3}x^3\right)\right]dx &= \frac{1}{\sqrt{3}}K_{2/3}(\xi)\\ \int_0^\infty \cos\left[\frac{3}{2}\xi\left(x+\frac{1}{3}x^3\right)\right]dx &= \frac{1}{\sqrt{3}}K_{1/3}(\xi)\end{aligned} \qquad [4.131]$$

Consequently, the energy radiated per unit frequency and per unit solid angle is equal to:

$$\frac{dI(\omega)}{d\Omega} = \frac{q^2\mu_0 c}{12\pi^3}\left(\frac{\omega\rho}{c}\right)^2\left(\frac{1}{\gamma^2}+\theta^2\right)^2\left[K_{2/3}^2(\xi)+\frac{\theta^2}{\frac{1}{\gamma^2}+\theta^2}K_{1/3}^2(\xi)\right] \qquad [4.132]$$

The first term in brackets corresponds to the radiation polarized in the plane of the orbit, and the second to the radiation polarized perpendicular to that plane.

Let us now examine this somewhat complex result. By integrating over all frequencies, we obtain the following:

$$\begin{aligned}\int_0^\infty \frac{dI(\omega)}{d\Omega}d\omega &= \frac{q^2\mu_0 c}{12\pi^3}\left(\frac{\rho}{c}\right)^2\left(\frac{1}{\gamma^2}+\theta^2\right)^2\left[\int_0^\infty \omega^2 K_{2/3}^2(\xi)\,d\omega+\right.\\ &\left.\frac{\theta^2}{\frac{1}{\gamma^2}+\theta^2}\int_0^\infty \omega^2 K_{1/3}^2(\xi)\,d\omega\right] = \frac{9q^2\mu_0 c}{4\pi^3}\frac{c}{\rho}\frac{1}{\left(\frac{1}{\gamma^2}+\theta^2\right)^{5/2}}\\ &\times\left[\int_0^\infty \xi^2 K_{2/3}^2(\xi)\,d\xi+\frac{\theta^2}{\frac{1}{\gamma^2}+\theta^2}\int_0^\infty \xi^2 K_{1/3}^2(\xi)\,d\xi\right]\\ &= \frac{7q^2}{64\pi\varepsilon_0\rho}\frac{1}{\left(\frac{1}{\gamma^2}+\theta^2\right)^{5/2}}\left[1+\frac{5}{7}\frac{\theta^2}{\frac{1}{\gamma^2}+\theta^2}\right]\end{aligned} \qquad [4.133]$$

where we have used the identities:

$$\begin{aligned}\int_0^\infty \xi^2 K_{2/3}^2(\xi)\,d\xi &= \frac{7\pi^2}{144}\\ \int_0^\infty \xi^2 K_{1/3}^2(\xi)\,d\xi &= \frac{5\pi^2}{144}\end{aligned} \qquad [4.134]$$

Equation [4.133] can be obtained directly by integrating [4.53] over time. As in [4.132], the first term in [4.133] corresponds to the parallel polarization to the plane of the orbit and the second to the perpendicular polarization. By integrating over all angles, it is found that the energy radiated parallel to the plane of the orbit is greater than that radiated perpendicularly.

The properties of the modified Bessel functions show that the radiation intensity is negligible for $\xi \gg 1$. From the definition of ξ, we see that this occurs at large angles: the higher the frequency, the smaller the critical angle beyond which the radiation becomes negligible. This shows that the radiation is largely confined to the plane that contains the motion, as illustrated by [4.133], and the confinement becomes tighter as the frequency increases relative to $\frac{c}{\rho}$. When the frequency ω becomes too high, ξ also becomes large at all angles, and the total energy emitted at that frequency becomes negligible. The critical frequency ω_c, beyond which there is negligible radiation in any direction, can be defined by setting $\xi(\omega_c) = 1$ for $\theta = 0$. Then we find:

$$\omega_c = 3\gamma^3 \left(\frac{c}{\rho}\right) = 3\left(\frac{E}{mc^2}\right)^3 \frac{c}{\rho} \qquad [4.135]$$

If the charge undergoes a truly circular motion, then $\frac{c}{\rho}$ is the fundamental rotational frequency, ω_0. Then it is possible to define a critical harmonic frequency $\omega_c = n_c\omega_0$ with a harmonic number:

$$n_c = 3\left(\frac{E}{mc^2}\right)^3 \qquad [4.136]$$

Since radiation is confined mainly to the orbital plane for $\gamma \gg 1$, it is instructive to evaluate the angular distribution [4.132] at $\theta = 0$. For frequencies well below the critical frequency ($\omega \ll \omega_c$), we get:

$$\begin{aligned} \left.\frac{dI(\omega)}{d\Omega}\right|_{\theta=0} &= \frac{q^2}{12\pi^3\varepsilon_0 c}\left(\frac{\omega\rho}{c}\right)^2 \frac{1}{\gamma^4} K_{2/3}^2\left(\frac{\omega\rho}{3c\gamma^3}\right) \\ &\simeq \frac{q^2}{4\pi\varepsilon_0 c}\left(\frac{\omega\rho}{c}\right)^{2/3}\left(\frac{3}{4}\right)^{1/3}\left[\frac{\Gamma(2/3)}{\pi}\right]^2 \end{aligned} \qquad [4.137]$$

where the approximation $K_\nu(x) \simeq \frac{\Gamma(\nu)}{2}\left(\frac{2}{x}\right)^\nu$ has been used for $x \ll 1$. For the opposite limit ($\omega \gg \omega_c$), the result is:

$$\left.\frac{dI(\omega)}{d\Omega}\right|_{\theta=0} \simeq \frac{3q^2}{8\pi^2\varepsilon_0 c}\frac{\omega}{\omega_c}\gamma^2 e^{-2\omega/\omega_c} \qquad [4.138]$$

where $K_\nu(x) \simeq \sqrt{\frac{\pi}{2x}}e^{-x}$ for $x \gg 1$. These limiting forms show that the spectrum at $\theta = 0$ increases with frequency approximately as $\omega^{2/3}$ well below the critical frequency, reaches a maximum in the neighborhood of ω_c and then tends toward zero (exponentially) above that frequency.

To determine the critical angle θ_c (for a fixed frequency) beyond which the radiation becomes negligible, we can approximately use the expression $\xi(\theta_c) - \xi(0) \simeq 1$. In the low-frequency range, $\xi(0)$ is very small (i.e. $\xi(0) \simeq 0$) and, therefore, $\xi(\theta_c) = \frac{\omega}{\omega_c}\left(1+\theta_c^2\gamma^2\right)^{3/2} \simeq \frac{\omega}{\omega_c}\theta_c^3\gamma^3 \simeq 1$. This gives:

$$\theta_c \simeq \frac{1}{\gamma}\left(\frac{\omega_c}{\omega}\right)^{1/3} = \left(\frac{3c}{\omega\rho}\right)^{1/3} \qquad [4.139]$$

We note that low-frequency components are emitted at angles much wider than the average $\langle\theta^2\rangle^{1/2} \sim \gamma^{-1}$. In the high-frequency limit ($\omega > \omega_c$), $\xi(0)$ is greater than one. As a function of angle θ, the intensity is written as follows:

$$\frac{dI(\omega)}{d\Omega} \simeq \frac{3q^2}{8\pi^2\varepsilon_0 c}\frac{\omega}{\omega_c}\gamma^2 e^{-2\omega/\omega_c} e^{-3\theta^2\gamma^2\omega/\omega_c} = \left.\frac{dI(\omega)}{d\Omega}\right|_{\theta=0} e^{-3\theta^2\gamma^2\omega/\omega_c} \qquad [4.140]$$

We see that the intensity decreases exponentially as ω increases. Thus, the critical angle, defined by $\left.\frac{dI(\omega)}{d\Omega}\right|_{\theta_c} = \left.\frac{dI(\omega)}{d\Omega}\right|_{\theta=0} e^{-1}$, is given by:

$$\theta_c \simeq \frac{1}{\gamma}\left(\frac{\omega_c}{3\omega}\right)^{1/2} \qquad [4.141]$$

which shows that the high-frequency components are confined within an angle much smaller than the average. Figure 4.8 shows the qualitative angular distribution of energy. For frequencies comparable to the critical frequency ω_c, the radiation is confined to angles on the order of γ^{-1}. For much lower (respectively, higher) frequencies, the angular spread is larger (respectively, smaller).

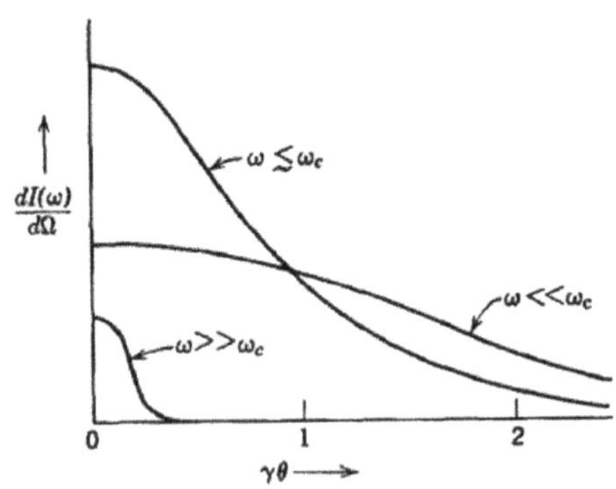

Figure 4.8. *Radiated energy per unit of solid angle and per unit frequency as a function of angle*

The total energy per unit frequency (frequency distribution of the total emitted energy) during the particle's passage can be found by integrating [4.132] over the angles:

$$I(\omega) = 2\pi \int_{-\pi/2}^{\pi/2} \frac{dI(\omega)}{d\Omega} \cos\theta d\theta \simeq 2\pi \int_{-\infty}^{\infty} \frac{dI(\omega)}{d\Omega} d\theta \qquad [4.142]$$

We can estimate the integral in the low-frequency range by using the value of the angular distribution [4.137] at $\theta = 0$ and the critical angle θ_c [4.139]. We then obtain:

$$I(\omega) = 2\pi\theta_c \left.\frac{dI(\omega)}{d\Omega}\right|_{\theta=0} \simeq \frac{q^2}{2\varepsilon_0 c}\left(\frac{\omega\rho}{c}\right)^{1/3}\left(\frac{9}{4}\right)^{1/3}\left[\frac{\Gamma(2/3)}{\pi}\right]^2 \qquad [4.143]$$

We observe that $I(\omega)$ increases like $\omega^{1/3}$ for $\omega \ll \omega_c$. In the high-frequency limit ($\omega \gg \omega_c$), we can directly integrate [4.140] over angles (using the Gaussian integral $\int_{-\infty}^{\infty} e^{-\alpha x^2} dx = \sqrt{\frac{\pi}{\alpha}}$):

$$I(\omega) \simeq \frac{3q^2}{4\pi\varepsilon_0 c}\frac{\omega}{\omega_c}\gamma^2 e^{-2\omega/\omega_c}\int_{-\infty}^{\infty} e^{-3\theta^2\gamma^2\omega/\omega_c} d\theta = \sqrt{\frac{3}{\pi}}\frac{q^2\gamma}{4\varepsilon_0 c}\left(\frac{\omega}{\omega_c}\right)^{1/2} e^{-2\omega/\omega_c}$$

[4.144]

4.8. Particle in circular motion

Let us perform the same calculation for a particle in truly circular motion. According to equation [4.123], the energy emitted per unit of solid angle during one cycle is equal to:

$$\frac{dW}{d\Omega} = \sum_{n=1}^{\infty}\frac{dW_n}{d\Omega} = \sum_{n=1}^{\infty}\frac{n^2\omega_0^3 q^2}{16\pi^3\varepsilon_0 c}\left|\int_0^{\frac{2\pi}{\omega_0}} \vec{\mathbf{n}} \times \left(\vec{\mathbf{n}} \times \vec{\beta}\right) e^{in\omega_0\left(t' - \frac{\vec{\mathbf{n}}\cdot\vec{\mathbf{r}}'}{c}\right)} dt'\right|^2$$

[4.145]

$\frac{dW_n}{d\Omega}$ represents the energy radiated at the frequency $n\omega_0$ during one cycle per unit of solid angle:

$$\begin{aligned}\frac{dW_n}{d\Omega} &= \frac{n^2\omega_0^3 q^2}{16\pi^3\varepsilon_0 c}\left|\int_0^{\frac{2\pi}{\omega_0}} \vec{\mathbf{n}} \times \left(\vec{\mathbf{n}} \times \vec{\beta}\right) e^{in\omega_0\left(t' - \frac{\vec{\mathbf{n}}\cdot\vec{\mathbf{r}}'}{c}\right)} dt'\right|^2 \\ &= \frac{n^2\omega_0^3 q^2}{16\pi^3\varepsilon_0 c}\int_0^{\frac{2\pi}{\omega_0}} dt' \int_0^{\frac{2\pi}{\omega_0}} dt'' \left[\vec{\mathbf{n}} \times \left(\vec{\mathbf{n}} \times \vec{\beta}'\right)\right] \cdot \left[\vec{\mathbf{n}} \times \left(\vec{\mathbf{n}} \times \vec{\beta}''\right)\right]\end{aligned}$$

$$
\begin{aligned}
&e^{in\omega_0\left[t'-t''-\frac{\vec{n}\cdot(\vec{r}'-\vec{r}'')}{c}\right]} = \frac{n^2\omega_0^3 q^2}{16\pi^3\varepsilon_0 c}\int_0^{\frac{2\pi}{\omega_0}} dt' \int_0^{\frac{2\pi}{\omega_0}} dt'' \left(\vec{n}\times\vec{\beta}''\right)\cdot \\
&\left\{\left[\vec{n}\times\left(\vec{n}\times\vec{\beta}'\right)\right]\times\vec{n}\right\} e^{in\omega_0\left[t'-t''-\frac{\vec{n}\cdot(\vec{r}'-\vec{r}'')}{c}\right]} \\
&= \frac{n^2\omega_0^3 q^2}{16\pi^3\varepsilon_0 c}\int_0^{\frac{2\pi}{\omega_0}} dt' \int_0^{\frac{2\pi}{\omega_0}} dt'' \left(\vec{n}\times\vec{\beta}''\right)\cdot\left(\vec{n}\times\vec{\beta}'\right) \\
&e^{in\omega_0\left[t'-t''-\frac{\vec{n}\cdot(\vec{r}'-\vec{r}'')}{c}\right]} = \frac{n^2\omega_0^3 q^2}{16\pi^3\varepsilon_0 c}\left|\int_0^{\frac{2\pi}{\omega_0}} \left(\vec{n}\times\vec{\beta}\right) e^{in\omega_0\left(t'-\frac{\vec{n}\cdot\vec{r}'}{c}\right)} dt'\right|^2
\end{aligned}
\quad [4.146]
$$

Dividing by the time period $\frac{2\pi}{\omega_0}$, we arrive at the time-averaged power at frequency $n\omega_0$ per unit of solid angle:

$$
\frac{dP_n}{d\Omega} = \frac{n^2\omega_0^4 q^2}{32\pi^4\varepsilon_0 c}\left|\int_0^{\frac{2\pi}{\omega_0}} \left(\vec{n}\times\vec{\beta}\right) e^{in\omega_0\left(t'-\frac{\vec{n}\cdot\vec{r}'}{c}\right)} dt'\right|^2 \quad [4.147]
$$

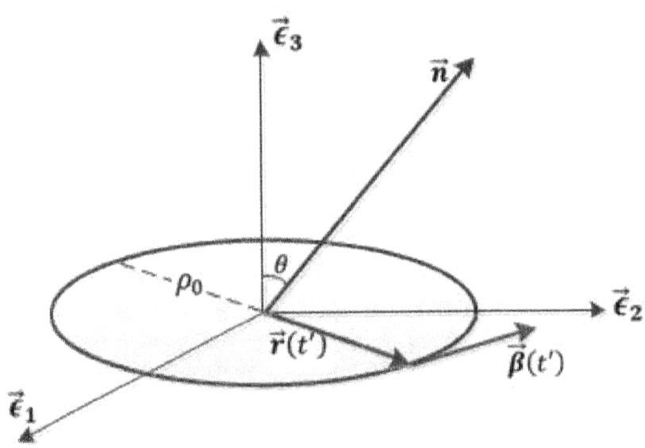

Figure 4.9. *Radiation of a charged particle in circular motion. For a color version of this figure, see www.iste.co.uk/abdesselam/electrodynamics3.zip*

Suppose that the charged particle undergoes a circular motion of radius ρ_0 in the Oxy plane around the origin O, as illustrated in Figure 4.9. Thus, we write:

$$
\begin{aligned}
\vec{r}\,(t') &= \rho_0\left[\vec{\epsilon}_1\cos\omega_0 t' + \vec{\epsilon}_2\sin\omega_0 t'\right] \\
\vec{\beta}\,(t') &= \frac{\rho_0\omega_0}{c}\left[-\vec{\epsilon}_1\sin\omega_0 t' + \vec{\epsilon}_2\cos\omega_0 t'\right]
\end{aligned}
\quad [4.148]
$$

We can also express $\vec{n}$ in terms of spherical coordinates as $\vec{n} = \vec{\epsilon}_1\sin\theta\cos\varphi + \vec{\epsilon}_2\sin\theta\sin\varphi + \vec{\epsilon}_3\cos\theta$. It then follows that:

$$
\begin{aligned}
\vec{\mathbf{n}} \cdot \vec{\mathbf{r}}\,(t') &= \rho_0 \left[\sin\theta\cos\varphi\cos\omega_0 t' + \sin\theta\sin\varphi\sin\omega_0 t'\right] \\
&= \rho_0 \sin\theta\cos\left(\omega_0 t' - \varphi\right) \\
\vec{\mathbf{n}} \times \vec{\beta}\,(t') &= \frac{\rho_0\omega_0}{c}\left[-\vec{\epsilon}_1\cos\theta\cos\omega_0 t' - \vec{\epsilon}_2\cos\theta\sin\omega_0 t' + \right. \\
&\quad \left. \vec{\epsilon}_3\sin\theta\left(\cos\varphi\cos\omega_0 t' + \sin\varphi\sin\omega_0 t'\right)\right] \\
&= \frac{\rho_0\omega_0}{c}\left[-\vec{\epsilon}_1\cos\theta\cos\omega_0 t' - \vec{\epsilon}_2\cos\theta\sin\omega_0 t'\right. \\
&\quad \left. + \vec{\epsilon}_3\sin\theta\cos\left(\omega_0 t' - \varphi\right)\right]
\end{aligned} \qquad [4.149]
$$

Taking into account the expressions [4.149], we get:

$$
\begin{aligned}
I &= \int_0^{2\pi/\omega_0} \left(\vec{\mathbf{n}} \times \vec{\beta}\right) e^{in\omega_0\left(t' - \frac{\vec{\mathbf{n}}\cdot\vec{\mathbf{r}}'}{c}\right)} dt' \\
&= \frac{\beta}{\omega_0}\left[-\vec{\epsilon}_1 K\cos\theta - \vec{\epsilon}_2 J\cos\theta + \vec{\epsilon}_3 L'\sin\theta\right]
\end{aligned} \qquad [4.150]
$$

where:

$$
\left\{\begin{matrix} K \\ J \\ L' \end{matrix}\right\} = \omega_0 \int_0^{2\pi/\omega_0} \left\{\begin{matrix} \cos\omega_0 t' \\ \sin\omega_0 t' \\ \cos\left(\omega_0 t' - \varphi\right) \end{matrix}\right\} e^{in\omega_0\left[t' - \frac{\rho_0}{c}\sin\theta\cos\left(\omega_0 t' - \varphi\right)\right]} dt' \qquad [4.151]
$$

with $\beta - \frac{\rho_0\omega_0}{c}$. Let us perform the change of the integration variable $y = \omega_0 t' - \varphi$:

$$
\left\{\begin{matrix} K \\ J \\ L' \end{matrix}\right\} = \int_0^{2\pi} \left\{\begin{matrix} \cos\left(y + \varphi\right) \\ \sin\left(y + \varphi\right) \\ \cos y \end{matrix}\right\} e^{i[ny - n\beta\sin\theta\cos y]} e^{in\varphi} dy \qquad [4.152]
$$

where the integration limits change from $[-\varphi, -\varphi + 2\pi]$ to $[0, 2\pi]$ due to periodicity. Thus, we see that:

$$
\begin{aligned}
\left\{\begin{matrix} J \\ K \end{matrix}\right\} &= e^{in\varphi} \int_0^{2\pi} \left\{\begin{matrix} \sin\left(y + \varphi\right) \\ \cos\left(y + \varphi\right) \end{matrix}\right\} e^{i(ny - n\beta\sin\theta\cos y)} dy \\
&= e^{in\varphi} \int_0^{2\pi} \left[\sin\varphi \left\{\begin{matrix} \cos y \\ -\sin y \end{matrix}\right\} + \cos\varphi \left\{\begin{matrix} \sin y \\ \cos y \end{matrix}\right\}\right] e^{i(ny - n\beta\sin\theta\cos y)} dy \\
&= e^{in\varphi} \left[\sin\varphi \left\{\begin{matrix} L \\ -M \end{matrix}\right\} + \cos\varphi \left\{\begin{matrix} M \\ L \end{matrix}\right\}\right]
\end{aligned} \qquad [4.153]
$$

where:

$$M = \int_0^{2\pi} \sin y e^{\mathrm{i}(ny - n\beta \sin\theta \cos y)} dy = \frac{1}{\mathrm{i}n\beta\sin\theta} \int_0^{2\pi} e^{\mathrm{i}ny} \frac{d}{dy} e^{-\mathrm{i}n\beta\sin\theta\cos y} dy$$

$$= -\frac{1}{\beta\sin\theta} \int_0^{2\pi} e^{\mathrm{i}ny} e^{-\mathrm{i}n\beta\sin\theta\cos y} dy = -\frac{2\pi}{\beta\sin\theta} \frac{J_n(n\beta\sin\theta)}{\mathrm{i}^n}$$

$$L = \int_0^{2\pi} \cos y e^{\mathrm{i}(ny - n\beta\sin\theta\cos y)} dy = -\frac{1}{\mathrm{i}} \frac{d}{d(n\beta\sin\theta)} \int_0^{2\pi} e^{\mathrm{i}(ny - n\beta\sin\theta\cos y)} dy$$

$$= -\frac{2\pi}{\mathrm{i}} \frac{J_n'(n\beta\sin\theta)}{\mathrm{i}^n} \qquad [4.154]$$

It therefore follows that:

$$|I|^2 = \frac{\beta^2}{\omega_0^2} \left[\cos^2\theta \left(|K|^2 + |J|^2\right) + \sin^2\theta |L|^2\right] \frac{\beta^2}{\omega_0^2} \left[\cos^2\theta \left(|M|^2 + |L|^2\right)\right.$$

$$\left. + \sin^2\theta |L|^2\right] = \frac{\beta^2}{\omega_0^2} \left[\cos^2\theta |M|^2 + |L|^2\right]$$

$$= \frac{4\pi^2\beta^2}{\omega_0^2} \left[\cos^2\theta \frac{J_n^2(n\beta\sin\theta)}{\beta^2\sin^2\theta} + J_n'^2(n\beta\sin\theta)\right]$$

$$= \frac{4\pi^2\beta^2}{\omega_0^2} \left[\frac{\cot^2\theta}{\beta^2} J_n^2(n\beta\sin\theta) + J_n'^2(n\beta\sin\theta)\right] \qquad [4.155]$$

Substituting [4.155] into [4.147], we finally get:

$$\frac{dP_n}{d\Omega} = \frac{n^2\omega_0^2 q^2\beta^2}{8\pi^2\varepsilon_0 c} \left[\frac{\cot^2\theta}{\beta^2} J_n^2(n\beta\sin\theta) + J_n'^2(n\beta\sin\theta)\right] \qquad [4.156]$$

Let us examine some limiting cases:

1) In the non-relativistic limit ($\beta \ll 1$), only the term $n = 1$ is expected to contribute appreciably. Using the approximations for Bessel functions $J_1(x) \simeq \frac{x}{2}$ and $J_1'(x) \simeq \frac{1}{2}$ for $x \ll 1$, we find:

$$\frac{dP_1}{d\Omega} = \frac{\omega_0^2 q^2\beta^2}{8\pi^2\varepsilon_0 c} \left[\frac{\cot^2\theta\sin^2\theta}{4} + \frac{1}{4}\right] = \frac{\omega_0^2 q^2\beta^2}{32\pi^2\varepsilon_0 c} \left[\cos^2\theta + 1\right] \qquad [4.157]$$

2) In the ultra-relativistic limit ($\beta \simeq 1$), most of the radiation is concentrated near the orbital plane (i.e. $\theta \simeq \frac{\pi}{2}$). By introducing the angular variable $\alpha = \frac{\pi}{2} - \theta \ll 1$, we deduce that:

$$\frac{dP_n}{d\Omega} = \frac{n^2\omega_0^2 q^2\beta^2}{8\pi^2\varepsilon_0 c} \left[\frac{\tan^2\alpha}{\beta^2} J_n^2(n\beta\cos\alpha) + J_n'^2(n\beta\cos\alpha)\right] \qquad [4.158]$$

The argument of the Bessel function is comparable to n (but always less than n). For this particular range of arguments, we know that:

$$J_n(x) \sim \frac{1}{\pi}\sqrt{\frac{2(n-x)}{3x}}K_{1/3}\left(\frac{2\sqrt{2}(n-x)^{3/2}}{3\sqrt{x}}\right) \qquad [4.159]$$

or simply:

$$J_n(x) \sim \frac{1}{\pi}\frac{y^{1/3}}{3^{1/6}x^{1/3}}K_{1/3}(y) \qquad [4.160]$$

with $y = \frac{2\sqrt{2}(n-x)^{3/2}}{3\sqrt{x}}$. Thus, the derivative in [4.158] can be expressed as:

$$\begin{aligned} J_n'(x) &\sim -\frac{1}{\pi}\frac{y^{1/3}}{3^{7/6}x^{4/3}}K_{1/3}(y) + \frac{1}{\pi}\frac{1}{3^{1/6}x^{1/3}}\frac{d}{dx}\left(y^{1/3}K_{1/3}(y)\right) \\ &= -\frac{1}{\pi}\frac{y^{1/3}}{3^{7/6}x^{4/3}}K_{1/3}(y) + \frac{1}{\pi}\frac{1}{3^{1/6}x^{1/3}}\frac{d}{dy}\left(y^{1/3}K_{1/3}(y)\right)\frac{dy}{dx} \\ &= -\frac{1}{\pi}\frac{y^{1/3}}{3^{7/6}x^{4/3}}K_{1/3}(y) + \frac{1}{\pi}\frac{y^{1/3}K_{2/3}(y)}{3^{1/6}x^{1/3}}\frac{\sqrt{2}(n-x)^{1/2}(2x+n)}{3x^{3/2}} \\ &\sim \frac{1}{\pi}\frac{y^{1/3}K_{2/3}(y)}{3^{1/6}x^{1/3}}\frac{\sqrt{2}(n-x)^{1/2}(2x+n)}{3x^{3/2}} \end{aligned} \qquad [4.161]$$

where we have only retained the dominant term. To arrive at the second term, we have used the property of derivation $\frac{d}{dx}(x^\nu K_\nu(x)) = -x^\nu K_{\nu-1}(x) = -x^\nu K_{1-\nu}(x)$. By inserting the results [4.160] and [4.161] into [4.158], we get:

$$\begin{aligned} \frac{dP_n}{d\Omega} &= \frac{n^2\omega_0^2q^2\beta^2}{8\pi^2\varepsilon_0 c}\frac{y^{2/3}}{\pi^2 3^{1/3}x^{2/3}} \\ &\times\left[\frac{\tan^2\alpha}{\beta^2}K_{1/3}^2(y) + \frac{2(n-x)(2x+n)^2}{9x^3}K_{2/3}^2(y)\right] \end{aligned} \qquad [4.162]$$

By expanding y and x in powers of α and $\frac{1}{\gamma}$, we obtain:

$$\begin{aligned} x &\simeq n\left(1 - \frac{1}{2\gamma^2} - \frac{\alpha^2}{2}\right) \\ y &\simeq \frac{n}{3}\left(\frac{1}{\gamma^2} + \alpha^2\right)^{3/2} \end{aligned} \qquad [4.163]$$

which gives approximately $\frac{2(n-x)(2x+n)^2}{9x^3} \simeq \frac{1}{\gamma^2} + \alpha^2$, $\frac{\tan^2 \alpha}{\beta^2} \simeq \alpha^2$ and $\frac{y^{2/3}}{x^{2/3}} \simeq \left(\frac{1}{\gamma^2} + \alpha^2\right) \frac{1}{3^{2/3}}$. It follows that:

$$\frac{dP_n}{d\Omega} = \frac{n^2 \omega_0^2 q^2}{24\pi^4 \varepsilon_0 c} \left(\frac{1}{\gamma^2} + \alpha^2\right)^2 \left[K_{2/3}^2 (y) + \frac{\alpha^2}{\frac{1}{\gamma^2} + \alpha^2} K_{1/3}^2 (y)\right] \quad [4.164]$$

This term describes the radiation at the particular frequency $\omega = n\omega_0$. We can therefore also write this in the form:

$$\frac{dP_n}{d\Omega} = \frac{\omega_0^2}{2\pi} \frac{q^2 \left(\omega^2 \omega_0^2\right)}{12\pi^3 \varepsilon_0 c \gamma^6} \left(1 + \alpha^2 \gamma^2\right)^2 \gamma^2 \left[K_{2/3}^2 (y) + \frac{\alpha^2 \gamma^2}{1 + \alpha^2 \gamma^2} K_{1/3}^2 (y)\right] \quad [4.165]$$

with:

$$y = \frac{\omega}{3\omega_0 \gamma^3} \left(1 + \alpha^2 \gamma^2\right)^{3/2} = \frac{\omega}{\omega_c} \left(1 + \alpha^2 \gamma^2\right)^{3/2} \quad [4.166]$$

In terms of ω_c, equation [4.166] becomes:

$$\frac{dP_n}{d\Omega} = \frac{\omega_0^2}{2\pi} \frac{3q^2}{4\pi^3 \varepsilon_0 c} \left(\frac{\omega}{\omega_c}\right)^2 \left(1 + \alpha^2 \gamma^2\right)^2 \gamma^2 \left[K_{2/3}^2 (y) + \frac{\alpha^2 \gamma^2}{1 + \alpha^2 \gamma^2} K_{1/3}^2 (y)\right]$$

[4.167]

This is the time-averaged power received at latitude α, at frequency $\omega = n\omega_0$, per unit solid angle. The energy per cycle received at this frequency is obtained by multiplying by $\frac{2\pi}{\omega_0}$. Finally, the energy per unit of frequency is found by multiplying by the inverse spacing factor of the harmonics, that is, $\frac{1}{\omega_0}$. We therefore conclude that for a single pulse, that is, one cycle of the particle:

$$\frac{dI(\omega)}{d\Omega} = \frac{2\pi}{\omega_0^2} \frac{dP_n}{d\Omega} = \frac{3q^2}{4\pi^3 \varepsilon_0 c} \left(\frac{\omega}{\omega_c}\right)^2 \left(1 + \alpha^2 \gamma^2\right)^2 \gamma^2 \times \left[K_{2/3}^2 (y) + \frac{\alpha^2 \gamma^2}{1 + \alpha^2 \gamma^2} K_{1/3}^2 (y)\right] \quad [4.168]$$

which is precisely what we found when calculating the radiation intensity of a particle in an instantaneous circular motion.

4.9. Thomson scattering of radiation

If a monochromatic electromagnetic plane wave is incident on a free particle of charge q and mass m, the particle will be accelerated and emit radiation. This radiation is emitted in directions other than that of the incident wave, but for non-relativistic motion of the particle, it will have the same frequency as the incident radiation. The entire process can be described as the scattering of incident radiation; it is known as *Thomson scattering*. Thomson scattering must be distinguished from *Compton scattering*, in which the same phenomenon is treated using quantum theory. Quantum theory must be used when the wavelength of the radiation λ is comparable to the Compton wavelength of the scatterer, $\lambda = \frac{2\pi c}{\omega} \sim \frac{\hbar}{mc} \equiv \lambda_c$ (λ_c is the Compton wavelength). This condition can also be written as $\hbar\omega \sim 2\pi mc^2$, which means that if the photon energy $\hbar\omega$ is comparable to the rest energy of the particle mc^2, then the calculation must be carried out using quantum theory. For example, for visible light, the photon energy is much smaller than the rest energy of an electron, which implies that the calculation using Thomson (classical) scattering is entirely appropriate.

The instantaneous power radiated by a particle (such as an electron) with charge e and mass m in non-relativistic motion is given by the Larmor formula:

$$\frac{dP}{d\Omega} = \frac{e^2 a^2}{16\pi^2 \varepsilon_0 c^3} \sin^2 \Theta \qquad [4.169]$$

where Θ is the angle between the direction of observation and the acceleration. The acceleration is produced by the incident electromagnetic wave. To determine the acceleration of the particle, we must solve the equation of motion of the scatterer. The force acting on it is determined from the incident field:

$$\vec{\mathbf{E}}_0 = E_0 \vec{\epsilon} e^{-\mathrm{i}\left(\omega t - \vec{\mathrm{k}} \cdot \vec{\mathbf{r}}\right)} \qquad [4.170]$$

where, for practical reasons, we take $\vec{\mathrm{k}} = \mathrm{k}\vec{\epsilon}_3$ and assume a linearly polarized incident plane wave $\vec{\epsilon} = \cos\psi \vec{\epsilon}_1 + \sin\psi \vec{\epsilon}_2$, as illustrated in Figure 4.10. Similarly, for simplicity, we assume that the velocity of the charged particle has no component along the direction of $\vec{\mathrm{k}}$. The magnetic force is also neglected, which is reasonable as long as the particle remains non-relativistic. If the particle is located in the plane $z = 0$, it is then subject to the incident field:

$$\vec{\mathbf{E}}_0 = E_0 \vec{\epsilon} e^{-\mathrm{i}\omega t} \qquad [4.171]$$

Thus, the force acting on the particle (with charge q and mass m) is $\vec{\mathbf{F}} = e\vec{\mathbf{E}}_0 = m\vec{\mathbf{a}}$.

This gives:

$$\vec{\mathbf{a}} = \frac{eE_0}{m} \vec{\epsilon} e^{-\mathrm{i}\omega t} \qquad [4.172]$$

and:

$$\vec{\mathrm{a}} \cdot \vec{\epsilon} = \frac{eE_0}{m} \sin\theta \left(\cos\psi\cos\varphi + \sin\psi\sin\varphi\right) e^{-i\omega t} = \frac{eE_0}{m} \sin\theta \cos\left(\psi - \varphi\right) e^{-i\omega t} \qquad [4.173]$$

where θ and φ indicate, in spherical coordinates, the direction of the observation point.

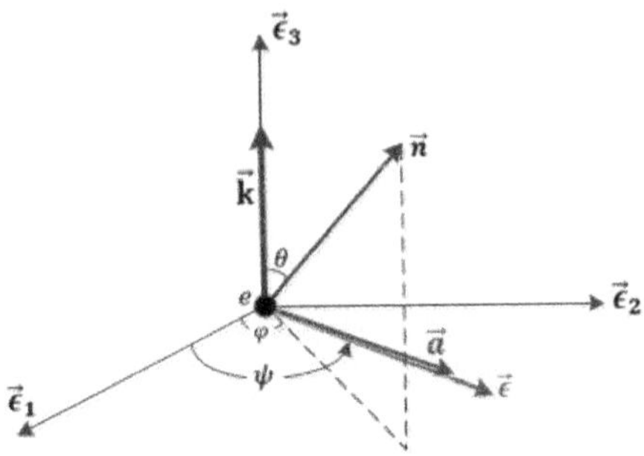

Figure 4.10. *Thomson scattering of a monochromatic plane wave with polarization $\vec{\epsilon}$ by a particle of charge e and mass m. For a color version of this figure, see www.iste.co.uk/abdesselam/electrodynamics3.zip*

The time-averaged power emitted in a particular direction is proportional to:

$$\begin{aligned}\langle a^2 \sin^2\Theta\rangle &= \langle a^2\left(1-\cos^2\Theta\right)\rangle = \langle a^2 - \left(\vec{\mathrm{a}}\cdot\vec{\epsilon}\right)^2\rangle \\ &= \frac{e^2\left|E_0\right|^2}{2m^2}\left[1-\sin^2\theta\cos^2\left(\psi-\varphi\right)\right]\end{aligned} \qquad [4.174]$$

where the angle brackets $\langle\cdots\rangle$ denote the average time. When incident radiation is unpolarized, an average over ψ must be performed.

$$\begin{aligned}\langle a^2 \sin^2\Theta\rangle &= \frac{1}{2\pi}\int_0^{2\pi} \frac{e^2\left|E_0\right|^2}{2m^2}\left[1-\sin^2\theta\cos^2\left(\psi-\varphi\right)\right] d\psi \\ &= \frac{e^2\left|E_0\right|^2}{2m^2}\left[1-\frac{1}{2}\sin^2\theta\right] = \frac{e^2\left|E_0\right|^2}{4m^2}\left[1+\cos^2\theta\right]\end{aligned} \qquad [4.175]$$

By substituting [4.175] into [4.169], we obtain:

$$\langle\frac{dP}{d\Omega}\rangle = \frac{e^4}{64\pi^2\varepsilon_0 c^3 m^2}\left|E_0\right|^2\left[1+\cos^2\theta\right] \qquad [4.176]$$

This is the power distribution on time average when the incident wave is unpolarized.

The *differential cross-section of scattering* is defined as the time-averaged power per unit solid angle divided by the time-averaged incident power per unit area (calculated from the Poynting vector of the incident wave, that is, $\langle \vec{\Pi} \cdot \vec{e}_3 \rangle = \frac{1}{2}\Re\left[\vec{e}_3 \cdot \left(\vec{\mathbf{E}} \times \vec{\mathbf{B}}^\star\right)\right] = \frac{1}{2}\varepsilon_0 c |E_0|^2$):

$$\langle \frac{d\sigma}{d\Omega} \rangle = \frac{\langle \frac{d\sigma}{d\Omega} \rangle}{\frac{1}{2}\varepsilon_0 c |E_0|^2} = \frac{1}{2}\left(\frac{e^2}{4\pi\varepsilon_0 mc^2}\right)^2 \left[1 + \cos^2\theta\right] \quad [4.177]$$

This is the *Thomson equation* for the scattering of light by a charged particle. The angular distribution of the scattering cross section is shown in Figure 4.11. The total cross-section, called the *Thomson cross-section*, is:

$$\sigma = \int \langle \frac{d\sigma}{d\Omega} \rangle d\Omega = 2\pi \int_0^\pi \frac{1}{2}\left(\frac{e^2}{4\pi\varepsilon_0 mc^2}\right)^2 \left[1 + \cos^2\theta\right] d\theta$$
$$= \frac{8\pi}{3}\left(\frac{e^2}{4\pi\varepsilon_0 mc^2}\right)^2 = \frac{8\pi}{3} r_e^2 \quad [4.178]$$

where r_e is the classical radius of the electron.

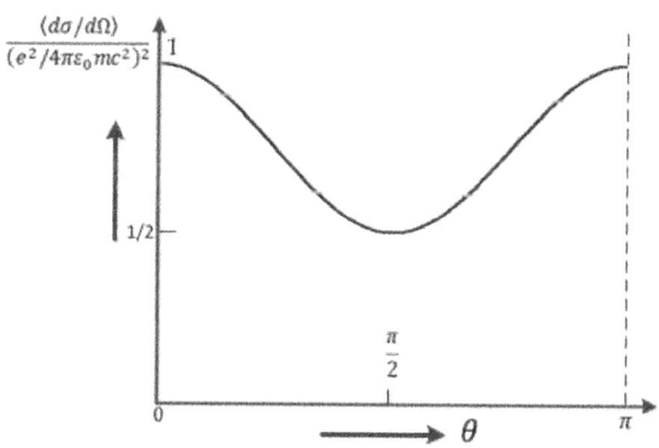

Figure 4.11. *Differential scattering cross-section of unpolarized radiation by a free electron*

EXAMPLE 4.6. BOUND ELECTRON.– We can also extend the previous calculations to a bound charge by using the damped harmonic oscillator model (elastic electron). Consider a charge $-e$ of mass m bound to the origin of coordinates (at its equilibrium position) by a restoring force $-m\omega_0 \overrightarrow{\mathbf{r}}$ (i.e. with a natural oscillation frequency ω_0) and a damping force $-\frac{m}{\tau}\overrightarrow{\mathbf{v}}$ (τ is the collision time). When an electric field $\overrightarrow{\mathbf{E}}_0\left(\overrightarrow{\mathbf{r}},t\right) = \overrightarrow{\epsilon}E_0 e^{-\mathrm{i}\left(\omega t - \overrightarrow{\mathrm{k}}\cdot\overrightarrow{\mathbf{r}}\right)}$ is applied, with $\overrightarrow{\mathrm{k}} = \mathrm{k}\overrightarrow{\epsilon}_3$, the charge responds with a displacement $\overrightarrow{\mathbf{r}}(t) = \overrightarrow{\mathbf{r}}_0 e^{-\mathrm{i}\omega t}$, where $\overrightarrow{\mathbf{r}}_0 = \frac{-eE_0/m}{\omega_0^2-\omega^2-\mathrm{i}\frac{\omega}{\tau}}$, assuming the approximation $\overrightarrow{\mathbf{E}}_0\left(\overrightarrow{\mathbf{r}},t\right) = \overrightarrow{\mathbf{E}}_0(0,t)$, which is reasonable when the particle's displacement is small compared to the wavelength of incident radiation. This is certainly true for visible light and an atomic electron. From $\overrightarrow{\mathbf{r}}(t)$, it is straightforward to calculate the acceleration:

$$\overrightarrow{\mathbf{a}}(t) = \overrightarrow{\epsilon}\frac{eE_0\omega^2/m}{\omega_0^2-\omega^2-\mathrm{i}\frac{\omega}{\tau}}e^{-\mathrm{i}\omega t} \qquad [4.179]$$

and then to find the time-averaged radiated (or scattered) power per unit of solid angle:

$$\begin{aligned}\langle\frac{dP}{d\Omega}\rangle &= \frac{e^2}{16\pi^2\varepsilon_0 c^3}\frac{1}{2\pi}\int_0^{2\pi}\frac{|a|^2}{2}\left[1-\sin^2\theta\cos^2(\psi-\varphi)\right]d\psi \\ &= \frac{e^2}{64\pi^2\varepsilon_0 c^3}|a|^2\left(1+\cos^2\theta\right) \\ &= \frac{e^2}{64\pi^2\varepsilon_0 c^3}\frac{\left(eE_0\omega^2/m\right)^2}{\left(\omega_0^2-\omega^2\right)^2+\frac{\omega^2}{\tau^2}}\left(1+\cos^2\theta\right) \\ &= \frac{e^2}{64\pi^2\varepsilon_0 c^3}\left(\frac{eE_0}{m}\right)^2\frac{\left(1+\cos^2\theta\right)}{\left(1-\frac{\omega_0^2}{\omega^2}\right)^2+\frac{1}{\omega^2\tau^2}}\end{aligned} \qquad [4.180]$$

It follows that:

$$\langle\frac{d\sigma}{d\Omega}\rangle = \frac{1}{2}r_e^2\frac{\left(1+\cos^2\theta\right)}{\left(1-\frac{\omega_0^2}{\omega^2}\right)^2+\frac{1}{\omega^2\tau^2}} \qquad [4.181]$$

When $\omega \gg \omega_0$ and $\omega \gg \tau^{-1}$, the previous equation reduces to the Thomson equation [4.177] (as expected), since at very high frequencies ω, the particle behaves as if it were free with respect to the field. Similarly, when $\omega_0 \gg \omega_0 \gg \tau^{-1}$, we obtain the result of *Rayleigh scattering*:

$$\langle\frac{d\sigma}{d\Omega}\rangle = \frac{1}{2}r_e^2\left(\frac{\omega}{\omega_0}\right)^4\left(1+\cos^2\theta\right) \qquad [4.182]$$

with characteristic ω^4 or $\frac{1}{\lambda^4}$ behavior, indicating dipole scattering. This means that short wavelengths (e.g. blue) are scattered much more than long wavelengths (e.g.

red). Note that in both cases (Thomson and Rayleigh), the unpolarized differential cross-section has the same angular dependence of $\left(1+\cos^2\theta\right)$. In both scenarios, it is assumed that the scattered wave has the same frequency as the incident wave (oscillating dipole). However, this assumption breaks down when the frequency of the incident wave becomes comparable to $\frac{mc^2}{\hbar}$, where mc^2 is the rest energy of the electron. In this high-frequency regime, the particle-like nature of light becomes increasingly apparent, and the incident photon can transfer a significant portion of its energy to the scattering electron. The scattered photon will then have a lower frequency. Thomson scattering thus becomes Compton scattering.

EXAMPLE 4.7. MACROSCOPIC STUDY OF LIGHT SCATTERING.– We will provide here a very general overview of the theory of light scattering caused by an inhomogeneity in the medium through which it propagates. This includes the particular case of scattering by a single electron. In a linear and inhomogeneous dielectric medium, the propagation of electromagnetic waves is governed by the Maxwell equations: (1) $0=\vec{\nabla}\cdot\left(\varepsilon_0\varepsilon_r\vec{\mathbf{E}}\right)\Longrightarrow\vec{\nabla}\cdot\vec{\mathbf{E}}=-\frac{\vec{\nabla}\varepsilon_r\cdot\vec{\mathbf{E}}}{\varepsilon_r}$; (2) $\vec{\nabla}\cdot\vec{\mathbf{B}}=0$; (3) $\vec{\nabla}\times\vec{\mathbf{E}}=-\frac{\partial\vec{\mathbf{B}}}{\partial t}$; (4) $\vec{\nabla}\times\vec{\mathbf{B}}=\varepsilon_0\mu_0\varepsilon_r\frac{\partial\vec{\mathbf{E}}}{\partial t}$. This gives:

$$
\begin{aligned}
\Delta\vec{\mathbf{E}}-\frac{\varepsilon_r}{c^2}\frac{\partial^2\vec{\mathbf{E}}}{\partial t^2}&=-\vec{\nabla}\left(\frac{\vec{\nabla}\varepsilon_r\cdot\vec{\mathbf{E}}}{\varepsilon_r}\right)\\
\Delta\vec{\mathbf{B}}-\frac{\varepsilon_r}{c^2}\frac{\partial^2\vec{\mathbf{B}}}{\partial t^2}&=-\frac{1}{c^2}\vec{\nabla}\varepsilon_r\times\frac{\partial\vec{\mathbf{E}}}{\partial t}
\end{aligned}
\qquad [4.183]
$$

Now suppose that ε_r fluctuates around an average value $\bar{\varepsilon}_r$, so that we can write, in the first approximation, $\varepsilon_r=\bar{\varepsilon}_r+\delta\varepsilon_r$, where $\delta\varepsilon_r\ll\bar{\varepsilon}_r$. Thus, the previous equations become:

$$
\begin{aligned}
\Delta\vec{\mathbf{E}}-\frac{\bar{\varepsilon}_r}{c^2}\frac{\partial^2\vec{\mathbf{E}}}{\partial t^2}&=-\vec{\nabla}\left(\frac{\vec{\nabla}\delta\varepsilon_r\cdot\vec{\mathbf{E}}}{\bar{\varepsilon}_r}\right)+\frac{\delta\varepsilon_r}{c^2}\frac{\partial^2\vec{\mathbf{E}}}{\partial t^2}\\
\Delta\vec{\mathbf{B}}-\frac{\bar{\varepsilon}_r}{c^2}\frac{\partial^2\vec{\mathbf{B}}}{\partial t^2}&=-\frac{1}{c^2}\vec{\nabla}\delta\varepsilon_r\times\frac{\partial\vec{\mathbf{E}}}{\partial t}+\frac{\delta\varepsilon_r}{c^2}\frac{\partial^2\vec{\mathbf{B}}}{\partial t^2}
\end{aligned}
\qquad [4.184]
$$

where only the linear terms in $\delta\varepsilon_r$ have been retained. Setting $\vec{\mathbf{E}}\left(\vec{\mathbf{r}},t\right)=\Re\left[\vec{\hat{\mathbf{E}}}\left(\vec{\mathbf{r}}\right)e^{-\mathrm{i}\omega t}\right]$ and $\vec{\mathbf{B}}\left(\vec{\mathbf{r}},t\right)=\Re\left[\vec{\hat{\mathbf{B}}}\left(\vec{\mathbf{r}}\right)e^{-\mathrm{i}\omega t}\right]$, we obtain:

$$
\begin{aligned}
\Delta\vec{\hat{\mathbf{E}}}+\mathsf{k}^2\vec{\hat{\mathbf{E}}}&=-\vec{\nabla}\left(\frac{\vec{\nabla}\delta\varepsilon_r\cdot\vec{\hat{\mathbf{E}}}}{\bar{\varepsilon}_r}\right)-\frac{\delta\varepsilon_r}{\bar{\varepsilon}_r}\mathsf{k}^2\vec{\hat{\mathbf{E}}}\\
\Delta\vec{\hat{\mathbf{B}}}+\mathsf{k}^2\vec{\hat{\mathbf{B}}}&=\frac{\mathrm{i}\omega}{c^2}\vec{\nabla}\delta\varepsilon_r\times\vec{\hat{\mathbf{E}}}-\frac{\delta\varepsilon_r}{\bar{\varepsilon}_r}\mathsf{k}^2\vec{\hat{\mathbf{B}}}
\end{aligned}
\qquad [4.185]
$$

where $\mathsf{k}^2 = \frac{\bar{\varepsilon}_r \omega^2}{c^2}$. These are the Helmholtz equations, whose solutions are well known. The first equation has the following solution:

$$\vec{\hat{\mathbf{E}}}\left(\vec{\mathbf{r}}\right) = \vec{\hat{\mathbf{E}}}_i\left(\vec{\mathbf{r}}\right) + \frac{1}{4\pi\bar{\varepsilon}_r}\int d\tau' \frac{e^{\mathsf{i}\mathsf{k}\left|\vec{\mathbf{r}}-\vec{\mathbf{r}}'\right|}}{\left|\vec{\mathbf{r}}-\vec{\mathbf{r}}'\right|}\left[\vec{\nabla}'\left(\vec{\hat{\mathbf{E}}}'\cdot\vec{\nabla}'\delta\varepsilon_r'\right) + \delta\varepsilon_r'\mathsf{k}^2\vec{\hat{\mathbf{E}}}'\right] \quad [4.186]$$

where $\vec{\nabla}'$ is the gradient with respect to $\vec{\mathbf{r}}'$. The first term is a solution of the homogeneous Helmholtz equation and describes the incident wave. The second term describes the scattered wave; it is an integral equation that is difficult to solve. To address this, we will assume that the scattered wave has a small amplitude compared to the incident wave. In the first approximation, we must replace $\vec{\hat{\mathbf{E}}}$ by $\vec{\hat{\mathbf{E}}}_0$ in the integral (*Born approximation*). The scattered wave then becomes:

$$\vec{\hat{\mathbf{E}}}_d\left(\vec{\mathbf{r}}\right) = \frac{\hat{E}_0}{4\pi\bar{\varepsilon}_r}\int d\tau' \frac{e^{\mathsf{i}\mathsf{k}\left|\vec{\mathbf{r}}-\vec{\mathbf{r}}'\right|}}{\left|\vec{\mathbf{r}}-\vec{\mathbf{r}}'\right|}\left[\vec{\nabla}'\left(\vec{\epsilon}\cdot\vec{\nabla}'\delta\varepsilon_r' e^{\mathsf{i}\mathsf{k}z'}\right) + \delta\varepsilon_r'\mathsf{k}^2\vec{\epsilon}\, e^{\mathsf{i}\mathsf{k}z'}\right] \quad [4.187]$$

where we set $\vec{\hat{\mathbf{E}}}_i = E_0\vec{\epsilon}\, e^{\mathsf{i}\mathsf{k}z}$. The vector $\vec{\epsilon}$ indicates the polarization of the incident wave ($\vec{\epsilon}$ lies in the Oxy plane). In the radiation zone, the previous equation reduces to:

$$\vec{\hat{\mathbf{E}}}_d\left(\vec{\mathbf{r}}\right) = \frac{\hat{E}_0}{4\pi\bar{\varepsilon}_r}\frac{e^{\mathsf{i}\mathsf{k}r}}{r}\int d\tau'\left[\vec{\nabla}'\left(\vec{\epsilon}\cdot\vec{\nabla}'\delta\varepsilon_r' e^{\mathsf{i}\mathsf{k}z'}\right) + \delta\varepsilon_r'\mathsf{k}^2\vec{\epsilon}\, e^{\mathsf{i}\mathsf{k}z'}\right] e^{-\mathsf{i}\vec{\mathsf{k}}_d\cdot\vec{\mathbf{r}}'} \quad [4.188]$$

where $\vec{\mathsf{k}}_d = \mathsf{k}\vec{\mathbf{n}}$ is the wave vector of the scattered wave in the direction $\vec{\mathbf{n}}$ (the incident wave vector is denoted $\vec{\mathsf{k}}_i = \mathsf{k}\vec{\epsilon}_3$). To simplify this expression, let us integrate the first term by parts:

$$\begin{aligned}&\int d\tau'\vec{\nabla}'\left(\vec{\epsilon}\cdot\vec{\nabla}'\delta\varepsilon_r' e^{\mathsf{i}\mathsf{k}z'}\right) e^{-\mathsf{i}\vec{\mathsf{k}}_d\cdot\vec{\mathbf{r}}'} = \mathsf{i}\vec{\mathsf{k}}_d\int d\tau'\left(\vec{\epsilon}\cdot\vec{\nabla}'\delta\varepsilon_r' e^{\mathsf{i}\mathsf{k}z'}\right)\\ &e^{-\mathsf{i}\vec{\mathsf{k}}_d\cdot\vec{\mathbf{r}}'} = -\mathsf{i}\vec{\mathsf{k}}_d\int d\tau'\delta\varepsilon_r'\vec{\epsilon}\cdot\vec{\nabla}'\left(e^{\mathsf{i}\mathsf{k}z'}e^{-\mathsf{i}\vec{\mathsf{k}}_d\cdot\vec{\mathbf{r}}'}\right)\\ &= -\vec{\mathsf{k}}_d\int d\tau'\delta\varepsilon_r'\vec{\epsilon}\cdot\left(\vec{\mathsf{k}}_d - \mathsf{k}\vec{\epsilon}_3\right) e^{\mathsf{i}\mathsf{k}z'}e^{-\mathsf{i}\vec{\mathsf{k}}_d\cdot\vec{\mathbf{r}}'}\end{aligned} \quad [4.189]$$

where it is assumed that the fields vanish rapidly at infinity so that the surface terms cancel out. Since $\vec{\epsilon}\cdot\vec{\epsilon}_3 = 0$, we find:

$$\vec{\hat{\mathbf{E}}}_d\left(\vec{\mathbf{r}}\right) = \frac{\hat{E}_0}{4\pi\bar{\varepsilon}_r}\frac{e^{\mathsf{i}\mathsf{k}r}}{r}\left[\mathsf{k}^2\vec{\epsilon} - \left(\vec{\mathsf{k}}_d\cdot\vec{\epsilon}\right)\vec{\mathsf{k}}_d\right]\int d\tau'\delta\varepsilon_r\left(\vec{\mathbf{r}}'\right) e^{-\mathsf{i}\vec{\mathbf{q}}\cdot\vec{\mathbf{r}}'} \quad [4.190]$$

where $\vec{\mathbf{q}} = \mathsf{k}\left(\vec{\mathbf{n}} - \vec{\epsilon}_3\right) = \vec{\mathsf{k}}_d - \vec{\mathsf{k}}_i$ is the wave vector difference between incident and scattered waves. Note also that $\vec{\hat{\mathbf{E}}}_d \cdot \vec{\mathsf{k}}_d = 0$ ($\vec{\hat{\mathbf{E}}}_d$ is perpendicular to the direction of scattering), which can easily be verified. In fact, we have $\left[\mathsf{k}^2\vec{\epsilon} - \left(\vec{\mathsf{k}}_d \cdot \vec{\epsilon}\right)\vec{\mathsf{k}}_d\right] \cdot \vec{\mathsf{k}}_d = 0$. The integral:

$$\delta\varepsilon_r\left(\vec{\mathbf{q}}\right) = \int d\tau' \delta\varepsilon_r\left(\vec{\mathbf{r}}'\right) e^{-\mathrm{i}\vec{\mathbf{q}}\cdot\vec{\mathbf{r}}'} \quad [4.191]$$

is simply the Fourier transform of $delta\varepsilon_r\left(\vec{\mathbf{r}}'\right)$. It follows that:

$$\vec{\hat{\mathbf{E}}}_d\left(\vec{\mathbf{r}}\right) = \frac{\hat{E}_0}{4\pi\bar{\varepsilon}_r}\frac{e^{\mathrm{i}\mathsf{k}r}}{r}\left[\mathsf{k}^2\vec{\epsilon} - \left(\vec{\mathsf{k}}_d \cdot \vec{\epsilon}\right)\vec{\mathsf{k}}_d\right]\delta\varepsilon_r\left(\vec{\mathbf{q}}\right) \quad [4.192]$$

The incident and scattered magnetic fields (in the far-field region) are given by:

$$\begin{aligned} \vec{\hat{\mathbf{B}}}_i &= \frac{E_0\sqrt{\hat{\varepsilon}_r}}{c}e^{\mathrm{i}\mathsf{k}z}\vec{\epsilon}_3 \times \vec{\epsilon} \\ \vec{\hat{\mathbf{B}}}_d &= \frac{E_0}{4\pi\sqrt{\hat{\varepsilon}_r}c}\frac{e^{\mathrm{i}\mathsf{k}r}}{r}\vec{\mathbf{n}} \times \left[\mathsf{k}^2\vec{\epsilon} - \left(\vec{\mathsf{k}}_d \cdot \vec{\epsilon}\right)\vec{\mathsf{k}}_d\right]\delta\varepsilon_r\left(\vec{\mathbf{q}}\right) \end{aligned} \quad [4.193]$$

Consequently, the incident and scattered time-averaged power fluxes are:

$$\begin{aligned} \langle\vec{\mathbf{\Pi}}\rangle_i &= \frac{\Re\left[\vec{\hat{\mathbf{E}}}_i \times \vec{\hat{\mathbf{B}}}_i^\star\right]}{2\mu_0} = \frac{\left|E_0\right|^2\sqrt{\bar{\varepsilon}_r}}{2\mu_0 c}\vec{\epsilon}_3 \\ \langle\vec{\mathbf{\Pi}}\rangle_d &= \frac{\Re\left[\vec{\hat{\mathbf{E}}}_d \times \vec{\hat{\mathbf{B}}}_d^\star\right]}{2\mu_0} = \frac{\left|E_0\right|^2\sqrt{\bar{\varepsilon}_r}}{2\mu_0 c}\frac{\left|\mathsf{k}^2\vec{\epsilon} - \left(\vec{\mathsf{k}}_d \cdot \vec{\epsilon}\right)\vec{\mathsf{k}}_d\right|^2}{16\pi^2\bar{\varepsilon}_r^2 r^2}\left|\delta\varepsilon_r\left(\vec{\mathbf{q}}\right)\right|^2\vec{\mathbf{n}} \end{aligned} \quad [4.194]$$

The differential scattering cross-section is defined as follows:

$$\begin{aligned} \frac{d\sigma}{d\Omega} &= \frac{r^2\langle\vec{\mathbf{\Pi}}\rangle_d \cdot \vec{\mathbf{n}}}{\langle\vec{\mathbf{\Pi}}\rangle_i \cdot \vec{\epsilon}_3} = \frac{\left|\mathsf{k}^2\vec{\epsilon} - \left(\vec{\mathsf{k}}_d \cdot \vec{\epsilon}\right)\vec{\mathsf{k}}_d\right|^2}{16\pi^2\bar{\varepsilon}_r^2}\left|\delta\varepsilon_r\left(\vec{\mathbf{q}}\right)\right|^2 \\ &= \frac{\left[\mathsf{k}^4 - \mathsf{k}^2\left(\vec{\mathsf{k}}_d \cdot \vec{\epsilon}\right)^2\right]}{16\pi^2\bar{\varepsilon}_r^2}\left|\delta\varepsilon_r\left(\vec{\mathbf{q}}\right)\right|^2 \\ &= \frac{\mathsf{k}^4}{16\pi^2\bar{\varepsilon}_r^2}\left[1 - \sin^2\theta\cos^2\left(\psi - \varphi\right)\right]\left|\delta\varepsilon_r\left(\vec{\mathbf{q}}\right)\right|^2 \end{aligned} \quad [4.195]$$

where ψ is the angle measured between $\vec{\epsilon}$ and $\vec{\epsilon}_1$ (i.e. the Ox axis). The angles θ and φ are the spherical coordinates indicating the direction of scattering $\vec{\mathbf{n}}$. For an unpolarized wave, we must average over ψ, which gives:

$$\frac{d\sigma}{d\Omega} = \frac{1}{2}\frac{\mathrm{k}^4}{16\pi^2\bar{\varepsilon}_r^2}\left[1+\cos^2\theta\right]\left|\delta\varepsilon_r\left(\vec{\mathbf{q}}\right)\right|^2 \qquad [4.196]$$

Let us now consider the scattering by a bound electron (assumed to be located at the origin). The relative permittivity function associated with this electron is:

$$\delta\epsilon_r\left(\vec{\mathbf{r}}\right) = \frac{e^2}{m\varepsilon_0\left(\omega_0^2-\omega^2-\mathrm{i}\omega\Gamma\right)}\delta^3\left(\vec{\mathbf{r}}\right) \qquad [4.197]$$

The Fourier transform $\delta\bar{\epsilon}_r\left(\vec{\mathbf{q}}\right)$ is a constant independent of $\vec{\mathbf{q}}$:

$$\delta\hat{\epsilon}_r\left(\vec{\mathbf{q}}\right) = \frac{e^2}{m\varepsilon_0\left(\omega_0^2-\omega^2-\mathrm{i}\frac{\omega}{\tau}\right)} \qquad [4.198]$$

Applying formula [4.196], we recover the Thomson or Rayleigh differential cross-section, depending on the case. Now suppose that the light is scattered by several electrons located at positions $\vec{\mathbf{r}}_i$. We then write:

$$\delta\epsilon_r\left(\vec{\mathbf{r}}\right) = \frac{e^2}{m\varepsilon_0\left(\omega_0^2-\omega^2-\mathrm{i}\frac{\omega}{\tau}\right)}\sum_i\delta^3\left(\vec{\mathbf{r}}-\vec{\mathbf{r}}_i\right) \qquad [4.199]$$

The Fourier transform $\delta\bar{\epsilon}_r\left(\vec{\mathbf{q}}\right)$ now depends on $\vec{\mathbf{q}}$:

$$\delta\hat{\epsilon}_r\left(\vec{\mathbf{q}}\right) = \frac{e^2}{m\varepsilon_0\left(\omega_0^2-\omega^2-\mathrm{i}\frac{\omega}{\tau}\right)}F\left(\vec{\mathbf{q}}\right) \qquad [4.200]$$

with:

$$F\left(\vec{\mathbf{q}}\right) = \sum_i e^{-\mathrm{i}\vec{\mathbf{q}}\cdot\vec{\mathbf{r}}_i} \qquad [4.201]$$

$F\left(\vec{\mathbf{q}}\right)$ is called the *form factor.*

4.10. Cherenkov radiation

A charged particle moving uniformly in a straight line through free space does not radiate. However, a particle traveling at constant speed through a material medium can radiate if its speed exceeds the speed of light in that medium. This radiation is called

Cherenkov radiation. The emission of Cherenkov radiation is a collective phenomenon that involves a large number of atoms in the medium: when the particle passes through the medium, the electrons (of the atoms) are accelerated by the field of the particles, and thus emit radiation. These radiations are added coherently to produce Cherenkov radiation when the incident particle moves faster than the speed of light in the medium, but not when it moves slower. The speed of light in the medium is denoted by $c' = \frac{c}{\sqrt{\varepsilon_r}}$, where ε_r is the relative permittivity of the medium, and the speed of the particle is v. Figure 4.12 shows a succession of spherical field wavelets for $v < c'$ and for $v > c'$. The wavelets interfere constructively only for $v > c'$ to form a radiation cone behind the particle. The normal of the cone forms an angle θ_c with the direction of motion:

$$\cos\theta_c = \frac{c'}{v} = \frac{c}{\sqrt{\varepsilon_r} v} \quad [4.202]$$

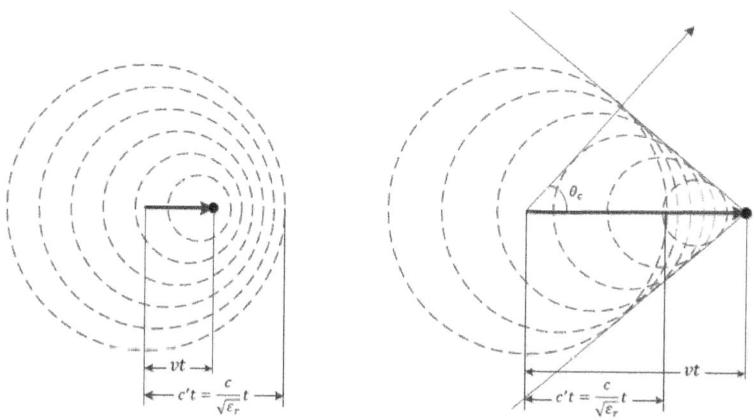

Figure 4.12. *Cherenkov radiation: spherical waves created by a particle moving at a speed lower (respectively, higher) than the speed of light in the medium*

This is the direction of emission of Cherenkov radiation.

The discussion presented here is an appropriate microscopic description of the origin of the radiation. However, if we are only interested in the angular and frequency distribution of the radiation, and not in the mechanism itself, it is possible to provide a simple derivation based on the results developed earlier. Recall that the angular frequency distribution of the radiation emitted by a moving charged particle is given by [4.107]. For a particle moving in a dielectric medium with relative permittivity ε_r, equation [4.110] must be modified as follows:

$$\frac{dI(\omega)}{d\Omega} = \frac{q^2\omega^2\sqrt{\varepsilon_r}}{16\pi^3\varepsilon_0 c^3}\left|\int_{-\infty}^{\infty}\vec{\mathbf{n}}\times\left(\vec{\mathbf{n}}\times\vec{\mathbf{v}}\right)e^{i\omega\left(t-\frac{\sqrt{\varepsilon_r}(\vec{\mathbf{n}}\cdot\vec{\mathbf{r}}')}{c}\right)}dt'\right|^2 \quad [4.203]$$

where substitutions $\varepsilon_0 \longrightarrow \varepsilon_0\varepsilon_r$ and $c \longrightarrow \frac{c}{\sqrt{\varepsilon_r}}$ have been made. For uniform rectilinear motion, that is, $\vec{\mathbf{r}}(t') = \vec{\mathbf{v}}t'$, we obtain:

$$\frac{dI(\omega)}{d\Omega} = \frac{q^2\sqrt{\varepsilon_r}}{4\pi\varepsilon_0 c^3}\left|\vec{\mathbf{n}}\times\vec{\mathbf{v}}\right|^2\left|\frac{\omega}{2\pi}\int_{-\infty}^{\infty}e^{i\omega t'\left[1-\frac{\sqrt{\varepsilon_r}}{c}\left(\vec{\mathbf{n}}\cdot\vec{\mathbf{v}}\right)\right]}dt'\right|^2 \quad [4.204]$$

The integral is a Dirac delta function. It therefore follows that:

$$\frac{dI(\omega)}{d\Omega} = \frac{q^2\sqrt{\varepsilon_r}\beta^2\sin^2\theta}{4\pi\varepsilon_0 c}\left|\delta\left(1-\sqrt{\varepsilon_r}\beta\cos\theta\right)\right|^2 \quad [4.205]$$

where θ is the angle measured between $\vec{\mathbf{n}}$ and $\vec{\mathbf{v}}$. The presence of the Dirac delta function ensures that radiation is emitted only in the direction corresponding to the angle θ_c (the *Cherenkov angle*):

$$\cos\theta_c = \frac{1}{\beta\sqrt{\varepsilon_r}} \quad [4.206]$$

The presence of the square of the Dirac delta function in expression [4.205] implies that the total energy radiated per unit frequency interval is infinite. This divergence arises because the particle is assumed to have been moving through the medium indefinitely. To obtain a meaningful result, we assume that the particle crosses a dielectric slab in a time interval $2T$ (not to be confused with the time period associated with the frequency). The infinite integral in [4.204] is therefore replaced by:

$$\frac{\omega}{2\pi}\int_{-T}^{T}e^{i\omega t'\left[1-\frac{\sqrt{\varepsilon_r}}{c}\left(\vec{\mathbf{n}}\cdot\vec{\mathbf{v}}\right)\right]}dt' = \frac{\omega T}{\pi}\frac{\sin\left[\omega T\left(1-\sqrt{\varepsilon_r}\beta\cos\theta\right)\right]}{\left[\omega T\left(1-\sqrt{\varepsilon_r}\beta\cos\theta\right)\right]} \quad [4.207]$$

The square of this function exhibits a peak at θ_c, provided that $\omega T \gg 1$. Assuming that $\beta > \frac{1}{\sqrt{\varepsilon_r}}$, so that the angle θ_c exists, the integral over the entire solid angle yields:

$$\int_{4\pi}d\Omega\left(\frac{\omega T}{\pi}\right)^2\frac{\sin^2\left[\omega T\left(1-\sqrt{\varepsilon_r}\beta\cos\theta\right)\right]}{\left[\omega T\left(1-\sqrt{\varepsilon_r}\beta\cos\theta\right)\right]^2} = \frac{2\omega T}{\pi\sqrt{\varepsilon_r}\beta}\times$$

$$\int_{1-\sqrt{\varepsilon_r}\beta}^{1+\sqrt{\varepsilon_r}\beta}\frac{\sin^2 x}{x^2}dx \approx \frac{2\omega T}{\sqrt{\varepsilon_r}\beta} \quad [4.208]$$

which shows that the amount of radiation is proportional to the time interval. From [4.204], we find that the total energy radiated per unit of frequency interval across the slab is:

$$I(\omega) = \frac{q^2\omega}{4\pi^2\varepsilon_0 c^2}\sin^2\theta_c\,(2cT\beta) \qquad [4.209]$$

To deduce the energy radiated per unit frequency interval and per unit length of the path, we must divide the previous expression by $(2c\beta T)$. Thus, we obtain:

$$\frac{dI(\omega)}{dx} = \frac{q^2\omega}{4\pi^2\varepsilon_0 c^2}\left[1-\cos^2\theta_c\right] = \frac{q^2\omega}{4\pi^2\varepsilon_0 c^2}\left[1-\frac{1}{\beta^2\varepsilon_r(\omega)}\right] \qquad [4.210]$$

where ω is chosen so that $\varepsilon_r(\omega) > \beta^{-2}$. In other words, radiation is emitted only in the frequency range where $\varepsilon_r(\omega) > \beta^{-2}$.

Chapter 5

Kirchhoff Diffraction

5.1. Initial value problem: Kirchhoff's integral representation

5.1.1. *Green's function for the time-dependent wave equation*

Wave equations have the following basic structure:

$$\Delta\psi - \frac{1}{c^2}\frac{\partial^2\psi}{\partial t^2} = -f\left(\overrightarrow{\mathbf{r}}, t\right) \qquad [5.1]$$

where $f\left(\overrightarrow{\mathbf{r}}, t\right)$ is a known function (called the *source*). The factor c is the propagation velocity of the wave in the medium. Here, we will present a different but equivalent resolution method. To solve equation [5.1], we need to find the Green function associated with this equation, as in electrostatics. The Green function depends on the variables $\left(\overrightarrow{\mathbf{r}}, \overrightarrow{\mathbf{r}}', t, t'\right)$ and satisfies the following equation:

$$\left(\Delta - \frac{1}{c^2}\frac{\partial^2}{\partial t^2}\right) G\left(\overrightarrow{\mathbf{r}}, t; \overrightarrow{\mathbf{r}}', t'\right) = -\delta^3\left(\overrightarrow{\mathbf{r}} - \overrightarrow{\mathbf{r}}'\right)\delta\left(t - t'\right) \qquad [5.2]$$

In terms of G, the solutions of equation [5.1] are written as:

$$\psi\left(\overrightarrow{\mathbf{r}}, t\right) = \int d\tau' \int dt' G\left(\overrightarrow{\mathbf{r}}, t; \overrightarrow{\mathbf{r}}', t'\right) f\left(\overrightarrow{\mathbf{r}}', t'\right) \qquad [5.3]$$

Obviously, G must satisfy certain boundary conditions imposed by physical considerations and depends only on the differences in coordinates $\left(\overrightarrow{\mathbf{r}} - \overrightarrow{\mathbf{r}}'\right)$ and time $(t - t')$. To find G, we must compute the Fourier transform for both sides in equation [5.2]. The Dirac delta functions on the right-hand side can be can be represented as:

$$\delta^3\left(\overrightarrow{\mathbf{r}} - \overrightarrow{\mathbf{r}}'\right)\delta\left(t - t'\right) = \frac{1}{(2\pi)^4}\int \mathrm{d}\mathrm{k}_x \mathrm{d}\mathrm{k}_y \mathrm{d}\mathrm{k}_z \int d\omega e^{\mathrm{i}\overrightarrow{\mathrm{k}}\cdot\left(\overrightarrow{\mathbf{r}} - \overrightarrow{\mathbf{r}}'\right)} e^{-\mathrm{i}\omega(t-t')} \qquad [5.4]$$

We thus express the representation of G as follows:

$$G\left(\vec{\mathbf{r}},t;\vec{\mathbf{r}}',t'\right)=\int \mathrm{d}k_x \mathrm{d}k_y \mathrm{d}k_z \int d\omega \mathsf{G}\left(\vec{\mathsf{k}},\omega\right) e^{\mathrm{i}\vec{\mathsf{k}}\cdot\left(\vec{\mathbf{r}}-\vec{\mathbf{r}}'\right)} e^{-\mathrm{i}\omega(t-t')} \qquad [5.5]$$

The Fourier transform $\mathsf{G}\left(\vec{\mathsf{k}},\omega\right)$ can now be easily determined. Substituting equations [5.4] and [5.5] into equation [5.2], we obtain:

$$\mathsf{G}\left(\vec{\mathsf{k}},\omega\right)=\frac{1}{(2\pi)^4}\frac{1}{\mathsf{k}^2-\frac{\omega^2}{c^2}} \qquad [5.6]$$

By substituting equation [5.6] into [5.5], we observe singularities at $k=\pm\frac{\omega}{c}$. To evaluate this integral, we require that G satisfy the following properties: (1) $G=0$ everywhere for $t<t'$; (2) G represents outgoing spherical waves (caused by a point source located at $\vec{\mathbf{r}}'$) for $t>t'$. Let us begin by integrating over ω. The singularities of $\mathsf{G}\left(\vec{\mathsf{k}},\omega\right)$ occur at the points $\omega=\pm c\mathsf{k}$. This integral can be evaluated in the complex ω-plane using the Cauchy theorem. For $t>t'$, the integral along the real axis in equation [5.5] can be extended over the full contour C in the lower half-plane, since the contribution over the semicircle at infinity vanishes exponentially. On the other hand, for $t<t'$, the contour must be closed in the upper half-plane, as illustrated in Figure 5.1 by the path C'. To obtain $G=0$ for $t<t'$, the poles $\pm c\mathsf{k}$ must be shifted below the real axis, as in Figure 5.1. In this case, the integral on C for $t>t'$ will give a nonzero contribution, while the integral on C for $t<t'$ cancels out. The poles can be displaced by replacing ω by $\omega+\mathrm{i}\epsilon$ in equation [5.6]. Then, the function G is given by:

$$G\left(\vec{\mathbf{r}},t;\vec{\mathbf{r}}',t'\right)=\frac{1}{(2\pi)^4}\int \mathrm{d}k_x \mathrm{d}k_y \mathrm{d}k_z \int d\omega \frac{e^{\mathrm{i}\vec{\mathsf{k}}\cdot\vec{\mathbf{R}}} e^{-\mathrm{i}\omega\tau}}{\mathsf{k}^2-\frac{(\omega+\mathrm{i}\epsilon)^2}{c^2}} \qquad [5.7]$$

where $\vec{\mathbf{R}}=\vec{\mathbf{r}}-\vec{\mathbf{r}}'$ and $\tau=t-t'$. ϵ is here a positive infinitesimal parameter. Integration over ω for $\tau>0$ can be performed using the Cauchy theorem (applied around the contour C shown in Figure 5.1).

This yields:

$$\begin{aligned} G &= \frac{c^2}{(2\pi)^4}\int \mathrm{d}k_x \mathrm{d}k_y \mathrm{d}k_z \left[\int_{-\infty}^{\infty} d\omega \frac{e^{\mathrm{i}\vec{\mathsf{k}}\cdot\vec{\mathbf{R}}} e^{-\mathrm{i}\omega\tau}}{c^2\mathsf{k}^2-(\omega+\mathrm{i}\epsilon)^2}\right] \\ &= \frac{c^2}{(2\pi)^4}\int \mathrm{d}k_x \mathrm{d}k_y \mathrm{d}k_z \left[\int_{\infty}^{-\infty} d\omega \frac{e^{\mathrm{i}\vec{\mathsf{k}}\cdot\vec{\mathbf{R}}} e^{-\mathrm{i}\omega\tau}}{(\omega+\mathrm{i}\epsilon)^2-c^2\mathsf{k}^2}\right] \end{aligned}$$

$$
\begin{aligned}
&= \frac{\mathrm{i}c^2}{(2\pi)^3}\int dk_x dk_y dk_z e^{\mathrm{i}\vec{\mathsf{k}}\cdot\vec{\mathbf{R}}}\left[\left.\frac{(\omega - c\mathsf{k} + \mathrm{i}\epsilon)\,e^{-\mathrm{i}\omega\tau}}{(\omega+\mathrm{i}\epsilon)^2 - c^2\mathsf{k}^2}\right|_{\omega=c\mathsf{k}-\mathrm{i}\epsilon}\right.\\
&\quad\left.+\left.\frac{(\omega + c\mathsf{k} + \mathrm{i}\epsilon)\,e^{-\mathrm{i}\omega\tau}}{(\omega+\mathrm{i}\epsilon)^2 - c^2\mathsf{k}^2}\right|_{\omega=-c\mathsf{k}-\mathrm{i}\epsilon}\right]\\
&= \frac{\mathrm{i}c^2}{8\pi^3}\int dk_x dk_y dk_z e^{\mathrm{i}\vec{\mathsf{k}}\cdot\vec{\mathbf{R}}}\left[\frac{e^{-\mathrm{i}c\mathsf{k}\tau}}{2c\mathsf{k}} + \frac{e^{\mathrm{i}c\mathsf{k}\tau}}{-2c\mathsf{k}}\right]\\
&= \frac{c}{8\pi^3}\int \frac{\sin(c\mathsf{k}\tau)}{\mathsf{k}} e^{\mathrm{i}\vec{\mathsf{k}}\cdot\vec{\mathbf{R}}} d^3\vec{\mathsf{k}}
\end{aligned} \quad [5.8]
$$

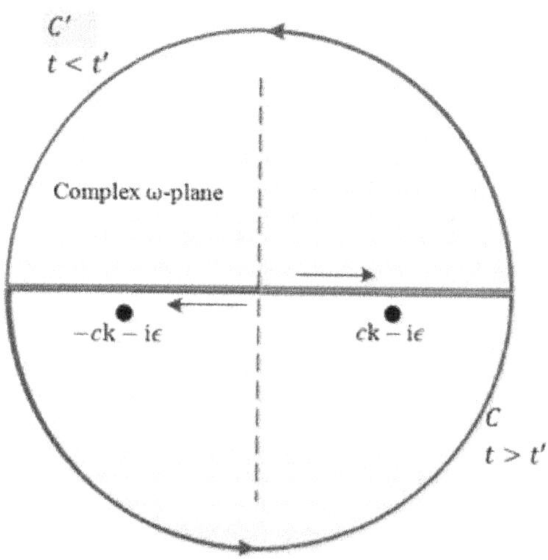

Figure 5.1. *Complex ω-plane with contour C for $t > t'$ and contour C' for $t < t'$. For a color version of this figure, see www.iste.co.uk/abdesselam/electrodynamics3.zip*

Integration over $dk_x dk_y dk_z$ can be carried out by first integrating over the angles (by switching to spherical coordinates). This gives:

$$
\begin{aligned}
G &= \frac{c}{8\pi^3}\int_{\mathsf{k}=0}^{\infty}\int_{\theta=0}^{\pi}\int_{\varphi=0}^{\pi} \sin(c\mathsf{k}\tau)\, e^{\mathrm{i}\mathsf{k}R\cos\theta}\mathsf{k}\sin\theta d\mathsf{k}d\theta d\varphi\\
&= \frac{c}{2\pi^2 R}\int_0^{\infty} \sin(c\mathsf{k}\tau)\sin(\mathsf{k}R)\,d\mathsf{k}
\end{aligned} \quad [5.9]
$$

Since the function $\sin(c\mathsf{k}\tau)\sin(\mathsf{k}R)$ is even in k, the integral can be extended over the whole interval, $-\infty < \mathsf{k} < +\infty$. Performing the change of variable $\xi = c\mathsf{k}$, the integral [5.9] can also be rewritten in the following form:

$$G = \frac{1}{8\pi^2 R}\int_{-\infty}^{\infty}\left[\cos\xi\left(\tau-\frac{R}{c}\right)-\cos\xi\left(\tau+\frac{R}{c}\right)\right]d\xi$$
$$= \frac{1}{8\pi^2 R}\int_{-\infty}^{\infty}\left[e^{i\xi\left(\tau-\frac{R}{c}\right)}-e^{i\xi\left(\tau+\frac{R}{c}\right)}\right]d\xi \qquad [5.10]$$

These are precisely the integral representations of the Dirac delta function. The argument of the second integral never vanishes, since $\tau > 0$. Therefore, only the first integral contributes to the expression of G. Consequently, we find:

$$G\left(\vec{\mathbf{r}},t;\vec{\mathbf{r}}',t'\right) = \frac{1}{4\pi R}\delta\left(\tau-\frac{R}{c}\right) = \frac{\delta\left(t-t'-\frac{\left|\vec{\mathbf{r}}-\vec{\mathbf{r}}'\right|}{c}\right)}{4\pi\left|\vec{\mathbf{r}}-\vec{\mathbf{r}}'\right|} \qquad [5.11]$$

This Green function is sometimes called the *retarded Green function*. It describes the causal nature of a wave disturbance. The effect observed at point $\vec{\mathbf{r}}$ at time t is due to a disturbance that originated at time t' at point $\vec{\mathbf{r}}'$. The solution to equation [5.1] in the absence of boundaries is therefore given by:

$$\psi\left(\vec{\mathbf{r}},t\right) = \int \frac{\delta\left(t-t'-\frac{\left|\vec{\mathbf{r}}-\vec{\mathbf{r}}'\right|}{c}\right)}{4\pi\left|\vec{\mathbf{r}}-\vec{\mathbf{r}}'\right|} f\left(\vec{\mathbf{r}}',t'\right) d\tau' dt' \qquad [5.12]$$

By integrating over t', we find the solution, known as the *retarded* solution:

$$\psi\left(\vec{\mathbf{r}},t\right) = \int \frac{\left[f\left(\vec{\mathbf{r}}',t'\right)\right]_{\text{ret.}}}{4\pi\left|\vec{\mathbf{r}}-\vec{\mathbf{r}}'\right|} d^3\vec{\mathbf{r}}' \qquad [5.13]$$

The square brackets $[\ ]_{\text{ret.}}$ mean that $f\left(\vec{\mathbf{r}}',t'\right)$ must be evaluated at the retarded time, $t' = t - \frac{\left|\vec{\mathbf{r}}-\vec{\mathbf{r}}'\right|}{c}$.

5.1.2. *Cauchy problem*

The expression [5.13] is the particular solution (with source term) of the inhomogeneous wave equation [5.1]. If we now want to find a solution $\psi\left(\vec{\mathbf{r}},t\right)$ that satisfies equation [5.1] and takes specific values at a fixed time $t = t_0$, namely, $\psi\left(\vec{\mathbf{r}},t_0\right)$ and $\frac{\partial\psi}{\partial t}\left(\vec{\mathbf{r}},t_0\right)$, we must add to expression [5.13] a solution of the homogeneous wave equation (without source term) to fulfill these boundary conditions. This is known as the *Cauchy initial value problem*. The values $\psi\left(\vec{\mathbf{r}},t_0\right)$ and $\frac{\partial\psi}{\partial t}\left(\vec{\mathbf{r}},t_0\right)$, given for all $\vec{\mathbf{r}}$, are called the *Cauchy initial conditions*. We aim to determine $\psi\left(\vec{\mathbf{r}},t\right)$ for $t > t_0$ inside a volume τ bounded by a closed surface S. To do this, we use the second Green identity, integrated over time from $t' = t_0$ to $t' = t_1 > t$:

$$\int_{t_0}^{t_1} dt' \int_{\tau} d^3\vec{\mathbf{r}}' \left(\phi\Delta\psi - \psi\Delta\phi\right) = \int_{t_0}^{t_1} dt' \oint_S \left(\phi\frac{\partial\psi}{\partial n'} - \psi\frac{\partial\phi}{\partial n'}\right) dS' \quad [5.14]$$

Setting $\phi = G$ and taking into account equations [5.1] and [5.2], the left-hand side of equation [5.14] is written as:

$$\begin{aligned}
&\int_{t_0}^{t_1} dt' \int_{\tau} d^3\vec{\mathbf{r}}' \left(G\Delta\psi - \psi\Delta G\right) \\
&= \int_{t_0}^{t_1} dt' \int_{\tau} d^3\vec{\mathbf{r}}' \left[\psi\left(\vec{\mathbf{r}}', t'\right)\delta^3\left(\vec{\mathbf{r}} - \vec{\mathbf{r}}'\right)\delta\left(t - t'\right) - \right. \\
&\left. Gf\left(\vec{\mathbf{r}}', t'\right) + \frac{1}{c^2}\left(G\frac{\partial^2\psi}{\partial t'^2} - \psi\frac{\partial^2 G}{\partial t'^2}\right)\right] \\
&= \psi\left(\vec{\mathbf{r}}, t\right) - \int_{\tau} \frac{\left[f\left(\vec{\mathbf{r}}', t'\right)\right]_{\text{ret.}}}{4\pi\left|\vec{\mathbf{r}} - \vec{\mathbf{r}}'\right|} d^3\vec{\mathbf{r}}' + \frac{1}{c^2}\int_{\tau} d^3\vec{\mathbf{r}}' \int_{t_0}^{t_1} \left(G\frac{\partial^2\psi}{\partial t'^2} - \psi\frac{\partial^2 G}{\partial t'^2}\right) dt' \\
&= \psi\left(\vec{\mathbf{r}}, t\right) - \int_{\tau} \frac{\left[f\left(\vec{\mathbf{r}}', t'\right)\right]_{\text{ret.}}}{4\pi\left|\vec{\mathbf{r}} - \vec{\mathbf{r}}'\right|} d^3\vec{\mathbf{r}}' + \frac{1}{c^2}\int_{\tau} d^3\vec{\mathbf{r}}' \int_{t_0}^{t_1} \frac{\partial}{\partial t'}\left(G\frac{\partial\psi}{\partial t'} - \psi\frac{\partial G}{\partial t'}\right) dt' \\
&= \psi\left(\vec{\mathbf{r}}, t\right) - \int_{\tau} \frac{\left[f\left(\vec{\mathbf{r}}', t'\right)\right]_{\text{ret.}}}{4\pi\left|\vec{\mathbf{r}} - \vec{\mathbf{r}}'\right|} d^3\vec{\mathbf{r}}' + \frac{1}{c^2}\int_{\tau} d^3\vec{\mathbf{r}}' \left(G\frac{\partial\psi}{\partial t'} - \psi\frac{\partial G}{\partial t'}\right)\Bigg|_{t_0}^{t_1} \quad [5.15]
\end{aligned}$$

Since $G = 0$ at $t' = t_1 > t$, the upper limit vanishes. Therefore, we can combine equations [5.14] and [5.15] to give the integral representation of $\psi\left(\vec{\mathbf{r}}, t\right)$ inside a volume τ bounded by the surface S, for times $t > t_0$:

$$\begin{aligned}
\psi\left(\vec{\mathbf{r}}, t\right) = &\int_{\tau} \frac{\left[f\left(\vec{\mathbf{r}}', t'\right)\right]_{\text{ret.}}}{4\pi\left|\vec{\mathbf{r}} - \vec{\mathbf{r}}'\right|} d^3\vec{\mathbf{r}}' + \frac{1}{c^2}\int_{\tau} d^3\vec{\mathbf{r}}' \left(G\frac{\partial\psi}{\partial t'} - \psi\frac{\partial G}{\partial t'}\right)\Bigg|_{t_0} + \\
&\int_{t_0}^{t_1} dt' \oint_S \left(G\frac{\partial\psi}{\partial n'} - \psi\frac{\partial G}{\partial n'}\right) dS' \quad [5.16]
\end{aligned}$$

5.1.3. *Poisson solution*

If τ extends to the entire space (i.e. the surface S is now located at infinity), then the integral over the surface in equation [5.16] can be omitted (since both ψ and G vanish at infinity). For simplicity, we choose $t_0 = 0$ and write that (the initial values are known at $t = 0$):

$$\begin{aligned}\psi\left(\vec{\mathbf{r}},0\right) &= F\left(\vec{\mathbf{r}}\right)\\ \frac{\partial\psi}{\partial t}\left(\vec{\mathbf{r}},0\right) &= D\left(\vec{\mathbf{r}}\right)\end{aligned} \qquad [5.17]$$

We also take the observation point at the origin (using spherical coordinates in the integrals). We thus obtain:

$$\begin{aligned}\psi\left(0,t\right) &= \frac{1}{4\pi}\int_{4\pi} d\Omega' \int_0^\infty r'dr' f\left(r',\theta',\varphi',t'=t-\frac{r'}{c}\right)\\ &+\frac{1}{4\pi c^2}\int_{4\pi} d\Omega' \int_0^\infty r'dr' \times\\ &\left[D\left(r',\theta',\varphi'\right)\delta\left(t-\frac{r'}{c}\right) - F\left(r',\theta',\varphi'\right)\frac{\partial}{\partial t'}\delta\left(t-t'-\frac{r'}{c}\right)\bigg|_{t'=0}\right]\\ &= \frac{1}{4\pi}\int_{4\pi} d\Omega' \int_0^\infty r'dr' f\left(r',\theta',\varphi',t'=t-\frac{r'}{c}\right)\\ &+\frac{1}{4\pi c^2}\int_{4\pi} d\Omega' \int_0^\infty dr' \times\\ &\left[r'D\left(r',\theta',\varphi'\right)\delta\left(t-\frac{r'}{c}\right) - r'F\left(r',\theta',\varphi'\right)\delta'\left(t-\frac{r'}{c}\right)\right]\end{aligned} \qquad [5.18]$$

Using the integration properties of the Dirac delta function listed in Appendix 3,[1] it finally follows that:

$$\begin{aligned}\psi\left(0,t\right) &= \frac{1}{4\pi}\int_{4\pi} d\Omega' \int_0^\infty r'dr' f\left(r',\theta',\varphi',t'=t-\frac{r'}{c}\right) +\\ &\frac{1}{4\pi}\int_{4\pi} d\Omega' \left[tD\left(ct,\theta',\varphi'\right) + \frac{\partial}{\partial t}\left(tF\left(ct,\theta',\varphi'\right)\right)\right]\end{aligned} \qquad [5.19]$$

This is the *Poisson solution* to the Cauchy problem. In the absence of sources ($f = 0$), only the values of the initial field at distances ct from the origin contribute at time t.

5.1.4. *Kirchhoff representation of the field*

The other result that can be obtained from equation [5.16] is the so-called *Kirchhoff representation* of the field within the volume τ in terms of the values of ψ and its

1 For online appendices, please see: www.iste.co.uk/abdesselam/electrodynamics1.zip.

derivatives on the boundary surface S. It is assumed that there are no sources within the volume τ and that the initial values of ψ and $\frac{\partial \psi}{\partial t}$ are zero. Alternatively, we can assume that the initial time is in the distant past, so that the solution of the initial value [5.19] within volume τ will not contribute. Thus, the field inside τ is given by:

$$\psi\left(\vec{\mathbf{r}}, t\right) = \int_{t_0}^{t_1} dt' \oint_S \left(G \frac{\partial \psi}{\partial n'} - \psi \frac{\partial G}{\partial n'} \right) dS' \qquad [5.20]$$

Using the expression for G, we can compute $\frac{\partial G}{\partial n'}$:

$$\begin{aligned} \vec{\nabla}' G &= \frac{\partial G}{\partial R} \vec{\nabla}' R = -\frac{\vec{\mathbf{R}}}{R} \frac{\partial}{\partial R} \left[\frac{\delta\left(t - t' - \frac{R}{c}\right)}{4\pi R} \right] \\ &= -\frac{\vec{\mathbf{R}}}{4\pi R} \left[-\frac{\delta\left(t - t' - \frac{R}{c}\right)}{R^2} + \frac{\delta'\left(t - t' - \frac{R}{c}\right)}{Rc} \right] \end{aligned} \qquad [5.21]$$

The term involving the derivative of the Dirac delta function can be integrated by parts with respect to time t'. Consequently, we obtain:

$$\psi\left(\vec{\mathbf{r}}, t\right) = \frac{1}{4\pi} \oint_S \vec{\mathbf{n}} \cdot \left[\frac{\vec{\nabla}' \psi\left(\vec{\mathbf{r}}', t'\right)}{R} - \frac{\vec{\mathbf{R}}}{R} \psi\left(\vec{\mathbf{r}}', t'\right) - \frac{\vec{\mathbf{R}}}{cR^2} \frac{\partial \psi\left(\vec{\mathbf{r}}', t'\right)}{\partial t'} \right]_{\text{ret.}} dS' \qquad [5.22]$$

where $\vec{\mathbf{R}} = \vec{\mathbf{r}} - \vec{\mathbf{r}}'$, and $\vec{\mathbf{n}}$ is the normal unit vector to the surface S. It should be emphasized that equation [5.22] is not a solution for the field, but merely an integral representation in terms of its value and the values of its spatial and temporal derivatives on the surface S. The integral [5.22], known as the *Kirchhoff integral*, forms the basis for the Huygens principle (the mathematical formulation of the Huygens principle). It is used as a starting point in all diffraction problems.

5.2. Kirchhoff integral for diffraction

The problem of diffraction concerns the study of the behavior of a wave incident on one or more obstacles or apertures on absorbing or conducting surfaces. The wave is scattered and may even be absorbed, leading to radiation propagating in directions other than the incidence. The purpose of all diffraction theories is to calculate the radiation emerging from a diffracting system. The first theory developed in this regard was that of Kirchhoff based on the idea of the superposition of elementary wavelets introduced by Huygens. In this section, we discuss the Kirchhoff method in detail. In the following section, we will introduce a vector version of the Kirchhoff integral.

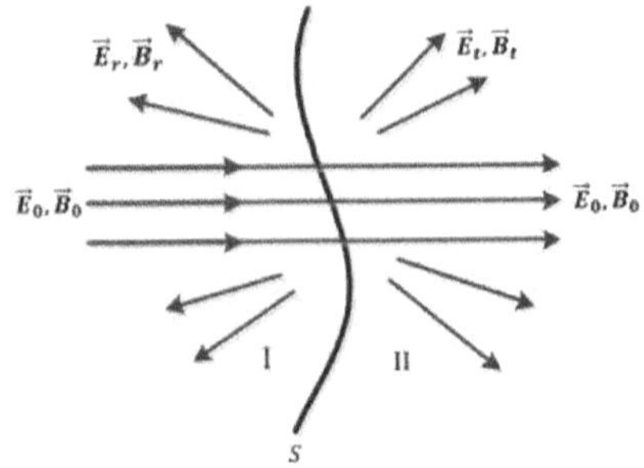

Figure 5.2. *Diffraction system: the surface S, containing apertures, gives rise, in addition to the incident fields, to reflected and transmitted fields in regions I and II*

In a diffraction process, there are two spatial regions, I and II, separated by a boundary surface S, as illustrated in Figure 5.2. For example, S may be an infinite metallic sheet with certain apertures. The incident wave, generated by sources in region I, approaches S from one side and is diffracted on the boundary surface, resulting in scattered waves, one transmitted and one reflected. If the incident wave is described by the fields $\left(\vec{\mathbf{E}}_0, \vec{\mathbf{B}}_0\right)$, the reflected wave by the fields $\left(\vec{\mathbf{E}}_r, \vec{\mathbf{B}}_r\right)$ and the transmitted wave by $\left(\vec{\mathbf{E}}_t, \vec{\mathbf{B}}_t\right)$, then the fields in regions I and II are, respectively, $\left(\vec{\mathbf{E}}_I = \vec{\mathbf{E}}_0 + \vec{\mathbf{E}}_r, \vec{\mathbf{B}}_I = \vec{\mathbf{B}}_0 + \vec{\mathbf{B}}_r\right)$ and $(\vec{\mathbf{E}}_{II} = \vec{\mathbf{E}}_0 + \vec{\mathbf{E}}_t, \vec{\mathbf{B}}_{II} = \vec{\mathbf{B}}_0 + \vec{\mathbf{B}}_t)$. By placing another screen behind the obstacle (i.e. in region II at a sufficiently large distance D, as shown in Figure 5.3), a *diffraction pattern* is observed, consisting of several spots: a central bright spot and smaller, dimmer lateral spots, separated by dark zones, in the case of a rectangular slit (aperture). For a circular aperture, we observe a central (circular) bright spot and circular rings (progressively smaller and dimmer). To get an approximate idea of the diffraction phenomenon, let us consider a scalar plane wave $\psi_I\left(\vec{\mathbf{r}}, t\right) = \psi_0 e^{-\mathrm{i}\left(\omega t - \vec{\mathrm{k}}_0 \cdot \vec{\mathbf{r}}\right)}$ incident on a planar screen with an aperture. Suppose that the screen is located at $z = 0$. According to the Huygens principle, the wave transmitted on the other side of the screen is the superposition of secondary waves emitted only by the aperture, and each of these secondary waves propagates spherically, with an initial phase given by the incident wave at each point of the aperture, as shown in Figure 5.4. According to this reasoning, the wave $\psi_{II}\left(\vec{\mathbf{r}}, t\right)$ on the other side of the screen is:

$$\begin{aligned}\psi_{II}\left(\vec{\mathbf{r}}, t\right) &= \int_{\text{aper.}} dS' \int dt' \psi_I\left(\vec{\mathbf{r}}', t'\right) G\left(\vec{\mathbf{r}}, t; \vec{\mathbf{r}}', t'\right) \\ &= \psi_0 \int_{\text{aper.}} \frac{e^{\mathrm{ik}\left|\vec{\mathbf{r}} - \vec{\mathbf{r}}'\right|}}{4\pi\left|\vec{\mathbf{r}} - \vec{\mathbf{r}}'\right|} e^{-\mathrm{i}\left(\omega t - \vec{\mathrm{k}}_0 \cdot \vec{\mathbf{r}}'\right)} dS' \end{aligned} \qquad [5.23]$$

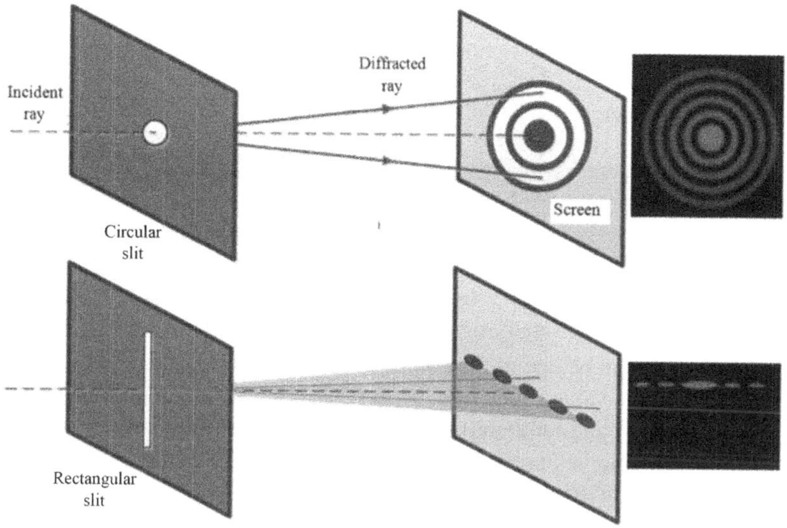

Figure 5.3. *Experimental setup for observing diffraction. For a color version of this figure, see www.iste.co.uk/abdesselam/electrodynamics3.zip*

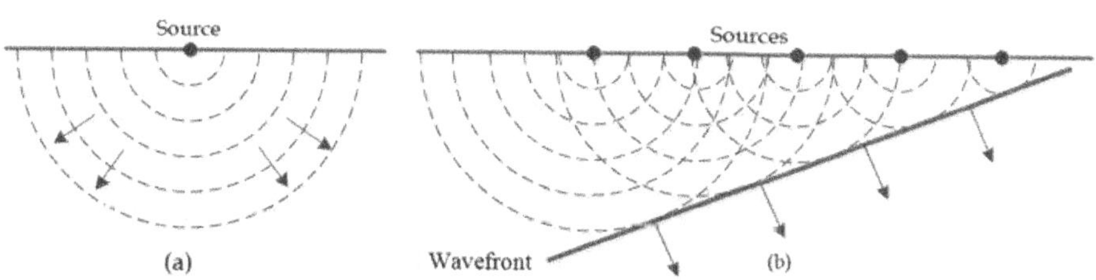

Figure 5.4. *The Huygens principle: (a) spherical wave in the case of a single source; (b) array of sources forming a plane wavefront*

The integral over $\overrightarrow{\mathbf{r}}'$ is carried out only over the aperture. By positioning ourselves now very far from the diffraction screen and assuming that the dimensions of the aperture are much smaller than the wavelength (i.e. in the radiation zone), we can use the approximation $\frac{e^{\mathrm{ik}|\overrightarrow{\mathbf{r}}-\overrightarrow{\mathbf{r}}'|}}{|\overrightarrow{\mathbf{r}}-\overrightarrow{\mathbf{r}}'|} \approx \frac{e^{\mathrm{ik}r}}{r} e^{-\mathrm{i}\overrightarrow{\mathrm{k}}\cdot\overrightarrow{\mathbf{r}}'}$, where $\overrightarrow{\mathrm{k}} = \mathrm{k}\frac{\overrightarrow{\mathbf{r}}}{r}$. Consequently, we obtain:

$$\psi_{II}\left(\overrightarrow{\mathbf{r}}, t\right) = \psi_0 \frac{e^{-\mathrm{i}(\omega t - \mathrm{k}r)}}{4\pi r} \int_{\text{aper.}} e^{-\mathrm{i}\left(\overrightarrow{\mathrm{k}} - \overrightarrow{\mathrm{k}}_0\right)\cdot\overrightarrow{\mathbf{r}}'} dS' = \psi_0 \frac{e^{-\mathrm{i}(\omega t - \mathrm{k}r)}}{4\pi r} I\left(\overrightarrow{\mathrm{k}} - \overrightarrow{\mathrm{k}}_0\right)$$

[5.24]

where:

$$I\left(\vec{\mathbf{q}}\right) = \int_{\text{aper.}} e^{-\mathrm{i}\vec{\mathbf{q}}\cdot\vec{\mathbf{r}}'} dS' \qquad [5.25]$$

$I\left(\vec{\mathbf{q}}\right)$ appears here as the Fourier transform of a function $f\left(\vec{\mathbf{r}}'\right)$ (called the *aperture function*), equal to 1 within the aperture and 0 elsewhere. The method presented here is entirely intuitive, but yields approximately correct results. Later, we will attempt to introduce the theoretical foundations that lead to an improved version of the Huygens formula [5.24]. Consider a scalar field $\psi\left(\vec{\mathbf{r}},t\right)$ defined on and within a closed surface S, satisfying the source-free wave equation within this region. The field $\psi\left(\vec{\mathbf{r}},t\right)$ can be one of the components of $\vec{\mathbf{E}}$ or $\vec{\mathbf{B}}$. In the previous section, it was shown that thc value of the field within S can be written in terms of the values of ψ and its (normal and temporal) derivatives on the surface, as given by equation [5.22]. If we assume a harmonic time dependence $e^{-\mathrm{i}\omega t}$, the integral form of $\psi\left(\vec{\mathbf{r}},t\right)$ also takes the form:

$$\psi\left(\vec{\mathbf{r}}\right) = \frac{1}{4\pi}\oint_S \frac{e^{\mathrm{i}kR}}{R}\vec{\mathbf{n}}\cdot\left[\vec{\nabla}'\psi + \mathrm{i}k\left(1+\frac{\mathrm{i}}{kR}\right)\frac{\vec{\mathbf{R}}}{R}\psi\right] dS' \qquad [5.26]$$

To apply equation [5.26] to the case of diffraction, consider that the surface S is made up of two surfaces S_1 and S_2: S_1 is the diffracting surface according to the problem to be solved (e.g. the conductive screen pierced by apertures), while S_2 is a sphere or hemisphere of very large radius (extending to infinity) in region II, as shown in Figure 5.5. Since the fields in part II originate from the apertures, they must appear as spherical (outgoing) waves in the vicinity of S_2. This means that:

$$\begin{aligned} &\psi \longrightarrow f(\theta,\varphi)\frac{e^{\mathrm{i}kr}}{r} \\ &\frac{1}{\psi}\frac{\partial\psi}{\partial r} \longrightarrow \left(\mathrm{i}k - \frac{1}{r}\right) \end{aligned} \qquad [5.27]$$

when $r \longrightarrow \infty$ (in the far-field zone). With these conditions (called *radiation conditions*) on ψ, it is easy to see that the integral [5.26] over S_2 is inversely proportional to the radius and will therefore vanish as this radius tends to infinity. Then, we obtain the Kirchhoff integral representation of ψ in region II:

$$\psi\left(\vec{\mathbf{r}}\right) = -\frac{1}{4\pi}\int_{S_1} \frac{e^{\mathrm{i}kR}}{R}\vec{\mathbf{n}}\cdot\left[\vec{\nabla}'\psi + \mathrm{i}k\left(1+\frac{\mathrm{i}}{kR}\right)\frac{\vec{\mathbf{R}}}{R}\psi\right] dS' \qquad [5.28]$$

where $\vec{\mathbf{n}}$ is now a unit normal vector to S_1, pointing into region II.

The Kirchhoff formula [5.28] shows that it is necessary to know the values of ψ and $\frac{\partial\psi}{\partial n}$ on the surface S_1. However, these values cannot be known in advance (unless the problem has already been solved exactly). For example, if S_1 is a perfectly conducting screen with an aperture and ψ represents the tangential component of the electric field on the surface S_1, then we know that ψ vanishes everywhere on S_1 except in the aperture. However, the value of ψ in the aperture remains undetermined. Without additional information, only approximate solutions can be found by making certain assumptions about ψ and $\frac{\partial \psi}{\partial n}$ on S_1. The Kirchhoff approximation consists of making the following assumptions:

1) ψ and $\frac{\partial \psi}{\partial n}$ vanish everywhere on S_1 except in the apertures.

2) The values of ψ and $\frac{\partial \psi}{\partial n}$ in the apertures are equal to those of the incident wave in the absence of any screen or obstacle.

Figure 5.5. *Different diffraction systems: region I contains the sources of radiation; region II is the diffraction zone*

The standard calculations used in diffraction theory (from classical optics) are all based on the Kirchhoff approximation. It is evident that the results obtained under this approximation have only approximate validity. Two important observations should be made:

– There is a fundamental mathematical inconsistency in the Kirchhoff assumptions. As demonstrated in Chapter 6 of Volume 1 for the Laplace equation (and similarly for the Helmholtz wave equation), the solution within a closed volume is uniquely determined by specifying ψ (Dirichlet boundary condition) or $\frac{\partial \psi}{\partial n}$ (Neumann boundary condition) on the surface. We cannot simultaneously specify both ψ and $\frac{\partial \psi}{\partial n}$ on the surface.

– The Kirchhoff approximation works better for short wavelengths, where the dimensions of the apertures are much larger compared to the wavelength. For medium and long wavelengths, the Kirchhoff approximation fails and yields incorrect results.

– Since this is a scalar theory, the Kirchhoff method cannot account for the polarization of the diffracted radiation, as the diffraction of electromagnetic radiation constitutes a boundary-value problem for vector fields. In the following section, we will introduce a vectorial version of the Kirchhoff integral [5.26].

5.3. Vector form of the Kirchhoff integral

To derive a vectorial form of the Kirchhoff integral, let us first express [5.26] as:

$$\psi\left(\vec{\mathbf{r}}\right) = \oint_S \left[G\vec{\mathbf{n}}\cdot\vec{\nabla}'\psi - \psi\vec{\mathbf{n}}\cdot\vec{\nabla}'G\right] dS' \qquad [5.29]$$

where:

$$G\left(\vec{\mathbf{r}},\vec{\mathbf{r}}'\right) = \frac{e^{\mathrm{i}kR}}{4\pi R} \qquad [5.30]$$

By writing a similar equation to that of [5.29] for the components E_x, E_y and E_z of the electric (or magnetic) field and then substituting them into the expression for $\vec{\mathbf{E}}$, we obtain:

$$\vec{\mathbf{E}}\left(\vec{\mathbf{r}}\right) = \oint_S \left[G\left(\vec{\mathbf{n}}\cdot\vec{\nabla}'\right)\vec{\mathbf{E}} - \vec{\mathbf{E}}\left(\vec{\mathbf{n}}\cdot\vec{\nabla}'G\right)\right] dS' \qquad [5.31]$$

with an analogous equation for the magnetic field $\vec{\mathbf{B}}$. Equation [5.31] represents the vectorial form of the Kirchhoff integral. However, this formulation is not practical for calculations. We can derive a more useful form through a series of transformations. First, let us express the bracketed term in the integral [5.31] as follows:

$$\left[G\left(\vec{\mathbf{n}}\cdot\vec{\nabla}'\right)\vec{\mathbf{E}} - \vec{\mathbf{E}}\left(\vec{\mathbf{n}}\cdot\vec{\nabla}'G\right)\right] = \left(\vec{\mathbf{n}}\cdot\vec{\nabla}'\right)\left(G\vec{\mathbf{E}}\right) - 2\vec{\mathbf{E}}\left(\vec{\mathbf{n}}\cdot\vec{\nabla}'G\right) \qquad [5.32]$$

Now using the identities:

$$\begin{aligned} \vec{\mathbf{n}}\times\left(\vec{\mathbf{E}}\times\vec{\nabla}'G\right) &= \vec{\mathbf{E}}\left(\vec{\mathbf{n}}\cdot\vec{\nabla}'G\right) - \left(\vec{\mathbf{n}}\cdot\vec{\mathbf{E}}\right)\vec{\nabla}'G \\ \vec{\nabla}'G\times\left(\vec{\mathbf{n}}\times\vec{\mathbf{E}}\right) &= \vec{\mathbf{n}}\left(\vec{\mathbf{E}}\cdot\vec{\nabla}'G\right) - \vec{\mathbf{E}}\left(\vec{\mathbf{n}}\cdot\vec{\nabla}'G\right) \end{aligned} \qquad [5.33]$$

we can eliminate the last term in [5.32]:

$$\left[G\left(\vec{\mathbf{n}}\cdot\vec{\nabla}'\right)\vec{\mathbf{E}} - \vec{\mathbf{E}}\left(\vec{\mathbf{n}}\cdot\vec{\nabla}'G\right)\right] = \left(\vec{\mathbf{n}}\cdot\vec{\nabla}'\right)\left(G\vec{\mathbf{E}}\right) - \vec{\mathbf{n}}\times\left(\vec{\mathbf{E}}\times\vec{\nabla}'G\right) - \left(\vec{\mathbf{n}}\cdot\vec{\mathbf{E}}\right)\vec{\nabla}'G - \vec{\mathbf{n}}\left(\vec{\mathbf{E}}\cdot\vec{\nabla}'G\right) - \left(\vec{\mathbf{n}}\times\vec{\mathbf{E}}\right)\times\vec{\nabla}'G \qquad [5.34]$$

Noting also that $\vec{\nabla}' \times \left(G\vec{\mathbf{E}}\right) = \vec{\nabla}' G \times \vec{\mathbf{E}} + G\vec{\nabla}' \times \vec{\mathbf{E}}$ and $\vec{\nabla}' \cdot \left(G\vec{\mathbf{E}}\right) = \vec{\nabla}' G \cdot \vec{\mathbf{E}} + G\vec{\nabla}' \cdot \vec{\mathbf{E}} = \vec{\nabla}' G \cdot \vec{\mathbf{E}}$ (since $\vec{\nabla}' \cdot \vec{\mathbf{E}} = 0$), the previous equation becomes:

$$\left[G\left(\vec{\mathbf{n}} \cdot \vec{\nabla}'\right)\vec{\mathbf{E}} - \vec{\mathbf{E}}\left(\vec{\mathbf{n}} \cdot \vec{\nabla}' G\right)\right] = \left(\vec{\mathbf{n}} \cdot \vec{\nabla}'\right)\left(G\vec{\mathbf{E}}\right) + \vec{\mathbf{n}} \times \left[\vec{\nabla}' \times \left(G\vec{\mathbf{E}}\right) - G\left(\vec{\nabla}' \times \vec{\mathbf{E}}\right)\right] - \left(\vec{\mathbf{n}} \cdot \vec{\mathbf{E}}\right)\vec{\nabla}' G - \vec{\mathbf{n}}\left[\vec{\nabla}' \cdot \left(G\vec{\mathbf{E}}\right)\right] - \left(\vec{\mathbf{n}} \times \vec{\mathbf{E}}\right) \times \vec{\nabla}' G$$

$$= \left(\vec{\mathbf{n}} \cdot \vec{\nabla}'\right)\left(G\vec{\mathbf{E}}\right) + \vec{\mathbf{n}} \times \left[\vec{\nabla}' \times \left(G\vec{\mathbf{E}}\right)\right] - \vec{\mathbf{n}}\left[\vec{\nabla}' \cdot \left(G\vec{\mathbf{E}}\right)\right] - G\vec{\mathbf{n}} \times \left(\vec{\nabla}' \times \vec{\mathbf{E}}\right) - \left(\vec{\mathbf{n}} \cdot \vec{\mathbf{E}}\right)\vec{\nabla}' G - \left(\vec{\mathbf{n}} \times \vec{\mathbf{E}}\right) \times \vec{\nabla}' G \qquad [5.35]$$

By substituting [5.35] into [5.31], we find that the integral of the first three terms involving the product $\left(G\vec{\mathbf{E}}\right)$ vanishes.

To verify this, we use the identities provided in Appendix 1 (these identities relate surface integrals on a closed surface to volume integrals within the enclosed region):[2]

$$\begin{aligned} \oint_S \vec{\mathbf{n}} \cdot \vec{\mathbf{A}} dS &= \int_\tau \vec{\nabla} \cdot \vec{\mathbf{A}} d^3\vec{\mathbf{r}} \\ \oint_S \vec{\mathbf{n}} \times \vec{\mathbf{A}} dS &= \int_\tau \vec{\nabla} \times \vec{\mathbf{A}} d^3\vec{\mathbf{r}} \\ \oint_S \phi \vec{\mathbf{n}} dS &= \int_\tau \vec{\nabla} \phi d^3\vec{\mathbf{r}} \end{aligned} \qquad [5.36]$$

where $\vec{\mathbf{A}}$ and ϕ are arbitrary vector and scalar functions, respectively.

Using these identities, the surface integral of the first three terms becomes:

$$\oint_S \left[\left(\vec{\mathbf{n}} \cdot \vec{\nabla}'\right)\left(G\vec{\mathbf{E}}\right) + \vec{\mathbf{n}} \times \left(\vec{\nabla}' \times \left(G\vec{\mathbf{E}}\right)\right) - \vec{\mathbf{n}}\vec{\nabla}' \cdot \left(G\vec{\mathbf{E}}\right)\right] dS$$

$$= \int_\tau \left[\Delta'\left(G\vec{\mathbf{E}}\right) + \vec{\nabla}' \times \left(\vec{\nabla}' \times \left(G\vec{\mathbf{E}}\right)\right) - \vec{\nabla}'\left(\vec{\nabla}' \cdot \left(G\vec{\mathbf{E}}\right)\right)\right] d^3\vec{\mathbf{r}} \qquad [5.37]$$

According to the identity $\vec{\nabla}' \times \left(\vec{\nabla}' \times \vec{\mathbf{A}}\right) = \vec{\nabla}'\left(\vec{\nabla}' \cdot \vec{\mathbf{A}}\right) - \Delta'\vec{\mathbf{A}}$, it is clear that the integral over the volume vanishes.

What remains, then, are the three other terms:

$$\vec{\mathbf{E}}\left(\vec{\mathbf{r}}\right) = -\oint_S \left[G\vec{\mathbf{n}} \times \left(\vec{\nabla}' \times \vec{\mathbf{E}}\right) + \left(\vec{\mathbf{n}} \cdot \vec{\mathbf{E}}\right)\vec{\nabla}' G + \left(\vec{\mathbf{n}} \times \vec{\mathbf{E}}\right) \times \vec{\nabla}' G\right] dS \qquad [5.38]$$

2 For online appendices, please see: www.iste.co.uk/abdesselam/electrodynamics1.zip.

From the Maxwell equations, we know that $\vec{\nabla}' \times \vec{\mathbf{E}} = \mathrm{i}k\vec{\mathbf{B}}$. This implies that:

$$\vec{\mathbf{E}}\left(\vec{\mathbf{r}}\right) = -\oint_S \left[\mathrm{i}\omega\left(\vec{\mathbf{n}} \times \vec{\mathbf{B}}\right)G + \left(\vec{\mathbf{n}} \cdot \vec{\mathbf{E}}\right)\vec{\nabla}'G + \left(\vec{\mathbf{n}} \times \vec{\mathbf{E}}\right) \times \vec{\nabla}'G\right] dS \quad [5.39]$$

This is the expression of the electric field at any point within the volume bounded by the surface S. An analogous expression to equation [5.39] is obtained for the magnetic field:

$$\vec{\mathbf{B}}\left(\vec{\mathbf{r}}\right) = -\oint_S \left[-\frac{\mathrm{i}\omega}{c^2}\left(\vec{\mathbf{n}} \times \vec{\mathbf{E}}\right)G + \left(\vec{\mathbf{n}} \cdot \vec{\mathbf{B}}\right)\vec{\nabla}'G + \left(\vec{\mathbf{n}} \times \vec{\mathbf{B}}\right) \times \vec{\nabla}'G\right] dS \quad [5.40]$$

In equations [5.39] and [5.40], the unit vector $\vec{\mathbf{n}}$ is the usual normal directed outwards. These integrals have a straightforward interpretation in terms of equivalent charge and current sources. The normal component of $\vec{\mathbf{E}}$ in equation [5.36] is clearly a surface charge density. Similarly, the tangential component of the magnetic field, $\left(\vec{\mathbf{n}} \times \vec{\mathbf{B}}\right)$, acts as a surface current. The other terms, $\left(\vec{\mathbf{n}} \cdot \vec{\mathbf{B}}\right)$ and $\left(\vec{\mathbf{n}} \times \vec{\mathbf{E}}\right)$, are, respectively, the surface density of magnetic charges and the magnetic current density. The expressions [5.39] and [5.40] are sometimes called the *Kirchhoff–Huygens integrals* (as they now take into account the vector nature of electromagnetic fields). Since the fields $\vec{\mathbf{E}}$ and $\vec{\mathbf{B}}$ must appear as outgoing spherical waves at infinity and obey the relation $\vec{\mathbf{E}} = c\left(\vec{\mathbf{B}} \times \frac{\vec{\mathbf{r}}}{r}\right)$, it is easy to show that the surface integral vanishes at infinity, that is, on S_2. Consequently, the electric field $\vec{\mathbf{E}}$ reduces to the expression:

$$\vec{\mathbf{E}}\left(\vec{\mathbf{r}}\right) = \int_{S_1} \left[\mathrm{i}\omega\left(\vec{\mathbf{n}} \times \vec{\mathbf{B}}\right)G + \left(\vec{\mathbf{n}} \cdot \vec{\mathbf{E}}\right)\vec{\nabla}'G + \left(\vec{\mathbf{n}} \times \vec{\mathbf{E}}\right) \times \vec{\nabla}'G\right] dS \quad [5.41]$$

where S_1 is the diffracting surface and the normal vector $\vec{\mathbf{n}}$ is now directed toward region II. To compute the diffracted fields, it is still necessary to know the values of $\vec{\mathbf{E}}$ and $\vec{\mathbf{B}}$ on the surface S_1.

EXAMPLE 5.1.– Imagine that the surface S_1, which contains the sources, is an infinite disk (a perfectly conducting one in equilibrium), with thickness e, and pierced with a few apertures, as shown in Figure 5.6. Region II now splits into two regions, II and II', which are connected only through the apertures in the disk. We denote the two sides of the disk by S_1 and S_1'. The normal vectors $\vec{\mathbf{n}}$ and $\vec{\mathbf{n}}' = -\vec{\mathbf{n}}$ are directed toward II and II', respectively. Our goal is to find an integral expression for the fields in region II in terms of the fields on the surface S_1 (on the right-hand side). This situation is analogous to the geometric configuration shown in Figure 5.5(a). If the fields in regions II and II' are, respectively, $\left(\vec{\mathbf{E}}, \vec{\mathbf{B}}\right)$ and $\left(\vec{\mathbf{E}}', \vec{\mathbf{B}}'\right)$, then it is clear that when the thickness e of the disk becomes vanishingly small, the integral [5.40] is written as:

$$\vec{\mathbf{E}}\left(\vec{\mathbf{r}}\right) = \int_{S_1} \left[\left(\vec{\mathbf{n}} \times \left(\vec{\mathbf{E}} - \vec{\mathbf{E}}'\right)\right) \times \vec{\nabla}' G + \vec{\mathbf{n}} \cdot \left(\vec{\mathbf{E}} - \vec{\mathbf{E}}'\right) \vec{\nabla}' G \right.$$
$$\left. + \mathrm{i}\omega \vec{\mathbf{n}} \times \left(\vec{\mathbf{B}} - \vec{\mathbf{B}}'\right) G\right] dS \qquad [5.42]$$

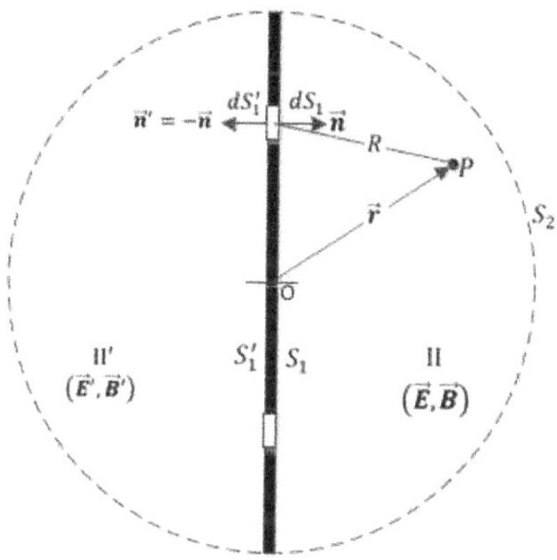

Figure 5.6. *Sources confined within an infinite disk (perfect conductor in equilibrium) of infinitesimal thickness e and pierced with apertures*

The electric field $\vec{\mathbf{E}}\left(\vec{\mathbf{r}}\right)$ on the left-hand side is $\vec{\mathbf{E}}$ or $\vec{\mathbf{E}}'$, depending on the location of the point $\vec{\mathbf{r}}$. However, the integral is only computed on the right surface S_1. Since the disk is a perfect conductor, the fields $\vec{\mathbf{E}}$ and $\vec{\mathbf{B}}$ must satisfy the following boundary conditions: $\vec{\mathbf{n}} \times \vec{\mathbf{E}} = 0$, $\vec{\mathbf{n}} \cdot \vec{\mathbf{B}} = 0$, while $\vec{\mathbf{n}} \cdot \vec{\mathbf{E}} \neq 0$, $\vec{\mathbf{n}} \times \vec{\mathbf{B}} \neq 0$. For physical considerations, the fields $\left(\vec{\mathbf{E}}, \vec{\mathbf{B}}\right)$ and $\left(\vec{\mathbf{E}}', \vec{\mathbf{B}}'\right)$ obey the symmetry rules:

$$\begin{aligned}
&\vec{\mathbf{n}} \times \vec{\mathbf{E}}\left(\vec{\mathbf{r}}\right) = -\vec{\mathbf{n}} \times \vec{\mathbf{E}}'\left(\vec{\mathbf{r}}'\right) \\
&\vec{\mathbf{n}} \cdot \vec{\mathbf{E}}\left(\vec{\mathbf{r}}\right) = \vec{\mathbf{n}} \cdot \vec{\mathbf{E}}'\left(\vec{\mathbf{r}}'\right) \\
&\vec{\mathbf{n}} \times \vec{\mathbf{B}}\left(\vec{\mathbf{r}}\right) = \vec{\mathbf{n}} \times \vec{\mathbf{B}}'\left(\vec{\mathbf{r}}'\right) \\
&\vec{\mathbf{n}} \cdot \vec{\mathbf{B}}\left(\vec{\mathbf{r}}\right) = -\vec{\mathbf{n}} \cdot \vec{\mathbf{B}}'\left(\vec{\mathbf{r}}'\right)
\end{aligned} \qquad [5.43]$$

where $\vec{\mathbf{r}}$ (in the II region) is the image of $\vec{\mathbf{r}}$ (in the II region) with respect to the S_1 plane, as shown in Figure 5.7. Obviously, these symmetry relations are compatible with the continuity equations quoted above, that is:

$$\begin{aligned}\left(\vec{\mathrm{n}} \cdot \vec{\mathrm{E}}\right)_{S_1} &= \left(\vec{\mathrm{n}} \cdot \vec{\mathrm{E}}'\right)_{S_1'} \\ \left(\vec{\mathrm{n}} \times \vec{\mathrm{B}}\right)_{S_1} &= \left(\vec{\mathrm{n}} \times \vec{\mathrm{B}}'\right)_{S_1'}\end{aligned} \qquad [5.44]$$

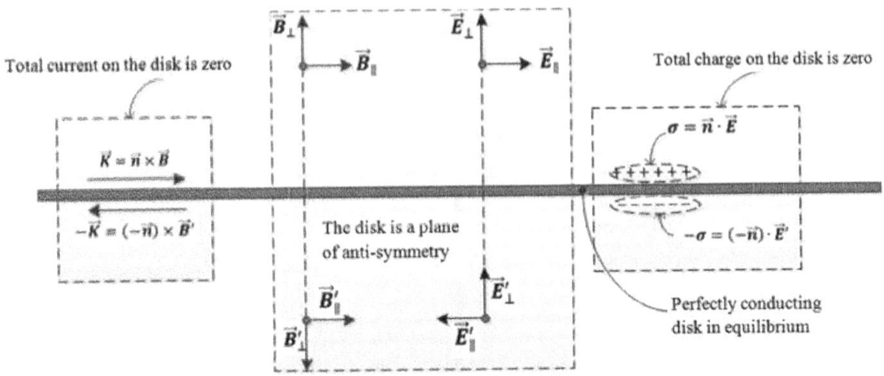

Figure 5.7. *Symmetry relation of the fields on either side of an infinite perfectly conducting disk of negligible thickness containing sources inside (the disk behaves like a plane of antisymmetry). For a color version of this figure, see www.iste.co.uk/abdesselam/electrodynamics3.zip*

Taking into account conditions [5.43], the expression [5.42] reduces to a single integral:

$$\vec{\mathrm{E}}\left(\vec{\mathrm{r}}\right) = 2\int_{S_1} \left(\vec{\mathrm{n}} \times \vec{\mathrm{E}}\right) \times \vec{\nabla}' G dS \qquad [5.45]$$

where $\vec{\mathrm{n}} \times \vec{\mathrm{E}}$ is the tangential electric field at S_1, $\vec{\mathrm{n}}$ is the normal directed toward region II and G is the Green function [5.30]. Since $\vec{\nabla}' G = -\vec{\nabla} G$, equation [5.45] can also be written as:

$$\vec{\mathrm{E}}\left(\vec{\mathrm{r}}\right) = 2\vec{\nabla} \times \int_{S_1} \vec{\mathrm{n}} \times \vec{\mathrm{E}}\left(\vec{\mathrm{r}}'\right) G\left(\vec{\mathrm{r}}, \vec{\mathrm{r}}'\right) dS \qquad [5.46]$$

The integral is computed only on the apertures, because the tangential component $\vec{\mathrm{n}} \times \vec{\mathrm{E}}\left(\vec{\mathrm{r}}'\right)$ is zero everywhere on the screen except on these apertures:

$$\vec{\mathrm{E}}\left(\vec{\mathrm{r}}\right) = \frac{1}{2\pi} \vec{\nabla} \times \int_{\text{aper.}} \frac{e^{ik\left|\vec{\mathrm{r}} - \vec{\mathrm{r}}'\right|}}{\left|\vec{\mathrm{r}} - \vec{\mathrm{r}}'\right|} \vec{\mathrm{n}} \times \vec{\mathrm{E}}\left(\vec{\mathrm{r}}'\right) dS' \qquad [5.47]$$

In the same way, it can be shown that:

$$\vec{\mathbf{B}} = \int_{S_1}\left[-2\frac{\mathrm{i}\omega}{c^2}\vec{\mathbf{n}}\times\vec{\mathbf{E}}G + 2\vec{\mathbf{n}}\cdot\vec{\mathbf{B}}\vec{\nabla}'G\right]dS' = -2\frac{\mathrm{i}\omega}{c^2}\int_{S_1}\vec{\mathbf{n}}\times\vec{\mathbf{E}}GdS'$$

$$= -\frac{\mathrm{i}\omega}{2\pi c^2}\int_{\text{aper.}}\frac{e^{\mathrm{ik}\left|\vec{\mathbf{r}}-\vec{\mathbf{r}}'\right|}}{\left|\vec{\mathbf{r}}-\vec{\mathbf{r}}'\right|}\vec{\mathbf{n}}\times\vec{\mathbf{E}}dS' \qquad [5.48]$$

since the tangential component $\vec{\mathbf{n}} \times \vec{\mathbf{E}}$ is zero except in the apertures, while the normal component $\vec{\mathbf{n}} \cdot \vec{\mathbf{B}}$ is zero everywhere on the disk. By combining equations [5.46] and [5.48], it is clearly verified that $\vec{\nabla} \times \vec{\mathbf{B}} = -\frac{\mathrm{i}\omega}{c^2}\vec{\mathbf{E}}$.

5.4. Babinet's principle of complementary screens

Before discussing examples of diffraction, we first aim to establish a useful relationship known as the Babinet principle. The Babinet principle relates the diffraction fields of a diffracting screen to those of its complementary screen. We first examine this principle within the Kirchhoff scalar approximation. It is assumed that the diffracting screen is located on a surface S that divides the space into regions I and II, as described in section 5.2. The screen occupies the entire surface S except for certain apertures. The complementary screen is the diffracting screen obtained by replacing the apertures with the screen material and the screen material with apertures. If the surface area of the original screen is S_a and that of the complementary screen is S_b, then $S = S_a + S_b$, as illustrated in Figure 5.8.

If there are sources inside S (in region I) that produce a field $\psi\left(\vec{\mathbf{r}}\right)$, then in the absence of any screen (original or complementary), the field $\psi\left(\vec{\mathbf{r}}\right)$ in region II is given by the Kirchhoff formula [5.26], where the integral is taken over the entire surface S. With the screen S_a in place, the field $\psi_a\left(\vec{\mathbf{r}}\right)$ in region II is given under the Kirchhoff approximation by equation [5.26], with the unobstructed field $\psi\left(\vec{\mathbf{r}}\right)$ inside the integral and the surface integration now limited to S_b (the apertures). Similarly, for the complementary screen S_b, the field $\psi_b\left(\vec{\mathbf{r}}\right)$ is derived under the same approximation by integrating over S_a. We thus obtain the obvious relation between the diffraction fields $\psi_a\left(\vec{\mathbf{r}}\right)$ and $\psi_b\left(\vec{\mathbf{r}}\right)$:

$$\psi\left(\vec{\mathbf{r}}\right) = \psi_a\left(\vec{\mathbf{r}}\right) + \psi_b\left(\vec{\mathbf{r}}\right) \qquad [5.49]$$

This is the *Babinet principle* as it is commonly formulated in optics. If ψ represents an incident plane wave, the Babinet principle states that the diffraction pattern far from the incident direction is identical for both the original screen and its complement.

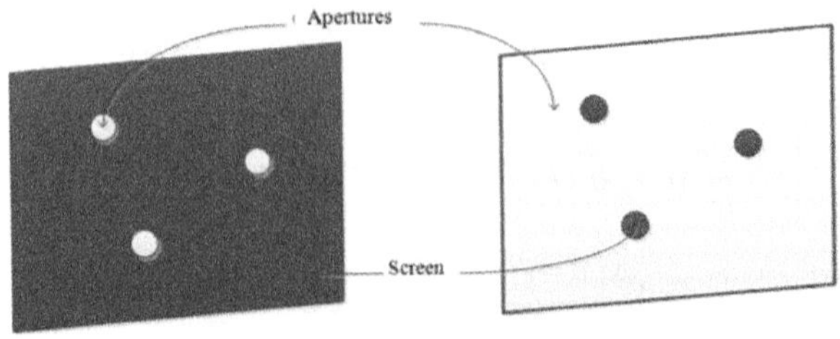

Figure 5.8. *Diffraction screen S_a and its complementary diffraction screen S_b*

Let us apply the Babinet principle to electromagnetic fields. Suppose that the screen S_a is thin, planar and perfectly conducting and consider incident fields $\vec{\mathbf{E}}_0$, $\vec{\mathbf{B}}_0$. The presence of the screen gives rise to transmitted and reflected fields, as shown in Figure 5.2. These transmitted and reflected fields are also called *scattered fields*. For a perfectly conducting screen, the tangential components of the resulting electric fields on both sides of the screen must vanish. Consequently, the incident field $\vec{\mathbf{E}}_0$ and the reflected field $\vec{\mathbf{E}}_r$ (on the side of region I) must satisfy $\vec{\mathbf{n}} \times \vec{\mathbf{E}}_r = -\vec{\mathbf{n}} \times \vec{\mathbf{E}}_0$ at every point on the surface S_a. Similarly, the incident field $\vec{\mathbf{E}}_0$ and the transmitted field $\vec{\mathbf{E}}_t$ (on the side of region II) must satisfy $\vec{\mathbf{n}} \times \vec{\mathbf{E}}_t = -\vec{\mathbf{n}} \times \vec{\mathbf{E}}_0$ at every point on S_a. For a thin and planar surface, the symmetry of the problem implies that the tangential components of the scattered magnetic fields $\vec{\mathbf{B}}_t$ and $\vec{\mathbf{B}}_r$ at the surface must be equal and opposite:

$$\vec{\mathbf{n}} \times \vec{\mathbf{B}}_t = -\vec{\mathbf{n}} \times \vec{\mathbf{B}}_r = \frac{\mu_0}{2}\vec{\mathbf{K}} \qquad [5.50]$$

where we have used the continuity equation $\vec{\mathbf{n}} \times \left(\vec{\mathbf{B}}_t - \vec{\mathbf{B}}_r\right) = \mu_0 \vec{\mathbf{K}}$. Here, the normal vector $\vec{\mathbf{n}}$ points into region II. Using arguments similar to those leading from [5.43] to [5.44], we can establish that at every point $\vec{\mathbf{r}}$ in region II and its image point $\vec{\mathbf{r}}'$ in region I, the scattered fields satisfy the following symmetry conditions:

$$\begin{aligned}
&\vec{\mathbf{n}} \times \vec{\mathbf{E}}_r(\vec{\mathbf{r}}') = \vec{\mathbf{n}} \times \vec{\mathbf{E}}_t(\vec{\mathbf{r}}) \\
&\vec{\mathbf{n}} \cdot \vec{\mathbf{E}}_r(\vec{\mathbf{r}}') = -\vec{\mathbf{n}} \cdot \vec{\mathbf{E}}_t(\vec{\mathbf{r}}) \\
&\vec{\mathbf{n}} \times \vec{\mathbf{B}}_r(\vec{\mathbf{r}}') = -\vec{\mathbf{n}} \times \vec{\mathbf{B}}_t(\vec{\mathbf{r}}) \\
&\vec{\mathbf{n}} \cdot \vec{\mathbf{B}}_r(\vec{\mathbf{r}}') = \vec{\mathbf{n}} \cdot \vec{\mathbf{B}}_t(\vec{\mathbf{r}})
\end{aligned} \qquad [5.51]$$

Note that these relations differ from those in [5.43], as the signs of $\vec{\mathbf{E}}_r$ and $\vec{\mathbf{B}}_r$ are reversed. In fact, the fields in equation [5.43] correspond to a double current sheet. The fields in equation [5.51] exhibit the appropriate symmetries for a single planar current sheet radiating in both directions.

An integral expression for the scattered magnetic field can be obtained in terms of the surface current $\overrightarrow{\mathbf{K}}$. In fact, the field $\overrightarrow{\mathbf{B}}_t$ is calculated from the vector potential $\overrightarrow{\mathbf{A}}$ generated by the current sheet:

$$\overrightarrow{\mathbf{B}}_t = \mu_0 \overrightarrow{\nabla} \times \int_{S_a} \overrightarrow{\mathbf{K}} G dS' \qquad [5.52]$$

where G is the Green function [5.30]. The integration here is carried out on the metallic surface S_a of the screen. Equation [5.52] can also be derived from equation [5.40]. Applying equation [5.39], we obtain the transmitted electric field as:

$$\overrightarrow{\mathbf{E}}_t\left(\overrightarrow{\mathbf{r}}\right) = 2i\omega \int_{S_a} \left(\overrightarrow{\mathbf{n}} \times \overrightarrow{\mathbf{B}}_t\right) G dS' \qquad [5.53]$$

Substituting $\overrightarrow{\mathbf{K}}$ from equation [5.50] into [5.52], the transmitted magnetic field can be written as:

$$\overrightarrow{\mathbf{B}}_t\left(\overrightarrow{\mathbf{r}}\right) = 2\overrightarrow{\nabla} \times \int_{S_a} \overrightarrow{\mathbf{n}} \times \overrightarrow{\mathbf{B}}_t\left(\overrightarrow{\mathbf{r}}'\right) G\left(\overrightarrow{\mathbf{r}}, \overrightarrow{\mathbf{r}}'\right) dS' \qquad [5.54]$$

This result is identical to that of equation [5.46], except that (1) the roles of $\overrightarrow{\mathbf{E}}$ and $\overrightarrow{\mathbf{B}}$ have been interchanged; (2) the integration in equation [5.53] is carried out over the metallic part of the screen, while in equation [5.46] it is only over the apertures; (3) the total electric field appears in equation [5.46], while only the scattered field appears in equation [5.53]. The comparison between equations [5.46] and [5.53] is the basis of the Babinet principle. Applying results [5.46] and [5.48] to the complementary screen of S_a, we obtain:

$$\begin{aligned} \overrightarrow{\mathbf{E}}'\left(\overrightarrow{\mathbf{r}}\right) &= 2\overrightarrow{\nabla} \times \int_{S_a} \overrightarrow{\mathbf{n}} \times \overrightarrow{\mathbf{E}}'\left(\overrightarrow{\mathbf{r}}'\right) G\left(\overrightarrow{\mathbf{r}}, \overrightarrow{\mathbf{r}}'\right) dS' \\ \overrightarrow{\mathbf{B}}'\left(\overrightarrow{\mathbf{r}}\right) &= -\frac{2i\omega}{c^2} \int_{S_a} \overrightarrow{\mathbf{n}} \times \overrightarrow{\mathbf{E}}'\left(\overrightarrow{\mathbf{r}}'\right) G\left(\overrightarrow{\mathbf{r}}, \overrightarrow{\mathbf{r}}'\right) dS' \end{aligned} \qquad [5.55]$$

The integration is carried out only on S_a, since it represents the aperture of the complementary screen. The field $\overrightarrow{\mathbf{E}}'$ in region II is the sum:

$$\overrightarrow{\mathbf{E}}' = \overrightarrow{\mathbf{E}}'_0 + \overrightarrow{\mathbf{E}}'_t \qquad [5.56]$$

where $\overrightarrow{\mathbf{E}}'_0$ is the incident electric field in the complementary diffraction problem and $\overrightarrow{\mathbf{E}}'_t$ is the corresponding transmitted (or diffracted) field. To shift from expressions [5.53] to [5.54], we must make the following correspondences:

$$\begin{aligned} \overrightarrow{\mathbf{B}}_t &\longrightarrow \pm\left(\overrightarrow{\mathbf{E}}'_0 + \overrightarrow{\mathbf{E}}'_t\right) \\ \overrightarrow{\mathbf{E}}_t &\longrightarrow \mp\left(\overrightarrow{\mathbf{B}}'_0 + \overrightarrow{\mathbf{B}}'_t\right) \end{aligned} \qquad [5.57]$$

Beginning with the complementary screen, it is clear that equations [5.57] must remain valid by interchanging the primes:

$$\begin{aligned} \vec{\mathbf{B}}'_t &\longrightarrow \mp\left(\vec{\mathbf{E}}_0 + \vec{\mathbf{E}}_t\right) \\ \vec{\mathbf{E}}'_t &\longrightarrow \pm\left(\vec{\mathbf{B}}_0 + \vec{\mathbf{B}}_t\right) \end{aligned} \qquad [5.58]$$

By comparing the correspondences [5.57] and [5.58], we verify that the incident fields in the original and complementary diffraction problems must be related as follows:

$$\begin{aligned} \vec{\mathbf{E}}'_0 &= \mp\vec{\mathbf{B}}_0 \\ \vec{\mathbf{B}}'_0 &= \pm\vec{\mathbf{E}}_0 \end{aligned} \qquad [5.59]$$

We see that to go from the original diffraction problem to the complementary one, the roles of the incident fields $\vec{\mathbf{E}}_0$ and $\vec{\mathbf{B}}_0$ must also be interchanged. The Babinet principle can be stated as follows: a diffracting system consists of a source that produces incident fields $\left(\vec{\mathbf{E}}_0, \vec{\mathbf{B}}_0\right)$ on a thin, planar, perfectly conducting screen with certain apertures. The complementary diffracting system consists of a source producing incident fields $\left(\vec{\mathbf{E}}'_0 = \mp\vec{\mathbf{B}}_0, \vec{\mathbf{B}}'_0 = \pm\vec{\mathbf{E}}_0\right)$ on the complementary screen. The transmitted fields are, respectively, $\left(\vec{\mathbf{E}}_t, \vec{\mathbf{B}}_t\right)$ and $\left(\vec{\mathbf{E}}'_t, \vec{\mathbf{B}}'_t\right)$ for the original and complementary systems. They are related by the following expressions:

$$\begin{aligned} \vec{\mathbf{E}}_t \pm \vec{\mathbf{B}}'_t &= -\vec{\mathbf{E}}_0 = \mp\vec{\mathbf{B}}'_0 \\ \vec{\mathbf{E}}'_t \mp \vec{\mathbf{B}}_t &= \pm\vec{\mathbf{B}}_0 = -\vec{\mathbf{E}}'_0 \end{aligned} \qquad [5.60]$$

These equations are the vector equivalents of the scalar equation [5.49]. The polarization of the incident wave on the complementary screen must, of course, be rotated according to equation [5.59].

5.5. Diffraction by a circular aperture

Diffraction has been extensively studied since the Kirchhoff pioneering work. It has been demonstrated that Kirchhoff's scalar theory, based on equation [5.28], is generally sufficient to describe the phenomenon. We will present here several examples to illustrate the application of the scalar and vector theorems [5.28] and [5.46]. We will also try to compare the precision of the approximation schemes.

Historically, diffraction models were classified into *Fresnel diffraction* and *Fraunhofer diffraction* depending on the distance between the observation point and the diffracting system. The diffraction phenomenon occurs when a wave encounters an obstacle or an aperture whose dimensions are on the order of, or smaller than, the wavelength of the wave. On the other hand, if the dimensions of the obstacle or aperture are greater than λ, diffraction does not occur. The smaller the aperture or obstacle, the more significant the diffraction effect. The observation point may lie in the near zone at a distance less than a wavelength from the diffracting system. In the near zone, the fields have a somewhat complicated structure and are of limited interest. Points located several wavelengths away from the diffracting system, but still relatively close to the obstacle or aperture, are said to lie in the *Fresnel zone*. At greater distances, large compared to both the size of the diffracting system and the wavelength, lies the *Fraunhofer zone*, which corresponds to the radiation zone.

If the observation point is far from the diffracting system, we can use the approximation $\left|\vec{\mathbf{r}} - \vec{\mathbf{r}}'\right| \approx r - \frac{\vec{\mathbf{r}}}{r} \cdot \vec{\mathbf{r}}'$. By keeping only the terms of order $\frac{1}{r}$, expression [5.28] becomes:

$$\psi\left(\vec{\mathbf{r}}\right) = -\frac{e^{\mathrm{i}kr}}{4\pi r} \int_{S_1} e^{-\mathrm{i}\,\vec{\mathsf{k}} \cdot \vec{\mathbf{r}}'} \left[\vec{\mathbf{n}} \cdot \vec{\nabla}'\psi\left(\vec{\mathbf{r}}'\right) + \mathrm{i}\,\vec{\mathbf{n}} \cdot \vec{\mathsf{k}}\,\psi\left(\vec{\mathbf{r}}'\right)\right] dS' \qquad [5.61]$$

where $\vec{\mathbf{r}}'$ denotes the position of the surface element dS', r is the magnitude of the vector $\vec{\mathbf{r}}$ from the origin to the observation point P and $\vec{\mathsf{k}} = \mathsf{k}\frac{\vec{\mathbf{r}}}{r}$ is the wave vector in the observation direction, as shown in Figure 5.9. For a plane surface, expression [5.46] reduces in this limit to:

$$\vec{\mathbf{E}}\left(\vec{\mathbf{r}}\right) = \frac{\mathrm{i}e^{\mathrm{i}kr}}{2\pi r}\vec{\mathsf{k}} \times \int_{S_1} \vec{\mathbf{n}} \times \vec{\mathbf{E}}\left(\vec{\mathbf{r}}'\right) e^{-\mathrm{i}\,\vec{\mathsf{k}} \cdot \vec{\mathbf{r}}'} dS' \qquad [5.62]$$

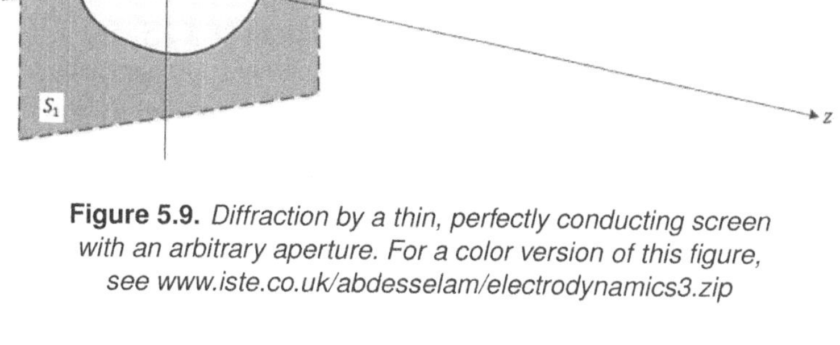

Figure 5.9. *Diffraction by a thin, perfectly conducting screen with an arbitrary aperture. For a color version of this figure, see www.iste.co.uk/abdesselam/electrodynamics3.zip*

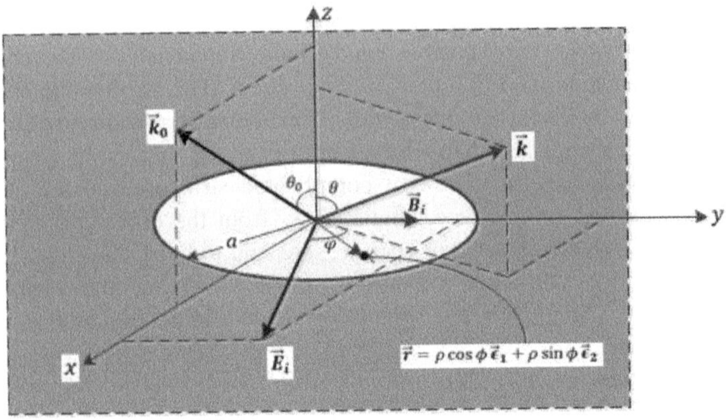

Figure 5.10. *Diffraction by an aperture of radius a. For a color version of this figure, see www.iste.co.uk/abdesselam/electrodynamics3.zip*

Consider a plane wave incident on a thin and perfectly conducting screen with a circular aperture of radius a. The incident wave makes an angle θ_0 with respect to the normal to the screen. Figure 5.10 shows an appropriate coordinate system:

– The screen is located in the Oxy plane at $z = 0$.

– The aperture is centered at point O.

– The incident wave arrives from the left side ($z < 0$) and propagates toward the $z > 0$ side (the diffraction region), passing through the screen via the aperture.

– The incident wave is polarized in the Oxz plane.

The incident electric and magnetic fields are written as:

$$\vec{\mathbf{E}}_i\left(\vec{\mathbf{r}}\right) = E_0\left(\vec{\epsilon}_1\cos\theta_0 - \vec{\epsilon}_3\sin\theta_0\right)e^{\mathrm{i}k(\cos\theta_0 z+\sin\theta_0 x)}$$
$$\vec{\mathbf{B}}_i\left(\vec{\mathbf{r}}\right) = B_0\vec{\epsilon}_2 e^{\mathrm{i}k(\cos\theta_0 z+\sin\theta_0 x)} \qquad [5.63]$$

where we have used the notation $\vec{\epsilon}_1$, $\vec{\epsilon}_2$ and $\vec{\epsilon}_3$ instead of $\vec{\mathbf{i}}$, $\vec{\mathbf{j}}$ and $\vec{\mathbf{k}}$ to denote the polarization directions. To compute the diffraction field, we apply the following approximation in equation [5.62]: the field in the surface integral $\vec{\mathbf{n}} \times \vec{\mathbf{E}}\left(\vec{\mathbf{r}}'\right)$ is replaced by the incident field $\vec{\mathbf{n}} \times \vec{\mathbf{E}}_i\left(\vec{\mathbf{r}}'\right)$.

In that case, we get:

$$\left(\vec{\mathbf{n}} \times \vec{\mathbf{E}}_i\right)_{z=0} = E_0\vec{\epsilon}_2\cos\theta_0 e^{\mathrm{i}k\sin\theta_0 x'} \qquad [5.64]$$

Introducing now polar coordinates for integration over the aperture, we obtain:

$$\begin{aligned}\vec{\mathbf{E}}\left(\vec{\mathbf{r}}\right) &= \frac{\mathrm{i}e^{\mathrm{i}\mathrm{k}r}E_0\cos\theta_0}{2\pi r}\left(\vec{\mathrm{k}}\times\vec{\epsilon}_2\right)\\ &\times\int_0^a \rho d\rho\int_0^{2\pi} d\phi\exp\left\{\mathrm{i}\mathrm{k}\rho\left[\sin\theta_0\cos\phi-\sin\theta\cos\left(\varphi-\phi\right)\right]\right\}\\ &= \frac{\mathrm{i}e^{\mathrm{i}\mathrm{k}r}E_0\cos\theta_0}{2\pi r}\left(\vec{\mathrm{k}}\times\vec{\epsilon}_2\right)\int_0^a\rho d\rho\\ &\times\int_0^{2\pi} d\phi\exp\left\{\mathrm{i}\mathrm{k}\rho\left[\left(\sin\theta_0-\sin\theta\cos\varphi\right)\cos\phi\right.\right.\\ &\left.\left.-\sin\theta\sin\varphi\sin\phi\right]\right\}\end{aligned} \quad [5.65]$$

where θ and φ are the spherical angles of $\vec{\mathrm{k}}$. We set $\vec{\mathbf{r}}' = \rho\cos\phi\,\vec{\epsilon}_1 + \rho\sin\phi\,\vec{\epsilon}_2$. To simplify [5.65], we introduce the notation:

$$\begin{aligned}&\xi\cos\phi_0 = -\sin\theta\sin\varphi\\ &\xi\sin\phi_0 = \sin\theta\cos\varphi - \sin\theta_0\end{aligned} \quad [5.66]$$

where $\xi = \sqrt{\sin^2\theta_0 + \sin^2\theta - 2\sin\theta_0\sin\theta\cos\varphi}$. Thus, we find:

$$\begin{aligned}\vec{\mathbf{E}}\left(\vec{\mathbf{r}}\right) &= \frac{\mathrm{i}e^{\mathrm{i}\mathrm{k}r}E_0\cos\theta_0}{2\pi r}\left(\vec{\mathrm{k}}\times\vec{\epsilon}_2\right)\int_0^a\rho d\rho\\ &\times\int_0^{2\pi} d\phi e^{\mathrm{i}\mathrm{k}\rho\xi\sin(\phi-\phi_0)} = \frac{\mathrm{i}e^{\mathrm{i}\mathrm{k}r}E_0\cos\theta_0}{2\pi r}\left(\vec{\mathrm{k}}\times\vec{\epsilon}_2\right)\\ &\int_0^a\rho d\rho\int_0^{2\pi} d\phi e^{\mathrm{i}\mathrm{k}\rho\xi\sin\phi} = \frac{\mathrm{i}e^{\mathrm{i}\mathrm{k}r}E_0\cos\theta_0}{r}\left(\vec{\mathrm{k}}\times\vec{\epsilon}_2\right)\\ &\times\int_0^a\rho d\rho J_0\left(\mathrm{k}\rho\xi\right)\\ &= \frac{\mathrm{i}e^{\mathrm{i}\mathrm{k}r}a^2E_0\cos\theta_0}{r}\left(\vec{\mathrm{k}}\times\vec{\epsilon}_2\right)\frac{J_1\left(\mathrm{k}a\xi\right)}{\mathrm{k}a\xi}\end{aligned} \quad [5.67]$$

where J_0 and J_1 are *Bessel functions*. This is the expression for the electric field in the Kirchhoff approximation. To go from the first line to the second in the previous calculation, we used the periodicity property of the function $e^{\mathrm{i}\mathrm{k}\rho\xi\sin\phi}$. In the last two lines, we used the following integral properties:

$$\begin{aligned}&J_0\left(x\right) = \frac{1}{2\pi}\int_0^{2\pi} e^{\mathrm{i}x\sin\phi}d\phi\\ &\int xJ_0\left(x\right)dx = xJ_1\left(x\right)\end{aligned} \quad [5.68]$$

The time-averaged diffracted power per unit of solid angle is thus given by:

$$\begin{aligned}\frac{dP}{d\Omega} &= \frac{r^2}{2\mu_0 c}\left|\vec{\mathbf{E}}\right|^2 = \frac{a^4 E_0^2 \cos^2\theta_0}{2\mu_0 c}\left|\vec{\mathrm{k}}\times\vec{\epsilon}_2\right|^2\left|\frac{J_1(\mathrm{k}a\xi)}{\mathrm{k}a\xi}\right|^2 \\ &= \frac{a^4 E_0^2\cos^2\theta_0 \mathrm{k}^2}{2\mu_0 c}\left(\cos^2\theta + \sin^2\theta\cos^2\varphi\right)\left|\frac{J_1(\mathrm{k}a\xi)}{\mathrm{k}a\xi}\right|^2 \\ &= P_i\cos\theta_0\frac{(a\mathrm{k})^2}{4\pi}\left(\cos^2\theta+\sin^2\theta\cos^2\varphi\right)\left|\frac{2J_1(\mathrm{k}a\xi)}{\mathrm{k}a\xi}\right|^2 \\ &= P_i\cos\theta_0\frac{(a\mathrm{k})^2}{4\pi}\left(\cos^2\varphi+\cos^2\theta\sin^2\varphi\right)\left|\frac{2J_1(\mathrm{k}a\xi)}{\mathrm{k}a\xi}\right|^2\end{aligned} \qquad [5.69]$$

where:

$$P_i = \frac{E_0^2\pi a^2\cos\theta_0}{2\mu_0 c} \qquad [5.70]$$

represents the total power normally incident on the aperture. If the aperture is large compared to the wavelength ($\mathrm{k}a \gg 1$), the factor $\left|\frac{2J_1(\mathrm{k}a\xi)}{\mathrm{k}a\xi}\right|^2$ peaks at 1 for $\xi = 0$ and rapidly falls to zero (with minor secondary maxima) within a region $\Delta\xi \sim \frac{1}{\mathrm{k}a}$ on either side of $\xi = 0$. This indicates that nearly the entire wave passes through the aperture following geometric optics, with only slight diffraction effects occurring. For $\mathrm{k}a \approx 1$, the Bessel function varies relatively slowly with angle; the transmitted wave is distributed in directions significantly different from the incident direction. For $\mathrm{k}a \ll 1$, the angular distribution is entirely governed by the factor $\left|\vec{\mathrm{k}}\times\vec{\epsilon}_2\right|^2$ in equation [5.69]. However, in this limit, the assumption of an unperturbed field in the aperture breaks down.

The total transmitted power can be obtained by integrating equation [5.69] over the entire front hemisphere. The ratio between transmitted and incident power is called the *transmission coefficient* T:

$$T = \frac{\cos\theta_0}{\pi}\int_0^{2\pi}d\varphi\int_0^{\pi/2}\left(\cos^2\varphi+\cos^2\theta\sin^2\varphi\right)\left|\frac{J_1(\mathrm{k}a\xi)}{\xi}\right|^2\sin\theta d\theta \qquad [5.71]$$

In particular, we have:

1) For $\mathrm{k}a \ll 1$, we have approximately $J_1(\mathrm{k}a\xi) \sim \frac{1}{2}\mathrm{k}a\xi$, which gives:

$$\begin{aligned}T &\simeq \frac{\cos\theta_0}{\pi}\int_0^{2\pi}d\varphi\int_0^{\pi/2}\left(\cos^2\varphi+\cos^2\theta\sin^2\varphi\right)\left(\frac{\mathrm{k}a}{2}\right)^2\sin\theta d\theta \\ &= \cos\theta_0\left(\frac{\mathrm{k}a}{2}\right)^2\int_0^{\pi/2}\left(1+\cos^2\theta\right)\sin\theta d\theta = \frac{1}{3}(\mathrm{k}a)^2\cos\theta_0\end{aligned} \qquad [5.72]$$

2) For $\text{k}a \gg 1$, the incident wave is only weakly diffracted (the diffracted radiation is confined between 0 and $\Delta\theta$ around the Oz axis). We thus obtain:

$$T \simeq \frac{\cos\theta_0}{\pi}\int_0^{2\pi} d\varphi \int_0^{\Delta\theta} \left(\frac{\text{k}a}{2}\right)^2 \theta d\theta = \cos\theta_0 \left(\frac{\text{k}a}{2}\right)^2 (\Delta\theta)^2 \qquad [5.73]$$

because the factor $\left|\frac{2J_1(\text{k}a\xi)}{\text{k}a\xi}\right|$ is approximately equal to unity over a band of width $\frac{2}{\text{k}a}$ around $\xi = 0$ and vanishes elsewhere. Since $\xi \simeq \sin\theta_0 - \theta\cos\varphi$ (when θ is very small), we see that $\Delta\theta \simeq \frac{2}{\text{k}a}$ (for fixed φ), which leads to the following:

$$T \simeq \cos\theta_0 \qquad [5.74]$$

When $\theta_0 = 0$ (the wave is normally incident on the screen), the transmission coefficient [5.71] becomes:

$$T = \int_0^{\pi/2} J_1^2(\text{k}a\sin\theta)\left(\frac{2}{\sin\theta} - \sin\theta\right) d\theta \qquad [5.75]$$

According to integral relations:

$$\begin{aligned} &\int_0^{\pi/2} J_1^2(\text{k}a\sin\theta)\frac{d\theta}{\sin\theta} = \int_0^{2\text{k}a} \frac{J_2(t)}{t} dt \\ &\int_0^{\pi/2} J_1^2(\text{k}a\sin\theta)\sin\theta d\theta = \frac{1}{2\text{k}a}\int_0^{2\text{k}a} J_2(t)\, dt \end{aligned} \qquad [5.76]$$

Equation [5.75] reduces to:

$$\begin{aligned} T &= \int_0^{2\text{k}a}\left[\frac{2J_2(t)}{t} - \frac{1}{2\text{k}a}J_2(t)\right] dt = \int_0^{2\text{k}a}\left[-2\frac{d}{dt}\left(\frac{J_1(t)}{t}\right) - \right. \\ &\left.\frac{1}{2\text{k}a}\left(J_0(t) - 2\frac{dJ_1(t)}{dt}\right)\right] dt = -2\left.\frac{J_1(t)}{t}\right|_0^{2\text{k}a} \\ &+\frac{1}{2\text{k}a}\left.J_1(t)\right|_0^{2\text{k}a} - \frac{1}{2\text{k}a}\int_0^{2\text{k}a} J_0(t)\, dt \\ &= 2\lim_{t\longrightarrow 0}\frac{J_1(t)}{t} - \frac{1}{2\text{k}a}\int_0^{2\text{k}a} J_0(t)\, dt = 1 - \frac{1}{2\text{k}a}\int_0^{2\text{k}a} J_0(t)\, dt \end{aligned} \qquad [5.77]$$

where the following properties have been used: $\frac{J_2(t)}{t} = -\frac{d}{dt}\left[\frac{J_1(t)}{t}\right]$, $J_2(t) = J_0(t) - 2\frac{dJ_1(t)}{dt}$ (obtained from recurrence relations) and $J_1(t) \simeq \frac{t}{2}$ for

$t \ll 1$. Noting now that $J_0(t) = 2\sum_{m=0}^{\infty} J'_{2m+1}(t)$ (also derived from recurrence relations), the previous equation can alternatively be written as:

$$T = 1 - \frac{1}{\mathrm{k}a}\sum_{m=0}^{\infty} J_{2m+1}(2\mathrm{k}a) \qquad [5.78]$$

We can see that the transmission coefficient T increases almost monotonically (but with small superposed oscillations) as $sfka$ increases. For $sfka \gg 1$, the first form [5.79] can be used to obtain an asymptotic expression:

$$\begin{aligned} T &= 1 - \frac{1}{2\mathrm{k}a}\int_0^{\infty} J_0(t)\,dt + \frac{1}{2\mathrm{k}a}\int_{2\mathrm{k}a}^{\infty} J_0(t)\,dt \\ &\simeq 1 - \frac{1}{2\mathrm{k}a} + \frac{1}{2\mathrm{k}a}\int_{2\mathrm{k}a}^{\infty}\sqrt{\frac{2}{\pi t}}\cos\left(t - \frac{\pi}{4}\right)dt \\ &\simeq 1 - \frac{1}{2\mathrm{k}a} + \frac{1}{\sqrt{2\pi}\mathrm{k}a}\left[\left.\int_{2\mathrm{k}a}^{\infty}\sqrt{\frac{1}{t}}\sin\left(t - \frac{\pi}{4}\right)\right|_{2\mathrm{k}a}^{\infty}\right. \\ &\quad \left. + \int_{2\mathrm{k}a}^{\infty}\frac{1}{2t^{3/2}}\sin\left(t - \frac{\pi}{4}\right)dt\right] \\ &\simeq 1 - \frac{1}{2\mathrm{k}a} - \frac{1}{2\sqrt{\pi}(\mathrm{k}a)^{3/2}}\sin\left(2\mathrm{k}a - \frac{\pi}{4}\right) + \cdots \end{aligned} \qquad [5.79]$$

which shows small oscillations explicitly.

We now wish to compare the above results (from the Kirchhoff vector approximation) with the standard scalar theory based on equation [5.61]. For an incident wave that makes an angle θ_0 with the normal to the screen, the immediate question is what to choose for the scalar function $\psi(\vec{\mathbf{r}})$. The simplest answer may be to choose the magnitude of the electric or magnetic field. If we take $\psi(\vec{\mathbf{r}}') = \left|\vec{\mathbf{E}}_i\right| e^{\mathrm{i}\vec{\mathrm{k}}\cdot\vec{\mathbf{r}}'} = E_0 e^{\mathrm{i}\vec{\mathrm{k}}\cdot\vec{\mathbf{r}}'}$, the diffracted intensity can be regarded as proportional to the absolute square of equation [5.61]. Taking $\psi(\vec{\mathbf{r}}') = E_0 e^{\mathrm{ik}\left(\cos\theta_0 z' + \sin\theta_0 x'\right)}$, we deduce that $\vec{\mathbf{n}}\cdot\vec{\nabla}'\psi = \vec{\epsilon}_3\cdot\vec{\nabla}'\psi = \frac{\partial\psi}{\partial z'} = \mathrm{ik}\cos\theta_0 E_0 e^{\mathrm{ik}\left(\cos\theta_0 z' + \sin\theta_0 x'\right)}$ and $\mathrm{i}\vec{\mathbf{n}}\cdot\vec{\mathrm{k}}\psi = \mathrm{ik}\cos\theta E_0 e^{\mathrm{ik}\left(\cos\theta_0 z' + \sin\theta_0 x'\right)}$. Substituting these results into equation [5.61], we obtain:

$$\psi(\vec{\mathbf{r}}) = -\frac{\mathrm{ik}e^{\mathrm{ik}r}E_0}{4\pi r}(\cos\theta_0 + \cos\theta)\int_{\text{ouv.}} e^{\mathrm{ik}\sin\theta_0 x'} e^{-\mathrm{i}\vec{\mathrm{k}}\cdot\vec{\mathbf{r}}'}dS'$$

$$= -\frac{\mathrm{ik}e^{\mathrm{ik}r}E_0}{4\pi r}(\cos\theta_0+\cos\theta)\int_{\rho'=0}^{a}$$

$$\times\int_{\phi'=0}^{2\pi} e^{\mathrm{ik}\rho'\left[(\sin\theta_0-\sin\theta\cos\varphi)\cos\phi'-\sin\theta\sin\varphi\sin\phi'\right]}\rho' d\rho' d\phi'$$

$$= -\frac{\mathrm{ik}e^{\mathrm{ik}r}E_0}{4\pi r}(\cos\theta_0+\cos\theta)\int_{\rho'=0}^{a}\rho' d\rho'\int_{\phi'=0}^{2\pi} d\phi' e^{\mathrm{ik}\rho'\xi\sin\phi'}$$

$$= -\frac{\mathrm{ik}e^{\mathrm{ik}r}E_0}{2r}(\cos\theta_0+\cos\theta)\int_{\rho'=0}^{a}\rho' d\rho' J_0(\mathrm{k}\rho'\xi)$$

$$= -\mathrm{ik}\frac{e^{\mathrm{ik}r}}{r}a^2E_0\left(\frac{\cos\theta_0+\cos\theta}{2}\right)\frac{J_1(\mathrm{k}a\xi)}{\mathrm{k}a\xi} \qquad [5.80]$$

This is the scalar equivalent form of equation [5.67]. The radiated power per unit solid angle in the Kirchhoff scalar approximation is equal to:

$$\frac{dP}{d\Omega}=\frac{r^2}{2\mu_0 r^2}\left|\psi\left(\vec{\mathbf{r}}\right)\right|^2=P_i\frac{(\mathrm{k}a)^2}{4\pi}\cos\theta_0\left(\frac{\cos\theta_0+\cos\theta}{2\cos\theta_0}\right)^2\left|\frac{2J_1(\mathrm{k}a\xi)}{\mathrm{k}a\xi}\right|^2 \qquad [5.81]$$

where P_i is given by [5.70].

A comparison between results [5.69] and [5.81] reveals some similarities and differences. Both formulas contain the same distribution factor $\left|\frac{2J_1(\mathrm{k}a\xi)}{\mathrm{k}a\xi}\right|^2$ and identical wavenumber dependence. The vector result [5.69] depends on the azimuthal angle φ (beyond what is contained in ξ), whereas the scalar result shows no such dependence (apart from ξ, of course). This azimuthal variation stems from the polarization properties of the field and must be absent in a scalar approximation. To better compare results [5.69] and [5.81], let us examine specific cases:

– $\theta_0=0$ (normal incidence), $\mathrm{k}a=4\pi$ and $\varphi=0$. In this case, expressions [5.69] and [5.81] reduce to:

$$\left.\frac{dP}{d\Omega}\right|_{\text{vector}}=\frac{P_i}{\pi}\left|\frac{J_1(4\pi\sin\theta)}{\sin\theta}\right|^2$$

$$\left.\frac{dP}{d\Omega}\right|_{\text{scalar}}=\frac{P_i}{\pi}\left(\frac{1+\cos\theta}{2}\right)^2\left|\frac{J_1(4\pi\sin\theta)}{\sin\theta}\right|^2 \qquad [5.82]$$

As seen from Figure 5.11, the curves are indistinguishable in the forward direction (at the diffraction peak). Although the difference is quite large toward the rear of the axis, this has little significance, since very little power is present in that region.

– For normal incidence ($\theta_0 = 0$) and k$a \gg 1$, the polarization dependence is negligible. The diffraction is confined at very small angles θ. The two approximations, scalar and vector, then reduce to the common expression:

$$\frac{dP}{d\Omega} \simeq P_i \frac{(\mathrm{k}a)^2}{\pi} \left| \frac{J_1(\mathrm{k}a \sin\theta)}{\mathrm{k}a \sin\theta} \right|^2 \qquad [5.83]$$

– Similarly, for $\theta_0 = \frac{\pi}{4}$, k$a = 4\pi$ and $\varphi = 0$, we arrive at:

$$\left.\frac{dP}{d\Omega}\right|_{\text{vector}} = \frac{P_i}{\sqrt{2}\pi} \left| \frac{J_1\left(\pi \left|\frac{\sqrt{2}}{2} - \sin\theta\right|\right)}{\left|\frac{\sqrt{2}}{2} - \sin\theta\right|} \right|^2$$

$$\left.\frac{dP}{d\Omega}\right|_{\text{scalar}} = \frac{P_i}{\sqrt{2}\pi} \left(\frac{\frac{\sqrt{2}}{2} + \cos\theta}{\sqrt{2}} \right)^2 \left| \frac{J_1\left(\pi \left|\frac{\sqrt{2}}{2} - \sin\theta\right|\right)}{\left|\frac{\sqrt{2}}{2} - \sin\theta\right|} \right|^2 \qquad [5.84]$$

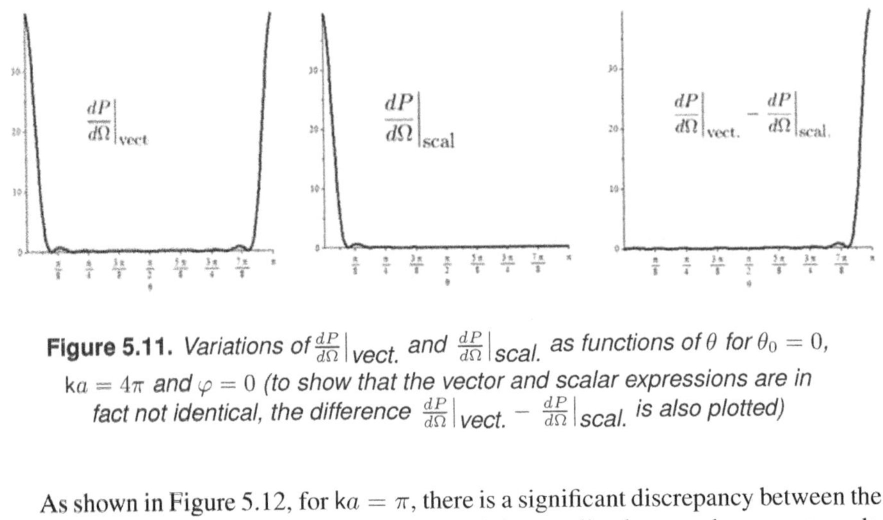

Figure 5.11. *Variations of $\left.\frac{dP}{d\Omega}\right|_{vect.}$ and $\left.\frac{dP}{d\Omega}\right|_{scal.}$ as functions of θ for $\theta_0 = 0$, k$a = 4\pi$ and $\varphi = 0$ (to show that the vector and scalar expressions are in fact not identical, the difference $\left.\frac{dP}{d\Omega}\right|_{vect.} - \left.\frac{dP}{d\Omega}\right|_{scal.}$ is also plotted)*

As shown in Figure 5.12, for k$a = \pi$, there is a significant discrepancy between the two approximations. The Kirchhoff vector result is actually closer to the correct result, even though the approximation begins to break down seriously for k$a \leq \pi$. For a rectangular aperture, the vector approximation and exact calculations yield consistent results, even down to k$a \sim 1$.

EXAMPLE 5.2. RECTANGULAR APERTURE.– Consider a thin, flat, perfectly conducting screen located in the Oxy plane. The screen contains a rectangular aperture with side lengths a and $b > a$, defined by $x = \pm\frac{a}{2}$ and $y = \pm\frac{b}{2}$. A plane wave normally arrives on the screen from the $z < 0$ region with its polarization vector at an angle to the Oy axis. We will compute the diffracted fields and the power radiated per unit solid angle, assuming that the tangential electric field within the aperture is equal to the undisturbed incident field. The diffracted electric field is calculated using the formula [5.62], where the undisturbed incident field (for a normally incident plane wave) can be taken as:

$$\vec{\mathbf{E}}\left(\vec{\mathbf{r}}\right) \simeq E_0\left(\vec{\epsilon}_1 \sin\alpha + \vec{\epsilon}_2 \cos\alpha\right) e^{\mathrm{i}kz'} \qquad [5.85]$$

where $\vec{\epsilon}_1 = \vec{\mathbf{i}}$ and $\vec{\epsilon}_2 = \vec{\mathbf{j}}$. For the rectangular screen, the surface S_1 is the rectangle located at $z' = 0$ with $|x'| \leq \frac{a}{2}$ and $|y'| \leq \frac{b}{2}$. The normal vector $\vec{n}$ to the surface S_1 is $\vec{\epsilon}_3 = \vec{\mathbf{k}}$. The resulting integral is then:

$$\vec{\mathbf{E}}\left(\vec{\mathbf{r}}\right) = \frac{\mathrm{i}E_0 e^{ikr}}{2\pi r}\vec{\mathsf{k}} \times \int_{S_1} \vec{\epsilon}_3 \times \left(\vec{\epsilon}_1 \sin\alpha + \vec{\epsilon}_2 \cos\alpha\right) e^{-\mathrm{i}\vec{\mathsf{k}}\,\vec{\mathbf{r}}'}$$

$$= \frac{\mathrm{i}E_0 e^{ikr}}{2\pi r}\vec{\mathsf{k}} \times \int_{-a/2}^{a/2} dx' \int_{-b/2}^{b/2} dy' \left(\vec{\epsilon}_2 \sin\alpha - \vec{\epsilon}_1 \cos\alpha\right) e^{-\mathrm{i}\vec{\mathsf{k}}\,\vec{\mathbf{r}}'}$$

$$= \frac{\mathrm{i}E_0 e^{ikr}}{2\pi r}\vec{\mathsf{k}} \times \left(\vec{\epsilon}_2 \sin\alpha - \vec{\epsilon}_1 \cos\alpha\right) \int_{-a/2}^{a/2} dx' e^{-\mathrm{i}\mathsf{k}_x x'} \int_{-b/2}^{b/2} dy' e^{-\mathrm{i}\mathsf{k}_y y'} \qquad [5.86]$$

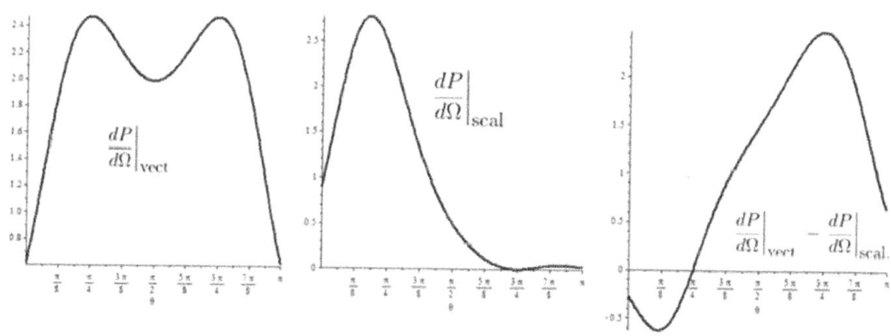

Figure 5.12. *Variations of $\left.\frac{dP}{d\Omega}\right|_{vect.}$ and $\left.\frac{dP}{d\Omega}\right|_{scal.}$ as functions of θ for $\theta_0 = \frac{\pi}{4}$, $\mathsf{k}a = 4\pi$ and $\varphi = 0$*

The integrals are straightforward to compute. They yield:

$$\vec{\mathbf{E}}\left(\vec{\mathbf{r}}\right) = \frac{2\mathrm{i}E_0 e^{ikr}}{\pi r}\left[-\mathsf{k}_z \sin\alpha\,\vec{\epsilon}_1 - \mathsf{k}_z \cos\alpha\,\vec{\epsilon}_2 + \vec{\epsilon}_3\left(\mathsf{k}_x \sin\alpha + \mathsf{k}_y \cos\alpha\right)\right] \times \frac{\sin\left(\mathsf{k}_x a/2\right)\sin\left(\mathsf{k}_y b/2\right)}{\mathsf{k}_x \mathsf{k}_y} \qquad [5.87]$$

Due to the rectangular geometry, this expression is simpler when written in Cartesian components. Expressing $\vec{\mathrm{k}}$ in terms of its spherical components, we get:

$$\vec{\mathbf{E}}\left(\vec{\mathbf{r}}\right) = \frac{2\mathrm{i}E_0 e^{ikr}}{\pi r \mathrm{k}}\left[-\vec{\epsilon}_1 \cos\theta \sin\alpha - \vec{\epsilon}_2 \cos\theta\cos\alpha + \vec{\epsilon}_3 \sin\theta \sin\left(\varphi+\alpha\right)\right] \times \frac{\sin\left(\frac{ka}{2}\sin\theta\cos\varphi\right)}{\sin\theta\cos\varphi}\frac{\sin\left(\frac{kb}{2}\sin\theta\sin\varphi\right)}{\sin\theta\sin\varphi} \tag{5.88}$$

The dissipated power can be expressed as follows:

$$\frac{dP}{d\Omega} = \frac{r^2}{2\mu_0 c}\left|\vec{\mathbf{E}}\right|^2 = \frac{1}{2\mu_0 c}\frac{4\left|E_0\right|^2}{\pi^2 \mathrm{k}^2}\left[\cos\theta^2 + \sin^2\theta\sin^2\left(\varphi+\alpha\right)\right] \times \frac{\sin^2\left(\frac{ka}{2}\sin\theta\cos\varphi\right)}{\sin^2\theta\cos^2\varphi}\frac{\sin^2\left(\frac{kb}{2}\sin\theta\sin\varphi\right)}{\sin^2\theta\sin^2\varphi} \tag{5.89}$$

In terms of the power normally incident on the aperture:

$$P_i = \frac{\left|E_0\right|^2}{2\mu_0 c}ab \tag{5.90}$$

the previous equation can be written as:

$$\frac{dP}{d\Omega} = \frac{P_i}{\pi^2}\left[\cos^2\theta + \sin^2\theta\sin^2\left(\varphi+\alpha\right)\right]\frac{\sin^2\left(\frac{ka}{2}\sin\theta\, cos\varphi\right)}{\frac{ka}{2}\sin^2\theta\cos^2\varphi}\frac{\sin^2\left(\frac{kb}{2}\sin\theta\sin\varphi\right)}{\frac{kb}{2}\sin^2\theta\sin^2\varphi} \tag{5.91}$$

For small apertures, equation [5.91] reduces to:

$$\frac{dP}{d\Omega} = \frac{P_i}{\pi^2}\frac{\mathrm{k}a}{2}\frac{\mathrm{k}b}{2}\left[\cos\theta^2 + \sin^2\theta\sin^2\left(\varphi+\alpha\right)\right] \tag{5.92}$$

Let us now compute the corresponding results within the Kirchhoff scalar approximation. Taking $\psi\left(\vec{\mathbf{r}}\right) \simeq E_0 e^{\mathrm{i}kz'}$, we get $\vec{\mathbf{n}} \cdot \vec{\nabla}'\psi = \vec{\epsilon}_3 \cdot \vec{\nabla}'\psi = \frac{\partial\psi}{\partial z'} = \mathrm{i}\mathrm{k}E_0 e^{\mathrm{i}kz'}$ and $\mathrm{i}\vec{\mathbf{n}} \cdot \vec{\mathrm{k}}'\psi = \mathrm{i}\mathrm{k}\cos\theta E_0 e^{\mathrm{i}kz'}$. Substituting these results into equation [5.61], we find:

$$\begin{aligned}\psi\left(\vec{\mathbf{r}}\right) &= -\frac{e^{ikr}}{4\pi r}\int_{\text{aper.}}\left(\mathrm{i}\mathrm{k}E_0 + \mathrm{i}\mathrm{k}\cos\theta E_0\right)e^{-\mathrm{i}\vec{\mathrm{k}}\cdot\vec{\mathbf{r}}'}dS' \\ &= -\frac{\mathrm{i}E_0 \mathrm{k} e^{ikr}}{4\pi r}\left(1+\cos\theta\right)\int_{-a/2}^{a/2} e^{-\mathrm{i}\mathrm{k}_x x'}dx'\int_{-b/2}^{b/2} e^{-\mathrm{i}\mathrm{k}_y y'}dy' \\ &= -\frac{\mathrm{i}E_0 e^{ikr}}{\pi \mathrm{k} r}\left(1+\cos\theta\right)\frac{\sin\left(\frac{ka}{2}\sin\theta\cos\varphi\right)}{\sin\theta\cos\varphi}\frac{\sin\left(\frac{kb}{2}\sin\theta\sin\varphi\right)}{\sin\theta\sin\varphi}\end{aligned} \tag{5.93}$$

Using the formulas $\frac{dP}{d\Omega} = \frac{r^2}{2\mu_0 c}|\psi|^2$ and $P_i = \frac{|\psi|^2}{2\mu_0 c}ab$, the scalar expression for the diffracted power becomes:

$$\frac{dP}{d\Omega} = \frac{P_i}{\pi^2}\left[\cos^4\frac{\theta}{2}\right]\frac{\sin^2\left(\frac{ka}{2}\sin\theta\cos\varphi\right)}{\frac{ka}{2}\sin^2\theta\cos^2\varphi}\frac{\sin^2\left(\frac{kb}{2}\sin\theta\sin\varphi\right)}{\frac{kb}{2}\sin^2\theta\sin^2\varphi} \qquad [5.94]$$

By comparing [5.94] (obtained using the Kirchhoff scalar approximation) to the vectorial expression [5.91], we see that the only difference lies in the polarization factors within the brackets. To better compare the two expressions, let us take, for example, $b = a$, $\alpha = \frac{\pi}{4}$ and $\varphi = 0$. For these parameters, expressions [5.91] and [5.94] reduce to:

$$\left.\frac{dP}{d\Omega}\right|_{\text{vector}} = \frac{P_i}{\pi^2}\left[\frac{1+\cos^2\theta}{2}\right]\frac{\sin^2\left(\frac{ka}{2}\sin\theta\right)}{\sin^2\theta}$$

$$\left.\frac{dP}{d\Omega}\right|_{\text{scalar}} = \frac{P_i}{\pi^2}\left[\cos^4\frac{\theta}{2}\right]\frac{\sin^2\left(\frac{ka}{2}\sin\theta\right)}{\sin^2\theta} \qquad [5.95]$$

which variations are illustrated in Figure 5.13 for $\mathrm{k}a = 4\pi$, $\mathrm{k}a = \pi$ and $\mathrm{k}a = \frac{\pi}{2}$. We can see that the agreement between the two approximations is perfect in the forward direction.

5.6. Scattering by a conducting sphere in the short-wavelength limit

Another type of problem that essentially falls into the category of diffraction is the scattering of waves by an obstacle. We will now study the scattering of an electromagnetic wave by a perfectly conducting obstacle whose dimensions are large compared to the wavelength. For a thin and flat obstacle, the techniques in section 5.5 can be used, possibly in conjunction with the Babinet principle. However, for other types of obstacles, we rely on the vector expression [5.41] established for scattered fields. If we restrict ourselves to the radiation zone, the integral [5.41] for the scattered field $\vec{\mathbf{E}}_d$ reduces to:

$$\vec{\mathbf{E}}_d = \frac{e^{\mathrm{i}kr}}{4\pi \mathrm{i}r}\int_{S_1}\left[\left(\vec{\mathrm{n}}\cdot\vec{\mathbf{E}}_d\right)\vec{\mathrm{k}} + \left(\vec{\mathrm{n}}\times\vec{\mathbf{E}}_d\right)\times\vec{\mathrm{k}} - \mathrm{k}c\vec{\mathrm{n}}\times\vec{\mathbf{B}}_d\right]e^{-\mathrm{i}\vec{\mathrm{k}}\cdot\vec{\mathrm{r}}'}dS' \qquad [5.96]$$

where $\vec{\mathrm{k}}$ is the wave vector of the scattered wave, and S_1 is the obstacle surface. In the radiation zone, the scattered magnetic field is computed as follows:

$$\vec{\mathbf{B}}_d = \frac{1}{\mathrm{k}c}\vec{\mathrm{k}}\times\vec{\mathbf{E}}_d = \frac{e^{\mathrm{i}kr}}{4\pi \mathrm{i}cr}\vec{\mathrm{k}}\times\int_{S_1}\left[\left(\vec{\mathrm{n}}\times\vec{\mathbf{E}}_d\right)\times\frac{\vec{\mathrm{k}}}{\mathrm{k}} - c\vec{\mathrm{n}}\times\vec{\mathbf{B}}_d\right]e^{-\mathrm{i}\vec{\mathrm{k}}\cdot\vec{\mathrm{r}}'}dS' \qquad [5.97]$$

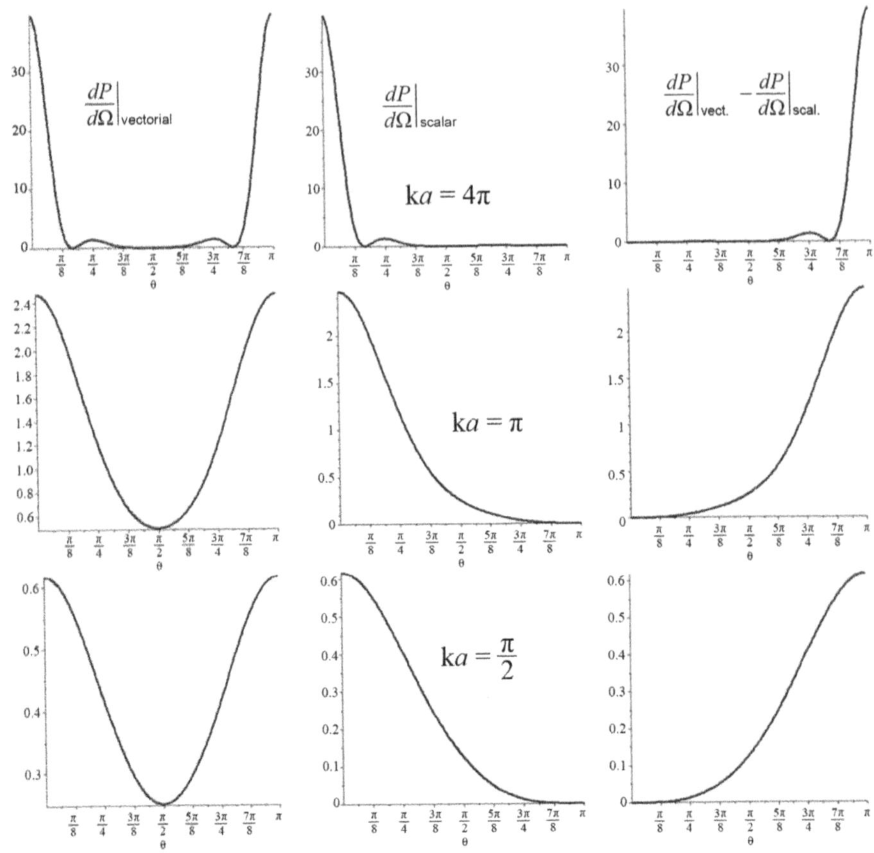

Figure 5.13. *Variations of $\frac{dP}{d\Omega}\big|_{vect.}$ and $\frac{dP}{d\Omega}\big|_{scal.}$ for a rectangular slit when $\alpha = \frac{\pi}{4}$, $\varphi = 0$ and ka $= 4\pi, \pi, \frac{\pi}{2}$*

For now, the fields on the surface of the obstacle are not known. To overcome this problem, we will make some approximations. If the wavelength is small compared to the dimensions of the obstacle, the surface can be roughly divided into an illuminated region and a shadowed region. In the shadowed region, the scattered fields on the surface of the obstacle must be nearly equal and opposite to the incident fields. In the illuminated region, the tangential electric field and the normal magnetic field of the scattered wave must be equal to and opposite to the corresponding incident fields to satisfy the boundary conditions on the surface of a perfectly conducting obstacle. However, the normal electric field and the tangential magnetic field in the illuminated region will be approximately equal to their incident values, as in the case of an infinite flat conducting sheet. This leads to the following approximate values for the scattered fields on the surface of the obstacle:

$$\text{Shadowed region:}\begin{cases}\vec{\mathbf{E}}_d \simeq -\vec{\mathbf{E}}_i \\ \vec{\mathbf{B}}_d \simeq -\vec{\mathbf{B}}_i\end{cases}$$

$$\text{Illuminated region:}\begin{cases}\vec{\mathbf{n}}\times\vec{\mathbf{E}}_d \simeq -\vec{\mathbf{n}}\times\vec{\mathbf{E}}_i \\ \vec{\mathbf{n}}\cdot\vec{\mathbf{B}}_d \simeq -\vec{\mathbf{n}}\cdot\vec{\mathbf{B}}_i \\ \vec{\mathbf{n}}\times\vec{\mathbf{B}}_d \simeq \vec{\mathbf{n}}\times\vec{\mathbf{B}}_i \\ \vec{\mathbf{n}}\cdot\vec{\mathbf{E}}_d \simeq \vec{\mathbf{n}}\cdot\vec{\mathbf{E}}_i\end{cases} \qquad [5.98]$$

where $\vec{\mathbf{E}}_i$ and $\vec{\mathbf{B}}_i$ are the fields of the incident wave. Using these boundary values, the scattered magnetic field given in equation [5.97] can be written as follows:

$$\vec{\mathbf{B}}_d = \frac{e^{\mathrm{i}kr}}{4\pi\mathrm{i}cr}\vec{\mathsf{k}}\times\left(\vec{\mathbf{F}}_{\text{shad.}}+\vec{\mathbf{F}}_{\text{illum.}}\right) \qquad [5.99]$$

where:

$$\vec{\mathbf{F}}_{\text{shad.}} = \int_{\text{shad.}}\left[\frac{\vec{\mathsf{k}}}{\mathsf{k}}\times\left(\vec{\mathbf{n}}\times\vec{\mathbf{E}}_i\right)+c\vec{\mathbf{n}}\times\vec{\mathbf{B}}_i\right]e^{-\mathrm{i}\vec{\mathsf{k}}\cdot\vec{\mathbf{r}}'}dS' \qquad [5.100]$$

is the integral over the shadowed region, and:

$$\vec{\mathbf{F}}_{\text{illum.}} = \int_{\text{illum.}}\left[\frac{\vec{\mathsf{k}}}{\mathsf{k}}\times\left(\vec{\mathbf{n}}\times\vec{\mathbf{E}}_i\right)-c\vec{\mathbf{n}}\times\vec{\mathbf{B}}_i\right]e^{-\mathrm{i}\vec{\mathsf{k}}\cdot\vec{\mathbf{r}}'}dS' \qquad [5.101]$$

is the integral over the illuminated region.

If the incident wave is a plane wave (with wave vector $\vec{\mathsf{k}}_0$):

$$\vec{\mathbf{E}}_i\left(\vec{\mathbf{r}}\right) = \vec{\mathbf{E}}_0 e^{\mathrm{i}\vec{\mathsf{k}}_0\cdot\vec{\mathbf{r}}}$$

$$\vec{\mathbf{B}}_i\left(\vec{\mathbf{r}}\right) = \frac{1}{\mathsf{k}c}\vec{\mathsf{k}}_0\times\vec{\mathbf{E}}_i\left(\vec{\mathbf{r}}\right) \qquad [5.102]$$

the integrals on the shadowed and illuminated regions of the obstacle surface are written as:

$$\begin{aligned}\vec{\mathbf{F}}_{\text{shad.}} &= \frac{1}{\mathsf{k}}\int_{\text{shad.}}\left[\vec{\mathsf{k}}\times\left(\vec{\mathbf{n}}\times\vec{\mathbf{E}}_0\right)+\vec{\mathbf{n}}\times\left(\vec{\mathsf{k}}_0\times\vec{\mathbf{E}}_0\right)\right]e^{\mathrm{i}\left(\vec{\mathsf{k}}_0-\vec{\mathsf{k}}\right)\cdot\vec{\mathbf{r}}'}dS' \\ &= \frac{1}{\mathsf{k}}\int_{\text{shad.}}\left[\left(\vec{\mathsf{k}}+\vec{\mathsf{k}}_0\right)\times\left(\vec{\mathbf{n}}\times\vec{\mathbf{E}}_0\right)+\vec{\mathbf{n}}\times\left(\vec{\mathsf{k}}_0\times\vec{\mathbf{E}}_0\right)-\right. \\ &\qquad \left.\vec{\mathsf{k}}_0\times\left(\vec{\mathbf{n}}\times\vec{\mathbf{E}}_0\right)\right]e^{\mathrm{i}\left(\vec{\mathsf{k}}_0-\vec{\mathsf{k}}\right)\cdot\vec{\mathbf{r}}'}dS' \qquad [5.103] \\ &= \frac{1}{\mathsf{k}}\int_{\text{shad.}}\left[\left(\vec{\mathsf{k}}+\vec{\mathsf{k}}_0\right)\times\left(\vec{\mathbf{n}}\times\vec{\mathbf{E}}_0\right)+\left(\vec{\mathbf{n}}\cdot\vec{\mathbf{E}}_0\right)\vec{\mathsf{k}}_0\right]e^{\mathrm{i}\left(\vec{\mathsf{k}}_0-\vec{\mathsf{k}}\right)\cdot\vec{\mathbf{r}}'}dS'\end{aligned}$$

and:

$$\overrightarrow{\mathbf{F}}_{\text{illum.}} = \frac{1}{\mathsf{k}} \int_{\text{illum.}} \left[\left(\overrightarrow{\mathsf{k}} - \overrightarrow{\mathsf{k}}_0 \right) \times \left(\overrightarrow{\mathbf{n}} \times \overrightarrow{\mathbf{E}}_0 \right) - \left(\overrightarrow{\mathbf{n}} \cdot \overrightarrow{\mathbf{E}}_0 \right) \overrightarrow{\mathsf{k}}_0 \right] e^{\mathrm{i}\left(\overrightarrow{\mathsf{k}}_0 - \overrightarrow{\mathsf{k}}\right)\cdot \overrightarrow{\mathbf{r}}'} dS' \qquad [5.104]$$

because $\overrightarrow{\mathsf{k}}_0 \cdot \overrightarrow{\mathbf{E}}_0 = 0$. These integrals behave very differently depending on the scattering angle. In the short-wavelength limit, both $\overrightarrow{\mathsf{k}} \cdot \overrightarrow{\mathbf{r}}'$ and $\overrightarrow{\mathsf{k}}_0 \cdot \overrightarrow{\mathbf{r}}'$ are large compared to unity. As a result, the exponential factors in equation [5.103] oscillate rapidly and cause the integrals to cancel out, except in the forward direction where $\overrightarrow{\mathsf{k}} \simeq \overrightarrow{\mathsf{k}}_0$. In that direction, the second terms of $\overrightarrow{\mathbf{F}}_{\text{shad.}}$ and $\overrightarrow{\mathbf{F}}_{\text{illum.}}$ have no contribution (in the expression for the field $\overrightarrow{\mathbf{B}}_d$), since the scattered field [5.199] is proportional to $\overrightarrow{\mathbf{F}} \times \overrightarrow{\mathsf{k}}$. Only the first terms in equation [5.103] contribute to the scattered field (at least in the forward direction). We see that $\overrightarrow{\mathbf{F}}_{\text{shad.}}$ and $\overrightarrow{\mathbf{F}}_{\text{illum.}}$ are proportional to $\overrightarrow{\mathsf{k}} \pm \overrightarrow{\mathsf{k}}_0$, respectively. The integral over the shadowed surface grows increasingly greater, and that over the illuminated surface tends toward zero as we approach the forward direction. When the scattering angle deviates from the forward direction, the integral over the shadowed surface quickly vanishes due to the oscillations of the exponential factor. However, the integral over the illuminated surface is small in the forward direction and is expected to remain small at all angles. The integral over the shadowed surface is clearly the contribution from diffraction, while the integral over the illuminated surface corresponds to the reflected wave. To evaluate the integrals in equation [5.103], we must specify the shape of the obstacle. Let us assume that it is a perfectly conducting sphere of radius a. Since the integral on the shadowed surface is significant only in the forward direction, we approximate it by taking $\overrightarrow{\mathsf{k}} \simeq \overrightarrow{\mathsf{k}}_0$ everywhere except in the exponential. By neglecting the second term of equation [5.103] and using the spherical coordinates on the surface of the sphere (shadowed region), we obtain:

$$\begin{aligned}
\overrightarrow{\mathbf{F}}_{\text{shad.}} &\simeq \frac{2}{\mathsf{k}} \int_{\text{shad.}} \overrightarrow{\mathsf{k}}_0 \times \left(\overrightarrow{\mathbf{n}} \times \overrightarrow{\mathbf{E}}_0 \right) e^{\mathrm{i}\left(\overrightarrow{\mathsf{k}}_0 - \overrightarrow{\mathsf{k}}\right)\cdot \overrightarrow{\mathbf{r}}'} dS' \\
&= -\frac{2}{\mathsf{k}} \int_{\text{shad.}} \left(\overrightarrow{\mathsf{k}}_0 \cdot \overrightarrow{\mathbf{n}} \right) \overrightarrow{\mathbf{E}}_0 e^{\mathrm{i}\left(\overrightarrow{\mathsf{k}}_0 - \overrightarrow{\mathsf{k}}\right)\cdot \overrightarrow{\mathbf{r}}'} dS' \\
&= -2\overrightarrow{\mathbf{E}}_0 a^2 \int_0^{\pi/2} d\alpha \int_0^{2\pi} d\beta \sin\alpha \cos\alpha e^{\mathrm{ik}(1-\cos\theta)\cos\alpha - \mathrm{ik}\sin\theta\sin\alpha\cos(\beta-\varphi)} \\
&= -2\overrightarrow{\mathbf{E}}_0 a^2 \int_0^{\pi/2} \sin\alpha \cos\alpha d\alpha e^{\mathrm{ik}(1-\cos\theta)\cos\alpha} \int_0^{2\pi} d\beta e^{-\mathrm{ik}\sin\theta\sin\alpha\cos(\beta-\varphi)}
\end{aligned} \qquad [5.105]$$

where θ, φ and α, β are the spherical angles of $\overrightarrow{\mathsf{k}}$ and $\overrightarrow{\mathbf{n}}$ with respect to $\overrightarrow{\mathsf{k}}_0$, that is, $\overrightarrow{\mathsf{k}}_0 = \mathsf{k}\overrightarrow{\mathbf{k}}$, $\overrightarrow{\mathsf{k}} = \mathsf{k}\left[\sin\theta\left(\cos\varphi\overrightarrow{\mathbf{i}} + \sin\varphi\overrightarrow{\mathbf{j}}\right) + \cos\theta\overrightarrow{\mathbf{k}}\right]$ and

$\overrightarrow{\mathbf{n}} = \sin\alpha\left(\cos\beta\,\overrightarrow{\mathbf{i}} + \sin\beta\,\overrightarrow{\mathbf{j}}\right) + \cos\alpha\,\overrightarrow{\mathbf{k}}$. For small angles θ, the exponent of the first exponential is of the order $\frac{\theta^2}{2}$, while that of the second exponential is of the order of θ. Therefore, the first exponential can be approximated as unity, yielding approximately:

$$\begin{aligned}
\overrightarrow{\mathbf{F}}_{\text{shad.}} &\simeq -2\overrightarrow{\mathbf{E}}_0 a^2 \int_0^{\pi/2} d\alpha \sin\alpha\cos\alpha \int_0^{2\pi} d\beta e^{-\mathrm{i}\mathrm{k}\theta\sin\alpha\cos(\beta-\varphi)} \\
&= -2\overrightarrow{\mathbf{E}}_0 a^2 \int_0^{\pi/2} d\alpha \sin\alpha\cos\alpha \int_0^{2\pi} d\beta e^{-\mathrm{i}\mathrm{k}\theta\sin\alpha\cos\beta} \\
&= -4\pi\overrightarrow{\mathbf{E}}_0 a^2 \int_0^{\pi/2} J_0\left(\mathrm{k}a\theta\sin\alpha\right)\sin\alpha\cos\alpha d\alpha \\
&\equiv -\frac{4\pi\overrightarrow{\mathbf{E}}_0 a^2}{\left(\mathrm{k}a\theta\right)^2}\int_0^{\mathrm{k}a\theta} xJ_0\left(x\right)dx \\
&= -4\pi\overrightarrow{\mathbf{E}}_0 a^2\frac{J_1\left(\mathrm{k}a\theta\right)}{\mathrm{k}a\theta}
\end{aligned} \qquad [5.106]$$

where we used $\int_0^{2\pi} d\beta e^{-\mathrm{i}\mathrm{k}\theta\sin\alpha\cos\beta} = 2\pi J_0\left(\mathrm{k}a\theta\sin\alpha\right)$ and $\int_0^{\mathrm{k}a\theta} xJ_0\left(x\right)dx = \mathrm{k}a\theta J_1\left(\mathrm{k}a\theta\right)$. We see that this is essentially the diffraction field created by a circular aperture [5.69].

The integral of the illuminated region [5.101], which produces the reflected wave, is a little harder to calculate. We must consider arbitrary scattering angles to evaluate this integral, as it disappears as we approach the forward direction (it does not have an essential contribution in this region at small angles θ). Since the function to be integrated in equation [5.101] consists of a vector function of angles with relatively slow variation multiplied by an exponential with rapid variation, the dominant contribution of the integral comes from the integration region where the phase of the exponential is stationary. The phase factor is:

$$f\left(\alpha,\beta\right) = \left(\overrightarrow{\mathrm{k}}_0 - \overrightarrow{\mathrm{k}}\right)\cdot\overrightarrow{\mathbf{r}}' = \mathrm{k}a\left[\left(1-\cos\theta\right)\cos\alpha - \sin\theta\sin\alpha\cos\left(\beta-\varphi\right)\right] \qquad [5.107]$$

It is easy to verify that the stationary point lies at the angles $\alpha_0 = \frac{\pi}{2} + \frac{\theta}{2}$ and $\beta_0 = \varphi$. At this point, the unit vector $\overrightarrow{\mathbf{n}}$ points in the direction of $\overrightarrow{\mathrm{k}} - \overrightarrow{\mathrm{k}}_0$. In fact, we have $\overrightarrow{\mathrm{k}} - \overrightarrow{\mathrm{k}}_0 = 2\mathrm{k}\sin\frac{\theta}{2}\left[\cos\frac{\theta}{2}\left(\cos\varphi\,\overrightarrow{\mathbf{i}} + \sin\varphi\,\overrightarrow{\mathbf{j}}\right) - \sin\frac{\theta}{2}\overrightarrow{\mathbf{k}}\right]$ and $\overrightarrow{\mathbf{n}} = \left[\cos\frac{\theta}{2}\left(\cos\varphi\,\overrightarrow{\mathbf{i}} + \sin\varphi\,\overrightarrow{\mathbf{j}}\right) - \sin\frac{\theta}{2}\overrightarrow{\mathbf{k}}\right]$. This gives $\overrightarrow{\mathrm{k}} - \overrightarrow{\mathrm{k}}_0 = 2\mathrm{k}\sin\frac{\theta}{2}\overrightarrow{\mathbf{n}}$. Expanding the phase factor around $\alpha = \alpha_0$ and $\beta = \beta_0$, we obtain:

$$f\left(\alpha,\beta\right) \simeq -2\mathrm{k}a\sin\frac{\theta}{2}\left[1 - \frac{1}{2}\left(x^2 + \cos^2\frac{\theta}{2}y^2\right) + \cdots\right] \qquad [5.108]$$

where $x = \alpha - \alpha_0$ and $y = \beta - \beta_0$. The integral [5.101] can be approached by evaluating the quantity between square brackets at α_0 and β_0:

$$\vec{\mathbf{F}}_{\text{illum.}} \simeq \frac{e^{-\mathrm{i}2\mathrm{k}a\sin\frac{\theta}{2}}}{\mathrm{k}} a^2 \cos\frac{\theta}{2}\left[2\mathrm{k}\sin\frac{\theta}{2}\vec{\mathbf{n}}_0 \times \left(\vec{\mathbf{n}}_0 \times \vec{\mathbf{E}}_0\right) - \left(\vec{\mathbf{n}}_0 \cdot \vec{\mathbf{E}}_0\right)\vec{\mathrm{k}}_0\right]$$
$$\int e^{\mathrm{i}\mathrm{k}a\sin\frac{\theta}{2}x^2}dx \int e^{\mathrm{i}\mathrm{k}a\sin\frac{\theta}{2}\cos^2\frac{\theta}{2}y^2}dy \qquad [5.109]$$

where $\vec{\mathbf{n}}_0$ is a unit vector in the direction of $\vec{\mathrm{k}} - \vec{\mathrm{k}}_0$. Since θ is not too small, the phase factors oscillate rapidly for large x or y. Consequently, the integration can be extended to $\pm\infty$ in each integral.

Using the result:

$$\int_{-\infty}^{\infty} e^{\mathrm{i}\lambda x^2}dx = \left(\frac{\pi}{\lambda}\right)^{1/2} e^{\mathrm{i}\frac{\pi}{4}} \qquad [5.110]$$

we obtain:

$$\vec{\mathbf{F}}_{\text{illum.}} \simeq \mathrm{i}\frac{\pi a e^{-\mathrm{i}2\mathrm{k}a\sin\frac{\theta}{2}}}{\mathrm{k}^2\sin\frac{\theta}{2}}\left[2\mathrm{k}\sin\frac{\theta}{2}\vec{\mathbf{n}}_0 \times \left(\vec{\mathbf{n}}_0 \times \vec{\mathbf{E}}_0\right) - \left(\vec{\mathbf{n}}_0 \cdot \vec{\mathbf{E}}_0\right)\vec{\mathrm{k}}_0\right]$$
$$= \mathrm{i}\frac{\pi a e^{-\mathrm{i}2\mathrm{k}a\sin\frac{\theta}{2}}}{\mathrm{k}^2\sin\frac{\theta}{2}}\left[2\mathrm{k}\sin\frac{\theta}{2}\left(\vec{\mathbf{n}}_0 \cdot \vec{\mathbf{E}}_0\right)\vec{\mathbf{n}}_0 - 2\mathrm{k}\sin\frac{\theta}{2}\vec{\mathbf{E}}_0 - \left(\vec{\mathbf{n}}_0 \cdot \vec{\mathbf{E}}_0\right)\vec{\mathrm{k}}_0\right]$$
$$= \mathrm{i}\frac{\pi a e^{-\mathrm{i}2\mathrm{k}a\sin\frac{\theta}{2}}}{\mathrm{k}^2\sin\frac{\theta}{2}}\left[\left(\vec{\mathbf{n}}_0 \cdot \vec{\mathbf{E}}_0\right)\left(\vec{\mathrm{k}} - 2\vec{\mathrm{k}}_0\right) - 2\mathrm{k}\sin\frac{\theta}{2}\vec{\mathbf{E}}_0\right] \qquad [5.111]$$

Thus, the scattered magnetic field from the illuminated part of the sphere is:

$$\vec{\mathbf{B}}_d^{\text{illum.}} \simeq \frac{e^{\mathrm{i}\mathrm{k}r}}{r}\frac{a e^{-\mathrm{i}2\mathrm{k}a\sin\frac{\theta}{2}}}{4c\mathrm{k}^2\sin\frac{\theta}{2}}\vec{\mathrm{k}} \times \left[\left(\vec{\mathbf{n}}_0 \cdot \vec{\mathbf{E}}_0\right)\left(\vec{\mathrm{k}} - 2\vec{\mathrm{k}}_0\right) - 2\mathrm{k}\sin\frac{\theta}{2}\vec{\mathbf{E}}_0\right]$$
$$= \frac{e^{\mathrm{i}\mathrm{k}r}}{r}\frac{a e^{-\mathrm{i}2\mathrm{k}a\sin\frac{\theta}{2}}}{4c\mathrm{k}^2\sin\frac{\theta}{2}}\left[2\left(\vec{\mathbf{n}}_0 \cdot \vec{\mathbf{E}}_0\right)\vec{\mathrm{k}}_0 \times \vec{\mathrm{k}} - 2\mathrm{k}\sin\frac{\theta}{2}\vec{\mathrm{k}} \times \vec{\mathbf{E}}_0\right] \qquad [5.112]$$

The scattered electric field $\vec{\mathbf{E}}_d^{\text{illum.}}$ is obtained in turn as follows:

$$\vec{\mathbf{E}}_d^{\text{illum.}} \simeq -\frac{c}{\mathrm{k}}\vec{\mathrm{k}} \times \vec{\mathbf{B}}_d^{\text{illum.}} = -\frac{e^{\mathrm{i}\mathrm{k}r}}{r}\frac{a e^{-\mathrm{i}2\mathrm{k}a\sin\frac{\theta}{2}}}{4\mathrm{k}^3\sin\frac{\theta}{2}}\left[2\left(\vec{\mathbf{n}}_0 \cdot \vec{\mathbf{E}}_0\right)\vec{\mathrm{k}} \times \left(\vec{\mathrm{k}}_0 \times \vec{\mathrm{k}}\right)\right.$$
$$\left. - 2\mathrm{k}\sin\frac{\theta}{2}\vec{\mathrm{k}} \times \left(\vec{\mathrm{k}} \times \vec{\mathbf{E}}_0\right)\right]$$

$$= -\frac{e^{ikr}}{r}\frac{ae^{-i2ka\sin\frac{\theta}{2}}}{2k^2}\left[2\left(\vec{n}_0\cdot\vec{E}_0\right)\vec{k}\times\left(\vec{k}_0\times\vec{n}_0\right)-\vec{k}\times\left(\vec{k}\times\vec{E}_0\right)\right]$$

$$= -\frac{e^{ikr}}{r}\frac{ae^{-i2ka\sin\frac{\theta}{2}}}{2k^2}\left[2\left(\vec{n}_0\cdot\vec{E}_0\right)\left(\vec{k}\cdot\vec{n}_0\right)\vec{k}_0-2\left(\vec{n}_0\cdot\vec{E}_0\right)\left(\vec{k}\cdot\vec{k}_0\right)\vec{n}_0\right.$$
$$\left.-\left(\vec{k}\cdot\vec{E}_0\right)\vec{k}+k^2\vec{E}_0\right]$$

$$= \frac{e^{ikr}}{r}\frac{ae^{-i2ka\sin\frac{\theta}{2}}}{2}\left[2\left(\vec{n}_0\cdot\vec{E}_0\right)\vec{n}_0-\vec{E}_0\right] \quad [5.113]$$

since $\vec{k}-\vec{k}_0 = 2k\sin\frac{\theta}{2}\vec{n}_0$, $\vec{n}_0\cdot\vec{k}_0 = -k\sin\frac{\theta}{2}$, $\vec{n}_0\cdot\vec{k} = k\sin\frac{\theta}{2}$ and $\vec{k}\cdot\vec{k}_0 = k^2\left(1-2\sin^2\frac{\theta}{2}\right)$. If $\vec{E}_0$ makes an angle γ with the Ox axis, then:

$$\vec{n}_0\cdot\vec{E}_0 = E_0\cos\frac{\theta}{2}\cos(\gamma-\varphi) \quad [5.114]$$

which gives:

$$\vec{E}_d^{\text{illum.}} \simeq -\frac{ae^{ikr}}{2r}e^{-i2ka\sin\frac{\theta}{2}}E_0\left[\frac{\vec{E}_0}{E_0}-2\cos\frac{\theta}{2}\cos(\gamma-\varphi)\vec{n}_0\right]$$
$$= -\frac{ae^{ikr}}{2r}e^{-i2ka\sin\frac{\theta}{2}}E_0\vec{\epsilon}_{\text{ilm.}} \quad [5.115]$$

where:

$$\vec{\epsilon}_{\text{illum.}} = \frac{\vec{E}_0}{E_0}-2\cos\frac{\theta}{2}\cos(\gamma-\varphi)\vec{n}_0 \quad [5.116]$$

is the polarization vector of the electric field originating from the illuminated region. It is easy to verify that $|\vec{\epsilon}_{\text{illum.}}| = 1$. We observe that the magnitude of the reflected field is constant, but its phase varies rapidly. The power scattered from the illuminated region, per unit of solid angle, is equal to:

$$\frac{dP_d^{\text{illum.}}}{d\Omega} = \frac{P_i}{4\pi} \quad [5.117]$$

where $P_i = \frac{\pi a^2 E_0^2}{2\mu_0 c}$ is the incident power per unit area multiplied by the projected area πa^2 of the sphere (on the Oxy plane), and consequently:

$$P_d^{\text{illum.}} = P_i \quad [5.118]$$

According to equations [5.106] and [5.99], the magnetic field scattered by the shadowed region is:

$$\vec{\mathbf{B}}_d^{\text{shad.}} \simeq -\frac{a^2 e^{ikr}}{icr}\frac{J_1(ka\theta)}{ka\theta}\vec{\mathsf{k}} \times \vec{\mathbf{E}}_0 \tag{5.119}$$

and we can therefore derive from this that:

$$\begin{aligned}\vec{\mathbf{E}}_d^{\text{shad.}} &= -\frac{c}{\mathsf{k}}\vec{\mathsf{k}} \times \vec{\mathbf{B}}_d^{\text{shad.}} = -i\mathsf{k}a^2\frac{J_1(ka\theta)}{ka\theta}\frac{e^{ikr}}{r}\frac{\vec{\mathsf{k}} \times \left(\vec{\mathsf{k}} \times \vec{\mathbf{E}}_0\right)}{\mathsf{k}^2} \\ &= i\mathsf{k}a^2\frac{J_1(ka\theta)}{ka\theta}\frac{e^{ikr}}{r}\left[\frac{\mathsf{k}^2\vec{\mathbf{E}}_0 - \left(\vec{\mathsf{k}} \cdot \vec{\mathbf{E}}_0\right)\vec{\mathsf{k}}}{\mathsf{k}^2}\right] \\ &= i\mathsf{k}a^2 E_0\frac{J_1(ka\theta)}{ka\theta}\frac{e^{ikr}}{r}\left[\frac{\vec{\mathbf{E}}_0}{E_0} - \sin\theta\cos(\gamma-\varphi)\frac{\vec{\mathsf{k}}}{\mathsf{k}}\right] \\ &= i\mathsf{k}a^2 E_0\frac{J_1(ka\theta)}{ka\theta}\frac{e^{ikr}}{r}\vec{\epsilon}_{\text{shad.}}\end{aligned} \tag{5.120}$$

where:

$$\vec{\epsilon}_{\text{shad.}} = \left[\frac{\vec{\mathbf{E}}_0}{E_0} - \sin\theta\cos(\gamma-\varphi)\frac{\vec{\mathsf{k}}}{\mathsf{k}}\right] \tag{5.121}$$

It can be easily verified that $|\vec{\epsilon}_{\text{shad.}}| \simeq 1$.

The comparison between the two contributions [5.115] and [5.120] shows that in the forward direction the field originating from the shadowed region is larger by a factor of $\mathsf{k}a \gg 1$. However, for much larger angles $\theta \sim \frac{1}{\mathsf{k}a}$, the field $\vec{\mathbf{E}}_d^{\text{shad.}}$ becomes very small, and the isotropic reflected field dominates. The power scattered (from the shadowed region), per unit solid angle, can be expressed as:

$$\frac{dP_d^{\text{shad.}}}{d\Omega} = P_i\frac{(\mathsf{k}a)^2}{4\pi}\left|\frac{2J_1(ka\theta)}{ka\theta}\right|^2\left[1-\sin^2\theta\cos^2(\gamma-\varphi)\right] \simeq \frac{(\mathsf{k}a)^2}{4\pi}\left|\frac{2J_1(ka\theta)}{ka\theta}\right|^2 \tag{5.122}$$

Let us summarize the results obtained: the power dissipated by the entire sphere, per unit solid angle, can now be:

$$\frac{dP_d}{d\Omega} = P_i\begin{cases}\dfrac{(\mathsf{k}a)^2}{4\pi}\left|\dfrac{2J_1(ka\theta)}{ka\theta}\right|^2 & \theta \lesssim \frac{1}{\mathsf{k}a} \\ \dfrac{1}{4\pi} & \theta \gg \frac{1}{\mathsf{k}a}\end{cases} \tag{5.123}$$

At small angles, the scattering resembles simple diffraction. At large angles, the scattering is isotropic. At intermediate angles, the two amplitudes interfere with each other. Due to this, the scattered power presents clear minimal values that are considerably smaller than the isotropic value at certain angles, as shown in Figure 5.14.

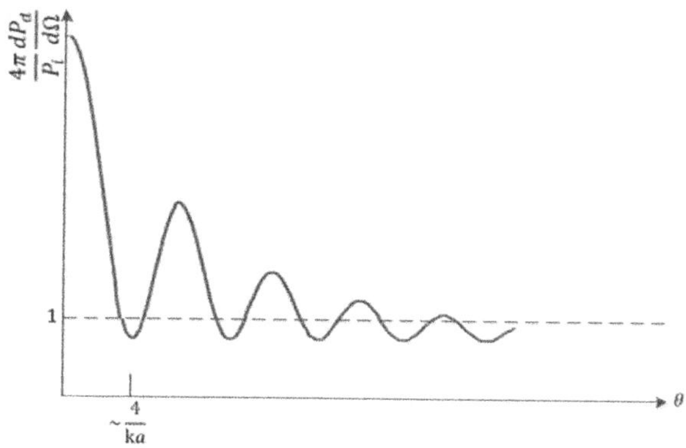

Figure 5.14. *Diffraction pattern of a conducting sphere, showing the forward peak caused by diffraction scattering from the shadowed surface, the contribution of isotropic reflected radiation and the interference radiation (maxima and minima)*

The total scattered power is obtained by performing integration over all angles. Neglecting the interference terms, the total scattered power is the sum of the integrals of the diffraction peak (localized at small angles) and the isotropic reflected part. It is easy to show that both integrals yield the same values. Consequently, we find:

$$P_d = P_i + P_i = 2P_i \qquad [5.124]$$

In fact, the first term P_i comes from direct reflection on the illuminated half of the sphere (πa^2), and the other one P_i comes from diffraction scattering by the shadowed half of the sphere.

Chapter 6

Theory of Special Relativity

6.1. Postulates of special relativity and Lorentz transformation

In this chapter, we will give an overview of Einstein's theory of special relativity. Before introducing the concepts of special relativity, let us first recall some essential features of Newtonian mechanics. Consider two Galilean reference frames, $Oxyz$ and $O'x'y'z'$, in translational motion relative to each other:

– In Newtonian mechanics, time is defined in an absolute manner ($t' = t$), independently of the reference frame. This means that the time interval between two events is an invariant.

– According to Galileo, the transformation, known as the *Galilean transformation*, of coordinates is written as:

$$\begin{aligned} t' &= t \\ \vec{\mathbf{r}}' &= \vec{\mathbf{r}} - \vec{\mathbf{V}}t \end{aligned} \qquad [6.1]$$

In particular, if $\vec{\mathbf{V}}$ is parallel to Oz, then:

$$\begin{aligned} t' &= t \\ x' &= x \\ y' &= y \\ z' &= z - Vt \end{aligned} \qquad [6.2]$$

– In transformation [6.1], distances are preserved, meaning that the length of a moving object does not change.

– Newtonian mechanics obeys Galileo's principle: the Newton law $\vec{\mathbf{F}} = m\vec{\gamma}$ remains invariant under transformation [6.1]. Therefore, this law can be applied to any Galilean reference frame. This principle is valid only for Newtonian forces.

– The interaction between material points is described by means of the potential energy (which is a function of the coordinates only). We see that this mode of description implies the instantaneous nature of the interactions. In fact, the forces exerted between these points of material depend (at every instant) only on their relative positions; the change in position of one is reflected simultaneously on the others.

It is well known that the electromagnetic force is not Newtonian, since it depends on the velocity of the charged particle. It should also be noted that Maxwell's equations do not obey Galileo's principle of relativity (that is, they are not invariant under the Galilean transformation). Therefore, there is an incompatibility between Newtonian mechanics and electromagnetic theory. Since Maxwell's equations are valid in all Galilean reference frames, Newtonian mechanics and Galilean transformation [6.1] must be modified. Other experimental results have confirmed this idea and led to the modification of the concepts of absolute time and distance. For example, if an electron with charge $-e$ and mass m_0 is accelerated by a potential difference U_0, Newtonian mechanics predicts the following:

$$\left(\frac{v}{c}\right)^2 = \frac{2eU_0}{m_0c^2} \qquad [6.3]$$

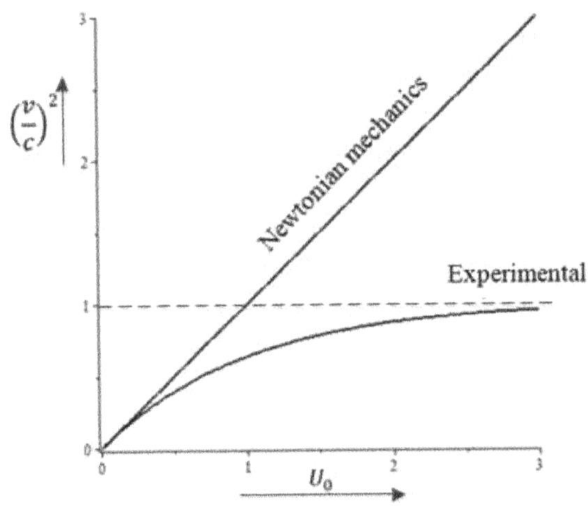

Figure 6.1. *Comparison between the speeds $v_{clas.}$ and $v_{rel.}$ of a particle accelerated by a potential difference U_0*

The experimental results, shown in Figure 6.1, show a profound disagreement with those predicted by the formula [6.3]. According to Figure 6.1, the electron cannot be accelerated to a speed exceeding the speed of light c in a vacuum. Therefore, Newtonian dynamics must be completely revised at high velocities; it only applies if $v \ll c$. Another example is the collision between a proton moving at high speed and a proton at rest. It has been experimentally observed that the collision produces other particles with non-negligible mass (in addition to the initial protons). The *Lavoisier principle* (conservation of matter) must therefore be abandoned. Furthermore, if one performs an energy balance before and after the collision (taking into account all classical forms of energy), an imbalance is also observed. Hence, the notion of energy must be reconsidered. Experiments have also shown that there are no instantaneous interactions in nature. This is why Newtonian mechanics, which is based on the concept of instantaneous propagation of interactions, contains a certain inaccuracy.

Experiments have demonstrated the existence of the *fundamental principle of relativity*, which states that all laws of nature are identical in all Galilean reference frames (*first postulate of special relativity*). In other words, the equations describing these laws must be invariant under a certain specific transformation of coordinates and time. From the principle of relativity, it follows that the propagation speed (denoted here by c) of the interactions is the same in all Galilean reference frames – thus it is a *universal constant*. This is the *second postulate of special relativity*: the speed of light (electromagnetic interaction) is the same in all Galilean reference frames. This speed does not depend on the state of motion of the source. It is evident that the existence of a limiting speed for the transmission of interactions also means that there are no motions in nature that exceed this speed. Indeed, if such a motion existed, it could be used to transport an interaction (at a speed higher than the limiting speed of interaction transmission).

If the speeds of moving bodies are very small compared to c, we can neglect the influence of the velocity limit of interactions on motion. Thus, relativistic mechanics reduces to Newtonian mechanics. Concretely, we pass from relativistic mechanics to Newtonian mechanics by taking $c \longrightarrow \infty$ in the equations of relativistic mechanics.

Time is absolute in Newtonian mechanics. In other words, the properties of time are independent of the reference frame – they are the same in all reference frames. This means that if two events are simultaneous for one observer, then they are also simultaneous for any other (in translational motion relative to the first). More generally, the time elapsed between two events must be the same in all frames of reference. However, it is easy to see that the concept of absolute time contradicts the postulates of relativity. Remember that in Newtonian mechanics, based on the notion of absolute time, the *principle of addition of velocities* plays a fundamental role. If this principle remains true, it must also apply to the propagation of interactions. It follows, then, that the speed of propagation of interactions must differ according to the chosen reference frame; this implies that time is not absolute. Time flows differently in Galilean reference frames. Consequently, the statement that two given events are separated by a given time interval only makes sense once the frame of reference in which the measurement was made is indicated. In particular, two events that are simultaneous in one frame of reference are not necessarily simultaneous in others.

EXAMPLE 6.1. SPHERICAL WAVEFRONT.– Velocity invariance means that a wavefront emitted from a point-like light source remains spherical in all reference frames in uniform relative motion. It is therefore impossible to determine the motion state of the source based on the shape of the wavefront. The goal here is to find a new coordinate transformation based on the geometry imposed by the invariance of c. Let $Oxyz$ and $O'x'y'z'$ be two Galilean reference frames in uniform relative motion. A wavefront emitted at time $t = 0$ by a source fixed at the origin of $Oxyz$ is described at $t > 0$ by an observer attached to the same frame by the sphere $x^2 + y^2 + z^2 = c^2t^2$. An observer in the frame $O'x'y'z'$ (which coincides with $Oxyz$ in $t = 0$ but moves uniformly with respect to $Oxyz$) sees the sphere $x'^2 + y'^2 + z'^2 = c^2t'^2$. If the velocity of $O'x'y'z'$ relative to $Oxyz$ is $\vec{\mathbf{V}} = V\vec{\mathbf{k}}$, the Galilean transformation [6.2] transforms the sphere from $O'x'y'z'$ into:

$$x^2 + y^2 + z^2 - 2zVt + V^2t^2 = c^2t^2 \qquad [6.4]$$

Clearly, the result obtained does not reproduce the sphere of $Oxyz$. The Galilean transformation is not compatible with the invariance of c. To find the relationship that transforms $x'^2 + y'^2 + z'^2 = c^2t'^2$ into $x^2 + y^2 + z^2 = c^2t^2$, let us make the following observations:

1) Since $\vec{\mathbf{V}} = V\vec{\mathbf{k}}$, we must always keep $x' = x$ and $y' = y$.

2) We cannot have $t' = t$ if we want to eliminate the $-2zVt + V^2t^2$ term in [6.4].

Let us try the linear transformation:

$$\begin{aligned} ct' &= a\,(ct + fz) \\ x' &= x \\ y' &= y \\ z' &= b\,(z - Vt) \end{aligned} \qquad [6.5]$$

Thus, we obtain:

$$\begin{aligned} &x'^2 + y'^2 + z'^2 - c^2t'^2 = x^2 + y^2 + \left(b^2 - a^2f^2\right)z^2 \\ &- \left(a^2c^2 - b^2V^2\right)t^2 - \left(b^2V + a^2fc\right)2tz \end{aligned} \qquad [6.6]$$

This equation corresponds to a spherical wave in $Oxyz$ if:

$$\begin{aligned} &b^2 - a^2f^2 = 1 \qquad \text{(i)} \\ &a^2 - b^2\frac{V^2}{c^2} = 1 \qquad \text{(ii)} \\ &b^2V + a^2fc = 0 \qquad \text{(iii)} \end{aligned} \qquad [6.7]$$

From equation [6.7iii], we obtain $f = -\frac{b^2}{a^2}\frac{V}{c}$. Substituting f into [6.7i] and taking into account [6.7ii], we find $\frac{b^2}{a^2}\left(a^2 - b^2\frac{V^2}{c^2}\right) = \frac{b^2}{a^2} = 1$; which gives $b = a = \frac{1}{\sqrt{1-\frac{V^2}{c^2}}}$ and $f = -\frac{V}{c}$. We therefore conclude that $x^2 + y^2 + z^2 = c^2t^2$ follows from $x'^2 + y'^2 + z'^2 = c^2t'^2$ if the following relations are satisfied:

$$\begin{aligned} t' &= \frac{-z\frac{V}{c^2} + t}{\sqrt{1 - \left(\frac{V}{c}\right)^2}} \\ x' &= x \\ y' &= y \\ z' &= \frac{z - Vt}{\sqrt{1 - \left(\frac{V}{c}\right)^2}} \end{aligned} \qquad [6.8]$$

This is the *Lorentz transformation.* ■

Let us try to give a more rigorous demonstration of [6.8]. An event is represented by a point (x, y, z, t) (called a *spacetime point*) in a four-dimensional affine space. This space is called *Minkowski spacetime.* Every material point has a corresponding worldline. Consider two Galilean reference frames $Oxyz$ and $O'x'y'z'$ (in translation with respect to each other). A signal propagating at the speed of light from point (x_1, y_1, z_1, t_1) to point (x_2, y_2, z_2, t_2) must satisfy the relation $c^2(t_2 - t_1)^2 - (x_2 - x_1)^2 - (y_2 - y_1)^2 - (z_2 - z_1)^2 = 0$. If these two events can also be observed in $O'x'y'z'$, we also write $c^2(t'_2 - t'_1)^2 - (x'_2 - x'_1)^2 - (y'_2 - y'_1)^2 - (z'_2 - z'_1)^2 = 0$, where (x'_1, y'_1, z'_1, t'_1) (respectively (x'_2, y'_2, z'_2, t'_2)) corresponds to (x_1, y_1, z_1, t_1) (respectively (x_2, y_2, z_2, t_2)). In general, if (x_1, y_1, z_1, t_1) and (x_2, y_2, z_2, t_2) represent the coordinates of any two events, the quantity:

$$s_{12}^2 = c^2(t_2 - t_1)^2 - (x_2 - x_1)^2 - (y_2 - y_1)^2 - (z_2 - z_1)^2 \qquad [6.9]$$

is called the *interval* between the two events. Thus, it follows from the constancy of the speed of light that if the interval between two events is zero in one Galilean reference frame, it is also zero in all others. If the two events are infinitesimally close, their intervals are written as:

$$\begin{aligned} ds^2 &= c^2dt^2 - dx^2 - dy^2 - dz^2 \quad \text{(reference frame } Oxyz) \\ ds'^2 &= c^2dt'^2 - dx'^2 - dy'^2 - dz'^2 \quad \text{(reference frame } O'x'y'z') \end{aligned} \qquad [6.10]$$

The quantities ds^2 and ds'^2 are infinitesimal and of the same order. Likewise, when $ds^2 = 0$, we must also have $ds'^2 = 0$. These two statements indicate that the two intervals ds^2 and ds'^2 must be proportional, that is:

$$ds^2 = \alpha ds'^2 \qquad [6.11]$$

The proportionality coefficient α depends only on the absolute value of the relative velocity between the two reference frames. It cannot depend on the coordinates or time because different points in space and different moments in time would then not be equivalent; this would contradict the uniformity of space and time. It also cannot depend on the direction of the relative velocity, as this would contradict the isotropy of space. This is the *third postulate of special relativity.* Consider three Galilean reference frames, $Oxyz$, $O_1x_1y_1z_1$, and $O_2x_2y_2z_2$. If $\vec{\mathbf{V}}_1$ (respectively $\vec{\mathbf{V}}_2$) represents the relative velocity of $O_1x_1y_1z_1$ (respectively $O_2x_2y_2z_2$) with respect to $Oxyz$, we write $ds^2 = \alpha\left(\left|\vec{\mathbf{V}}_1\right|\right) ds_1^2$ and $ds^2 = \alpha\left(\left|\vec{\mathbf{V}}_2\right|\right) ds_2^2$, which gives $ds_2^2 = \frac{\alpha\left(\left|\vec{\mathbf{V}}_1\right|\right)}{\alpha\left(\left|\vec{\mathbf{V}}_2\right|\right)} ds_1^2 = \alpha\left(\left|\vec{\mathbf{V}}_1 - \vec{\mathbf{V}}_2\right|\right) ds_1^2$. Therefore, it follows that $\frac{\alpha\left(\left|\vec{\mathbf{V}}_1\right|\right)}{\alpha\left(\left|\vec{\mathbf{V}}_2\right|\right)} = \alpha\left(\left|\vec{\mathbf{V}}_1 - \vec{\mathbf{V}}_2\right|\right)$, where $\left|\vec{\mathbf{V}}_1 - \vec{\mathbf{V}}_2\right|$ depends on the magnitudes of the vectors $\vec{\mathbf{V}}_1$, $\vec{\mathbf{V}}_2$ and the angle between them. However, the left-hand side $\frac{\alpha\left(\left|\vec{\mathbf{V}}_1\right|\right)}{\alpha\left(\left|\vec{\mathbf{V}}_2\right|\right)}$ does not depend on this angle. This equality can only hold if the coefficient $\alpha\left(\left|\vec{\mathbf{V}}\right|\right)$ is a constant (equal to unity according to the same formula). Thus, we obtain $ds^2 = ds'^2$ and, consequently:

$$s^2 = s'^2 \qquad [6.12]$$

The interval between two events is therefore the same in all Galilean reference frames (thus, s^2 is an invariant). According to the sign of s^2, three cases are distinguished:

1) $s^2 > 0$, the interval is of the time type: the two events can have a causal relationship.

2) $s^2 < 0$, the interval is of the space type: the two events are absolutely separated; there can be no causal relationship between the two events, because no signal from the first point can reach the second.

3) $s^2 = 0$: the events can be connected by a signal propagating at the speed of light (electromagnetic interaction).

Consider two Galilean reference frames $Oxyz$ and $O'x'y'z'$. If the coordinates x and y remain unchanged when moving from one frame to the other, equation [6.12] reduces to $c^2t^2 - z^2 = c^2t'^2 - z'^2$. Setting $\sigma = ict$, we obtain the equation $\sigma^2 + z^2 = \sigma'^2 + z'^2$, which expresses a conservation of distance. It follows that σ' and z' must be related to σ and z by a transformation of the form:

$$\begin{cases} \sigma' = \cos\psi\sigma + \sin\psi z \\ z' = -\sin\psi\sigma + \cos\psi z \end{cases} \qquad \begin{cases} \sigma = \cos\psi\sigma' - \sin\psi z' \\ z = \sin\psi\sigma' + \cos\psi z' \end{cases} \qquad [6.13]$$

In terms of the coordinates t and z, the previous transformations can be written as:

$$\begin{cases} ct' = \cosh\phi ct - \sinh\phi z \\ z' = -\sinh\phi ct + \cosh\phi z \end{cases} \qquad \begin{cases} ct = \cosh\phi ct' + \sinh\phi z' \\ z = \sinh\phi ct' + \cosh\phi z' \end{cases} \qquad [6.14]$$

where the change of variable $\mathrm{i}\psi = \phi$ has been made.

To determine the value of ϕ, we need to study the motion of the origin O' of the reference frame $O'x'y'z'$ relative to $Oxyz$. On the one hand, the equation of motion is obtained by setting $z' = 0$ in [6.14], which gives $z = ct\tanh\phi$. On the other hand, mechanics tells us that the trajectory of the origin O' is described by $z = Vt$. It follows that: $\tanh\phi = \frac{V}{c}$, $\cosh\phi = \frac{1}{\sqrt{1-\left(\frac{V}{c}\right)^2}}$, and $\sinh\phi = \frac{\frac{V}{c}}{\sqrt{1-\left(\frac{V}{c}\right)^2}}$.

Thus, the previous transformation equations can be written as:

$$\begin{cases} t' = \dfrac{-z\frac{V}{c^2} + t}{\sqrt{1-\left(\frac{V}{c}\right)^2}} \\ x' = x \\ y' = y \\ z' = \dfrac{z - Vt}{\sqrt{1-\left(\frac{V}{c}\right)^2}} \end{cases} \qquad \begin{cases} t = \dfrac{z'\frac{V}{c^2} + t'}{\sqrt{1-\left(\frac{V}{c}\right)^2}} \\ x = x' \\ y = y' \\ z = \dfrac{z' + Vt'}{\sqrt{1-\left(\frac{V}{c}\right)^2}} \end{cases} \qquad [6.15]$$

This is the *Lorentz transformation*. When $V \ll c$, this transformation reduces to the Galilean transformation. In matrix notation, the Lorentz transformation takes the following form:

$$\begin{pmatrix} ct \\ x \\ y \\ z \end{pmatrix} = \begin{pmatrix} \Gamma & 0 & 0 & \beta\Gamma \\ 0 & 1 & 0 & 0 \\ 0 & 0 & 1 & 0 \\ \beta\Gamma & 0 & 0 & \Gamma \end{pmatrix} \begin{pmatrix} ct' \\ x' \\ y' \\ z' \end{pmatrix} \qquad \begin{pmatrix} ct' \\ x' \\ y' \\ z' \end{pmatrix} = \begin{pmatrix} \Gamma & 0 & 0 & -\beta\Gamma \\ 0 & 1 & 0 & 0 \\ 0 & 0 & 1 & 0 \\ -\beta\Gamma & 0 & 0 & \Gamma \end{pmatrix} \begin{pmatrix} ct \\ x \\ y \\ z \end{pmatrix}$$

[6.16]

where $\beta = \frac{V}{c}$ and $\Gamma = \frac{1}{\sqrt{1-\beta^2}}$.

Transformation [6.15] represents a special case in which the relative motion between $Oxyz$ and $O'x'y'z'$ is parallel to the Oz axis. It is easy to generalize the result for an arbitrary velocity $\vec{\mathbf{V}}$ describing the translation of $O'x'y'z'$ with respect to $Oxyz$.

Transformation [6.15] clearly applies to the components of the coordinates that are parallel and perpendicular to $\vec{\mathbf{V}}$:

$$\begin{aligned} t' &= \frac{1}{\sqrt{1-\left(\frac{V}{c}\right)^2}}\left(t-\frac{\vec{\mathbf{V}}\cdot\vec{\mathbf{r}}}{c^2}\right) \\ \vec{\mathbf{r}}'_{\parallel} &= \frac{1}{\sqrt{1-\left(\frac{V}{c}\right)^2}}\left(\vec{\mathbf{r}}_{\parallel}-\vec{\mathbf{V}}t\right) \\ \vec{\mathbf{r}}'_{\perp} &= \vec{\mathbf{r}}_{\perp} \end{aligned} \qquad [6.17]$$

Noting that $\vec{\mathbf{r}}'_{\parallel} = \frac{(\vec{\mathbf{V}}\cdot\vec{\mathbf{r}})\vec{\mathbf{V}}}{V^2}$ and $\vec{\mathbf{r}}_{\perp} = \vec{\mathbf{r}} - \vec{\mathbf{r}}_{\parallel}$, we can combine equations [6.17] to obtain the general Lorentz transformation:

$$\begin{aligned} t' &= \frac{1}{\sqrt{1-\left(\frac{V}{c}\right)^2}}\left(t-\frac{\vec{\mathbf{V}}\cdot\vec{\mathbf{r}}}{c^2}\right) \\ \vec{\mathbf{r}}' &= \vec{\mathbf{r}} + \left(\frac{1}{\sqrt{1-\left(\frac{V}{c}\right)^2}}-1\right)\frac{\vec{\mathbf{V}}\cdot\vec{\mathbf{r}}}{V^2}\vec{\mathbf{V}} - \frac{1}{\sqrt{1-\left(\frac{V}{c}\right)^2}}\vec{\mathbf{V}}t \end{aligned} \qquad [6.18]$$

Note that [6.18] represents a single Lorentz transformation for a reference frame $O'x'y'z'$ moving at velocity $\vec{\mathbf{V}}$ relative to the reference frame $Oxyz$. In general, successive Lorentz transformations do not commute, except when the velocities (of the reference frames) are parallel.

6.2. FitzGerald–Lorentz length contraction, time dilation and proper time

The rule for the contraction of object dimensions was first introduced by FitzGerald and Lorentz using the invariance properties of Maxwell's equations under the Lorentz transformation. We will now show that length contraction in the direction of motion is a more general phenomenon that applies to any relative motion. Consider a rod of length L_0 at rest and oriented parallel to the $O'z'$ axis in the reference frame $O'x'y'z'$, as shown in Figure 6.2. By definition, $L_0 = z'_2 - z'_1$, where z'_1 and z'_2 are the coordinates of the ends of the rod in $O'x'y'z'$. In the $Oxyz$ reference frame, the length of the rod is again, by definition, $L = z_2 - z_1$, where z_1 and z_2 are the instantaneous coordinates of the ends of the rod in $Oxyz$, observed at the same instant t. According to [6.15], the length in $O'x'y'z'$ is given by $L_0 = z'_2 - z'_1 = \frac{z_2 - z_1}{\sqrt{1-\left(\frac{V}{c}\right)^2}} = \frac{L}{\sqrt{1-\left(\frac{V}{c}\right)^2}}$, which gives:

$$L = \sqrt{1-\left(\frac{V}{c}\right)^2}\, L_0 \qquad [6.19]$$

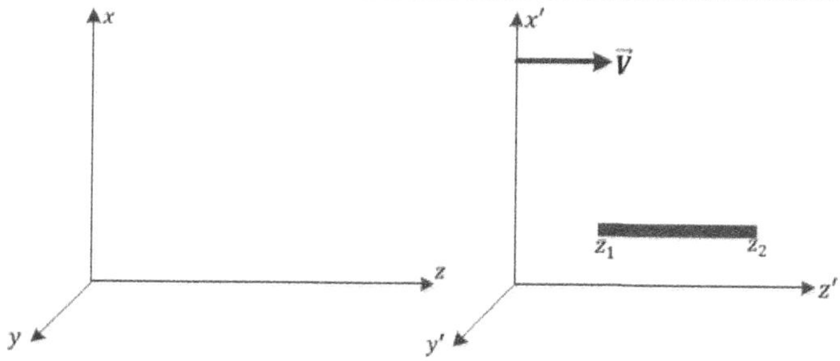

Figure 6.2. *FitzGerald–Lorentz contraction (moving rod)*

This is the *length contraction formula* given by FitzGerald and Lorentz. Note that length is defined at a common time t in reference $Oxyz$. The fact that the measurement does not occur at the same time in $O'x'y'z'$ is irrelevant to the definition of length in $Oxyz$ (simultaneity is simply a relative concept).

Another consequence of the theory of special relativity is *time dilation.* It is observed that a moving clock runs more slowly than an (identical) clock at rest. The most fundamental clocks are unstable elementary particles. Each type of particle decays at rest with a well-defined lifetime (mean lifetime). These particles can be used as clocks. Consider a meson with the lifetime τ_0 at rest in the reference frame $O'x'y'z'$, which moves at a constant velocity V relative to the frame $Oxyz$. Suppose that the meson is created at the origin of $O'x'y'z'$ at time $t' = t = 0$. From the perspective of the frame $Oxyz$, the position of the meson is given by $z = Vt$. If it lives for a time τ_0 in the reference frame $O'x'y'z'$, then at the moment of its decay, we find:

$$t' = \tau_0 = \frac{t - \frac{V}{c^2}z}{\sqrt{1-\left(\frac{V}{c}\right)^2}} = t\sqrt{1-\left(\frac{V}{c}\right)^2} \qquad [6.20]$$

τ is the lifetime observed in the reference frame $Oxyz$. Consequently:

$$\tau = \frac{\tau_0}{\sqrt{1-\left(\frac{V}{c}\right)^2}} \qquad [6.21]$$

Seen from $Oxyz$, the moving meson lives longer than a meson at rest in $O'x'y'z'$.

More generally, consider a clock with arbitrary motion in a Galilean reference frame $Oxyz$. This motion can be considered uniform at all times. Consequently, we can rigidly attach a coordinate system to the clock at each instant (a proper reference frame that will also be Galilean). During the time interval dt, the moving clock covers the distance $\sqrt{dx^2 + dy^2 + dz^2}$. According to the interval invariance, we write $c^2dt^2 - dx^2 - dy^2 - dz^2 = c^2d\tau^2$, where $d\tau$ is the proper time indicated by the clock.

Thus, we obtain:

$$d\tau = \sqrt{1 - \frac{\dot{x}^2 + \dot{y}^2 + \dot{z}^2}{c^2}}dt = \sqrt{1 - \frac{v^2}{c^2}}dt \qquad [6.22]$$

where v is the velocity of the clock relative to the reference frame $Oxyz$. After integration, we find:

$$\tau_2 - \tau_1 = \int_{t_1}^{t_2} \sqrt{1 - \left(\frac{v}{c}\right)^2} dt \qquad [6.23]$$

If v is constant, the preceding formula reduces to [6.21]. The proper time is therefore always shorter than the corresponding time in the fixed reference frame.

6.3. Law of addition of velocities

The Lorentz transformation [6.15] can be used to derive the law of addition of velocities.

Suppose that there exists a particle moving with velocity $\vec{v}'$ in the reference frame $O'x'y'z'$, described by the spherical angles θ' and φ'. We want to determine the components of the velocity $\vec{v}$ as seen from the reference frame $Oxyz$. To obtain the components of the velocity vector in the frame $Oxyz$, we write $\frac{dx}{dt} = \frac{dx'}{dt'}\frac{dt'}{dt}$, $\frac{dy}{dt} = \frac{dy'}{dt'}\frac{dt'}{dt}$, and $\frac{dz}{dt} = \left[1 - \left(\frac{V}{c}\right)^2\right]^{-1/2}\left(\frac{dz'}{dt'} + V\right)\frac{dt'}{dt}$, where $dt = \left[1 - \left(\frac{V}{c}\right)^2\right]^{-1/2}\left(1 + \frac{dz'}{dt'}\frac{V}{c^2}\right)dt'$.

Thus, we obtain:

$$\begin{aligned} v_x &= \frac{\sqrt{1 - \left(\frac{V}{c}\right)^2} v'_x}{1 + \frac{Vv'_z}{c^2}} \\ v_y &= \frac{\sqrt{1 - \left(\frac{V}{c}\right)^2} v'_y}{1 + \frac{Vv'_z}{c^2}} \\ v_z &= \frac{v'_z + V}{1 + \frac{Vv'_z}{c^2}} \end{aligned} \qquad [6.24]$$

The magnitude v and the angles θ, φ in the reference frame $Oxyz$ are easily obtained. Since $\frac{v'_x}{v'_y} = \frac{v_x}{v_y}$, the azimuthal angles are equal, $\varphi' = \varphi$.

Similarly:

$$\begin{aligned} \tan\theta &= \sqrt{1-\left(\frac{V}{c}\right)^2}\,\frac{v'\sin\theta'}{v'\cos\theta'+V} \\ v &= \frac{\sqrt{v'^2+V^2+2v'V\cos\theta'-\left[\frac{v'V\sin\theta'}{c}\right]^2}}{1+\frac{v'V}{c^2}\cos\theta'} \end{aligned} \qquad [6.25]$$

Inverse results expressing (v', θ') as functions of (v, θ') can be obtained by exchanging $v \longleftrightarrow v'$, $\theta \longleftrightarrow \theta'$, and also changing the sign of V.

For an arbitrary translational velocity $\vec{\mathbf{V}}$, we have:

$$\begin{aligned} v_\perp &= \sqrt{1-\left(\frac{V}{c}\right)^2}\,\frac{v'_\perp}{1+\frac{Vv'_\parallel}{c^2}} \\ v_\parallel &= \frac{v'_\parallel+V}{1+\frac{Vv'_\parallel}{c^2}} \end{aligned} \qquad [6.26]$$

In the same way, we have $\vec{\mathbf{v}}'_\parallel = \left(\frac{\vec{\mathbf{v}}'.\vec{\mathbf{V}}}{V^2}\right)\vec{\mathbf{V}}$ and $\vec{\mathbf{v}}'_\perp = \vec{\mathbf{v}}' - \vec{\mathbf{v}}'_\parallel$; this gives:

$$\vec{\mathbf{v}} = \frac{\Gamma^{-1}\vec{\mathbf{v}}' + \left[1+\left(1-\Gamma^{-1}\right)\left(\frac{\vec{\mathbf{v}}'.\vec{\mathbf{V}}}{V^2}\right)\right]\vec{\mathbf{V}}}{1+\frac{\vec{\mathbf{V}}.\vec{\mathbf{v}}'}{c^2}} \qquad [6.27]$$

We clearly see that if the velocities v' and V are very small compared to c, the velocity addition law reduces to that of Galilean relativity: $\vec{\mathbf{v}} = \vec{\mathbf{v}}' + \vec{\mathbf{V}}$. However, if v' or V are close to c, modifications appear. It is impossible to obtain a speed greater than that of light by adding two velocities, even if each is very close to c.

In fact, we have:

$$\begin{aligned} v^2 &= \frac{\left(1-\left(\frac{V}{c}\right)^2\right)v'^2_\perp + \left(v'_\parallel+V\right)^2}{\left(1+\frac{Vv'_\parallel}{c^2}\right)^2} = \frac{v'^2 - \frac{V^2}{c^2}v'^2_\perp + 2Vv'_\parallel + V^2}{\left(1+\frac{Vv'_\parallel}{c^2}\right)^2} \\ &= \frac{c^2\left(1+\frac{Vv'_\parallel}{c^2}\right)^2 - c^2\left(1-\frac{v'^2}{c^2}\right)\left(1-\frac{V^2}{c^2}\right)}{\left(1+\frac{Vv'_\parallel}{c^2}\right)^2} \\ &= c^2\left[1-\frac{\left(1-\frac{v'^2}{c^2}\right)\left(1-\frac{V^2}{c^2}\right)}{\left(1+\frac{Vv'_\parallel}{c^2}\right)^2}\right] \le c^2 \end{aligned} \qquad [6.28]$$

since $v' \leq c$ and $V \leq c$. In particular, if $v' = c$, then $v = c$ as well. We thus see that the speed of light is the same in all Galilean reference frames. This is a mathematical expression of the second postulate of special relativity.

EXAMPLE 6.2. STELLAR ABERRATION.– Bradley was the first to experimentally confirm the Earth's revolution around the Sun by showing that stars undergo the phenomenon of light aberration. A light source always appears slightly closer to the direction in which the observer is moving. Let us study this phenomenon. In the Galilean reference frame $Oxyz$ attached to the Sun, a light ray propagates in the Oxz plane in the direction of negative z and x, making an angle θ with $-Oz$. The speed of light is:

$$\vec{\mathbf{v}} = (-c\sin\theta, 0, -c\cos\theta) \qquad [6.29]$$

In the reference frame $O'x'y'z'$ (attached to the Earth), moving with velocity V relative to $Oxyz$ (in the $+Oz$ direction), the velocity becomes:

$$\vec{\mathbf{v}}' = \left(\frac{-c\sin\theta\sqrt{1-\beta^2}}{1+\beta\cos\theta}, 0, c\frac{-\cos\theta-\beta}{1+\beta\cos\theta}\right) \qquad [6.30]$$

Here, V is the velocity of the Earth and $\beta = \frac{V}{c}$. The angle that the light ray makes with $-O'z'$ in the moving reference frame $O'x'y'z'$ is then given by:

$$\tan\theta' = \frac{\sin\theta}{\cos\theta+\beta}\sqrt{1-\beta^2} \qquad [6.31]$$

The following are two remarks:

1) A ray propagating in $Oxyz$ along $-Ox$ (that is, toward the negative x-direction) makes a non-zero angle δ' with $-O'x'$ in the moving reference frame $O'x'y'z'$. Setting $\theta = \frac{\pi}{2}$, the previous relation leads to $\tan\delta = \tan\left(\frac{\pi}{2}-\theta'\right) = \frac{1}{\tan\theta'} \approx \beta$, and consequently $\delta \approx \beta$, since $\beta \ll 1$. Thus, we see that the light ray is deflected backward by a very small angle $\delta \approx \beta$ relative to $-Ox'$.

2) If now we are traveling aboard a rocket moving at a speed close to that of light, we easily see that $\lim_{\beta\to1} \tan\delta = \lim_{\beta\to1} \frac{\beta}{\sqrt{1-\beta^2}} = \infty$ and, consequently, $\delta = \frac{\pi}{2}$. The ray will bend completely backward. The entire celestial sphere will appear to be coming toward us. It feels as if we are in a tunnel with all of the stars located directly in front of us.

EXAMPLE 6.3. LIGHT EMISSION FROM A MOVING SOURCE.– A source moving with velocity $\vec{\mathbf{V}}$ parallel to Oz with respect to a reference frame $Oxyz$ emits light in a cone with an opening angle θ' in the reference frame $O'x'y'z'$ (attached to the source). Consider the light ray in the Oxz plane (Ox is the vertical axis). The propagation velocity of the ray is $\vec{v}' = (c\sin\theta', 0, c\cos\theta')$, where θ' is the angle measured between the ray and the $O'z'$ axis. In $Oxyz$, the velocity has the components: $\vec{v} = \left(\frac{c\sin\theta'\sqrt{1-\beta^2}}{1+\beta\cos\theta'}, 0, c\frac{\cos\theta'+\beta}{1+\beta\cos\theta'}\right)$. The direction of propagation therefore forms an angle θ with the Oz axis (in the $Oxyz$ reference frame):

$$\tan\theta = \frac{\sin\theta'}{\cos\theta' + \beta}\sqrt{1-\beta^2} \qquad [6.32]$$

In particular, if $\theta' = \frac{\pi}{2}$ (emission into the entire half-space in front of the source in the direction of motion), then we obtain:

$$\tan\theta = \frac{\sqrt{1-\beta^2}}{\beta} \qquad [6.33]$$

We clearly see that $\theta < \theta'$. It is observed that as $\beta \longrightarrow 1$, the angle θ tends toward zero. The light is then concentrated in the direction of the source's motion from the point of view of the observer seeing it move.

EXAMPLE 6.4. FIZEAU EXPERIMENT.– This experiment was first set up to measure the speed $v_{\|}$ of light in a liquid moving at speed V relative to the laboratory frame $Oxyz$. In the $O'x'y'z'$ reference frame of the liquid, the light moves with speed $v_{\|} = \frac{c}{\mathrm{n}}$, where n is the refractive index of the liquid. If we assume Galilean relativity to be true, we should have:

$$v_{\|} = v_{\|} + V = \frac{c}{\mathrm{n}} + V \qquad [6.34]$$

However, by analyzing the experimental results, Fizeau found a different formula:

$$v_{\|} = \frac{c}{\mathrm{n}} + V\left(1 - \frac{1}{\mathrm{n}^2}\right) \qquad [6.35]$$

Special relativity easily explains this difference. Indeed, the speed of light in the laboratory frame is equal to:

$$v_{\|} = \frac{\frac{c}{\mathrm{n}} + V}{1 + \frac{V(c/\mathrm{n})}{c^2}} = \frac{\frac{c}{\mathrm{n}} + V}{1 + \frac{V}{c\mathrm{n}}} \simeq \left(\frac{c}{\mathrm{n}} + V\right)\left(1 - \frac{V}{c\mathrm{n}} + \cdots\right) \simeq \frac{c}{\mathrm{n}} + V - \frac{V}{\mathrm{n}^2} + \cdots \qquad [6.36]$$

where terms in $\frac{1}{c}$, etc., have been neglected.

EXAMPLE 6.5. DOPPLER EFFECT.–

1) Let us start by reasoning within the framework of Galilean relativity and place ourselves in a fixed reference frame in which light propagates at speed c. A point source S emits periodic signals while moving in the direction Oz with velocity V. A receiver is placed at point $M\,(x_M, y_M, 0)$, as shown in Figure 6.3. At time t, the source S emits a first light signal that reaches M at time $t_1 = t + \frac{|\vec{\mathbf{r}}(t) - \vec{\mathbf{r}}_M|}{c}$, where $\vec{\mathbf{r}}(t) = \vec{\mathbf{V}}t$ and $\vec{\mathbf{r}}_M = x_M\,\vec{\mathbf{i}} + y_M\,\vec{\mathbf{j}}$ represent the positions of the source and the receiver relative to O, respectively. The source emits a second light signal at time $t + T_0$, which is received at M at time $t_2 = t + T_0 + \frac{|\vec{\mathbf{r}}(t+T_0) - \vec{\mathbf{r}}_M|}{c}$. The time interval T that separates the reception at M of the two signals is equal to:

$$
\begin{aligned}
T = t_2 - t_1 &= T_0 + \frac{|\vec{\mathbf{r}}\,(t+T_0) - \vec{\mathbf{r}}_M| - |\vec{\mathbf{r}}\,(t) - \vec{\mathbf{r}}_M|}{c} \\
&= T_0 \left(1 + \frac{1}{c}\left[\frac{|\vec{\mathbf{r}}\,(t+T_0) - \vec{\mathbf{r}}_M| - |\vec{\mathbf{r}}\,(t) - \vec{\mathbf{r}}_M|}{T_0}\right]\right)
\end{aligned}
\qquad [6.37]
$$

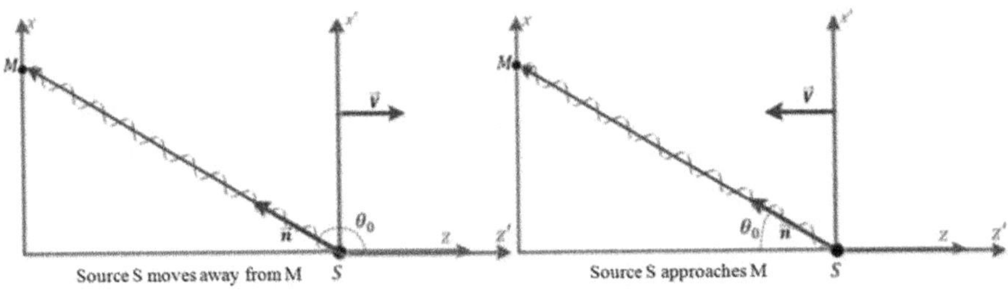

Figure 6.3. *Doppler effect: when the source is moving, the signals are received at a different frequency ν, which is lower (respectively higher) than ν_0 if the source S is moving away from (respectively approaching) the observer. For a color version of this figure, see www.iste.co.uk/abdesselam/electrodynamics3.zip*

If $\vec{\mathbf{r}}\,(t)$ varies little over a period T_0, we can make the approximation:

$$
T \simeq T_0 \left[1 + \frac{1}{c}\left(\frac{\vec{\mathbf{r}}\,(t) - \vec{\mathbf{r}}_M}{|\vec{\mathbf{r}}\,(t) - \vec{\mathbf{r}}_M|}\right)\cdot \vec{\mathbf{V}}\right] = T_0\left[1 - \frac{1}{c}\vec{\mathbf{n}}\cdot\vec{\mathbf{V}}\right] \qquad [6.38]
$$

The unit vector $\vec{\mathbf{n}}$ here points from S to M. By inverting the previous equation, we obtain:

$$
\nu \simeq \frac{\nu_0}{1 - \frac{\vec{\mathbf{n}}\cdot\vec{\mathbf{V}}}{c}} = \frac{\nu_0}{1 - \frac{V}{C}\cos\theta_0} \qquad [6.39]
$$

where θ_0 is the angle measured between $\vec{\mathbf{n}}$ and $\vec{\mathbf{V}}$. Signals emitted by a source S at frequency ν_0 are received at M at a different frequency ν. The frequency ν is lower (respectively higher) than ν_0 if S is moving away from (respectively approaching) M. This phenomenon is well known. It is the *Doppler effect*.

2) Let us examine the problem within the framework of special relativity. The reasoning is almost identical, but with one difference concerning T_0. In fact, T_0 is the proper period of signal emission (measured by an observer attached to S and its reference frame $O'x'y'z'$). For the observer at M, in the reference frame $Oxyz$, the source emits signals spaced by ΓT_0, due to the phenomenon of time dilation. Therefore, it is sufficient to replace T_0 with ΓT_0 in [6.38] to obtain:

$$T \simeq \Gamma T_0 \left(1 - \frac{\vec{\mathbf{n}} \cdot \vec{\mathbf{V}}}{c}\right) = T_0 \frac{1 - \frac{\vec{\mathbf{n}} \cdot \vec{\mathbf{V}}}{c}}{\sqrt{1-\beta^2}} \qquad [6.40]$$

Thus, we obtain:

$$\nu = \nu_0 \frac{\sqrt{1-\beta^2}}{1 - \frac{\vec{\mathbf{n}} \cdot \vec{\mathbf{V}}}{c}} = \nu_0 \frac{\sqrt{1-\beta^2}}{1 - \frac{V}{c}\cos\theta_0} \qquad [6.41]$$

6.4. Four-dimensional Formalism

6.4.1. *Brehme–Lorentz diagram*

We will briefly describe how the coordinates of a point M on a plane Oxy change during a rotation. An *active* rotation can be performed by turning the vector $\overrightarrow{\mathbf{OM}}$ by an angle $\theta \geq 0$ in the trigonometric direction, or a *passive* rotation by turning the axes Ox and Oy by an angle $\theta \geq 0$ in the clockwise direction, as shown in Figure 6.4. In the reference frames Oxy and $Ox'y'$, the vector $\overrightarrow{\mathbf{OM}}$ is written as $\overrightarrow{\mathbf{OM}} = x\,\vec{\mathbf{i}} + y\,\vec{\mathbf{j}} = x'\,\vec{\mathbf{i}}\,' + y'\,\vec{\mathbf{j}}\,'$, or the vectors $\left(\vec{\mathbf{i}}, \vec{\mathbf{j}}\right)$ of the old base are expressed in terms of the vectors $\left(\vec{\mathbf{i}}\,', \vec{\mathbf{j}}\,'\right)$ of the new base as follows: $\vec{\mathbf{i}} = cos\theta\,\vec{\mathbf{i}}\,' + \sin\theta\,\vec{\mathbf{j}}\,'$ and $\vec{\mathbf{j}} = -\sin\theta\,\vec{\mathbf{i}}\,' + \cos\theta\,\vec{\mathbf{j}}\,'$. Consequently, the coordinates (x', y') will be linked to the coordinates (x, y) through the matrix transformation:

$$\begin{pmatrix} x \\ y \end{pmatrix} = \begin{pmatrix} \cos\theta & -\sin\theta \\ \sin\theta & \cos\theta \end{pmatrix} \begin{pmatrix} x' \\ y' \end{pmatrix} = \mathbb{R}(\theta) \begin{pmatrix} x' \\ y' \end{pmatrix} \qquad [6.42]$$

The transformation from the old coordinates (x, y) to the new coordinates (x', y') is thus ensured by the rotation matrix $\mathbb{R}(\theta)$. Now, consider two vectors $\vec{\mathbf{a}}$ and $\vec{\mathbf{b}}$ represented in the reference frame Oxy by the components (a_x, a_y) and (b_x, b_y). The images of the components (a_x, a_y) and (b_x, b_y) in rotation $\mathbb{R}(\theta)$ are, respectively, denoted by (a'_x, a'_y) and (b'_x, b'_y). The dot product between $\vec{a}$ and $\vec{b}$ is equal to:

$$\vec{a} \cdot \vec{b} = a_x b_x + a_y b_y = \begin{pmatrix} b_x & b_y \end{pmatrix} \begin{pmatrix} a_x \\ a_y \end{pmatrix} = [b]^t [a] \qquad [6.43]$$

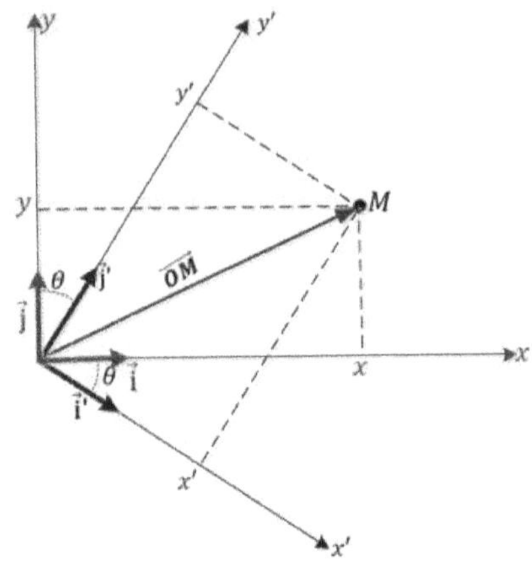

Figure 6.4. *Rotation of the axes of the frame Oxy by an angle θ in the clockwise direction. For a color version of this figure, see www.iste.co.uk/abdesselam/electrodynamics3.zip*

In the new reference frame, we have:

$$[b']^t [a'] = [b]^t \mathbb{R}(\theta)^t \mathbb{R}(\theta) [a] \qquad [6.44]$$

However, the rotation matrix $\mathbb{R}(\theta)$ satisfies the following relation:

$$\mathbb{R}(\theta)^t \mathbb{R}(\theta) = \begin{pmatrix} \cos\theta & \sin\theta \\ -\sin\theta & \cos\theta \end{pmatrix} \begin{pmatrix} \cos\theta & -\sin\theta \\ \sin\theta & \cos\theta \end{pmatrix} = \begin{pmatrix} 1 & 0 \\ 0 & 1 \end{pmatrix} \qquad [6.45]$$

The scalar (dot) product is therefore invariant under rotation:

$$\vec{a} \cdot \vec{b} = a_x b_x + a_y b_y = [b]^t [a] = [b']^t [a'] = a'_x b'_x + a'_y b'_y \qquad [6.46]$$

We find an analogy with the Lorentz transformation.

The Brehme–Lorentz diagram provides a graphical representation of Lorentz transformations, similar to how rotations are depicted. Consider two reference frames, $Oxyz$ and $O'x'y'z'$, in translational motion relative to each other with a velocity V (along the parallel axes Oz and $O'z'$). The coordinates x and y remain unchanged. Imagine two initially coincident planes, $Oz\,(ct)$ and $Oz'\,(ct')$ (note that O here is not the origin of the frame $Oxyz$), then rotate the perpendicular axes Oz' and $O\,(ct)$ by an angle $\frac{\alpha}{2}$ clockwise. Repeat the same operation with the axes Oz and $O\,(ct')$, but rotate them counterclockwise this time, as shown in Figure 6.5(a). The axes Oz and Oz' thus form an angle α between them, as do the axes $O\,(ct)$ and $O\,(ct')$. Note also that the axes Oz' and $O\,(ct)$ are perpendicular to each other. The same applies to the axes Oz and $O\,(ct')$. The coordinates of an event E are obtained by perpendicular projection on each axis:

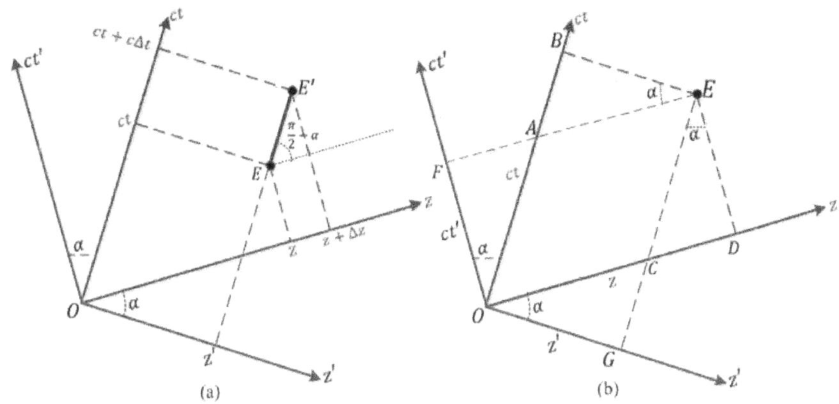

Figure 6.5. *Brehme–Lorentz diagram for passing from one reference frame to another (in translation relative to each other) under Lorentz transformation conditions. For a color version of this figure, see www.iste.co.uk/abdesselam/electrodynamics3.zip*

– Let us begin by relating the angle α to the velocity V of the reference frame $O'x'y'z'$. A traveler stationary along the Oz' axis is sitting at a fixed z' coordinate and observes a time interval $\Delta t'$ elapsing. This spacetime interval corresponds to the segment EE' in Figure 6.5(a). In the original reference frame $Oxyz$, the two events E and E' are separated by a spatial distance Δz and a time duration Δt. For an observer linked to the Oz axis, the traveler is moving at the velocity of the $O'x'y'z'$ frame:

$$V = \frac{\Delta z}{\Delta t} \qquad [6.47]$$

By construction, the segment EE', parallel to the axis $O\,(ct)$, has a length $c\Delta t$ and makes an angle $\frac{\pi}{2} - \alpha$ with the axis Oz. Thus, it is possible to express α as a function of V:

$$\sin\alpha = \cos\left(\frac{\pi}{2} - \alpha\right) = \frac{\Delta z}{c\Delta t} = \frac{V}{c} = \beta \qquad [6.48]$$

– Figure 6.5(b) illustrates the purely geometric way of expressing the coordinates z and ct of point E in terms of the coordinates z' and ct'. Note that the lengths OB

and OD correspond respectively to the coordinates ct and z. They can be decomposed as follows:

$$\begin{aligned} ct &\equiv OB = OA + AB \\ z &\equiv OD = OC + CD \end{aligned} \qquad [6.49]$$

Let us start with OB. Since AOF and AEB are right triangles with $\widehat{AOF} = \widehat{BEA} = \alpha$, then:

$$\begin{aligned} \frac{ct'}{OA} &= \frac{OF}{OA} = \cos\alpha \\ \frac{AB}{z'} &= \frac{AB}{BE} = \tan\alpha \end{aligned} \qquad [6.50]$$

and, consequently:

$$ct = OA + AB = \frac{ct'}{\cos\alpha} + \tan\alpha z' \qquad [6.51]$$

Now let us move on to segment OD. Since GOC and CED are also right-angled triangles, with $\widehat{GOC} = \widehat{CED} = \alpha$, we obtain:

$$z = OC + CD = \frac{z'}{\cos\alpha} + \tan\alpha ct' \qquad [6.52]$$

The transformation from the coordinates (z', ct') to the coordinates (z, ct) is given by:

$$\begin{aligned} ct &= \frac{ct' + \sin\alpha z'}{\cos\alpha} \\ z &= \frac{z' + \sin\alpha ct'}{\cos\alpha} \end{aligned} \qquad [6.53]$$

Now, we have shown that $\sin\alpha = \beta$, which implies that $\cos\alpha = \sqrt{1-\beta^2} = \Gamma^{-1}$. Thus, the coordinate transformation is written as:

$$\begin{aligned} ct &= \Gamma\left(ct' + \beta z'\right) \\ z &= \Gamma\left(z' + \beta ct'\right) \end{aligned} \qquad [6.54]$$

The Brehme–Lorentz diagram is indeed an elegant and simple geometric representation of the Lorentz transformation.

– Let us now verify geometrically that the quantity $(\Delta s)^2 = c^2 (\Delta t)^2 - (\Delta z)^2$ is a relativistic invariant. A spacetime interval EE' is shown in Figure 6.6. It has coordinates $(\Delta z, c\Delta t)$ in $Oxyz$ and $(\Delta z', c\Delta t')$ in $O'x'y'z'$. We can construct two right triangles EAE' and EBE' based on these coordinates. The distances EB and AE' are, respectively, equal to the absolute values $|\Delta z|$ and $|c\Delta t|$. Similarly, the distances EA and BE' are equal to $|\Delta z'|$ and $|c\Delta t'|$. Since the triangles are right triangles, it follows that $EE' = EA^2 + AE'^2 = EB^2 + BE'^2$, which gives $(\Delta z')^2 + (c\Delta t)^2 = (\Delta z)^2 + (c\Delta t')^2$. We have thus obtained:

$$(c\Delta t)^2 - (\Delta z)^2 = (c\Delta t')^2 - (\Delta z')^2 \qquad [6.55]$$

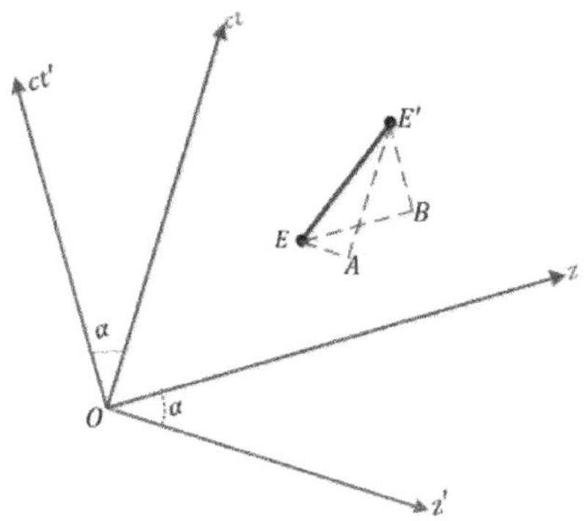

Figure 6.6. *Invariance of the element $(\Delta s)^2$ in the Brehme–Lorentz diagram. For a color version of this figure, see www.iste.co.uk/abdesselam/electrodynamics3.zip*

We thus see that $(\Delta s)^2$ is invariant under a change of Galilean reference frame. It behaves with respect to Lorentz transformations just as the scalar product $\vec{a} \cdot \vec{b}$ behaves with respect to rotations in the plane.

6.4.2. *Transformations and Lorentz group*

In the preceding discussion, we derived the Lorentz transformations from the postulate of the invariance of the speed of light and the principle of relativity. It is also possible to do so in an even more elegant way using group theory. At the foundation lies the Galilean transformation, which can be represented as follows:

$$\begin{pmatrix} t \\ z \end{pmatrix} = T_V \begin{pmatrix} t' \\ z' \end{pmatrix} = \begin{pmatrix} 1 & 0 \\ V & 1 \end{pmatrix} \begin{pmatrix} t' \\ z' \end{pmatrix} \qquad [6.56]$$

It is easy to see that this forms a group whose identity element is T_0. Each element T_V has an inverse given by $T_V^{-1} = T_{-V}$. Moreover, the property $T_V T_W = T_{V+W}$ ensures the associativity and commutativity of the group.

Let us now consider the most general transformation of z and ct that preserves the group structure. The most general transformation could, in principle, involve the coordinates x and y. However, invariance under translations perpendicular to the Oz axis clearly requires that x and y do not appear in the transformation law for z. Similarly, for given z and ct, the transformed time ct' should not depend on x and y. Ultimately, the most general transformation law can be expressed using a 2×2-matrix relation:

$$\begin{pmatrix} ct' \\ z' \end{pmatrix} = \mathcal{L}(V) \begin{pmatrix} ct \\ z \end{pmatrix} = \begin{pmatrix} a(V) & b(V) \\ e(V) & f(V) \end{pmatrix} \begin{pmatrix} ct \\ z \end{pmatrix} \qquad [6.57]$$

We can significantly refine the form of this transformation using a simple symmetry argument. Invert the Oz and $O'z'$ axes of the reference frames $Oxyz$ and $O'x'y'z'$. In other words, we perform the changes $(X = y, Y = x, Z = -z)$ and $(X' = y', Y' = x', Z' = -z')$ (the permutations $x \leftrightarrow y$ and $x' \leftrightarrow y'$ are necessary to maintain the right-handed orientation of the frames $Oxyz$ and $O'x'y'z'$). Then, we attempt to express the new transformation that relates the coordinates (ct, X, Y, Z) to (ct', X', Y', Z'). By symmetry, we observe that $X = X'$ and $Y = Y'$, while the coordinates (ct, Z) and (ct', Z') will be related by a transformation of the same form as [6.57]:

$$\begin{pmatrix} ct \\ Z \end{pmatrix} = \begin{pmatrix} a & b \\ e & f \end{pmatrix} \begin{pmatrix} ct' \\ Z' \end{pmatrix} \qquad [6.58]$$

which can also be written in the form:

$$\begin{pmatrix} ct \\ z \end{pmatrix} = \begin{pmatrix} a & -b \\ -e & f \end{pmatrix} \begin{pmatrix} ct' \\ z' \end{pmatrix} \qquad [6.59]$$

However, this latter relation is also the inverse transform of the transformation we are seeking:

$$\begin{pmatrix} ct \\ z \end{pmatrix} = \frac{1}{af - be} \begin{pmatrix} f & -b \\ -e & a \end{pmatrix} \begin{pmatrix} ct' \\ z' \end{pmatrix} \qquad [6.60]$$

By comparing the two formulas [6.59] and [6.60], we directly deduce that the determinant of the transformation must be equal to 1:

$$af - be = 1 \qquad [6.61]$$

and:

$$a = f \qquad [6.62]$$

We can also use the Abelian property of the group. By writing $\mathcal{L}(U)\mathcal{L}(V) = \mathcal{L}(V)\mathcal{L}(U) = \mathcal{L}(W)$, we find:

$$\begin{aligned} a(W) &= a(U)a(V) + b(U)e(V) = a(U)a(V) + b(V)e(U) \\ b(W) &= a(U)b(V) + a(V)b(U) \\ e(W) &= a(U)e(V) + a(V)e(U) \end{aligned} \qquad [6.63]$$

The first equation can only be satisfied for arbitrary velocities if $\frac{e(U)}{b(U)}$ is a constant, unless one of these functions vanishes identically. If this ratio is constant, a suitable choice of space and time units allows its value to be set to ± 1. We will have to distinguish four cases:

1) $e(V) = -b(V)$: the matrix is therefore antisymmetric and its two coefficients satisfy $a^2 + b^2 = 1$. We can thus set $a = \cos\theta$ and $b = \sin\theta$. The transformation matrix then becomes: $\begin{pmatrix} \cos\theta & \sin\theta \\ -\sin\theta & \cos\theta \end{pmatrix}$. This is a simple rotation about the origin in spacetime. The additive representation of this group is the rotation angle θ.

2) $b(V) = 0$: the value of the determinant, therefore, imposes that $a = 1$, and the transformation matrix takes the form: $\begin{pmatrix} 1 & 0 \\ e(V) & 1 \end{pmatrix}$. Since the motion of O' occurs at speed V (that is, $z' = 0$ implies $z = Vt$), we obtain $e(V) = -\beta = -\frac{V}{c}$. This group is simply the Galilean group, whose additive parameter is the velocity V.

3) Here too, we must have $a = 1$, and the matrix is $\begin{pmatrix} 1 & b(V) \\ 0 & 1 \end{pmatrix}$. This group resembles the Galilean group, but it transforms time rather than space. It is called the *Carroll group*.

4) $e(V) = b(V)$: we then have $a^2 - b^2 = 1$, and we can set $a = \cosh\phi(V)$ and $b = -\sinh\phi(V)$. The transformation matrix $\begin{pmatrix} \cosh\phi & -\sinh\phi \\ -\sinh\phi & \cosh\phi \end{pmatrix}$ is that of a hyperbolic rotation (a rotation about the origin by a purely imaginary angle). The additive representation of this group (called the *Lorentz group*) is the rotation angle ϕ.

We thus see that there are four possible groups, each parameterized by a single additive parameter. To determine the correct group, we must invoke additional arguments. Let us begin by requiring the transformation to preserve the temporal order of events (that is, to respect causality). Now, it is easy to see that the spacetime rotation group and the Carroll group include transformations that change the sign of t. Therefore, these groups must be excluded for simple causality reasons. This leaves us with only the Galilean and Lorentz groups. The former is ruled out because it leads to the ordinary law of the addition of velocities. Hence, the desired transformation must belong to the Lorentz group. All that remains is to determine the value of $\phi(V)$. This can be done by examining the motion of O'. From the transformation, we immediately deduce that $-\sinh\phi ct + \cosh\phi z = 0$, which leads to $\tanh\phi = \frac{V}{c} = \beta$. It follows that $\cosh\phi = \Gamma$ and $\sinh\phi = \beta\Gamma$, and thus we recover transformation [6.15].

Note that in the above, we made a very specific choice of axes. It is also possible to find Lorentz transformations corresponding to different directions of velocity. In that case, the Lorentz transformation is written as $\mathcal{L}\left(\vec{\mathbf{V}}\right)$, where $\vec{\mathbf{V}}$ is the velocity vector of $O'x'y'z'$ relative to $Oxyz$, with an arbitrary direction. To obtain such a transformation, we must compose the transformation we just derived with spatial rotations, spatial reflections, and even (if necessary) time reflections. This yields the full Lorentz group, which describes all possible changes of reference frames. Within the full group, several subgroups can be distinguished:

– The *proper* Lorentz group includes Lorentz transformations combined with spatial rotations. The determinant of the corresponding matrix is equal to 1.

– The *orthochronous* Lorentz group includes the Lorentz group possibly combined with spatial reflections. The determinant of the corresponding matrix can then be ± 1.

EXAMPLE 6.6. REVISITED LAW OF VELOCITY COMPOSITION.– Let us consider three reference frames in relative motion. We can write:

$$\phi(W) = \phi(U) + \phi(V) \qquad [6.64]$$

We can easily derive the law of composition of relativistic velocities by expressing W in terms of U and V. To do this, simply take cosh or sinh from the previous equation. We then obtain:

$$\begin{aligned} \sinh\phi(W) &= \beta(W)\Gamma(W) = \Gamma(U)\Gamma(V)\left[\beta(U)+\beta(V)\right] \\ \cosh\phi(W) &= \Gamma(W) = \Gamma(U)\Gamma(V)\left[1+\beta(U)\beta(V)\right] \end{aligned} \qquad [6.65]$$

which immediately gives:

$$\beta(W) = \frac{\beta(U)+\beta(V)}{1+\beta(U)\beta(V)} \qquad [6.66]$$

and, consequently:

$$W = \frac{U+V}{1+\frac{UV}{c^2}} \qquad [6.67]$$

6.4.3. *Contravariant coordinates*

An event is fully defined in a reference frame by the coordinates (ct, x, y, z). We therefore consider a vector space (the Minkowski spacetime $\mathcal{M}$), which is four-dimensional (hence the term four-vector), endowed with a basis $\{\mathbf{e}_\mu\}\ 0 \leq \mu \leq 3$. We then associate to this event the four-position vector $\mathbf{R} = \sum_\mu x^\mu \mathbf{e}_\mu$, where $x^0 = ct$, $x^1 = x$, $x^2 = y$, and $x^3 = z$. The components x^μ are called the *contravariant components* of the four-vector (with upper indices). To simplify notation, we adopt the *Einstein convention* for the summation of repeated indices. When the same index appears at the top and bottom of an expression, we must sum the expression over all possible values of this index. We therefore simply write $\mathbf{R} = \sum_\mu x^\mu \mathbf{e}_\mu \equiv x^\mu \mathbf{e}_\mu$.

Minkowski space is also equipped with a scalar product. Consider two four-vectors, $\mathbf{R} = x^\mu \mathbf{e}_\mu$ and $\mathbf{Q} = y^\mu \mathbf{e}_\mu$ (associated with two events). The scalar product is defined by:

$$\mathbf{R} \cdot \mathbf{Q} = x^\mu y^\nu \mathbf{e}_\mu \cdot \mathbf{e}_\nu \qquad [6.68]$$

For this definition to be compatible with that of the interval, we must have $\mathbf{R} \cdot \mathbf{R} = c^2t^2 - x^2 - y^2 - z^2$. By setting:

$$g_{\mu\nu} = \mathbf{e}_\mu \cdot \mathbf{e}_\nu \qquad [6.69]$$

the scalar product can also be written as:

$$\mathbf{R} \cdot \mathbf{Q} = x^\mu y^\nu g_{\mu\nu} \qquad [6.70]$$

We thus obtain a scalar product consistent with the definition of the interval by taking:

$$(g_{\mu\nu}) = \begin{pmatrix} 1 & 0 & 0 & 0 \\ 0 & -1 & 0 & 0 \\ 0 & 0 & -1 & 0 \\ 0 & 0 & 0 & -1 \end{pmatrix} \qquad [6.71]$$

This is the *metric tensor*.

6.4.4. *Covariant coordinates*

The components x_μ, defined as follows, are called the *covariant coordinates* of the event:

$$x_\mu = g_{\mu\nu} x^\nu \qquad [6.72]$$

The correspondence rules between the contravariant and covariant components are very simple: $x_0 = x^0$, $x_1 = -x^1$, $x_2 = -x^2$, and $x_3 = -x^3$. Using this notation, the scalar product [6.70] can be written as:

$$\mathbf{R} \cdot \mathbf{Q} = x^\mu y_\mu = x_\mu y^\mu \qquad [6.73]$$

We can also introduce the matrix of $g^{\mu\nu}$ (the inverse of the matrix $g_{\mu\nu}$):

$$(g^{\mu\nu}) = \begin{pmatrix} 1 & 0 & 0 & 0 \\ 0 & -1 & 0 & 0 \\ 0 & 0 & -1 & 0 \\ 0 & 0 & 0 & -1 \end{pmatrix} \quad [6.74]$$

$$g^{\mu\sigma} g_{\sigma\nu} = \delta^{\mu}_{\nu}$$

where δ^{μ}_{ν} are the *Kronecker delta* (1 if the two indices are equal; 0 otherwise). Using this new matrix, we can show that:

$$g^{\mu\nu} x_{\nu} = g^{\mu\nu} g_{\nu\sigma} x^{\sigma} = \delta^{\mu}_{\sigma} x^{\sigma} = x^{\mu} \quad [6.75]$$

The metric $g_{\mu\nu}$ is thus used to lower the indices, while $g^{\mu\nu}$ is used to raise them.

It is always possible to associate with any four-vector $\mathbf{R}$ a linear form $\widetilde{\mathbf{R}}$ in the following way: $\widetilde{\mathbf{R}}(\mathbf{Q}) = \mathbf{R} \cdot \mathbf{Q}$. Thus, we have $\widetilde{\mathbf{R}}(\mathbf{P} + \mathbf{Q}) = \mathbf{R} \cdot \mathbf{P} + \mathbf{R} \cdot \mathbf{Q}$, $\left(\widetilde{\mathbf{P}} + \widetilde{\mathbf{R}}\right)(\mathbf{Q}) = \mathbf{P} \cdot \mathbf{Q} + \mathbf{R} \cdot \mathbf{Q}$, and $\left(a\widetilde{\mathbf{R}}\right)(\mathbf{Q}) = a(\mathbf{R} \cdot \mathbf{Q})$. All of these relations endow the set of linear forms with a vector space structure; this is the dual space $\mathcal{M}^{\star}$ of Minkowski space $\mathcal{M}$ ($\mathcal{M}^{\star}$ and $\mathcal{M}$ have the same dimension, namely 4). In the space $\mathcal{M}^{\star}$, we can choose a basis. Due to the isomorphism with $\mathcal{M}$, it is natural to choose as a basis the linear forms associated with the elements of the basis of $\mathcal{M}$. Thus, we define $\widetilde{\mathbf{e}}^{\mu}(\mathbf{e}_{\nu}) = \delta^{\mu}_{\nu}$. Hence, from a four-vector $\mathbf{R} = x^{\mu}\mathbf{e}_{\mu}$, we can construct the linear form $x_{\mu}\widetilde{\mathbf{e}}^{\mu}$. The action of this form on the four-vector $\mathbf{Q} = u^{\mu}\mathbf{e}_{\mu}$ is written as $(x_{\mu}\widetilde{\mathbf{e}}^{\mu})(y^{\nu}\mathbf{e}_{\nu}) = x_{\nu}y^{\nu} = \mathbf{R} \cdot \mathbf{Q}$. If the contravariant components are the components of the four-vector, then the covariant components are those of the linear form associated with that four-vector on the dual basis. Due to the isomorphism between $\mathcal{M}$ and $\mathcal{M}^{\star}$ (mentioned above), we will use the notation $\overrightarrow{\mathbf{e}}^{\mu}$ instead of $\widetilde{\mathbf{e}}^{\mu}$. In fact, we have $\widetilde{\mathbf{e}}^{\mu}(\mathbf{e}_{\nu}) = \delta^{\mu}_{\nu} = g^{\mu\sigma} g_{\sigma\nu} = (g^{\mu\sigma}\mathbf{e}_{\sigma}) \cdot \mathbf{e}_{\nu} = (g^{\mu\sigma}\mathbf{e}_{\sigma})(\mathbf{e}_{\nu})$. Therefore, it is possible to take $\widetilde{\mathbf{e}}^{\mu} = g^{\mu\sigma}\mathbf{e}_{\sigma} = \mathbf{e}^{\mu}$.

6.4.5. *Change of basis*

We have already seen that a change of reference frame transforms the contravariant coordinates of an event via the Lorentz transformation:

1) For contravariant coordinates, the Lorentz transformation is written as:

$$x'^{\mu} = \mathcal{L}^{\mu}{}_{\nu} x^{\nu} \quad [6.76]$$

where the x^{ν} (respectively, x'^{ν}) are the contravariant components of the event in the frame $Oxyz$ (respectively, $O'x'y'z'$). In the case of transformation [6.15], we simply have:

$$(\mathcal{L}^\mu{}_\nu) = \begin{pmatrix} \Gamma & 0 & 0 & -\beta\Gamma \\ 0 & 1 & 0 & 0 \\ 0 & 0 & 1 & 0 \\ -\beta\Gamma & 0 & 0 & \Gamma \end{pmatrix} \qquad [6.77]$$

The inverse transformation is also given by:

$$x^\mu = \left(\mathcal{L}^{-1}\right)^\mu{}_\nu x'^\nu \qquad [6.78]$$

The inverse matrix $\mathcal{L}^{-1}$ is obtained simply by changing the sign of β.

2) Let us examine the transformation law for covariant coordinates (in the dual space). To do this, we write that:

$$x'_\mu = \mathcal{L}_\mu{}^\nu x_\nu \qquad [6.79]$$

We can find the relation between $\mathcal{L}^\mu{}_\nu$ and $\mathcal{L}_\mu{}^\nu$ from the invariance of the scalar product. In fact, we have $x'^\mu y'_\mu = x^\nu y_\nu$, with $x'^\mu = \mathcal{L}^\mu{}_\nu x^\nu$ and $y'_\mu = \mathcal{L}_\mu{}^\nu y_\nu$. We then deduce $\mathcal{L}^\mu{}_\rho x^\rho \mathcal{L}_\mu{}^\sigma y_\sigma = x^\nu y_\nu = x^\rho y_\sigma \delta^\sigma_\rho$. This last relation must hold for any four-vectors $\mathbf{R}$ and $\mathbf{Q}$. Therefore, we must have:

$$\mathcal{L}^\mu{}_\rho \mathcal{L}_\mu{}^\sigma = \delta^\sigma_\rho \qquad [6.80]$$

Since equation [6.77] is symmetric, the transformation matrix for covariant coordinates is simply the inverse of the transformation matrix for contravariant coordinates.

3) Let us introduce the metric tensor into the expression of the scalar product: $\mathbf{R} \cdot \mathbf{Q} = x^\rho g_{\rho\sigma} y^\sigma = x'^\mu g_{\mu\nu} y'^\nu = \mathcal{L}^\mu{}_\rho x^\rho g_{\mu\nu} \mathcal{L}^\nu{}_\sigma y^\sigma$. Since this relation must hold for all four-vectors $\mathbf{R}$ and $\mathbf{Q}$, we find:

$$g_{\rho\sigma} = \mathcal{L}^\mu{}_\rho g_{\mu\nu} \mathcal{L}^\nu{}_\sigma \qquad [6.81]$$

From this expression, we see that the determinant of the matrix representing the Lorentz transformation is ± 1 (recall that it is negative only if the transformation involves a space or time reflection).

4) Multiplying formula [6.81] by $g^{\tau\rho}$ and performing the summations, we obtain:

$$g^{\tau\rho} g_{\rho\sigma} = \delta^\tau_\sigma = g^{\tau\rho} \mathcal{L}^\mu{}_\rho g_{\mu\nu} \mathcal{L}^\nu{}_\sigma = \left(g^{\tau\rho} g_{\mu\nu} \mathcal{L}^\mu{}_\rho\right) \mathcal{L}^\nu{}_\sigma \qquad [6.82]$$

By comparing [6.82] with [6.80], it comes that:

$$\mathcal{L}_\nu{}^\tau = g_{\nu\mu} \mathcal{L}^\mu{}_\rho g^{\rho\tau} \qquad [6.83]$$

and conversely:

$$\mathcal{L}^\nu{}_\tau = g^{\nu\mu} \mathcal{L}_\mu{}^\rho g_{\rho\tau} \qquad [6.84]$$

5) It is also possible to introduce the quantities:

$$\begin{aligned} \mathcal{L}^{\mu\tau} &= \mathcal{L}^{\mu}{}_{\rho} g^{\rho\tau} \\ \mathcal{L}_{\mu\sigma} &= g_{\mu\nu} \mathcal{L}^{\nu}{}_{\sigma} \end{aligned} \qquad [6.85]$$

describing the simultaneous change from one reference frame to another and from one type of coordinates to another. Indeed, it is easy to verify that:

$$\begin{aligned} x'^{\mu} &= \mathcal{L}^{\mu}{}_{\rho} x^{\rho} = \mathcal{L}^{\mu}{}_{\rho} g^{\rho\nu} x_{\nu} = \mathcal{L}^{\mu\nu} x_{\nu} \\ x'_{\mu} &= \mathcal{L}_{\mu}{}^{\rho} x_{\rho} = \mathcal{L}_{\mu}{}^{\rho} g_{\rho\nu} x^{\nu} = \mathcal{L}_{\mu\nu} x^{\nu} \end{aligned} \qquad [6.86]$$

6) We can use relation [6.80] to express the inverse change of reference frame without involving the transformation $\mathcal{L}^{-1}$:

$$\mathcal{L}^{\mu}{}_{\rho} x'_{\mu} = \mathcal{L}^{\mu}{}_{\rho} \mathcal{L}_{\mu}{}^{\nu} x_{\nu} = \delta^{\nu}_{\rho} x_{\nu} = x_{\rho} \qquad [6.87]$$

7) Let us determine the transformation laws for the elements of the bases of $\mathcal{M}$ and $\mathcal{M}^{\star}$. Noting that $x_{\nu} = \mathbf{R} \cdot \mathbf{e}_{\nu}$ and $x'_{\mu} = \mathbf{R} \cdot \mathbf{e}'_{\mu}$, we have $\mathbf{R} \cdot \mathbf{e}'_{\mu} = \mathcal{L}_{\mu}{}^{\nu} \mathbf{R} \cdot \mathbf{e}_{\nu}$. Since this relation must hold for all $\mathbf{R}$, we deduce that:

$$\mathbf{e}'_{\mu} = \mathcal{L}_{\mu}{}^{\nu} \mathbf{e}_{\nu} \qquad [6.88]$$

The elements of the basis of $\mathcal{M}$ transform like covariant components. By similarly writing $x'^{\mu} = \mathbf{e}'^{\mu} \cdot \mathbf{R}$ and $x^{\nu} = \mathbf{e}^{\nu} \cdot \mathbf{R}$, we find $\mathbf{e}'^{\mu} \cdot \mathbf{R} = \mathcal{L}^{\mu}{}_{\nu} \mathbf{e}^{\nu} \cdot \mathbf{R}$. This implies that:

$$\mathbf{e}'^{\mu} = \mathcal{L}^{\mu}{}_{\nu} \mathbf{e}^{\nu} \qquad [6.89]$$

The elements of the dual basis transform like contravariant components.

6.4.6. *Four-velocity, four-momentum and four-acceleration*

Let us generalize the notion of four-vector to other physical quantities. A *four-vector* is defined as any quantity expressed by four coordinates (covariant or contravariant) that transforms like the position four-vector under a change in the reference frame (that is, under a Lorentz transformation). The contravariant components of any four-vector $\mathbf{A}$ can be written in terms of three spatial components that form a vector and one temporal component. We use the following notation: $\mathbf{A} = \left(a^0, \vec{a}\right)$. The covariant components are written as $\left(a^0, -\vec{a}\right)$. All previously introduced concepts (scalar product, basis change rules, etc.) apply to four-vectors, regardless of their physical nature.

Let us also define the notion of a *four-scalar* or simply *scalar*: it is a quantity that is independent of the reference frame (such as the rest mass, the charge of a particle, the scalar product of two four-vectors, etc.). The multiplication of a four-vector by a scalar yields another four-vector.

Here are a few examples:

1) *Four-velocity*: to define the velocity of a relativistic particle, we must introduce a time parameter to follow the particle along its worldline. This parameter must be a scalar. The velocity can thus be defined as the ratio of the change in the four-vector position (over a small time interval) to the change in this time parameter. The result is, of course, a four-vector. The absolute time of a reference frame is not suitable for parameterizing the trajectory. However, the proper time τ of the moving particle is independent of the reference frame and therefore makes it possible to define a four-velocity as:

$$\mathbf{U} = \frac{d\mathbf{R}}{d\tau} \qquad [6.90]$$

Let us express this quantity in terms of the spatial velocity $\vec{\mathbf{v}}$ of the particle in a given reference frame $Oxyz$. If the proper time changes by $d\tau$, the coordinate time in $Oxyz$ changes by $dt = \gamma dt$ (time dilation). The position of the particle changes by $d\vec{\mathbf{r}} = \vec{\mathbf{v}}dt$. We then immediately have $\frac{d\mathbf{R}}{dt} = (c, \vec{\mathbf{v}})$, and consequently:

$$\mathbf{U} = (c\gamma, \gamma\vec{\mathbf{v}}) \qquad [6.91]$$

Note that the square of the norm of $\mathbf{U}$, $\mathbf{U}^{\mu}\mathbf{U}_{\mu} = c^2\gamma^2 - \gamma\vec{\mathbf{v}}^2 = c^2$, is nothing other than the square of the speed of light.

2) *Four-momentum*: by multiplying the four-velocity by the particle's rest mass m_0, we obtain the energy-momentum four-vector or the four-momentum:

$$\mathbf{P} = m_0\mathbf{U} = (p^0, \vec{\mathbf{p}}) = (m_0 c\gamma, m_0\gamma\vec{\mathbf{v}}) \qquad [6.92]$$

We will return to this four-vector in detail later.

3) *Four-acceleration*: the four-acceleration (the acceleration four-vector) is defined as:

$$\mathbf{\Gamma} = \frac{d\mathbf{U}}{d\tau} = \gamma\frac{d\mathbf{U}}{dt} = (c\gamma\gamma', \gamma\gamma'\vec{\mathbf{v}} + \gamma^2\vec{\mathbf{a}}) \qquad [6.93]$$

where $\gamma' = \frac{d\gamma}{dt}$ is the derivative of the Lorentz factor γ, and $\vec{\mathbf{a}}$ is the three-dimensional spatial acceleration of the particle. Since $\gamma' = \frac{\gamma^3}{c^2}\vec{\mathbf{v}} \cdot \vec{\mathbf{a}}$, we have the following:

$$\mathbf{\Gamma} = \left(\frac{\gamma^4}{c}\vec{\mathbf{v}} \cdot \vec{\mathbf{a}}, \frac{\gamma^4}{c^2}\left(\vec{\mathbf{v}} \cdot \vec{\mathbf{a}}\right)\vec{\mathbf{v}} + \gamma^2\vec{\mathbf{a}}\right) \qquad [6.94]$$

It is easy to verify that the four-acceleration is orthogonal to the four-velocity:

$$\mathbf{\Gamma}^{\mu}\mathbf{U}_{\mu} = 0 \qquad [6.95]$$

4) *Four-current density*: we will now construct a four-vector from the charge density ρ and the current density $\overrightarrow{\mathbf{J}}$. To that end, consider a charge q modeled as a charge density ρ contained within a small volume element dV (defined in the reference frame $Oxyz$). Thus, we write $q = \rho dV$. The particle moves with velocity $\overrightarrow{\mathbf{v}}$ in $Oxyz$. The current density is given by $\overrightarrow{\mathbf{J}} = \rho\overrightarrow{\mathbf{v}}$ inside the volume (it is zero outside). In another reference frame $O'x'y'z'$, the quantities ρ and dV transform into ρ' and dV'. Only the total charge q, which is an intrinsic property of the particle, is a four-scalar. Thus, we have $\rho dV = \rho' dV'$. Now, consider a time interval dt in $Oxyz$. During this time, the particle moves by $dx^\mu = \left(cdt, d\overrightarrow{\mathbf{r}}\right)$, which is clearly a four-vector. The quantity $\rho dV dx^\mu = \rho dV dt \frac{dx^\mu}{dt} = \rho \frac{d\Omega}{c}\frac{dx^\mu}{dt}$ is therefore also a four-vector. Now, $d\Omega$ is a four-scalar. Indeed, we have:

$$d\Omega' = \left\| \frac{\partial\left(ct, x, y, z\right)}{\partial\left(ct', x', y', z'\right)} \right\| d\Omega = \left|\det\left(\mathcal{L}\right)\right| d\Omega = d\Omega \qquad [6.96]$$

and, consequently, we deduce that:

$$J^\mu = \rho\frac{dx^\mu}{dt} \equiv \left(\rho c, \overrightarrow{\mathbf{j}}\right) \qquad [6.97]$$

is a four-vector that combines the charge and current densities produced by a point charge. This reasoning can be extended to larger distributions (composed of multiple charges).

5) *Wave four-vector*: consider a plane wave of frequency ω and wave vector $\overrightarrow{\mathrm{k}}$, represented by $\psi\left(t, \overrightarrow{\mathbf{r}}\right) = A\cos\left(\omega t - \overrightarrow{\mathrm{k}}\cdot\overrightarrow{\mathbf{r}}\right)$. Here are the essential points:

a) The wavefront is the set of points where the phase is constant at a given instant. After a displacement $\Delta\overrightarrow{\mathbf{r}}$ (at a fixed time t), the phase becomes $\omega t - \overrightarrow{\mathrm{k}}\cdot\overrightarrow{\mathbf{r}} - \overrightarrow{\mathrm{k}}\cdot\Delta\overrightarrow{\mathbf{r}}$. It remains constant, at a given instant, if $\overrightarrow{\mathrm{k}}\cdot\Delta\overrightarrow{\mathbf{r}} = 0$. In other words, we must move perpendicularly to the wave vector to keep the phase constant. The wavefront is therefore perpendicular to $\overrightarrow{\mathrm{k}}$.

b) To follow a wave crest, the phase must remain constant. The displacement must satisfy $\omega dt - \overrightarrow{\mathrm{k}}\cdot d\overrightarrow{\mathbf{r}} = 0$, which gives $\omega - \overrightarrow{\mathrm{k}}\cdot\frac{d\overrightarrow{\mathbf{r}}}{dt} = 0$. If the displacement is in the direction of the wave vector, then the phase velocity is $c = \left|\frac{d\overrightarrow{\mathbf{r}}}{dt}\right| = \frac{\omega}{\mathrm{k}}$. The wave thus propagates in the direction of $\overrightarrow{\mathrm{k}}$, that is, perpendicular to the wavefront.

c) Two plane waves can interfere with each other: it is the phase difference that determines the illumination at a given point. If one observer sees destructive interference at a point, another observer moving relative to them will observe the same thing. This can easily be explained if the phase is Lorentz-invariant. The phase remains invariant under a change of reference frame if it can be written as the scalar product of two four-vectors. One is the four-vector position $\left(ct, \overrightarrow{\mathbf{r}}\right) = \left(ct, x, y, z\right)$, and the other is the wave four-vector $\left(\frac{\omega}{c}, \overrightarrow{\mathsf{k}}\right) = \left(\frac{\omega}{c}, \mathsf{k}_x, \mathsf{k}_y, \mathsf{k}_z\right)$. The transformation of these components follows the usual rules for any four-vector:

$$\begin{pmatrix} \frac{\omega'}{c} \\ \mathsf{k}'_x \\ \mathsf{k}'_y \\ \mathsf{k}'_z \end{pmatrix} = \begin{pmatrix} \Gamma & 0 & 0 & -\beta\Gamma \\ 0 & 1 & 0 & 0 \\ 0 & 0 & 1 & 0 \\ -\beta\Gamma & 0 & 0 & \Gamma \end{pmatrix} \begin{pmatrix} \frac{\omega}{c} \\ \mathsf{k}_x \\ \mathsf{k}_y \\ \mathsf{k}_z \end{pmatrix} \qquad [6.98]$$

The norm of the wave four-vector is zero, $\mathsf{k}_\mu \mathsf{k}^\mu = \frac{\omega^2}{c^2} - \overrightarrow{\mathsf{k}}^2 = 0$. This implies that $\omega = c\mathsf{k}$ in all reference frames.

EXAMPLE 6.7. REVISITING THE DOPPLER EFFECT.– A light source is moving at a constant velocity $\overrightarrow{\mathbf{V}} = V\overrightarrow{\mathbf{k}}$ with respect to a fixed reference frame $Oxyz$. If the source emits (in its own rest frame $O'x'y'z'$) a wave of frequency ω_0 (the proper frequency) in the direction $\overrightarrow{\mathsf{k}}_0$ (the proper wave vector), forming an angle θ_0 with the Oz'-axis, then:

$$\begin{aligned} \omega &= \omega_0 \frac{1 + \frac{V}{c}\cos\theta_0}{\sqrt{1-\beta^2}} \\ \cos\theta &= \omega_0 \frac{\cos\theta_0 + \frac{V}{c}}{1 + \beta\cos\theta_0} \end{aligned} \qquad [6.99]$$

where ω and θ are, respectively, the frequency and the emission angle of the wave as measured in $Oxyz$.

6.4.7. *Tensors*

For further details, please refer to Appendix 2.[1] Tensors are objects used to represent multilinear applications or multivectors. A tensor is a generalization of the concept of a matrix with n indices. The tensor has an arbitrary number of lower (covariant) and upper (contravariant) indices:

– *Contravariant tensors*: the tensor product allows us to associate with $\mathcal{M}$ a larger space $\mathcal{M} \otimes \mathcal{M}$. Any pair of elements $\mathbf{R}$ and $\mathbf{Q}$ of $\mathcal{M}$ is associated with an element $\mathbf{R} \otimes \mathbf{Q}$ of $\mathcal{M} \otimes \mathcal{M}$. A basis of $\mathcal{M} \otimes \mathcal{M}$ is formed by the 16 possible products formed with the 4 four-vectors of the basis of $\mathcal{M}$, $\mathbf{e}_\mu \otimes \mathbf{e}_\nu$ (the dimension of the tensor product space is $16 = 4^2$). The components of $\mathbf{R} \otimes \mathbf{Q}$ on this basis are the products of the components of $\mathbf{R}$ and $\mathbf{Q}$:

$$\mathbf{R} \otimes \mathbf{Q} = x^\mu y^\nu \mathbf{e}_\mu \otimes \mathbf{e}_\nu \qquad [6.100]$$

1 For online appendices, please see: www.iste.co.uk/abdesselam/electrodynamics1.zip.

It is also possible to define objects more general than tensor products of four-vectors: rank-2 tensors, completely contravariant elements of the space $\mathcal{M} \otimes \mathcal{M}$. We will denote $T^{\mu\nu}$ the components of such a tensor in the basis $\mathbf{e}_\mu \otimes \mathbf{e}_\nu$. We can also establish the rule for transforming a tensor in a change of basis (by Lorentz transformation): the components $T^{\mu\nu}$ transform as a product of contravariant components:

$$T'^{\mu\nu} = \mathcal{L}^{\mu}{}_{\rho}\mathcal{L}^{\nu}{}_{\sigma}T^{\rho\sigma} \qquad [6.101]$$

and, conversely:

$$T^{\mu\nu} = \mathcal{L}_{\rho}{}^{\mu}\mathcal{L}_{\sigma}{}^{\nu}T'^{\rho\sigma} \qquad [6.102]$$

We can also define the space $\mathcal{M}^{\otimes n}$ (n times the tensor product of $\mathcal{M}$ with itself). The elements of this space, of dimension 4^n, are the completely contravariant tensors of rank n whose components are written $T^{\mu_1\mu_2\cdots\mu_n}$. These components are transformed as follows:

$$T'^{\mu_1\mu_2\cdots\mu_n} = \mathcal{L}^{\mu_1}{}_{\nu_1}\mathcal{L}^{\mu_2}{}_{\nu_2}\cdots\mathcal{L}^{\mu_n}{}_{\nu_n}T^{\nu_1\nu_2\cdots\nu_n} \qquad [6.103]$$

A rank-n tensor describes a multilinear transformation that associates a four-vector to $n-1$ four-vectors.

– *Covariant tensors, mixed tensors*: the same can be done for $\mathcal{M}^\star$. It is possible to define second-order tensors that are fully covariant, whose components in the tensor basis $\mathbf{e}^\mu \otimes \mathbf{e}^\nu$ are written as $T_{\mu\nu}$. These quantities transform as follows:

$$T'_{\mu\nu} = \mathcal{L}_{\mu}{}^{\rho}\mathcal{L}_{\nu}{}^{\sigma}T_{\rho\sigma} \qquad [6.104]$$

In the same way, we define the tensor product between $\mathcal{M}$ and $\mathcal{M}^\star$. This gives rise to mixed tensors of rank 2 (or higher, if multiple copies of $\mathcal{M}$ and $\mathcal{M}^\star$ are used), whose components are written as $T^{\mu}_{\ \nu}$ (for $\mathcal{M} \otimes \mathcal{M}^\star$) or $T_{\mu}^{\ \nu}$ (for $\mathcal{M}^\star \otimes \mathcal{M}$). The transformation rule for such a mixed tensor is simply:

$$T'^{\mu}_{\ \nu} = \mathcal{L}^{\mu}{}_{\rho}\mathcal{L}_{\nu}{}^{\sigma}T^{\rho}_{\ \sigma} \qquad [6.105]$$

and it generalizes easily to any mixed tensor of arbitrary rank.

– *Metric tensor*: the metric tensor $g_{\mu\nu}$ or $g^{\mu\nu}$ can be used to raise or lower indices. For example, we can write $T^{\mu\nu} = g^{\mu\rho}g^{\nu\sigma}T_{\rho\sigma}$, $T^{\mu}_{\ \nu} = g^{\mu\rho}g_{\nu\sigma}T_{\rho}^{\ \sigma}$, $T^{\mu}_{\ \nu} = g_{\nu\rho}T^{\mu\rho}$, etc.

– *Rank-2 tensor*: rank-2 tensors can be viewed as linear mappings from the space $\mathcal{M}$ into itself (or from its dual into itself). A fully covariant rank-2 tensor, for instance, naturally defines a linear mapping from $\mathcal{M}$ to $\mathcal{M}^\star$. The image W of a four-vector (V^μ) is then written as:

$$W_\mu = T_{\mu\nu}V^\nu \qquad [6.106]$$

etc. The transformation rules for [6.106] can be derived automatically. Suppose that $W'_\rho = \mathcal{L}_\rho{}^\mu W_\mu$ and $V^\nu = \mathcal{L}_\sigma{}^\nu V'_\sigma$. It then follows that:

$$W'_\rho = \mathcal{L}_\rho{}^\mu W_\mu = \mathcal{L}_\rho{}^\mu T_{\mu\nu} V^\nu = \mathcal{L}_\rho{}^\mu T_{\mu\nu} \mathcal{L}_\sigma{}^\nu V'_\sigma = T'_{\rho\sigma} V'^\sigma \qquad [6.107]$$

– *Symmetric tensor*: a rank-2 tensor is said to be *symmetric* if:

$$T^{\mu\nu} = T^{\nu\mu} \qquad [6.108]$$

It follows immediately that $T_{\mu\nu} = T_{\nu\mu}$ and $T^\mu_\nu = T_\nu{}^\mu$.

– *Antisymmetric tensor*: a rank-2 tensor is said to be *antisymmetric* if:

$$T^{\mu\nu} = -T^{\nu\mu} \qquad [6.109]$$

The diagonal terms are obviously zero, that is, $T^{\mu\mu} = 0$. An antisymmetric tensor can be written in the form:

$$(T^{\mu\nu}) = \begin{pmatrix} 0 & a_x & a_y & a_z \\ -a_x & 0 & -b_z & b_y \\ -a_y & b_z & 0 & -b_x \\ -a_z & -b_y & b_x & 0 \end{pmatrix} \qquad [6.110]$$

– *Trace of a tensor*: the *trace* of a rank-2 tensor is the quantity $T^\mu_\mu = T_\mu{}^\mu$. More generally, the *contracted tensor* over the first two indices of a tensor $T^{\nu_1\nu_2\nu_3\cdots\nu_n}$ is the quantity $T^{\mu\,\nu_3\cdots\nu_n}_\mu$. Contracting a rank-$n$ tensor yields a tensor of rank $n-2$. Contracting a rank-2 tensor (that is, taking the trace) results in a rank-0 tensor, which is a scalar. Contracting a rank-3 tensor gives a result that transforms like a four-vector $T'^{\mu\,rho}_\mu = \mathcal{L}^\mu_\sigma \mathcal{L}^\nu_\mu \mathcal{L}^\sigma_\tau T^{\sigma\,\tau}_\nu = \delta^\nu_\sigma \mathcal{L}^\rho_\tau T^{\sigma\,\tau}_\nu = \mathcal{L}^\rho_\tau T^{\sigma\,\tau}_\sigma$.

– *Levi-Civita tensor*: in Minkowski space, we can define a contravariant rank-4 tensor, $\epsilon^{\mu\nu\rho\sigma}$, whose components are the *Levi-Civita symbols* with four indices. The metric tensor is used to raise and lower the indices. Thus, we write:

$$\epsilon_{\alpha\beta\eta\xi} = g_{\alpha\mu} g_{\beta\nu} g_{\eta\rho} g_{\xi\sigma} \epsilon^{\mu\nu\rho\sigma} \qquad [6.111]$$

Since the indices of the Levi-Civita tensor must be different and the metric tensor is diagonal, the product of the metric tensor elements on the right-hand side of [6.111] is $g_{00}g_{11}g_{22}g_{33} = -1$. By definition, we have $\epsilon^{0123} = 1$ (thus $\epsilon_{0123} = -1$). It is also possible to define the Levi-Civita tensor from a determinant:

$$\epsilon^{\mu\nu\rho\sigma} = \begin{vmatrix} \delta^\mu_0 & \delta^\mu_1 & \delta^\mu_2 & \delta^\mu_3 \\ \delta^\nu_0 & \delta^\nu_1 & \delta^\nu_2 & \delta^\nu_3 \\ \delta^\rho_0 & \delta^\rho_1 & \delta^\rho_2 & \delta^\rho_3 \\ \delta^\sigma_0 & \delta^\sigma_1 & \delta^\sigma_2 & \delta^\sigma_3 \end{vmatrix} \qquad [6.112]$$

This definition can be used to compute the contraction over multiple indices of a product of Levi-Civita tensors. We can easily verify the identities:

$$\begin{aligned} \epsilon^{\alpha\beta\mu\nu}\epsilon_{\alpha\beta\rho\sigma} &= -2\left(\delta^{\mu}_{\sigma}\delta^{\nu}_{\tau} - \delta^{\nu}_{\sigma}\delta^{\mu}_{\tau}\right) \\ \epsilon^{\alpha\beta\mu\nu}\epsilon_{\alpha\beta\mu\sigma} &= -6\delta^{\nu}_{\tau} \\ \epsilon^{\alpha\beta\mu\nu}\epsilon_{\alpha\beta\mu\nu} &= -24 \end{aligned} \qquad [6.113]$$

6.4.8. *Differentiation, vector analysis and integration*

1) For a function of the position four-vector, differentiation with respect to the contravariant coordinates is:

$$\partial_{\mu} = \frac{\partial}{\partial x^{\mu}} \qquad [6.114]$$

It is clear that the four quantities ∂_{μ} form the four covariant components of an operator that generalizes $\vec{\nabla}$ to Minkowski space. Indeed, we write:

$$df = (\partial_{\mu} f)\, dx^{\mu} \qquad [6.115]$$

Since df is a scalar and dx^{μ} is a contravariant four-vector, it follows that ∂_{μ} is a covariant four-vector. Under a Lorentz transformation, they transform as follows:

$$\partial'_{\mu} = \mathcal{L}_{\mu}^{\ \nu}\partial_{\nu} \qquad [6.116]$$

where ∂'_{μ} represents the derivatives with respect to the new contravariant coordinates. It is also possible to define differentiation with respect to the covariant coordinates:

$$\partial^{\mu} = \frac{\partial}{\partial x_{\mu}} \qquad [6.117]$$

These differential operators clearly form the contravariant components of a four-vector. The rules for raising and lowering indices can naturally be extended to these differential operators by writing:

$$\partial^{\mu} = g^{\mu\nu}\partial_{\mu} \qquad [6.118]$$

2) Starting from ∂_{μ}, it is possible to generalize the operators of three-dimensional vector analysis. If f is a scalar function, the gradient is written as:

$$\begin{aligned} \partial_{\mu} f &= \left(\frac{1}{c}\frac{\partial f}{\partial t}, \vec{\nabla} f\right) \\ \partial^{\mu} f &= \left(\frac{1}{c}\frac{\partial f}{\partial t}, -\vec{\nabla} f\right) \end{aligned} \qquad [6.119]$$

By considering a four-vector field $A^\mu = \left(a^0, \vec{a}\right)$, its divergence can be defined as follows:

$$\partial^\mu A_\mu = \partial_\mu A^\mu = \frac{1}{c}\frac{\partial a^0}{\partial t} + \vec{\nabla} \cdot \vec{a} \qquad [6.120]$$

The analogue of the *curl* is the completely antisymmetric rank-2 tensor $\partial^\mu A^\nu - \partial^\nu A^\mu$. The curl can also be written in a fully covariant form: $\partial_\mu A_\nu - \partial_\nu A_\mu$. The *Laplacian* (or the *d'Alembertian*) is the norm of the vector ∂^μ:

$$\partial^\mu \partial_\mu = \frac{1}{c^2}\frac{\partial^2}{\partial t^2} - \Delta \qquad [6.121]$$

3) A volume integral in space-time can be defined for any type of quantity $\int \cdot \, d\Omega$, where $d\Omega = cdtdxdydz$ is the integration element in space-time. We have already seen that $d\Omega$ is a scalar.

4) A surface element is a 3D object. It can be considered as being spanned by three four-vectors dx^μ, dy^μ, and dz^μ (by analogy with a 2D surface element spanned by two infinitesimal vectors). dS^μ must be orthogonal to any four-vector forming the element, and its magnitude must provide a measure of the *volume* of the surface element. We can obtain dS^μ using the usual procedure. First, we construct the rank-3 tensor $dS^{\mu\nu\rho}$:

$$dS^{\mu\nu\rho} = \begin{vmatrix} dx^\mu & dy^\mu & dz^\mu \\ dx^\nu & dy^\nu & dz^\nu \\ dx^\rho & dy^\rho & dz^\rho \end{vmatrix} \qquad [6.122]$$

The desired surface element is then obtained by contracting this tensor with the completely antisymmetric rank-4 tensor:

$$dS^\sigma = -\frac{1}{6}\epsilon^{\sigma\mu\nu\rho} dS_{\mu\nu\rho} \qquad [6.123]$$

5) We can establish for surface integrals a theorem that generalizes the divergence theorem (see Appendix 2)[2]:

$$\int_S A^\mu dS_\mu = \int_V \partial_\mu A^\mu d\Omega \qquad [6.124]$$

where V is a volume in spacetime and S its boundary surface.

6) We can also define a 2D integration. The integration element is an antisymmetric rank-2 tensor formed by the four-vectors dx^μ and dy^ν spanning the

2 For online appendices, please see: www.iste.co.uk/abdesselam/electrodynamics1.zip.

integration element, $df^{\mu\nu} = dx^\mu dy^\nu - dx^\nu dy^\mu$. Finally, a line integral can be defined along a worldline. The curl theorem relates the integral over a 2D object to the integral over its boundary (see Appendix 2)[3]:

$$\int A_\mu dx^\mu = \int df^{\mu\nu} \left(\partial_\mu A_\nu - \partial_\nu A_\mu\right) \tag{6.125}$$

6.5. Relativistic dynamics

6.5.1. *Relativistic Lagrangian for a free particle*

Let us now attempt to define an action or a Lagrangian for a free relativistic particle. The principle of least action states that for any system, there exists an integral S, called the *action*, which reaches a minimum for the actual motion (that is, the worldline of the particle). This integral does not depend on the choice of reference frame. In other words, it must be invariant under Lorentz transformations in the context of special relativity. Moreover, if the particle is free, the principle of least action becomes the principle of least proper time. In the rest frame of the particle, the action must be written as $S = \int_1^2 \alpha d\tau$. Switching to the Galilean frame $Oxyz$, we obtain:

$$S = \alpha \int_1^2 \sqrt{1 - \frac{v^2}{c^2}} dt \tag{6.126}$$

where v is the speed of the particle in $Oxyz$. Thus, the Lagrange function (or Lagrangian), called *relativistic*, is given by $L\left(\vec{v}\right) = \alpha\sqrt{1 - \frac{v^2}{c^2}}$. Similarly, we know that when $c \longrightarrow \infty$, the Lagrangian must reduce to its classical expression, that is, $\lim_{c \longrightarrow \infty} L\left(\vec{v}\right) = \frac{1}{2} m_0 v^2$, where m_0 is the rest mass of the particle. Since constant terms do not affect the equations of motion and can be omitted, we can retain only the term $L\left(\vec{v}\right) = -\frac{\alpha v^2}{2c^2}$, which gives $\alpha = -m_0 c^2$. Therefore, the relativistic Lagrangian of a free particle is:

$$L\left(\vec{v}\right) = -m_0 c^2 \sqrt{1 - \frac{v^2}{c^2}} \tag{6.127}$$

In terms of ds, the relativistic action [6.126] becomes:

$$S = -m_0 c \int_1^2 ds = -m_0 c \int_1^2 \sqrt{dx^\mu dx_\mu} \tag{6.128}$$

---

3 For online appendices, please see: www.iste.co.uk/abdesselam/electrodynamics1.zip.

The actual movement therefore corresponds to a minimal spacetime interval. In other words, a free particle evolves from (t_1, x_1, y_1, z_1) to (t_2, x_2, y_2, z_2) in such a way that the spacetime distance separating the two events is as small as possible.

We will now derive the equations of motion from this expression for the action. Consider two worldlines that are very close to each other between events 1 and 2 (which here serve as initial conditions). One of these lines is the actual worldline (which makes the action stationary). This is parameterized by the proper time of the particle τ. The events along this worldline can therefore be written as $x^\mu(\tau)$. The other worldline deviates from the actual one by infinitesimal quantities. The coordinates of the events on this second worldline are written as $x^\mu(\tau) + \delta x^\mu(\tau)$, where $\delta x^\mu(\tau)$ are infinitesimal quantities. The action along this second worldline is:

$$\begin{aligned} S + \delta S &= -m_0 c \int_1^2 \sqrt{(dx^\mu + d\delta x^\mu)(dx_\mu + d\delta x_\mu)} \\ &\simeq -m_0 c \int_1^2 \left[1 + \frac{dx_\mu d\delta x^\mu}{ds^2}\right] ds \\ &= -m_0 c \int_1^2 ds - m_0 c \int_1^2 \frac{dx_\mu}{ds} d\delta x^\mu \\ &= -m_0 c \int_1^2 ds - m_0 \int_1^2 \frac{dx_\mu}{d\tau} d\delta x^\mu \\ &= -m_0 c \int_1^2 ds - m_0 \int_1^2 U_\mu d\delta x^\mu \end{aligned} \qquad [6.129]$$

where U_μ is the four-velocity. By performing an integration by parts (on the last term of the previous expression), we obtain:

$$\delta S = -m_0 \int_1^2 U_\mu d\delta x^\mu = -m_0\ U_\mu \delta x^\mu|_1^2 + m_0 \int_1^2 \frac{dU_\mu}{d\tau} \delta x^\mu d\tau \qquad [6.130]$$

The first term on the right-hand side is zero, since the two worldlines coincide at the boundary events. The action S would therefore be extremal for the actual worldline if:

$$\frac{dU_\mu}{d\tau} = 0 \qquad [6.131]$$

This is the equation of motion of the free particle.

6.5.2. *Four-momentum*

To introduce the four-momentum, we use the results of analytical mechanics. The four-momentum is defined by the formula:

$$P_\mu = -\partial_\mu S \qquad [6.132]$$

To estimate P_μ, we consider two actual worldlines of the particle, one connecting t_1 to t_2 (the reference worldline), and the other from t_1 to $t_2 + dt_2$. The calculation of the variation of the action is similar to the previous one. The two actual lines are parametrized by the proper time of the first universe line. The only difference is that the variation does not disappear for event 2, $\delta x^\mu(t_2) = dx^\mu(t_2)$. Equation [6.130] is therefore still valid. Since both trajectories are real, the integral term is identically zero (the equation of motion is satisfied along both worldlines). However, the first term is nonzero and equals $U_\mu dx^\mu(t_2)$. We therefore deduce (by comparing the result obtained with definition [6.132]) that:

$$P_\mu = m_0 U_\mu = m_0 \frac{dx_\mu}{d\tau}$$
$$P^\mu = m_0 U^\mu = m_0 \frac{dx^\mu}{d\tau} \qquad [6.133]$$

Let us examine the physical meaning of the components of P^μ. We have $P^\mu = \left(-\frac{1}{c}\frac{\partial S}{\partial t}, \vec{\nabla} S\right)$. A simple comparison with the results of analytical mechanics indicates that the time component of P^μ is the mechanical energy (the Hamilton–Jacobi equation), while the spatial components represent the momentum of the particle:

$$P^\mu = \left(\frac{E}{c}, \vec{\mathbf{p}}\right) \qquad [6.134]$$

Energy and momentum thus appear as the temporal and spatial components of a four-vector. They must transform together under a change of reference frame. In a frame $O'x'y'z'$ moving at speed $\vec{\mathbf{V}}$ along the Oz axis relative to $Oxyz$, we obtain:

$$\frac{E'}{c} = \frac{\frac{E}{c} - \beta p_z}{\sqrt{1-\beta^2}}$$
$$p'_x = p_x$$
$$p'_x = p_x \qquad [6.135]$$
$$p'_z = \frac{p_z - \beta\frac{E}{c}}{\sqrt{1-\beta^2}}$$

The scalar product $P_\mu P^\mu$ is equal to $P_\mu P^\mu = m_0^2 \frac{dx_\mu dx^\mu}{d\tau^2} = m_0^2 \frac{ds^2}{d\tau^2} = m_0^2 c^2$, so it is indeed an invariant.

$$\frac{E^2}{c^2} - \vec{\mathbf{p}}^2 = m_0^2 c^2 \qquad [6.136]$$

EXAMPLE 6.8. TWO-PARTICLE SYSTEM.– Consider a system consisting of two particles (a) and (b). $P_\mu^a + P_\mu^b$ is a four-vector like P_μ^a and P_μ^b. Therefore, we can write for the two-particle system:

$$(E_a + E_b)^2 - \left(\vec{\mathrm{p}}_a + \vec{\mathrm{p}}_b\right)^2 c^2 = (E'_a + E'_b)^2 - \left(\vec{\mathrm{p}}'_a + \vec{\mathrm{p}}'_b\right)^2 c^2 \qquad [6.137]$$

This relation expresses the invariance of $E^2 - \vec{\mathrm{p}}^2 c^2$ for the system of two particles when switching from the reference frame $Oxyz$ to the frame $O'x'y'z'$. We can choose $O'x'y'z'$ as the center-of-mass frame of the two particles. In this reference frame, we have $\vec{\mathrm{p}}'_a + \vec{\mathrm{p}}'_b = 0$, which implies that $(E_a + E_b)^2 - \left(\vec{\mathrm{p}}_a + \vec{\mathrm{p}}_b\right)^2 c^2 = (E'_a + E'_b)^2$. ■

From the expression of four-velocity and the formulas [6.133], we also write:

$$E = m_0 \gamma c^2 = \frac{m_0 c^2}{\sqrt{1 - \frac{v^2}{c^2}}}$$
$$\vec{\mathrm{p}} = m_0 \gamma \vec{\mathrm{v}} = \frac{m_0 \vec{\mathrm{v}}}{\sqrt{1 - \frac{v^2}{c^2}}} \qquad [6.138]$$

Setting, in analogy with classical mechanics, $\vec{\mathrm{p}} = m\vec{\mathrm{v}}$, we define an apparent mass equal to $m = \frac{m_0}{\sqrt{1-\frac{v^2}{c^2}}}$, which depends on the velocity and, therefore, on the chosen reference frame. We can see that $m \longrightarrow \infty$ when $v \longrightarrow c$, which explains why a particle never reaches the speed of light. Particles that move at the speed of light therefore have a vanishing rest mass $m_0 = 0$. The relation $E = mc^2$ shows that a particle at rest has an internal energy $m_0 c^2$. Therefore, there is a close correlation between mass and energy. In other words, mass and energy are of the same nature, which means that they are merely different manifestations of the same thing. This is the *principle of mass–energy equivalence*. There is a very important formula that relates the momentum $\vec{\mathrm{p}}$, the velocity $\vec{\mathrm{v}}$ and the energy E of a particle, which is obtained by combining equations [6.138]:

$$\vec{\mathrm{p}} = \frac{E}{c^2}\vec{\mathrm{v}} \qquad [6.139]$$

If the particle thus moves at the speed of light $v = c$ (photon, neutrino, etc.), its momentum is related to its total energy by the relation $p = \frac{E}{c}$. In particular, for a photon of energy $\hbar\omega$, where $\hbar = \frac{h}{2\pi}$ is the reduced Planck constant and ω the photon frequency, we obtain $p = \frac{\hbar\omega}{c} = \frac{2\pi\hbar}{\lambda}$, where λ is the wavelength. In four-vector notation, we also write:

$$P^\mu = \left(\frac{E}{c}, \vec{\mathrm{p}}\right) = \hbar K^\mu = \left(\frac{\hbar\omega}{c}, \hbar\,\vec{\mathrm{k}}\right) \qquad [6.140]$$

The kinetic energy of a particle moving with speed $\vec{\mathrm{v}}$ is the difference between its relativistic energy E and its rest energy $m_0 c^2$, that is, $T = E - m_0 c^2 =$

$m_0\left[\left(1-\frac{v^2}{c^2}\right)^{-1/2}-1\right]c^2$. For low velocities ($v \ll c$), we can use a Taylor expansion to obtain $T \simeq \frac{1}{2}m_0v^2$. We recover the classical expression for the kinetic energy. We also write $\left(\frac{v}{c}\right)^2 = \frac{2T}{m_0c^2}$. Now, in relativistic mechanics, the relation between kinetic energy and velocity is $\left(\frac{v}{c}\right)^2 = 1-\left(\frac{m_0c^2}{T+m_0c^2}\right)^2$. This formula was experimentally verified by Bertozzi.

6.5.3. *Application to collisions*

In a collision, both the relativistic momentum and the energy of the particle system are conserved before and after impact. Indeed, global translational invariance in spacetime implies that any isolated system possesses an energy–momentum four-vector that is conserved. If the collision is elastic, that is, if the same particles are present after the impact, their rest energy is conserved in the collision, and consequently, the kinetic energy of the particle system is also conserved. To briefly illustrate the study of relativistic collisions, we examine two examples: the Compton effect and the calculation of a reaction threshold.

EXAMPLE 6.9. COMPTON EFFECT.– Compton scattering provides evidence for the momentum $\vec{\mathbf{p}}$ of a photon. A photon with frequency ω collides with an electron at rest (of mass m_0) in the reference frame $Oxyz$. After an elastic collision, the photon is scattered with a frequency ω', different from ω, in a direction that makes an angle θ with its original direction of motion (in the direction of Ox). The electron is set in motion in a direction specified by the angle ψ, with a velocity $\vec{\mathbf{v}}$, as shown in Figure 6.7. The four-momentum of the photon before and after the collision are denoted by P and P', respectively. Similarly, let Q and Q' be the electron's initial and final four-momentum:

$$\begin{aligned}
\mathbf{P} &= \left(\frac{\hbar\omega}{c}, \frac{\hbar\omega}{c}, 0, 0\right) \\
\mathbf{P}' &= \left(\frac{\hbar\omega'}{c}, \frac{\hbar\omega'}{c}\cos\theta, \frac{\hbar\omega'}{c}\sin\theta, 0\right) \\
\mathbf{Q} &= (m_0c, 0, 0, 0) \\
\mathbf{Q}' &= \left(\frac{E}{c}, m_0\gamma c\cos\psi, m_0\gamma c\sin\psi, 0\right)
\end{aligned} \qquad [6.141]$$

The law of conservation of energy–momentum gives:

$$\mathbf{P}+\mathbf{Q} = \mathbf{P}'+\mathbf{Q}' \qquad [6.142]$$

We can immediately deduce that:

$$\begin{aligned}
(\mathbf{P}-\mathbf{P}')^2 &= (\mathbf{Q}-\mathbf{Q}')^2 \\
\mathbf{P}^2+\mathbf{P}'^2-2\mathbf{P}\cdot\mathbf{P}' &= \mathbf{Q}^2+\mathbf{Q}'^2-2\mathbf{Q}\cdot\mathbf{Q}'
\end{aligned} \qquad [6.143]$$

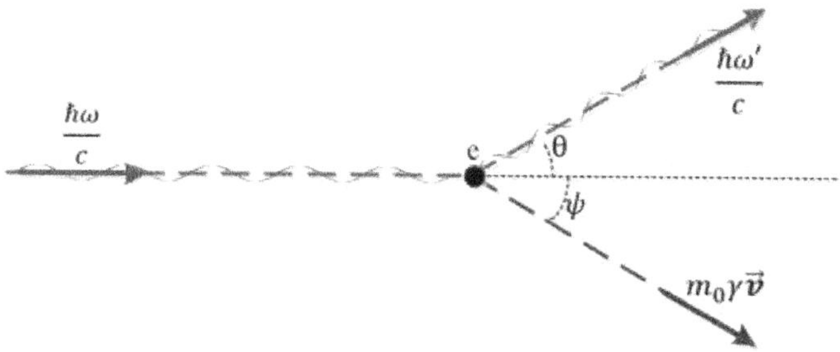

Figure 6.7. *Compton effect: scattering of a photon by an electron initially at rest. For a color version of this figure, see www.iste.co.uk/abdesselam/electrodynamics3.zip*

Since $\mathbf{P}^2 = \mathbf{P}'^2 = 0$ (the photon has zero mass), $\mathbf{Q}^2 = \mathbf{Q}'^2 = m_0 c^2$, $\mathbf{P} \cdot \mathbf{P}' = \frac{\hbar^2 \omega\omega'}{c^2}(1 - \cos\theta)$, and $\mathbf{Q} \cdot \mathbf{Q}' = m_0 E$, we obtain:

$$\frac{\hbar^2 \omega\omega'}{c^2}(1 - \cos\theta) = m_0 \left(E - m_0 c^2\right) \qquad [6.144]$$

To eliminate E from the previous expression, let us use the law of conservation of energy (the time component of [6.142]):

$$\hbar\omega + m_0 c^2 = \hbar\omega' + E \qquad [6.145]$$

It therefore follows that:

$$\lambda' - \lambda = \frac{h}{m_0 c}(1 - \cos\theta) = \lambda_c (1 - \cos\theta) \qquad [6.146]$$

where $\lambda = \frac{2\pi c}{\omega}$ and $\lambda' = \frac{2\pi c}{\omega'}$ are the wavelengths associated with the photon before and after the collision, respectively. λ_c is called the *Compton wavelength*. The electron is set in motion in a direction characterized by the angle ψ:

$$\tan\psi = \frac{-\sin\theta}{\frac{\omega}{\omega'} - \cos\theta} = -\frac{1}{1 + \frac{\lambda_c}{\lambda}} \frac{1}{\tan\frac{\theta}{2}} \qquad [6.147]$$

EXAMPLE 6.10. REACTION THRESHOLD.– Inelastic collisions between two particles A and B can lead to the production of a number of new particles $C + D + \cdots$. We assume that in the laboratory frame $Oxyz$, the target particle B is at rest. The problem is to determine the minimum energy of the incident particle A required for the reaction to occur. In passing, it is noted that from the same pair of initial particles, different reactions can occur, each with its own threshold energy. Let $O'x'y'z'$ be the center-of-mass frame of the particles A and B. In the collision, energy is conserved. $E'_A + E'_B$ is therefore equal to the sum of the energies of the formed particles. For these particles to exist, their energy must be at least equal to their rest energy. In the $O'x'y'z'$ frame, all of the formed particles can be at rest, since the total momentum is zero in this frame. Therefore, the following condition must hold:

$$E'_A + E'_B \geq \sum_i m_i c^2 \qquad [6.148]$$

where the sum runs over all of the particles produced in the reaction. We must therefore satisfy the condition for the minimum energy $E'_{A,\min}$:

$$E'_{A,\min} + E'_B = \sum_i m_i c^2 \qquad [6.149]$$

By writing the relativistic invariance $E^2 - p^2c^2$ for the system of the two particles A and B, we obtain:

$$\left(E'_{A,\min} + E'_B\right)^2 = \left(\sum_i m_i c^2\right)^2 = \left(E_{A,\min} + E_B\right)^2 - p_A^2 c^2 \qquad [6.150]$$

If we write this invariance for particle A alone, $E_{A,\min}^2 - p_A^2 c^2 = m_A^2 c^4$, we then deduce that $\left(\sum_i m_i c^2\right)^2 = \left(E_{A,\min} + E_B\right)^2 - E_{A,\min}^2 + m_A^2 c^4 = E_B(2E_{A,\min} + E_B) + m_A^2 c^4$ and consequently:

$$E_{A,\min} = \frac{\left(\sum_i m_i c^2\right)^2 - m_A^2 c^4 - E_B^2}{2E_B} \qquad [6.151]$$

Consider, for example, the creation of a proton–antiproton pair by the collision of two protons $p+p \longrightarrow p+p+(p+\overline{p})$. The minimum energy required for this reaction to occur is $E_{A,\min} = \frac{\left(4M_pc^2\right)^2 - M_p^2c^4 - M_p^2c^4}{2M_pc^2} = 7M_pc^2$. Thus, the energy threshold is $7M_pc^2$. The kinetic energy of the projectile must therefore be at least $6M_pc^2$.

6.5.4. *Minkowski four-force*

For a particle, the equation of motion is $\overrightarrow{\mathrm{f}} = \frac{d\overrightarrow{\mathrm{p}}}{dt}$, where $\overrightarrow{\mathrm{f}}$ is the result of the forces acting on the particle in the reference frame $Oxyz$. However, in this form, the law of motion is not invariant under Lorentz transformations. To obtain an invariant form, one must differentiate the four-momentum with respect to the particle's proper time τ instead of t. Thus, the four-vector (called the *Minkowski four-force*) is defined as:

$$G^{\mu} = \frac{dP^{\mu}}{d\tau} \qquad [6.152]$$

will play the role of a force. Since the norm of the four-momentum P^{μ} is constant, G^{μ} must necessarily be orthogonal to the momentum: $G^{\mu}P_{\mu} = G_{\mu}P^{\mu} = 0$. In terms of energy and spatial momentum, this can be written as:

$$G^{\mu} = \left(\frac{\gamma}{c}\frac{dE}{dt}, \gamma\frac{d\vec{\mathbf{p}}}{dt}\right) \qquad [6.153]$$

We can set $\vec{\mathbf{f}} = \frac{d\vec{\mathbf{p}}}{dt}$, thus defining the spatial force as the derivative of momentum with respect to ordinary time. The orthogonality of G_{μ} and P^{μ} simply results from the fact that $E^2 - \vec{\mathbf{p}}^2c^2$ is a constant. The quantity $c\gamma^{-1}G^0$ represents the power delivered by the force $\vec{\mathbf{f}}$ in the frame $Oxyz$. In fact, it is easily verified that $c\gamma^{-1}G^0 = \frac{dE}{dt} = \frac{c^2}{E}\vec{\mathbf{f}} \cdot \vec{\mathbf{p}} = \vec{\mathbf{f}} \cdot \vec{\mathbf{v}}$.

In a reference frame transformation (from $Oxyz$ to $O'x'y'z'$), the components of the four-force vector transform in the following way:

$$\begin{cases} \dfrac{\gamma'}{c}\dfrac{dE'}{dt'} = \Gamma\dfrac{\gamma}{c}\left(\dfrac{dE}{dt} - c\beta f_z\right) \\ \gamma' f'_x = \gamma f_x \\ \gamma' f'_y = \gamma f_y \\ \gamma' f'_z = \Gamma\gamma\left(f_z - \dfrac{\beta}{c}\dfrac{dE}{dt}\right) \end{cases} \qquad \begin{cases} \dfrac{\gamma'}{c}\vec{\mathbf{f}}' \cdot \vec{\mathbf{v}}' = \Gamma\dfrac{\gamma}{c}\left(\vec{\mathbf{f}} \cdot \vec{\mathbf{v}} - c\beta f_z\right) \\ \gamma' f'_x = \gamma f_x \\ \gamma' f'_y = \gamma f_y \\ \gamma' f'_z = \Gamma\gamma\left(f_z - \dfrac{\beta}{c}\vec{\mathbf{f}} \cdot \vec{\mathbf{v}}\right) \end{cases} \qquad [6.154]$$

where $V = \beta c$ is the translation velocity of $O'x'y'z'$ with respect to $Oxyz$.

EXAMPLE 6.11. INTERACTION FORCE BETWEEN TWO ELECTRONS.– Two electrons separated by a fixed distance a are moving with the same translational velocity v in the reference frame $Oxyz$ along the Oz axis, as illustrated in Figure 6.8. Suppose that both electrons are located along Oz. In the reference frame $O'x'y'z'$ attached to the two electrons, the components of the interaction force experienced between the two electrons are $f'_x = f'_y = 0$ and $f'_z = \frac{e^2}{4\pi\varepsilon_0 a^2}$. Using the inverse transformation of [6.154], we obtain $f_x = f_y = 0$ and $f_z = \frac{e^2}{4\pi\varepsilon_0 a^2}\sqrt{1 - \frac{v^2}{c^2}}$. If the two electrons are on the $O'x'$ axis, we find $f'_x = \frac{e^2}{4\pi\varepsilon_0 a^2}$ and $f'_y = f'_z = 0$; this also gives $f_x = \frac{e^2}{4\pi\varepsilon_0 a^2}\sqrt{1 - \frac{v^2}{c^2}}$ and $f_y = f_z = 0$. We see that the force depends on the velocity of the electrons in the fixed frame $Oxyz$.

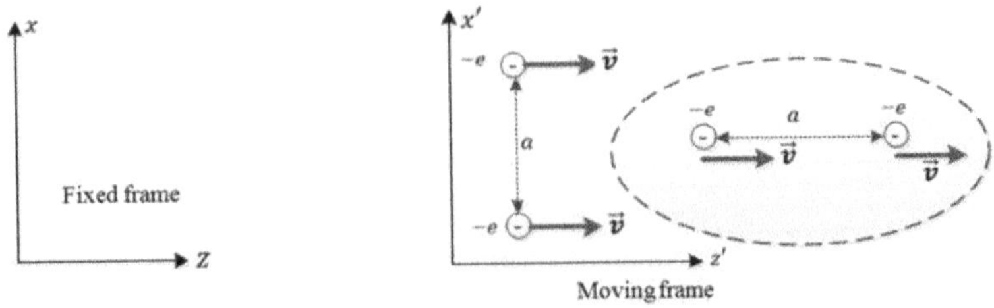

Figure 6.8. *Transformation of the electrostatic force between two electrons*

EXAMPLE 6.12. PARADOX OF THE CONDUCTING WIRE.– Consider an infinite cylindrical and conducting wire carrying an electric current and a negative charge q (located near the wire at a distance a) moving parallel to the wire in the same direction as the electrons, as illustrated in Figure 6.9. In the fixed reference frame $Oxyz$, the current I generates a magnetic field $\vec{\mathbf{B}}$, which exerts a force qvB on the charge q, directed along Ox (but downward, meaning an attractive force) given by:

$$F_x = \frac{\mu_0 q I v}{2\pi a} \qquad [6.155]$$

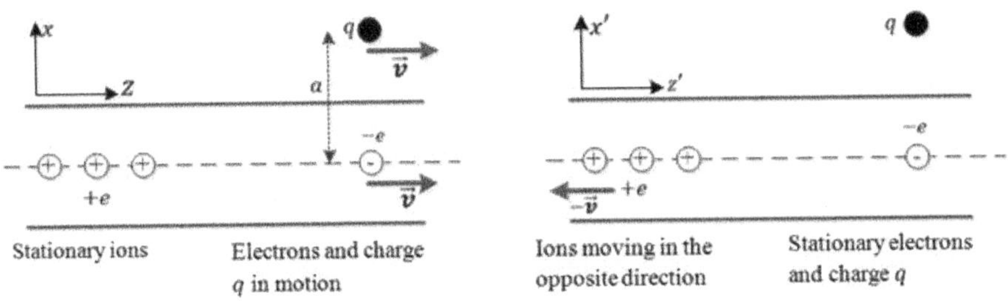

Figure 6.9. *Paradox of the conducting wire: the force experienced by the charge q is of electrostatic origin (due to the contraction of the wire)*

In the reference frame $O'x'y'z'$ (linked to the charge q and the electrons), there is a magnetic field due to the ionic current, but it does not act on the charge q, which is at rest. In classical mechanics, the charge does not experience any force; hence the paradox. In relativistic mechanics, the paradox is resolved because the FitzGerald–Lorentz contraction changes the volume of the wire and the wire appears positively charged in the $O'x'y'z'$ frame. Therefore, it still attracts the charge q. Let us examine this phenomenon in detail. Let us first place ourselves in the fixed reference frame $Oxyz$ (the wire is at rest and the electrons are moving). The density of the ionic charges is $\rho_+ = N_0 e$, where N_0 is the number of ions per unit volume (for simplicity, we assume that the ion charge is $+e$). The electron density is $\rho_- = -N_0 e$, since the wire is electrically neutral in $Oxyz$. In this frame, we therefore have $\rho_t = \rho_+ + \rho_- = 0$, the ion velocity is $\vec{\mathbf{v}}_+ = 0$, and the ionic current is $\vec{\mathbf{J}}_+ = \rho_+ \vec{\mathbf{v}}_+ = 0$. For electrons, we have $\vec{\mathbf{v}}_- = \vec{\mathbf{v}}$ and $\vec{\mathbf{J}}_- = \rho_- \vec{\mathbf{v}}_- = -N_0 e \vec{\mathbf{v}}$ (the electronic current). The total current carried by the wire is $I = J\_S = -SN_0 ev$, where S is the cross-sectional area of the wire. The attractive force experienced by the charge q is then given by:

$$F_x = \frac{\mu_0 q \rho_- S v^2}{2\pi a} \qquad [6.156]$$

Consider now the reference frame $O'x'y'z'$ where the electrons and charge q are at rest. The electron current is zero. However, there remains an ionic current I', which generates a magnetic field that has no effect on the charge q, since q is stationary. If a force acts on q, it must therefore be of electrostatic origin. To determine the new charge densities ρ'_+ and ρ'_-, we must take a proper length L_0 of the wire (that is, as measured in $Oxyz$) and compute, in $O'x'y'z'$, the new charge Q' and the new volume $V' = SL'$. The density is then $\rho' = \frac{Q'}{V}$. However, we know that volume V' differs from $V_0 = L_0 S$ due to length contraction. Specifically, we have $L' = L_0\sqrt{1-\frac{v^2}{c^2}}$ (as observed in $O'x'y'z'$), which implies that $V' = V_0\sqrt{1-\frac{v^2}{c^2}}$. On the other hand, the charge Q' remains unchanged, that is, $Q' = Q$. We can thus write, in general: $(\rho)_{\text{moving}} = \frac{(\rho)_{\text{stationary}}}{\sqrt{1-\frac{v^2}{c^2}}}$. In our case, the ions are stationary in $Oxyz$. However, the electrons are stationary in $O'x'y'z'$. This leads to $\rho'_+ = \frac{\rho_+}{\sqrt{1-\frac{v^2}{c^2}}}$ and $\rho_- = \frac{\rho'_-}{\sqrt{1-\frac{v^2}{c^2}}}$. The total charge density of the wire in $O'x'y'z'$ is then:

$$\rho'_t = \rho'_+ + \rho'_- = \frac{\rho_+}{\sqrt{1-\frac{v^2}{c^2}}} + \rho_-\sqrt{1-\frac{v^2}{c^2}} - \frac{\rho_+\frac{v^2}{c^2}}{\sqrt{1-\frac{v^2}{c^2}}} \qquad [6.157]$$

since $\rho_- = -\rho_+$. We observe that $\rho'_t \neq 0$. The electric field generated by ρ'_t is therefore radially outward and indeed attracts the negative charge. Let us compute this force in $O'x'y'z'$. The electric field E' produced by an infinite wire carrying a linear charge density $S\rho'_t$ (charge per unit length) is given by $E' = \frac{\rho'_t S}{2\pi\varepsilon_0 a}$. This yields:

$$F'_x = qE' = \frac{q}{2\pi\varepsilon_0}\frac{\rho_+ S}{a}\frac{v^2/c^2}{\sqrt{1-\frac{v^2}{c^2}}} \qquad [6.158]$$

By comparing [6.156] and [6.158], it is verified that:

$$F'_x = \frac{F_x}{\sqrt{1-\frac{v^2}{c^2}}} \qquad [6.159]$$

which is in agreement with the Lorentz transformation laws. We also see that E'_x is given by:

$$E'_x = \frac{vB}{\sqrt{1 - \frac{v^2}{c^2}}} \qquad [6.160]$$

Chapter 7

Covariant Formulation

7.1. A brief remark on field theory

We now want to work on the interaction between charged particles via an electromagnetic field in a relativistic framework. But before that, we will give a brief overview of the concepts of field theory.

7.1.1. *Transition from a discrete system to a continuous one*

We applied the Lagrange method to an infinite 1D solid (consisting of identical atoms of mass m, connected to each other by identical springs of stiffness k), undergoing small longitudinal vibrations, as illustrated in Figure 7.1. This modeling is valid around an equilibrium position since, in that case, any interaction potential can be approximated by a harmonic potential. At equilibrium, the distance between two neighboring atoms is equal to a. The generalized coordinate, denoted by $\phi_n = \phi(x_n, t)$, describes the displacement of the nth atom relative to its equilibrium position.

Thus, the Lagrangian of the solid is:

$$L = \frac{m}{2} \sum_n \dot{\phi}_n^2 - \frac{k}{2} \sum_n (\phi_{n+1} - \phi_n)^2 \qquad [7.1]$$

and, consequently, the equations of motion of the atoms are given by:

$$m\ddot{\phi}_n + k(2\phi_n - \phi_{n+1} - \phi_{n-1}) = 0 \qquad [7.2]$$

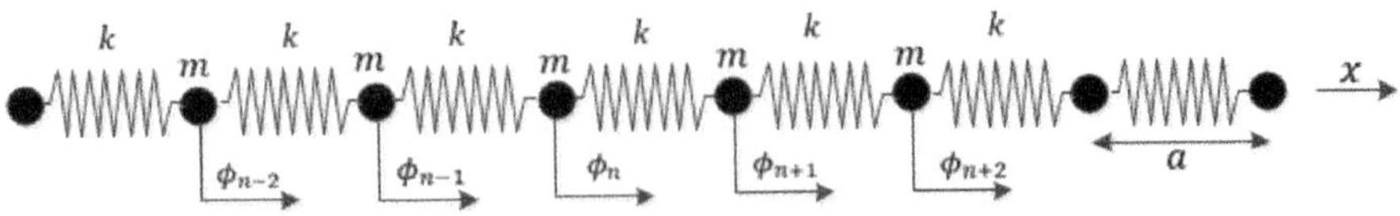

Figure 7.1. *Longitudinal oscillations in an infinitely long 1D solid*

We now aim to take the continuous limit in [7.1] and [7.2] by letting $a \longrightarrow 0$. To do so, let us first rewrite these equations in the form:

$$L = \frac{1}{2}\left[\sum_n a\frac{m}{a}\left(\frac{\partial\phi(x_n,t)}{\partial t}\right)^2 - \sum_n a(ka)\left(\frac{\phi(x_n+a,t)-\phi(x_n,t)}{a}\right)^2\right] \quad \text{(i)}$$
$$\frac{m}{a}\frac{\partial^2\phi(x_n,t)}{\partial t^2} + (ak)\left(\frac{2\phi(x_n,t)-\phi(x_n+a,t)-\phi(x_n-a,t)}{a^2}\right) = 0 \quad \text{(ii)}$$

[7.3]

By setting $\sigma = \frac{m}{a}$ (the mass density of the 1D solid), $Y = ka$, and taking the limit $a \longrightarrow 0$, the previous equations reduce to:

$$L = \frac{1}{2}\int\left[\sigma\left(\frac{\partial\phi(x,t)}{\partial t}\right)^2 - Y\left(\frac{\partial\phi(x,t)}{\partial x}\right)^2\right]dx \quad \text{(i)}$$
$$\sigma\frac{\partial^2\phi(x,t)}{\partial t^2} - Y\frac{\partial^2\phi(x,t)}{\partial x^2} = 0 \quad \text{(ii)} \qquad [7.4]$$

The summation over n is simply replaced by an integration over the position x in [7.3]. Equation [7.4ii] describes the propagation of a wave $\phi(x,t)$, where $v = \sqrt{\frac{Y}{\sigma}}$ is the phase velocity. The quantity:

$$\mathcal{L} = \frac{1}{2}\left[\sigma\left(\frac{\partial\phi(x,t)}{\partial t}\right)^2 - Y\left(\frac{\partial\phi(x,t)}{\partial x}\right)^2\right] \qquad [7.5]$$

is called the *Lagrangian density*. $\phi(x,t)$ is called the *longitudinal displacement field.*

7.1.2. *Continuous system*

Any continuous system, that is, one with an infinite number of degrees of freedom, can be described by a Lagrangian of the form:

$$L = \int_{\text{sys.}} \mathcal{L} d\tau \qquad [7.6]$$

where $\mathcal{L}$ is the *Lagrangian density* of the system (with $d\tau$ being the volume element $dxdydz$). The integral can be extended over the entire ordinary space $\mathbb{R}^3$, provided that $\mathcal{L} = 0$ outside the system. It should also be noted that the variables x, y and z are merely parameters used to identify a region in space, not variables that describe the evolution of the system itself.

7.1.3. *Euler–Lagrange equations*

If the system is described by the fields $\left\{\phi_i\left(\vec{\mathbf{r}}, t\right)\right\}_{1\leq i\leq N}$, its Lagrangian density depends on the functions ϕ_i and their first derivatives $\partial_t\phi_i$, $\partial_x\phi_i$, $\partial_y\phi_i$ and $\partial_z\phi_i$, that is:

$$\mathcal{L} = \mathcal{L}\left(\phi_i, \partial_t\phi_i, \partial_x\phi_i, \partial_y\phi_i, \partial_z\phi_i\right) \tag{7.7}$$

It cannot contain higher order derivatives of ϕ_i, because the fields interact only locally. To obtain the field equations (i.e. the equations of motion of the fields), we must use the Hamilton principle. Physical fields are those that make the action extremal:

$$\delta S = \delta\int_{t_1}^{t_2}\int_{x_1}^{x_2}\int_{y_1}^{y_2}\int_{z_1}^{z_2}\mathcal{L}dtd\tau = \delta\int_{t_1}^{t_2}\int_{x_1}^{x_2}\int_{y_1}^{y_2}\int_{z_1}^{z_2}\delta\mathcal{L}dtd\tau = 0 \tag{7.8}$$

By rejecting virtual variations in time and in the coordinates (x, y, z), and considering variations in the fields, we obtain:

$$\sum_i\int\left[\frac{\partial\mathcal{L}}{\partial\phi_i}\delta\phi_i + \sum_{\eta=x,y,z}\frac{\partial\mathcal{L}}{\partial\left[\partial_\eta\phi_i\right]}\delta\left[\partial_\eta\phi_i\right] + \frac{\partial\mathcal{L}}{\partial\left[\partial_t\phi_i\right]}\delta\left[\partial_t\phi_i\right]\right]dtd\tau = 0 \tag{7.9}$$

The second term is written as:

$$\begin{aligned}\sum\int\frac{\partial\mathcal{L}}{\partial\left[\partial_\eta\phi_i\right]}\delta\left[\partial_\eta\phi_i\right]dtd\tau &= \sum\int\frac{\partial\mathcal{L}}{\partial\left[\partial_\eta\phi_i\right]}\partial_\eta\left[\delta\phi_i\right]dtd\tau\\ &= \sum\int\left[\partial_\eta\left(\frac{\partial\mathcal{L}}{\partial\left[\partial_\eta\phi_i\right]}\delta\phi_i\right) - \partial_\eta\left(\frac{\partial\mathcal{L}}{\partial\left[\partial_\eta\phi_i\right]}\right)\delta\phi_i\right]dtd\tau\\ &= \sum\int\left[\int dS_\eta\frac{\partial\mathcal{L}}{\partial\left[\partial_\eta\phi_i\right]}\delta\phi_i - \int d\tau\partial_\eta\left(\frac{\partial\mathcal{L}}{\partial\left[\partial_\eta\phi_i\right]}\right)\delta\phi_i\right]dt\\ &= -\sum\int\partial_\eta\left(\frac{\partial\mathcal{L}}{\partial\left[\partial_\eta\phi_i\right]}\right)\delta\phi_i dtd\tau\end{aligned} \tag{7.10}$$

because the variations $\delta\phi_i$ must vanish at the boundaries. Similarly, the third term is transformed into the following:

$$\int \frac{\partial \mathcal{L}}{\partial [\partial_t \phi_i]} \delta [\partial_t \phi_i] \, dt d\tau = \int \frac{\partial \mathcal{L}}{\partial [\partial_t \phi_i]} \partial_t [\delta \phi_i] \, dt d\tau = \int \left[\partial_t \left(\frac{\partial \mathcal{L}}{\partial [\partial_t \phi_i]} \delta \phi_i \right) - \partial_t \left(\frac{\partial \mathcal{L}}{\partial [\partial_t \phi_i]} \right) \delta \phi_i \right] dt d\tau = \int d^3\tau \, \frac{\partial \mathcal{L}}{\partial [\partial_t \phi_i]} \delta \phi_i \Big|_{t_1}^{t_2} - \int \partial_t \left(\frac{\partial \mathcal{L}}{\partial [\partial_t \phi_i]} \right) \delta \phi_i dt d\tau$$
$$= - \int \partial_t \left(\frac{\partial \mathcal{L}}{\partial [\partial_t \phi_i]} \right) \delta \phi_i dt d\tau \quad [7.11]$$

because the virtual variations $\delta\phi_i$ also vanish at t_1 and t_2. By combining these results, we obtain:

$$\sum_i \int \left[\frac{\partial \mathcal{L}}{\partial \phi_i} - \sum_{\eta = x,y,z} \partial_\eta \left(\frac{\partial \mathcal{L}}{\partial [\partial_\eta \phi_i]} \right) - \partial_t \left(\frac{\partial \mathcal{L}}{\partial [\partial_t \phi_i]} \right) \right] \delta \phi_i dt d\tau = 0 \quad [7.12]$$

This equation must hold for all values of $\delta\phi_i$, which gives:

$$\sum_{\eta = x,y,z} \partial_\eta \left(\frac{\partial \mathcal{L}}{\partial [\partial_\eta \phi_i]} \right) + \partial_t \left(\frac{\partial \mathcal{L}}{\partial [\partial_t \phi_i]} \right) - \frac{\partial \mathcal{L}}{\partial \phi_i} = 0 \quad [7.13]$$

This is the *Euler–Lagrange equation* for a continuous system described by the fields ϕ_i. It can be rewritten in a more compact form:

$$\sum_{\eta = t,x,y,z} \partial_\eta \left(\frac{\partial \mathcal{L}}{\partial [\partial_\eta \phi_i]} \right) - \frac{\partial \mathcal{L}}{\partial \phi_i} = 0 \quad [7.14]$$

For a relativistic field theory, the Lagrangian density must be a scalar quantity, so that the Euler–Lagrange equations derived from it remain invariant under Lorentz transformations. In this case, equations [7.14] are written as follows:

$$\partial_\mu \left(\frac{\partial \mathcal{L}}{\partial [\partial_\mu \phi_i]} \right) - \frac{\partial \mathcal{L}}{\partial \phi_i} = 0 \quad [7.15]$$

7.1.4. *Hamiltonian formalism*

It is instructive to derive the Hamiltonian associated with [7.5]. Defining the conjugate momenta $\pi_k (x^\mu)$ of the canonical variables $\phi_k (x^\mu)$ as:

$$\pi_k = \frac{\partial \mathcal{L}}{\partial (\partial_0 \phi_k)} \equiv \frac{\partial \mathcal{L}}{\partial \dot{\phi}_k} \quad [7.16]$$

the *Hamiltonian density* is written, by analogy with the discrete case, as:

$$\mathcal{H} = \sum_k \pi_k \dot{\phi}_k - \mathcal{L} \tag{7.17}$$

The Hamiltonian density $\mathcal{H}$ therefore corresponds to the Legendre transform of the Lagrangian density with respect to $\dot{\phi}_k$. The Hamiltonian then appears as an integral of the density $\mathcal{H}$ over all space:

$$H(t) = \int \mathcal{H} d\tau = \int \left(\sum_k \pi_k \dot{\phi}_k - \mathcal{L} \right) d\tau \tag{7.18}$$

By substituting $\mathcal{L}$ with $\sum_k \pi_k \dot{\phi}_k - \mathcal{H}$ in the expression for the action, we find:

$$\begin{aligned} \delta S = \delta \int \left(\sum_k \pi_k \dot{\phi}_k - \mathcal{H} \right) dt d\tau = \sum_k \int \left[\left(\dot{\phi}_k - \frac{\partial \mathcal{H}}{\partial \pi_k} \right) \delta \pi_k + \right. \\ \left. \left(\pi_k \frac{\partial [\delta \phi_k]}{\partial t} - \frac{\partial \mathcal{H}}{\partial \phi_k} \delta \phi_k - \frac{\partial \mathcal{H}}{\partial [\partial_\mu \phi_k]} \partial_\mu [\delta \phi_k] \right) \right] dt d\tau \\ = \sum_k \int \left[\left(\dot{\phi}_k - \frac{\partial \mathcal{H}}{\partial \pi_k} \right) \delta \pi_k - \left(\dot{\pi}_k + \frac{\partial \mathcal{H}}{\partial \phi_k} - \partial_\mu \left(\frac{\partial \mathcal{H}}{\partial [\partial_\mu \phi_k]} \right) \right) \delta \phi_k \right] dt d\tau \end{aligned} \tag{7.19}$$

where we used $\delta \phi_k = \delta \pi_k = 0$ at the boundaries and at infinity. Assuming that $\delta \phi_i$ and $\delta \pi_i$ are independent and setting δS for the actual motion, we obtain:

$$\begin{aligned} \dot{\phi}_k &= \frac{\partial \mathcal{H}}{\partial \pi_k} \\ \dot{\pi}_k &= -\frac{\partial \mathcal{H}}{\partial \phi_k} + \partial_\mu \left(\frac{\partial \mathcal{H}}{\partial [\partial_\mu \phi_k]} \right) \end{aligned} \tag{7.20}$$

These equations of motion for the fields are called *functional canonical Hamilton equations*. Using the functional derivatives defined by:

$$\begin{aligned} \frac{\delta}{\delta \pi_k} &= \frac{\partial}{\partial \pi_k} \\ \frac{\delta}{\delta \phi_k} &= \frac{\partial}{\partial \phi_k} - \partial_\mu \left(\frac{\partial}{\partial [\partial_\mu \phi_k]} \right) \end{aligned} \tag{7.21}$$

Equations [7.20] can also be written as:

$$\begin{aligned} \dot{\phi}_k &= \frac{\delta \mathcal{H}}{\delta \pi_k} \\ \dot{\pi}_k &= -\frac{\delta \mathcal{H}}{\delta \phi_k} \end{aligned} \tag{7.22}$$

These are similar to the Hamiltonian equations for discrete systems.

EXAMPLE 7.1. (Vibrating string).– Returning to the Lagrangian density [7.5] of the vibrating string, it is clear that $\frac{\partial \mathcal{L}}{\partial \phi} = 0$, $\frac{\partial \mathcal{L}}{\partial[\partial_t \phi]} = \sigma \partial_t \phi$ and $\frac{\partial \mathcal{L}}{\partial[\partial_x \phi]} = -Y \partial_x \phi$. Substituting these results into [7.14] leads to the wave equation [7.4ii]. Definition [7.16] yields the conjugate momentum $\pi(x,t) = \sigma \dot{\phi}(x,t)$, which implies $\dot{\phi}(x,t) = \frac{\pi(x,t)}{\sigma}$ and leads to the Hamiltonian density:

$$\mathcal{H} = \frac{\Pi^2(x,t)}{2\sigma} + \frac{Y}{2}(\partial_x \phi(x,t))^2 \qquad [7.23]$$

The first term corresponds to the kinetic energy of the vibrating string and the second to its potential energy.

EXAMPLE 7.2. (Klein–Gordon equation).– In the case of the Lagrangian density of Klein–Gordon:

$$\mathcal{L} = \frac{1}{2}\left(g^{\mu\nu}\partial_\mu \phi \partial_\nu \phi - m^2\phi^2\right) = \frac{1}{2}\left(\dot{\phi}^2 - \left(\vec{\nabla}\phi\right)^2 - m^2\phi^2\right) \qquad [7.24]$$

we find:

$$\left(\partial_\mu \partial^\mu + m^2\right)\phi = 0 \qquad [7.25]$$

The second expression of $\mathcal{L}$ allows an easy computation of the conjugate momentum of ψ, $\pi = \dot{\psi}$, and the Hamiltonian density is:

$$\mathcal{H} = \frac{1}{2}\pi^2 + \left(\vec{\nabla}\phi\right)^2 + m^2\phi^2 \qquad [7.26]$$

The energy is calculated by integrating the preceding expression over all space.

7.1.5. *Gauge invariance*

In analytical mechanics, two Lagrangians L and L' describe the same mechanical system if they differ only by a total derivative of a function depending on position and time. For continuous systems, two Lagrangian densities $\mathcal{L}$ and $\mathcal{L}'$ describe the same system if:

$$\mathcal{L}' = \mathcal{L} + \partial_\mu f^\mu \qquad [7.27]$$

where $f^\mu \equiv f^\mu(\phi_i)$ is a function of the fields ϕ_i. It will be verified that $\delta S' = \delta S = 0$. Indeed, we have:

$$\delta S' = \delta S + \int \left[\partial_t \left(\delta f^0\right) + \sum_i \partial_i \left(\delta f^i\right) \right] dt d\tau = 0 + \int \delta f^0 \big|_{t_1}^{t_2} d\tau$$
$$+ \sum_i \int dt \oint_{S \longrightarrow \infty} \left(\delta f^i\right) dS_i = \sum_j \int d\tau \frac{\partial f^0}{\partial \phi_j} \delta\phi_j \big|_{t_1}^{t_2}$$
$$+ \textstyle\sum_{i,j} \int dt \oint_{S \longrightarrow \infty} \left(\frac{\partial f^i}{\partial \phi_j} \delta\phi_j \right) dS_i \qquad [7.28]$$

It is clear that the last two terms vanish, because $\delta\phi_i = 0$ at the boundaries and on the surface at infinity, which implies that $\delta S' = \delta S = 0$.

7.1.6. *Infinitesimal symmetry*

We consider an infinitesimal transformation of the fields ϕ_i: $\phi_i(x^\mu) \longrightarrow \phi_i'(x^\mu) = \phi_i(x^\mu) + \epsilon\delta\phi_i(x^\mu)$, where $\epsilon \ll 1$. The system is said to possess symmetry under this transformation if its Lagrangian density remains invariant up to a total derivative, which means that there exists a four-vector f^μ such that:

$$\mathcal{L}(x^\mu) \longrightarrow \mathcal{L}'(x^\mu) = \mathcal{L}(x^\mu) + \epsilon\partial_\mu f^\mu + \mathcal{O}\left(\epsilon^2\right) \qquad [7.29]$$

where $f^\mu = f^\mu(\phi_i)$ depends on the fields ϕ_i. Generally, first-order in ϵ, the transformation leads to:

$$\delta\mathcal{L} = \sum_i \left(\frac{\partial \mathcal{L}}{\partial \phi_i} \epsilon\delta\phi_i + \frac{\partial \mathcal{L}}{\partial\left[\partial_\mu \phi_i\right]} \epsilon\delta\left[\partial_\mu \phi_i\right] \right) + \mathcal{O}\left(\epsilon^2\right) = \sum_i \left[\frac{\partial \mathcal{L}}{\partial \phi_i} \delta\phi_i - \right.$$
$$\left. \partial_\mu \left(\frac{\partial \mathcal{L}}{\partial\left[\partial_\mu \phi_i\right]} \right) \right] \epsilon\delta\phi_i + \epsilon\partial_\mu \left(\sum_i \frac{\partial \mathcal{L}}{\partial\left[\partial_\mu \phi_i\right]} \delta\phi_i \right) + \mathcal{O}\left(\epsilon^2\right) \qquad [7.30]$$

If we assume that the equation of motion is satisfied, the first term multiplying $\epsilon\delta\phi_i$ in [7.30] vanishes. It therefore follows that:

$$\delta\mathcal{L} = \epsilon\partial_\mu \left(\sum_i \frac{\partial \mathcal{L}}{\partial\left[\partial_\mu \phi_i\right]} \delta\phi_i \right) + \mathcal{O}\left(\epsilon^2\right) = \epsilon\partial_\mu f^\mu \qquad [7.31]$$

Thus, we can derive the explicit form of the conserved current:

$$\partial_\mu J^\mu = 0 \qquad [7.32]$$

with:

$$J^{\mu} = \sum_{i} \frac{\partial \mathcal{L}}{\partial \left(\partial_{\mu} \phi_{i}\right)} \delta \phi_{i} - f^{\mu} \tag{7.33}$$

J^{μ} is called the *Noether current.* Equation [7.32] shows that the charge $Q = \int J^{0} d\tau$ associated with this symmetry is conserved over time (Noether charge). Indeed, we have:

$$\frac{dQ}{dt} = c \int \partial_{0} J^{0} d\tau = -c \int \vec{\nabla} \cdot \vec{\mathbf{J}}\, d\tau = -c \oint_{S \longrightarrow 0} \vec{\mathbf{J}} \cdot d\vec{\mathbf{S}} = 0 \tag{7.34}$$

7.1.7. *Energy–momentum tensor of a field*

We now wish to determine the conserved current corresponding to a spacetime translation. From classical mechanics, we know that invariance under spatial translation is associated with conservation of the total momentum of the system. Similarly, invariance under time translation is associated with conservation of total energy. It is now possible to unify these two invariants using the covariant formalism. We assume that the Lagrangian does not have explicit dependence on x. Let us consider an infinitesimal spacetime translation:

$$x^{\mu} \longrightarrow x'^{\mu} = x^{\mu} + \epsilon^{\mu} \tag{7.35}$$

where $\epsilon^{\mu} \longrightarrow 0$. The fields transform as follows:

$$\phi_{k}(x) \longrightarrow \phi_{k}(x') = \phi_{k}(x + \epsilon) = \phi_{k}(x) + \epsilon^{\mu} \partial_{\mu} \phi_{k} + \mathcal{O}\left(\epsilon^{2}\right) \tag{7.36}$$

that is, $\delta \phi_{k} = \epsilon^{\mu} \partial_{\mu} \phi_{k}$. The Lagrangian is affected by the transformation in the same way, since it does not explicitly depend on x. In fact, we have:

$$\mathcal{L}(x) \longrightarrow \mathcal{L}\left(\phi_{k}(x + \epsilon), \partial \phi_{k}(x + \epsilon)\right) = \mathcal{L}(x + \epsilon) = \mathcal{L}(x) + \epsilon^{\mu} \partial_{\mu} \mathcal{L} + \mathcal{O}\left(\epsilon^{2}\right) \tag{7.37}$$

Consequently, we obtain $\delta \mathcal{L} = \epsilon^{\mu} \partial_{\mu} \mathcal{L} = \partial_{\mu}\left(\epsilon^{\mu} \mathcal{L}\right)$, which implies that $f^{\mu} = \epsilon^{\mu} \mathcal{L}$. We can therefore use [7.33] to deduce the corresponding conserved current:

$$J^{\mu} = \sum_{k} \frac{\partial \mathcal{L}}{\partial \left[\partial_{\mu} \phi_{k}\right]} \epsilon^{\nu} \partial_{\nu} \phi_{k} - \epsilon^{\mu} \mathcal{L} = \epsilon^{\nu} \left(\sum_{k} \frac{\partial \mathcal{L}}{\partial \left[\partial_{\mu} \phi_{k}\right]} \partial_{\nu} \phi_{k} - \delta^{\mu}_{\nu} \mathcal{L} \right) \tag{7.38}$$

It is more convenient to express this current in the form $J^{\mu} = \epsilon^{\nu} \left(J^{\mu}\right)_{\nu}$ with:

$$\left(J^{\mu}\right)_{\nu} = \sum_{k} \frac{\partial \mathcal{L}}{\partial \left[\partial_{\mu} \phi_{k}\right]} \partial_{\nu} \phi_{k} - \delta^{\mu}_{\nu} \mathcal{L} = T^{\mu}{}_{\nu} \tag{7.39}$$

which shows the energy–momentum tensor $T^{\mu}_{;;\nu}$. Note that the four possible values of ν correspond to the four conserved quantities: the total energy E of the field and the three components of the total momentum P^i $(i = 1, 2, 3)$ of the field. Indeed, we have:

$$\partial_\mu T^\mu{}_\nu = 0 \qquad [7.40]$$

and the four conserved charges are:

$$\begin{aligned} E &= \int T^{00} d\tau \\ P^i &= \frac{1}{c} \int T^{0i} d\tau \qquad\qquad i - 1, 2, 3 \end{aligned} \qquad [7.41]$$

It is always possible to add to the energy–momentum tensor $T^{\mu\nu}$ a term $\partial_\rho W^{\mu\nu\rho}$, where $W^{\mu\nu\rho} = -W^{\rho\nu\mu}$ is an arbitrary tensor antisymmetric in its first and last indices, without affecting the conservation equation $\partial_\mu T^{\mu\nu} = 0$, nor altering the momentum $P^i = \frac{1}{c} \int T^{0i} d\tau$.

EXAMPLE 7.3. (Klein–Gordon equation).–We applied the above results to the case of the Klein–Gordon field [7.24]. The energy–momentum tensor is given by:

$$T^{\mu\nu} = \partial^\mu \phi \partial^\nu \phi - g^{\mu\nu} \mathcal{L} \qquad [7.42]$$

and takes a symmetric form $T^{\mu\nu} = T^{\nu\mu}$. We can then deduce that:

$$\begin{aligned} E &= \int \frac{1}{2} \left[\dot{\phi}^2 + \left(\vec{\nabla} \phi \right)^2 + m^2 \phi^2 \right] d\tau \\ \vec{\mathbf{P}} &= \int \frac{1}{c^2} \dot{\phi} \vec{\nabla} \phi d\tau \end{aligned} \qquad [7.43]$$

7.1.8. *Angular momentum tensor*

We now assume that the system is also invariant under the Lorentz transformation $x^\mu \longrightarrow x^\mu + \omega^{\mu\nu} x_\nu$, where $\omega^{\mu\nu} = -\omega^{\nu\mu}$ is an infinitesimal parameter. The Noether current associated with this transformation is given by $J^\mu_\omega = w_{\nu\rho} M^{\mu\nu\rho}$, where $M^{\mu\nu\rho} = x^\rho T^{\mu\nu} - x^\nu T^{\mu\rho}$ is the *angular momentum tensor*.

7.1.9. *Electromagnetic interaction*

We will now treat, in a relativistic framework, the interaction between charged particles via an electromagnetic field. Our goal is to express the non-trivial relativistic interaction between particles using a four-vector field. In other words, within the framework of special relativity, we will construct the simplest (non-trivial) field theory capable of reproducing the Maxwell results. We will postulate the form of the actions for both the particles and the field (choosing the simplest and most natural forms). From these actions, we will derive the Euler–Lagrange equations that describe the dynamics of the particles coupled to the field, as well as the dynamics of the field coupled to the particles. These equations will turn out to be the Maxwell equations, thus demonstrating that electromagnetism is the simplest field theory within the framework of relativity. Our system will thus consist of a set of particles that interact with a four-vector field. The total action for this system will therefore be written as:

$$S = S_{\text{free particles}} + S_{\text{free field}} + S_{\text{interaction}} \qquad [7.44]$$

where $S_{\text{free particles}}$, $S_{\text{free field}}$ and $S_{\text{interaction}}$ represent, respectively, the action of the particles in the absence of a field, the action that describes the field alone (in the absence of particles) and the action of the particle–field interaction. This problem will be approached in two steps (as in all elementary treatments of electromagnetism). We first consider a single particle in the presence of a given (imposed) field. We assume that this particle (called a test particle) is small enough not to perturb the imposed field (which is created by a larger source – the source particles). The only dynamical variables are therefore those of the free particle. We can then write the Euler–Lagrange equations and show that they exactly yield the expression of the Lorentz force. In the second step, we consider a field in interaction with particles whose motion is given. We represent the motion of these particles using charge and current densities and postulate a simple form for $S_{\text{free field}}$. The Euler–Lagrange equations that follow from this action are simply the Maxwell equations involving sources.

7.2. Free particle in a field

7.2.1. *Four-potential*

An electric charge is an intrinsic property of the particle (independent of its velocity); it is a Lorentz invariant. The experimental proof of this statement is the neutrality of atoms (an atom remains neutral regardless of the velocities of the electrons inside it). Consider a volume element $d\tau$ occupied by electric charges. Since observers moving at different velocities measure different volumes, the charge density within the volume must necessarily depend on the velocity of the volume. In fact, an observer will find that the length of the volume moving at relative velocity v is shortened by the factor γ^{-1} in the direction of motion. Therefore, the charge density measured by the observer is $\rho = \rho_0\gamma$, where ρ_0 is the proper charge density, that is, in the rest reference frame of the volume (the frame attached to the volume).

The *four-current density* vector, also called the *four-current* vector, is defined as the product of the proper density ρ_0 and the *four-velocity* vector $U^\mu = \left(\gamma c, \gamma \overrightarrow{\mathbf{v}}\right)$:

$$J^\mu = \rho_0 U^\mu = \left(\rho_0 \gamma c, \rho_0 \gamma \overrightarrow{\mathbf{v}}\right) = \left(\rho c, \overrightarrow{\mathbf{j}}\right) \tag{7.45}$$

The norm of this four-vector is given by $J^\mu J_\mu = \rho_0^2 c^2$. We also observe that the components of the four-current density appear in the expression of charge conservation, $\overrightarrow{\nabla} \cdot \overrightarrow{\mathbf{j}} + \frac{1}{c}\frac{\partial}{\partial t}(c\rho) = 0$, which thus allows us to write:

$$\partial_\mu J^\mu = 0 \tag{7.46}$$

This is the *covariant form of the charge conservation equation.*

According to the classical formulation of electromagnetism, the scalar and vector potentials are given by:

$$\begin{aligned} \frac{V\left(x^\mu\right)}{c} &= \frac{\mu_0}{4\pi}\int_{\text{source}} \frac{\rho\left(t', \overrightarrow{\mathbf{r}}'\right)\Big|_{t'=t-\frac{\left|\overrightarrow{\mathbf{r}}-\overrightarrow{\mathbf{r}}'\right|}{c}}}{\left|\overrightarrow{\mathbf{r}} - \overrightarrow{\mathbf{r}}'\right|} d\tau' \\ \overrightarrow{\mathbf{A}}\left(x^\mu\right) &= \frac{\mu_0}{4\pi}\int_{\text{source}} \frac{\overrightarrow{\mathbf{j}}\left(t', \overrightarrow{\mathbf{r}}'\right)\Big|_{t'=t-\frac{\left|\overrightarrow{\mathbf{r}}-\overrightarrow{\mathbf{r}}'\right|}{c}}}{\left|\overrightarrow{\mathbf{r}} - \overrightarrow{\mathbf{r}}'\right|} d\tau' \end{aligned} \tag{7.47}$$

where $\rho\left(t', \overrightarrow{\mathbf{r}}'\right)$ denotes the charge density and $\overrightarrow{\mathbf{j}}\left(t', \overrightarrow{\mathbf{r}}'\right)$ the current density in the volume τ' of the source. The time t' refers to the retarded time, or the time at the source, since the field propagates at the speed c; thus, the field emitted by the source at position $\overrightarrow{\mathbf{r}}'$ at time t' will be felt at position $\overrightarrow{\mathbf{r}}$ at time $t = t' + \frac{\left|\overrightarrow{\mathbf{r}}-\overrightarrow{\mathbf{r}}'\right|}{c}$. The propagation equations for the scalar and vector potentials are written as:

$$\begin{aligned} \left(\frac{1}{c^2}\frac{\partial^2}{\partial t^2} - \Delta\right)\left(\frac{V}{c}\right) &= -\mu_0 c\rho \\ \left(\frac{1}{c^2}\frac{\partial^2}{\partial t^2} - \Delta\right)\overrightarrow{\mathbf{A}} &= -\mu_0 \overrightarrow{\mathbf{j}} \end{aligned} \tag{7.48}$$

Since the d'Alembertian operator $\partial_\mu \partial^\mu = \frac{1}{c^2}\frac{\partial^2}{\partial t^2} - \Delta$ is covariant, it follows that the four components $\left(\frac{V}{c}, \overrightarrow{\mathbf{A}}\right)$ form, just like $\left(c\rho, \overrightarrow{\mathbf{j}}\right)$, a four-vector. This four-vector is called the *four-potential* and is denoted by $A^\mu = \left(\frac{V}{c}, \overrightarrow{\mathbf{A}}\right)$. The Lorentz gauge condition $\overrightarrow{\nabla} \cdot \overrightarrow{\mathbf{A}} + \frac{1}{c}\frac{\partial}{\partial t}\left(\frac{V}{c}\right) = 0$, which has the same form as the charge conservation law, shows that the four-potential satisfies an equation analogous to [7.46], that is:

$$\partial_\mu A^\mu = 0 \tag{7.49}$$

Finally, it should be noted that the expressions [7.47] can be combined into a single formula:

$$A^{\mu}\left(x^{\nu}\right)=\frac{\mu_{0}}{4\pi}\int_{\text{source}}\frac{\left[J^{\mu}\right]_{\text{retarded}}}{\left|\overrightarrow{\mathbf{r}}-\overrightarrow{\mathbf{r}}'\right|}d\tau' \qquad [7.50]$$

From this last equation, it follows that $\frac{d\tau'}{\left|\overrightarrow{\mathbf{r}}-\overrightarrow{\mathbf{r}}'\right|}$ is an invariant (a scalar).

7.2.2. *Equation of relativistic dynamics*

We have already seen that to obtain relativistic, and thus covariant, equations, the action must be a Lorentz scalar. The simplest possibility is a constant multiplied by the interval element ds, that is, $S_{\text{free particle}} = \int \alpha ds$. Moreover, we know that $ds = cdt\sqrt{1-\frac{v^2}{c^2}}$. Thus, the simplest action for a free particle can be written as:

$$S_{\text{free particle}}=\int \alpha c dt\sqrt{1-\frac{v^{2}}{c^{2}}} \qquad [7.51]$$

In the limit of speeds much less than c, we must recover the classical free particle action (up to a gauge), $S_{\text{conventional}} = \int \frac{1}{2}mv^2 dt$. By performing a Taylor expansion in $\beta = \frac{v}{c}$, we obtain, from [7.51], that $S_{\text{free particle}} = \int \alpha c dt - \int \frac{1}{2}\alpha\frac{v^2}{c}dt$. By eliminating the gauge term and identifying the second term with the classical action, we find $\alpha = -mc$. The simplest covariant action for a relativistic free particle is therefore:

$$S_{\text{free particle}}=-mc\int_{1}^{2}ds \qquad [7.52]$$

Let us now consider the case of a charged particle acted upon by an electromagnetic field represented by the four-vector $A^{\mu} = \left(\frac{V}{c}, \overrightarrow{\mathbf{A}}\right)$. The interaction between the field and the particle must be represented by an action that is invariant under a Lorentz transformation. The simplest nontrivial quantity that we can write is thus:

$$S_{\text{interaction}}=-q\int_{1}^{2}A_{\mu}dx^{\mu} \qquad [7.53]$$

where q is a quantity that describes the strength of the particle-field coupling called the *charge*. The minus sign is conventional. It should be noted that the worldline (trajectory in spacetime) of the particle is parameterized by its proper time τ. Due to the additivity property, the total action of a charged particle in an electromagnetic field is as follows:

$$S = -mc \int_1^2 ds - q \int_1^2 A_\mu dx^\mu \qquad [7.54]$$

We must now minimize this action to determine the actual trajectory. The optimization variable here is x^μ. The variation of the corresponding action is:

$$\delta S = -\int_1^2 [mc\delta ds + q\delta A_\mu dx^\mu + qA_\mu \delta dx^\mu] \qquad [7.55]$$

Since $ds = \sqrt{dx_\mu dx^\mu}$, it follows that $\delta ds = \frac{1}{2}\frac{\delta(dx_\mu dx^\mu)}{\sqrt{dx_\mu dx^\mu}} = \frac{1}{2}\frac{\delta(dx_\mu dx^\mu)}{ds} = \frac{dx^\mu}{ds}\delta dx_\mu = \frac{dx_\mu}{ds}\delta dx^\mu$, where we used the fact that $\delta(dx_\mu dx^\mu) = 2dx^\mu \delta dx_\mu = 2dx_\mu \delta dx^\mu$. Introducing the four-velocity $\frac{dx_\mu}{ds} = \frac{1}{c}\frac{dx_\mu}{d\tau} = \frac{1}{c}U_\mu$, we find that $\delta ds = \frac{1}{c}U_\mu \delta dx^\mu$. Thus, equation [7.55] becomes:

$$\delta S = -\int_1^2 [mU_\mu \delta dx^\mu + q\delta A_\mu dx^\mu + qA_\mu \delta dx^\mu] \qquad [7.56]$$

By interchanging the variations δ and d, and then integrating the first and third terms by parts, we obtain:

$$\begin{aligned} \delta S &= -\int_1^2 [mU_\mu d\delta x^\mu + q\delta A_\mu dx^\mu + qA_\mu d\delta x^\mu] \\ &= -\int_1^2 [-mdU_\mu \delta x^\mu + q\delta A_\mu dx^\mu - qdA_\mu \delta x^\mu] \end{aligned} \qquad [7.57]$$

where the boundary terms resulting from this integration vanish, since we must impose $\delta x^\mu(1) = \delta x^\mu(2) = 0$ at the endpoints. Finally, by writing $\delta A_\mu = \frac{\partial A_\mu}{\partial x^\nu}\delta x^\nu = A_{\mu,\nu}\delta x^\nu$ and $dA_\mu = \frac{\partial A_\mu}{\partial x^\nu}dx^\nu = A_{\mu,\nu}dx^\nu$, it follows that:

$$\delta S = -\int_1^2 [-mdU_\mu \delta x^\mu + qA_{\mu,\nu}\delta x^\nu dx^\mu - qA_{\mu,\nu}dx^\nu \delta x^\mu] \qquad [7.58]$$

Noting that $dx^\mu = U^\mu d\tau$ and by inverting the indices μ and ν in the middle term of [7.58], the variation of the action is written as:

$$\delta S = -\int_1^2 \left[-m\frac{dU_\mu}{d\tau} + qA_{\nu,\mu}U^\nu - qA_{\mu,\nu}U^\nu\right] d\tau \delta x^\mu \qquad [7.59]$$

For the action to reach its minimum, δS must vanish for any arbitrary variation δx^μ. The equation of motion thus reads:

$$m\frac{dU_\mu}{d\tau} = q\left(A_{\nu,\mu} - A_{\mu,\nu}\right)U^\nu \qquad [7.60]$$

By introducing the rank-2 tensor (called the *field tensor* or *Faraday tensor*):

$$F_{\mu\nu} = A_{\nu,\mu} - A_{\mu,\nu} = \partial_\mu A_\nu - \partial_\nu A_\mu = \frac{\partial A_\nu}{\partial x^\mu} - \frac{\partial_\nu A_\mu}{\partial x^\nu} \qquad [7.61]$$

the equation of motion of a relativistic charged particle takes the form:

$$m\frac{dU_\mu}{d\tau} = qF_{\mu\nu}U^\nu = G_\mu \qquad [7.62]$$

The quantity G^μ is precisely the covariant component of the four-force. It is simply the contraction of the field tensor with the velocity of the particle. Note that this four-vector equation encompasses both the change in the particle's energy and its momentum. Equation [7.62] represents a generalization of Newton's law of dynamics, but in four dimensions.

7.2.3. *Electromagnetic field tensor*

By definition, $F_{\mu\nu}$ is an antisymmetric rank-2 tensor; it is called the *four-curl of the potential* $\left(\frac{V}{c}, \overrightarrow{\mathbf{A}}\right)$. It has only six independent components. The three spacetime components form a spatial vector, while the other three purely spatial components form a pseudovector. Let us now express $F_{\mu\nu}$ in terms of the components of the four-potential $A^\mu = \left(\frac{V}{c}, \overrightarrow{\mathbf{A}}\right)$ (the covariant form of the four-potential is written $A_\mu = \left(\frac{V}{c}, -\overrightarrow{\mathbf{A}}\right)$), where V and $\overrightarrow{\mathbf{A}}$ are the scalar and vector potentials, respectively:

– *Temporal part*: the spacetime components can be written in the form:

$$F_{0i} = \partial_0 A_i - \partial_i A_0 = -\frac{\partial A^i}{\partial(ct)} - \frac{1}{c}\frac{\partial V}{\partial x^i} = -\frac{1}{c}\left(\frac{\partial A^i}{\partial t} + \frac{\partial V}{\partial x^i}\right) = \frac{E^i}{c} \qquad [7.63]$$

where E^i are the components of the electric field:

$$\overrightarrow{\mathbf{E}} = -\overrightarrow{\nabla}V - \frac{\partial \overrightarrow{\mathbf{A}}}{\partial t} \qquad [7.64]$$

We adopt the convention $E^1 = -E_1 = E_x$, $E^2 = -E_2 = E_y$, and $E^3 = -E_3 = E_z$ (the same applies to the fields $\overrightarrow{\mathbf{A}}$, $\overrightarrow{\mathbf{B}}$, $\overrightarrow{\mathbf{v}}$, etc.).

– *Spatial part*:

$$F_{ij} = \partial_i A_j - \partial_j A_i \qquad [7.65]$$

These are, up to a sign, the components of a 3D curl. By introducing the pseudovector (magnetic field):

$$\vec{\mathbf{B}} = \vec{\nabla} \times \vec{\mathbf{A}} \qquad [7.66]$$

we get:

$$\begin{aligned} F_{12} &= \partial_1 A_2 - \partial_2 A_1 = -\frac{\partial A_y}{\partial x} + \frac{\partial A_x}{\partial y} = -B_z \\ F_{13} &= \partial_1 A_3 - \partial_3 A_1 = -\frac{\partial A_z}{\partial x} + \frac{\partial A_x}{\partial z} = B_y \\ F_{23} &= \partial_2 A_3 - \partial_3 A_2 = -\frac{\partial A_z}{\partial y} + \frac{\partial A_y}{\partial z} = -B_x \end{aligned} \qquad [7.67]$$

The results obtained can be summarized in a matrix form:

$$F_{\mu\nu} = \begin{pmatrix} 0 & E_x/c & E_y/c & E_z/c \\ -E_x/c & 0 & -B_z & B_y \\ -E_y/c & B_z & 0 & -B_x \\ -E_z/c & -B_y & B_x & 0 \end{pmatrix} \qquad [7.68]$$

or alternatively, in fully contravariant form:

$$F^{\mu\nu} = \begin{pmatrix} 0 & -E_x/c & -E_y/c & -E_z/c \\ E_x/c & 0 & -B_z & B_y \\ E_y/c & B_z & 0 & -B_x \\ E_z/c & -B_y & B_x & 0 \end{pmatrix} \qquad [7.69]$$

Gauge invariance is inherent in the definition of the tensor $F^{\mu\nu}$. The equation of motion of the particle remains unchanged if an arbitrary four-gradient is added to the four-potential. In fact, let us set:

$$A'_\mu = A_\mu + \partial_\mu \phi \qquad [7.70]$$

where ϕ is an arbitrary four-scalar field. The new field tensor is then:

$$F_{\mu\nu} = \partial_\mu A'_\nu - \partial_\nu A'_\mu = \partial_\mu A_\nu - \partial_\nu A_\mu + \partial_\mu \partial_\nu \phi - \partial_\nu \partial_\mu \phi = \partial_\mu A_\nu - \partial_\nu A_\mu = F_{\mu\nu}$$

[7.71]

By expressing this transformation in terms of the spatial and temporal components of the potential, we recover the standard form of the gauge transformation: $\vec{\mathbf{A}} \longrightarrow \vec{\mathbf{A}} - \vec{\nabla}\phi$ and $V \longrightarrow V + \frac{\partial \phi}{\partial t}$. To fix the potential, an additional gauge condition must be imposed. To respect relativistic invariance, this gauge condition must be covariant. The simplest one (the Lorenz gauge) requires the four-divergence of the potential to vanish, that is, $\partial_\mu A^\mu = 0$, which can be written in terms of the spatial and temporal components as $\frac{1}{c^2}\frac{\partial V}{\partial t} + \vec{\nabla} \cdot \vec{\mathbf{A}} = 0$. Note that the electrostatic gauge or the Coulomb gauge, $\vec{\nabla} \cdot \vec{\mathbf{A}} = 0$, breaks the covariance. We also see that fields $\vec{\mathbf{E}}$ and $\vec{\mathbf{B}}$ are not affected by the gauge transformation [7.70]. The same observation applies to $S_{\text{interaction}}$.

7.2.4. *Lorentz force*

Let us return to the relativistic equation of motion [7.62]. The temporal component of this equation gives:

$$m\frac{d(\gamma c)}{d\tau} = qF_{0\nu}U^\nu = qF_{0i}U^i = \frac{q\gamma}{c}\vec{\mathbf{E}} \cdot \vec{\mathbf{v}} \qquad [7.72]$$

since $F_{00} = 0$. Introducing the relativistic energy $\mathcal{E} = m\gamma c^2$, we obtain:

$$\frac{d\mathcal{E}}{d\tau} = q\gamma\vec{\mathbf{E}} \cdot \vec{\mathbf{v}} \qquad [7.73]$$

We see that the variation in energy is purely of electric origin. The other spatial components are written as:

$$\begin{aligned} m\frac{d(\gamma v_x)}{d\tau} &= -qF_{10}U^0 - qF_{12}U^2 - qF_{13}U^3 = q\gamma\left(E_x + v_yB_z - v_zB_y\right) \\ m\frac{d(\gamma v_y)}{d\tau} &= -qF_{20}U^0 - qF_{21}U^1 - qF_{23}U^3 = q\gamma\left(E_y + v_zB_x - v_xB_z\right) \\ m\frac{d(\gamma v_z)}{d\tau} &= -qF_{30}U^0 - qF_{31}U^1 - qF_{32}U^2 = q\gamma\left(E_z + v_xB_y - v_yB_x\right) \end{aligned} \qquad [7.74]$$

They can also be combined in the following form:

$$m\frac{d(\gamma\vec{\mathbf{v}})}{dt} = q\left(\vec{\mathbf{E}} + \vec{\mathbf{v}} \times \vec{\mathbf{B}}\right) \qquad [7.75]$$

They express the force experienced by a charge q in an electromagnetic field; this is the *relativistic Lorentz force*. Like the electromagnetic field itself, this force is inherently relativistic. Of course, when $\gamma \longrightarrow 1$, we recover the empirical formula for the Lorentz force introduced earlier. In terms of momentum, $\overrightarrow{\mathbf{p}} = m\gamma\overrightarrow{\mathbf{v}}$, equation [7.75] can also be written as:

$$\frac{d\overrightarrow{\mathbf{p}}}{dt} = q\left(\overrightarrow{\mathbf{E}} + \overrightarrow{\mathbf{v}} \times \overrightarrow{\mathbf{B}}\right) \qquad [7.76]$$

EXAMPLE 7.4. (Constant magnetic field).– Consider the motion of a charged particle in a constant magnetic field. The magnetic field does not modify the total energy of the particle. This implies that γ is constant, and the equation of motion simply reads:

$$m\gamma\frac{d\overrightarrow{\mathbf{v}}}{dt} = q\overrightarrow{\mathbf{v}} \times \overrightarrow{\mathbf{B}} \qquad [7.77]$$

The same applies as in the non-relativistic case (with the simple replacement of m by $m\gamma$) (see section 4.1.3). The trajectory is a helix around the magnetic field with a radius:

$$R = \gamma\frac{mv_\perp}{qB} \qquad [7.78]$$

where $v_\perp$ is the magnitude of the velocity perpendicular to the magnetic field. The frequency of the circular motion in the plane perpendicular to $\overrightarrow{\mathbf{B}}$ (cyclotron frequency) is equal to:

$$\omega_c = \frac{qB}{m\gamma} \qquad [7.79]$$

EXAMPLE 7.5. (Constant electric field).– Now consider a charged particle moving in a constant electric field. The particle is assumed to be initially at rest, and the electric field is directed along the Ox-axis. The equation of motion projected onto the Ox-axis reads:

$$m\frac{d\left(\gamma\dot{x}\right)}{dt} = qE \qquad [7.80]$$

The energy of the particle and thus the γ factor are not constant here. From this equation, we deduce that $\gamma\dot{x} = \left(1 - \frac{\dot{x}^2}{c^2}\right)^{-1/2}\dot{x} = \frac{qEt}{m}$ and, consequently, $\dot{x} = \left[1 + \left(\frac{qEt}{mc}\right)^2\right]^{-1/2}\frac{qEt}{mc}$. By integrating this last equation, we obtain:

$$x\left(t\right) = \frac{mc^2}{qE}\left[\sqrt{1 + \left(\frac{qEt}{mc}\right)^2} - 1\right] \qquad [7.81]$$

provided that we take $x\left(0\right) = 0$. For low speeds, the expression [7.81] reduces to $x\left(t\right) = \frac{qEt^2}{2m}$.

7.2.5. *Change of reference frames for the field*

Since the tensor $F^{\mu\nu}$ is of rank 2, we can easily write the transformation law for the fields when switching from one reference frame $Oxyz$ to another $O'x'y'z'$:

$$F'^{\mu\nu}\left(x'^{\alpha}=\Lambda^{\alpha}{}_{\beta}x^{\beta}\right)=\Lambda^{\mu}{}_{\rho}\Lambda^{\nu}{}_{\sigma}F^{\rho\sigma}\left(x^{\beta}\right) \qquad [7.82]$$

If $Oxyz$ and $O'x'y'z'$ are in relative motion with velocity V along Oz, then:

$$\mathcal{L}^{\mu}{}_{\nu}=\begin{pmatrix}\Gamma & 0 & 0 & -\beta\Gamma\\ 0 & 1 & 0 & 0\\ 0 & 0 & 0 & 1\\ \beta\Gamma & 0 & 0 & \Gamma\end{pmatrix} \qquad [7.83]$$

and, consequently, the new values of the electric and magnetic fields are given by:

$$\begin{aligned}
F'^{01}&=\mathcal{L}^{0}{}_{0}\mathcal{L}^{1}{}_{1}F^{01}+\mathcal{L}^{0}{}_{3}\mathcal{L}^{1}{}_{1}F^{31}\\
F'^{02}&=\mathcal{L}^{0}{}_{0}\mathcal{L}^{2}{}_{2}F^{02}+\mathcal{L}^{0}{}_{3}\mathcal{L}^{2}{}_{2}F^{32}\\
F'^{03}&=\mathcal{L}^{0}{}_{0}\mathcal{L}^{3}{}_{3}F^{03}+\mathcal{L}^{0}{}_{3}\mathcal{L}^{3}{}_{0}F^{30}\\
F'^{12}&=\mathcal{L}^{1}{}_{1}\mathcal{L}^{2}{}_{2}F^{12}\\
F'^{13}&=\mathcal{L}^{1}{}_{1}\mathcal{L}^{3}{}_{0}F^{10}+\mathcal{L}^{1}{}_{1}\mathcal{L}^{3}{}_{3}F^{13}\\
F'^{23}&=\mathcal{L}^{2}{}_{2}\mathcal{L}^{3}{}_{0}F^{20}+\mathcal{L}^{2}{}_{2}\mathcal{L}^{3}{}_{3}F^{23}
\end{aligned} \qquad [7.84]$$

which yields:

$$\begin{aligned}
E'_x&=\Gamma\left(E_x-VB_y\right)\\
E'_y&=\Gamma\left(E_y+VB_x\right)\\
E'_z&=E_z\\
B'_x&=\Gamma\left(B_x+\frac{V}{c^2}E_y\right)\\
B'_y&=\Gamma\left(B_y-\frac{V}{c^2}E_x\right)\\
B'_z&=B_z
\end{aligned} \qquad [7.85]$$

The inverse transformation is obtained by changing the sign of V. In vector form, these equations can be grouped as follows:

$$\begin{aligned}
\vec{\mathbf{E}}'_{\parallel}&=\vec{\mathbf{E}}_{\parallel}\\
\vec{\mathbf{E}}'_{\perp}&=\Gamma\left(\vec{\mathbf{E}}_{\perp}+\vec{\mathbf{V}}\times\vec{\mathbf{B}}\right)\\
\vec{\mathbf{B}}'_{\parallel}&=\vec{\mathbf{B}}_{\parallel}\\
\vec{\mathbf{B}}'_{\perp}&=\Gamma\left(\vec{\mathbf{B}}_{\perp}-\frac{1}{c^2}\vec{\mathbf{V}}\times\vec{\mathbf{E}}\right)
\end{aligned} \qquad [7.86]$$

A Galilean approximation can be obtained by retaining only the lowest order in $\frac{V}{c}$ in the preceding equations:

$$\begin{aligned} \vec{\mathbf{E}}' &= \vec{\mathbf{E}} + \vec{\mathbf{V}} \times \vec{\mathbf{B}} \\ \vec{\mathbf{B}}' &= \vec{\mathbf{B}} - \frac{1}{c^2}\vec{\mathbf{V}} \times \vec{\mathbf{E}} \end{aligned} \qquad [7.87]$$

EXAMPLE 7.6. (Electromagnetic field of a moving charged particle).– Let us calculate the electric and magnetic fields produced by a point charge particle q moving at a velocity $\vec{\mathbf{v}} = v\,\vec{\mathbf{k}}$. In the rest frame of the particle ($O'x'y'z'$), the fields are given by:

$$\begin{aligned} \vec{\mathbf{E}}' &= \frac{q\,\vec{\mathbf{r}}'}{4\pi\varepsilon_0 r'^3} \\ \vec{\mathbf{B}}' &= 0 \end{aligned} \qquad [7.88]$$

In the fixed reference frame $Oxyz$, these fields transform as follows:

$$\begin{aligned} \vec{\mathbf{E}} &= \frac{q}{4\pi\varepsilon_0 r'^3}\left(\gamma x', \gamma y', z'\right) \\ \vec{\mathbf{B}} &= \frac{q}{4\pi\varepsilon_0 r'^3}\left(-\frac{v\gamma}{c^2}y', \frac{v\gamma}{c^2}x', 0\right) \end{aligned} \qquad [7.89]$$

or, in terms of coordinates x, y, z, and t:

$$\begin{aligned} \vec{\mathbf{E}} &= \frac{q}{4\pi\varepsilon_0\left[x^2 + y^2 + \gamma^2\left(z - vt\right)^2\right]^{3/2}}\left(\gamma x, \gamma y, z - vt\right) \\ \vec{\mathbf{B}} &= \frac{q}{4\pi\varepsilon_0\left[x^2 + y^2 + \gamma^2\left(z - vt\right)^2\right]^{3/2}}\left(-\frac{v\gamma}{c^2}y, \frac{v\gamma}{c^2}x, 0\right) = \frac{1}{c^2}\vec{\mathbf{v}} \times \vec{\mathbf{E}} \end{aligned} \qquad [7.90]$$

EXAMPLE 7.7. (Electromagnetic field generated by an infinite wire).– We consider an infinite wire along the Oz-axis (the cross-sectional area S of the wire is very small). We always remain outside the wire. In the reference frame $Oxyz$ attached to the wire, the electric charges within the wire move in a uniform motion with velocity $\vec{\mathbf{v}} = v\,\vec{\mathbf{k}}$, thus creating in the wire a current density $\vec{\mathbf{j}} = \rho\vec{\mathbf{v}}$, where ρ is the charge density. The electric current on the wire is simply the flux of the vector of current density through the cross-section of the wire, that is, $I = \rho v S$. In the rest frame $O'x'y'z'$ of the electric charges, the charges are at rest: the current density $\vec{\mathbf{j}}\,'$ and the current I' are zero. Let M be an arbitrary point in $O'x'y'z'$, specified by the cylindrical coordinates (a', θ', z'). We denote by (a, θ, z) the coordinates of M in the frame $Oxyz$. Since a' is orthogonal to the relative motion of the two frames, we have $a' = a$. Similarly, the angle θ is given by $\tan\theta = \frac{y}{x} = \frac{y'}{x'} = \tan\theta'$, because x and y are not affected here by the Lorentz transformation. This implies $\theta' = \theta$. Since the electric field $\vec{\mathbf{E}}'(M)$ in

$O'x'y'z'$ is purely radial, it can easily be calculated using Gauss's theorem. This field is simply:

$$\vec{\mathbf{E}}'(a) = \frac{\rho' S}{2\pi\varepsilon_0 a'}\vec{e}_r = \frac{\rho' S}{2\pi\varepsilon_0 a}\vec{e}_r \qquad [7.91]$$

In the reference frame $O'x'y'z'$, the tensor $F'^{\mu\nu}$ is written as:

$$F'^{\mu\nu} = \begin{pmatrix} 0 & -E'_x/c = \frac{E'(a)}{c}\cos\theta & -E'_y/c = \frac{E'(a)}{c}\sin\theta & 0 \\ E'_x/c = \frac{E'(a)}{c}\cos\theta & 0 & 0 & 0 \\ E'_y/c = \frac{E'(a)}{c}\sin\theta & 0 & 0 & 0 \\ 0 & 0 & 0 & 0 \end{pmatrix} \qquad [7.92]$$

When going from the reference frame $Oxyz$ to the frame $O'x'y'z'$, the contravariant components $F^{\mu\nu}$ of the electromagnetic tensor transform according to the law $F'^{\mu\nu} = \Lambda^\mu{}_\alpha \Lambda^\nu{}_\beta F^{\alpha\beta}$, and conversely, $F^{\mu\nu} = \left(\Lambda^{-1}\right)^\mu{}_\alpha \left(\Lambda^{-1}\right)^\nu{}_\beta F'^{\alpha\beta}$. Using the fact that the matrices (Λ) and $\left(\Lambda^{-1}\right)$ are symmetric, we obtain $F^{\mu\nu} = \left(\Lambda^{-1}\right)^\mu_\alpha F'^{\alpha\beta} \left(\Lambda^{-1}\right)^\nu_\beta$. The last expression can be interpreted as a matrix product (since the Lorentz matrix is symmetric):

$$F^{\mu\nu} = \begin{pmatrix} \Gamma & 0 & 0 & \beta\gamma \\ 0 & 1 & 0 & 0 \\ 0 & 0 & 1 & 0 \\ \beta\gamma & 0 & 0 & \gamma \end{pmatrix} \begin{pmatrix} 0 & -\frac{E'(a)}{c}\cos\theta & -\frac{E'(a)}{c}\sin\theta & 0 \\ \frac{E'(a)}{c}\cos\theta & 0 & 0 & 0 \\ \frac{E'(a)}{c}\sin\theta & 0 & 0 & 0 \\ 0 & 0 & 0 & 0 \end{pmatrix} \times \begin{pmatrix} \Gamma & 0 & 0 & \beta\gamma \\ 0 & 1 & 0 & 0 \\ 0 & 0 & 1 & 0 \\ \beta\gamma & 0 & 0 & \gamma \end{pmatrix} \qquad [7.93]$$

which yields:

$$F^{\mu\nu} = \begin{pmatrix} 0 & -\gamma\frac{E'(a)}{c}\cos\theta & -\gamma\frac{E'(a)}{c}\sin\theta & 0 \\ \gamma\frac{E'(a)}{c}\cos\theta & 0 & 0 & \beta\gamma\frac{E'(a)}{c}\cos\theta \\ \gamma\frac{E'(a)}{c}\sin\theta & 0 & 0 & \beta\gamma\frac{E'(a)}{c}\sin\theta \\ 0 & -\beta\gamma\frac{E'(a)}{c}\cos\theta & -\beta\gamma\frac{E'(a)}{c}\sin\theta & 0 \end{pmatrix} \qquad [7.94]$$

Comparing this result with [7.69], we can deduce that:

$$
\begin{aligned}
E_x &= \gamma E'(a)\cos\theta = \gamma\frac{\rho' S}{2\pi\varepsilon_0 a}\cos\theta = \frac{\rho S}{2\pi\varepsilon_0 a}\cos\theta \\
E_y &= \gamma E'(a)\sin\theta = \gamma\frac{\rho' S}{2\pi\varepsilon_0 a}\sin\theta = \frac{\rho S}{2\pi\varepsilon_0 a}\sin\theta \\
E_z &= 0
\end{aligned}
\qquad [7.95]
$$

or, equivalently (in vector form):

$$
\vec{\mathbf{E}}(a) = \frac{\rho S}{2\pi\varepsilon_0 a}\vec{\mathbf{e}}_r \qquad [7.96]
$$

Working in the $Oxyz$ frame and applying the Gauss theorem once again, we naturally obtain the same result as that given by the Lorentz transformation. For the magnetic field, we find that:

$$
\begin{aligned}
B_x &= -\beta\gamma\frac{E'(a)}{c}\sin\theta = -\beta\gamma\frac{\rho' S}{2\pi\varepsilon_0 ac}\sin\theta = -\beta\frac{\rho S}{2\pi\varepsilon_0 ac}\sin\theta = -\frac{\mu_0 I}{2\pi a}\sin\theta \\
B_x &= \beta\gamma\frac{E'(a)}{c}\cos\theta = \beta\gamma\frac{\rho' S}{2\pi\varepsilon_0 ac}\cos\theta = \beta\frac{\rho S}{2\pi\varepsilon_0 ac}\cos\theta = \frac{\mu_0 I}{2\pi a}\cos\theta \\
B_z &= 0
\end{aligned}
\qquad [7.97]
$$

or, equivalently:

$$
\vec{\mathbf{B}}(a) = \frac{\mu_0 I}{2\pi a}\vec{\mathbf{e}}_\theta \qquad [7.98]
$$

The AmpÃ¨re theorem in the $Oxyz$ frame leads to the same result. The magnetic field appears to depend on the reference frame: in a frame where the charges are in motion, $\vec{\mathbf{B}}$ is non-zero. In the frame moving with the charges, the field $\vec{\mathbf{B}}$ vanishes. We can therefore interpret $\vec{\mathbf{B}}$ as a relativistic manifestation linked to the choice of reference frame for observing the system.

7.2.6. *Invariants of the electromagnetic field*

We can form four-scalars from the field tensor (invariants of the electromagnetic field). There are only two (the simplest ones) that are of physical interest:

– The quantity $F_{\mu\nu}F^{\mu\nu}$ is a four-scalar and therefore an invariant of the field. In terms of the electric and magnetic fields, it is written as:

$$
F_{\mu\nu}F^{\mu\nu} = 2F_{0i}F^{0i} + 2F_{12}F^{12} + 2F_{13}F^{13} + 2F_{23}F^{23} = \frac{2}{c^2}\left(c^2B^2 - E^2\right) \qquad [7.99]
$$

The quantity $c^2B^2 - E^2$ is therefore an invariant of the field. Let us give a direct application of this invariant. Consider an electromagnetic plane wave in a reference frame $Oxyz$. The magnitudes of the electric and magnetic fields are related by the relation $E = cB$. This implies that the field invariant under consideration is zero and, consequently, we have $E' = cB'$ in all reference frames.

– The other invariant is written as $\epsilon^{\mu\nu\rho\sigma}F_{\mu\nu}F_{\rho\sigma}$, where $\epsilon^{\mu\nu\rho\sigma}$ is the totally antisymmetric rank-4 tensor; it is equal to $+1$ if the four indices are an even permutation of $(0, 1, 2, 3)$, -1 if they form an odd permutation, and 0 in all other cases. This invariant contains 24 non-zero terms. In addition, these terms are grouped into sets of 8 that are equal to each other. In fact, given a specific permutation (μ, ν, ρ, σ), we obtain an identical form by swapping μ and ν and/or ρ and σ. Furthermore, we obtain an identical term by exchanging the first pair (μ, ν) with the second pair (ρ, σ). For our calculation, we choose the permutations $(0, 1, 2, 3)$, $(0, 2, 1, 3)$ and $(0, 3, 1, 2)$ (all the others can be obtained by applying the above operations). Since $\epsilon^{0123} = +1$, $\epsilon^{0213} = -1$ and $\epsilon^{0312} = +1$, we find:

$$\epsilon^{\mu\nu\rho\sigma}F_{\mu\nu}F_{\rho\sigma} = 8F_{01}F_{23} - 8F_{02}F_{13} + 8F_{03}F_{12} = -8\frac{\vec{\mathbf{E}}\cdot\vec{\mathbf{B}}}{c} \qquad [7.100]$$

The scalar product between the electric and magnetic fields is therefore invariant under a change of reference frame. Let us give a direct application of this property. For a plane wave, $\vec{\mathbf{E}}$ and $\vec{\mathbf{B}}$ are perpendicular, and their magnitudes are in a ratio of c. This property is observed in all reference frames for a wave.

7.2.7. *First group of Maxwell equations*

We have previously stated that the tensor $F_{\mu\nu}$ is the four-curl of the potential. In 3D, this property implies a vanishing divergence. We will now attempt to extend this result to 4D. The four-divergence of $F_{\mu\nu}$ is a rank-3 tensor that can be naturally expressed (by analogy with the 3D case) as:

$$\begin{aligned}\partial_\rho F_{\mu\nu} + \partial_\nu F_{\rho\mu} + \partial_\mu F_{\nu\rho} &= \partial_\rho\partial_\mu A_\nu - \partial_\rho\partial_\nu A_\mu + \partial_\nu\partial_\rho A_\mu - \partial_\nu\partial_\mu A_\rho \\ &\quad +\partial_\mu\partial_\nu A_\rho - \partial_\mu\partial_\rho A_\nu = 0\end{aligned} \qquad [7.101]$$

because derivatives commute. This equation is a direct consequence of the fact that the field is derived from a potential. To understand its physical meaning, write this equation in terms of electric and magnetic fields. First, note that there are $4^3 = 64$ possible equations, but only four of them are non-trivial:

– If the three indices are equal $(\rho = \mu = \nu)$, equation [7.101] reduces to $0 = 0$, because the diagonal terms of $F_{\mu\nu}$ are all zero ($F_{\mu\nu}$ is antisymmetric).

– If two of the three indices are equal, $\mu = \nu \neq \rho$, equation [7.101] gives $\partial_\rho F_{\mu\mu} + \partial_\mu (F_{\rho\mu} + F_{\mu\rho}) = 0$. This is also a trivial equation because of the antisymmetry of $F_{\mu\nu}$.

– It is therefore necessary to have $\mu \neq \nu \neq \rho$ in order to obtain a non-trivial equation. Note that equation [7.101] is obtained for the ordered choice $[\rho, \mu, \nu]$. For $[\rho, \nu, \mu]$, we get the same result. In fact, we have $0 = \partial_\rho F_{\nu\mu} + \partial_\mu F_{\rho\nu} + \partial_\nu F_{\mu\rho} = -(\partial_\rho F_{\mu\nu} + \partial_\mu F_{\nu\rho} + \partial_\nu F_{\rho\mu})$. The order of the indices ρ, μ and ν does not matter. There are only four possibilities:

1) For $[\rho, \mu, \nu] = [1, 2, 3]$, we write $\partial_1 F_{23} + \partial_3 F_{12} + \partial_2 F_{31} = 0$, which implies $\partial_x (-B_x) + \partial_y (-B_y) + \partial_z (-B_z) = \vec{\nabla} \cdot \vec{\mathbf{B}} = 0$.

2) For $[\rho, \mu, \nu] = [0, 1, 2]$, we find $0 = \partial_0 F_{12} + \partial_2 F_{01} + \partial_1 F_{20} = \frac{\partial}{\partial(ct)} (-B_z) + \frac{\partial}{\partial x} (-E_y/c) + \frac{\partial}{\partial y} (E_x/c)$, which gives $\left(\vec{\nabla} \times \vec{\mathbf{E}}\right)_z = -\frac{\partial B_z}{\partial t}$.

3) For $[\rho, \mu, \nu] = [0, 1, 3]$, we similarly obtain $\left(\vec{\nabla} \times \vec{\mathbf{E}}\right)_y = -\frac{\partial B_y}{\partial t}$.

4) For $[\rho, \mu, \nu] = [0, 2, 3]$, we also get $\left(\vec{\nabla} \times \vec{\mathbf{E}}\right)_x = -\frac{\partial B_x}{\partial t}$.

Equation [7.101] is therefore equivalent to the two Maxwell equations (first group of Maxwell equations):

$$\begin{aligned} &\vec{\nabla} \cdot \vec{\mathbf{B}} = 0 \\ &\vec{\nabla} \times \vec{\mathbf{E}} = -\frac{\partial \vec{\mathbf{B}}}{\partial t} \end{aligned} \qquad [7.102]$$

7.3. Sources of the field: second group of Maxwell equations

In this section, we establish the equations that relate the field tensor to its sources, that is, to the motion of charged particles. In the previous section, we studied the motion of a single particle, so the dynamical variables of the problem were the position and momentum of the particle. In what follows, we assume that the particle dynamics are known (i.e. the current) and we will focus on the dynamics of the field. The dynamical variables are thus the values of the potential or the fields at all points in space and at every instant. Now consider a set of charged particles in motion (the current). It is known that these particles generate an additional field of the same nature as the one that causes their motion. To obtain a complete description, it is therefore necessary to generalize the action [7.53] to multiple charges, then add a term that describes the self-interaction of the field (i.e. the free field itself).

7.3.1. *Interaction between a field and a current*

Consider a set of charged particles whose motion is known in advance. Each particle is identified by an index (k). The four-current vector can thus be written as $J^{\mu} = \left(c\rho, \overrightarrow{\mathbf{v}}\right)$, with:

$$\begin{aligned} \rho &= \sum_{(k)} q_{(k)} \delta^3 \left(\overrightarrow{\mathbf{r}} - \overrightarrow{\mathbf{r}}_{(k)}(t)\right) \\ \overrightarrow{\mathbf{j}} &= \sum_{(k)} q_{(k)} \overrightarrow{\mathbf{v}}_{(k)} \delta^3 \left(\overrightarrow{\mathbf{r}} - \overrightarrow{\mathbf{r}}_{(k)}(t)\right) \end{aligned} \qquad [7.103]$$

where $q_{(k)}$, $\overrightarrow{\mathbf{r}}_{(k)}(t)$ and $\overrightarrow{\mathbf{v}}_{(k)}$ are, respectively, the charge, position and velocity of particle (k). The generalization of action [7.53] to a set of charges is straightforward. The superposition principle allows us to write:

$$S_{\text{interaction}} = -\sum_{(k)} q_{(k)} \int_1^2 A_{\mu}\left(x^{\nu}_{(k)}\right) U^{\mu}_{(k)} d\tau_{(k)} \qquad [7.104]$$

where $x^{\nu}_{(k)}$ is the position of the particle (k), and $\tau_{(k)}$ is the proper time associated with the particle. Due to time dilation, we write $d\tau_{(k)} = \frac{dt}{\gamma_{(k)}}$, where $\gamma_{(k)}$ is the *Lorentz factor*. Substituting this expression into the interaction action [7.104], we obtain:

$$S_{\text{interaction}} = -\int_1^2 dt \left[\sum_{(k)} A_{\mu}\left(x^{\nu}_{(k)}\right) \begin{Bmatrix} q_{(k)} c \\ q_{(k)} \overrightarrow{\mathbf{v}}_{(k)} \end{Bmatrix}\right] \qquad [7.105]$$

where $U^{\mu}_{(k)} = \left(\gamma_{(k)} c, \gamma_{(k)} \overrightarrow{\mathbf{v}}_{(k)}\right)$. This notation was used to show the expressions of ρ and $\overrightarrow{\mathbf{j}}$. To put the expression [7.105] into integral form, we use the Dirac delta function:

$$A_{\mu}\left(x^{\nu}_{(k)}\right) = \int d\tau A_{\mu}\left(ct, \overrightarrow{\mathbf{r}}\right) \delta^3\left(\overrightarrow{\mathbf{r}} - \overrightarrow{\mathbf{r}}_{(k)}(t)\right) \qquad [7.106]$$

The integral here is taken over all space. It follows that:

$$S_{\text{interaction}} = -\int_1^2 dt \int d\tau A_{\mu}\left(ct, \overrightarrow{\mathbf{r}}\right) \left[\sum_{(k)} \begin{Bmatrix} q_{(k)} c \\ q_{(k)} \overrightarrow{\mathbf{v}}_{(k)} \end{Bmatrix} \delta^3\left(\overrightarrow{\mathbf{r}} - \overrightarrow{\mathbf{r}}_{(k)}(t)\right)\right]$$

[7.107]

We recognize, within the brackets, the expression of the four-current vector. Finally, we deduce that:

$$S_{\text{interaction}} = -\int\int A_{\mu} J^{\mu} dt d\tau \qquad [7.108]$$

We therefore conclude that the interaction action takes into account all charged particles and their motions via the current density. The integration element here is $dtd\tau$. This is the volume element (up to a factor of c) in Minkowski spacetime (a relativistic invariant). The quantity $A_\mu J^\mu$ is the Lagrangian density of the field–source interaction.

7.3.2. *Free field action (field self-interaction)*

Imagine a region of spacetime in which there are no particles, that is, $J^\mu = 0$, but only a field: this is referred to as a free field, just as a particle is considered free when it is subject to no force and hence to no field. A simple way to construct the action of a free field is to base it on the tensor $F_{\mu\nu}$. Since $F_{\mu\nu}$ is gauge invariant, it immediately follows that such an action would also be gauge invariant. The action must also be a Lorentz scalar and quadratic in the field. Indeed, a Lagrangian density is homogeneous to an energy density, which must be a quadratic function of the dynamical variables. The two scalar field invariants given in section 7.2.6 satisfy these conditions, except that the scalar $\epsilon^{\mu\nu\alpha\beta}F_{\mu\nu}F_{\alpha\beta}$ can be written as a four-divergence: it thus falls within the gauge freedom of the action. In fact, we can write:

$$\begin{aligned}\epsilon^{\mu\nu\alpha\beta}F_{\mu\nu}F_{\alpha\beta} - 2\epsilon^{\mu\nu\alpha\beta}\partial_\mu A_\nu F_{\alpha\beta} &= \partial_\mu\left(2\epsilon^{\mu\nu\alpha\beta}A_\nu F_{\alpha\beta}\right) + 2\epsilon^{\nu\mu\alpha\beta}A_\nu\partial_\mu F_{\alpha\beta} \\ &= \partial_\mu\left(2\epsilon^{\mu\nu\alpha\beta}A_\nu F_{\alpha\beta}\right) + \frac{2}{3}\epsilon^{\nu\mu\alpha\beta}A_\nu\left[\partial_\mu F_{\alpha\beta} + \partial_\beta F_{\mu\alpha} + \partial_\alpha F_{\beta\mu}\right] \\ &= \partial_\mu\left(2\epsilon^{\mu\nu\alpha\beta}A_\nu F_{\alpha\beta}\right)\end{aligned} \qquad [7.109]$$

Thus, only $F^{\mu\nu}F_{\mu\nu}$ remains to construct an action. We will therefore try the quantity:

$$S_{\text{free field}} = \kappa\int\int F_{\mu\nu}F^{\mu\nu}dtd\tau \qquad [7.110]$$

where κ is a constant to be determined. The total action of electromagnetism is thus written as:

$$S = S_{\text{free particles}} + \int_{t_1}^{t_2} dt\int d\tau\left[-A_\mu J^\mu + \kappa F_{\mu\nu}F^{\mu\nu}\right] \qquad [7.111]$$

In this expression, the action of free particles is a simple constant, since the dynamics of the latter are known.

To find the equations governing the dynamics of a free particle, we fixed the boundary conditions of its motion, namely x^μ (1) and x^μ (2), and then looked for the minimum of the action under variations $x^\mu \longrightarrow x^\mu + \delta x^\mu$. Another type of variation is possible for the action [7.111]: the tensor $F_{\mu\nu}$ depends on (x^ν) only through the four-potential (A^α). For a given particle current, we can therefore consider calculating the variation of the action δS corresponding to a variation $A^\mu \longrightarrow A^\mu + \delta A^\mu$, while imposing fixed boundary conditions $A^\mu(1)$ and $A^\mu(2)$. The variation of the total action is therefore:

$$\delta S = 0 + \int_{t_1}^{t_2} dt \int d\tau \left[-(\delta A_\mu) J^\mu + \kappa\delta(F_{\mu\nu}F^{\mu\nu})\right] \quad [7.112]$$

However, we know that $\delta(F_{\mu\nu}F^{\mu\nu}) = 2F^{\mu\nu}\delta F_{\mu\nu}$ and $\delta F_{\mu\nu} = \partial_\mu(\delta A_\nu) - \partial_\nu(\delta A_\mu)$ (because the variations δ and ∂_μ commute). Thus, we obtain:

$$\begin{aligned}\delta(F_{\mu\nu}F^{\mu\nu}) &= 2F^{\mu\nu}\partial_\mu(\delta A_\nu) - 2F^{\mu\nu}\partial_\nu(\delta A_\mu) \\ &= 2F^{\nu\mu}\partial_\nu(\delta A_\mu) - 2F^{\mu\nu}\partial_\nu(\delta A_\mu) = -4F^{\mu\nu}\partial_\nu(\delta A_\mu)\end{aligned} \quad [7.113]$$

Substituting [7.113] into [7.112] yields:

$$\delta S = -\int d\tau \int_{t_1}^{t_2} dt \left[J^\mu \delta A_\mu + 4\kappa F^{\mu\nu}\partial_\nu(\delta A_\mu)\right] \quad [7.114]$$

Integrating the second term by parts and noting that the boundary term always vanishes, we remain with:

$$\delta S = -\int d\tau \int_{t_1}^{t_2} dt \left[J^\mu - 4\kappa\partial_\nu F^{\mu\nu}\right]\delta A_\mu = 0 \quad [7.115]$$

Since expression [7.115] must vanish for all variations δA_μ, it follows that the field satisfies the equation:

$$J^\mu = 4\kappa\partial_\nu F^{\mu\nu} \quad [7.116]$$

To identify the constant κ, we now express the previous equation in terms of the observables $\overrightarrow{\mathbf{E}}$ and $\overrightarrow{\mathbf{B}}$. In fact, the time component of this vector equation reads $J^0 = 4\kappa, \partial_\nu F^{0\nu}$, which gives $\rho c = 4\kappa\left(\frac{\partial 0}{\partial(ct)} + \frac{\partial(-E_x/c)}{\partial x} + \frac{\partial(-E_y/c)}{\partial y} + \frac{\partial(-E_z/c)}{\partial z}\right)$, or, equivalently, $\overrightarrow{\nabla} \cdot \overrightarrow{\mathbf{E}} = -\frac{\rho c^2}{4\kappa}$. By choosing $-\frac{1}{4\kappa} = \mu_0$, we recover the Maxwell–Gauss equation:

$$\overrightarrow{\nabla} \cdot \overrightarrow{\mathbf{E}} = \frac{\rho}{\varepsilon_0} \quad [7.117]$$

For $\mu = 1$, we obtain $J^1 = 4\kappa\partial_\nu F^{1\nu} = -\frac{1}{\mu_0}\left(\frac{\partial(E_x/c)}{\partial(ct)} + \frac{\partial 0}{\partial x} + \frac{\partial(-B_z)}{\partial y} + \frac{(B_y)}{\partial z}\right)$, which implies $\left(\vec{\nabla} \times \vec{\mathbf{B}}\right)_x = \left(\mu_0 \vec{\mathbf{j}} + \frac{1}{c^2}\frac{\partial \vec{\mathbf{E}}}{\partial t}\right)_x$. For $\mu = 2$ and 3, we find the two other components of the equation:

$$\vec{\nabla} \times \vec{\mathbf{B}} = \mu_0 \vec{\mathbf{j}} + \frac{1}{c^2}\frac{\partial \vec{\mathbf{E}}}{\partial t} \qquad [7.118]$$

The Euler–Lagrange equations that determine the field in terms of sources are thus simply:

$$\partial_\mu F^{\mu\nu} = \mu_0 J^\nu \qquad [7.119]$$

They encompass the second group of Maxwell equations (linking the fields $\vec{\mathbf{E}}$ and $\vec{\mathbf{B}}$ to the sources). Equation [7.119] can also be written in terms of the components of the four-potential. Using the expression for $F^{\mu\nu}$, it is found that:

$$\partial_\mu \partial^\mu A^\nu - \partial_\mu \partial^\nu A^\mu = \mu_0 J^\nu \qquad [7.120]$$

If we require the four-potential to satisfy the Lorenz gauge condition $\partial_\mu A^\mu = 0$, the second term vanishes (by swapping the partial derivatives). Equation [7.120] then becomes:

$$\partial_\mu \partial^\mu A^\nu = \mu_0 J^\nu \qquad [7.121]$$

The differential operator on the left is none other than the square of the four-gradient; it is the d'Alembert operator or the four-Laplacian. In terms of fields V and $\vec{\mathbf{A}}$, the previous equation yields:

$$\left(\frac{1}{c^2}\frac{\partial^2}{\partial t^2} - \Delta\right)\left\{\begin{matrix} \vec{\mathbf{A}} \\ V \end{matrix}\right\} = \left\{\begin{matrix} \mu_0 \vec{\mathbf{j}} \\ \frac{\rho}{\varepsilon_0} \end{matrix}\right\} \qquad [7.122]$$

7.3.3. *Magnetization and polarization*

Since the components of the fields $\vec{\mathbf{E}}$ and $\vec{\mathbf{B}}$ are those of the rank-2 tensor $F_{\mu\nu}$, we may expect that the components of $\vec{\mathbf{D}} = \varepsilon_0 \vec{\mathbf{E}} + \vec{\mathbf{P}}$ and $\vec{\mathbf{H}} = \frac{\vec{\mathbf{B}}}{\mu_0} - \vec{\mathbf{M}}$ are also those of another rank-2 tensor, which we will denote by $H_{\mu\nu}$. This tensor can be constructed by writing:

$$H_{\mu\nu} = \varepsilon_0 c^2 F_{\mu\nu} + P_{\mu\nu} \qquad [7.123]$$

where $P_{\mu\nu}$ is a second-rank tensor containing magnetization and polarization. The (01)-component gives $H_{01} = \varepsilon_0 c E_x + P_{01}$, which implies that $P_{01} = cP_x$ and $H_{01} = cD_x$. Similarly, the (13)-component gives $H_{13} = \varepsilon_0 c^2 B_y + P_{13} = \frac{B_y}{\mu_0} + P_{13} = H_y$, which allows us to deduce $P_{13} = -M_y$ and $H_{13} = H_y$. Continuing in the same manner for the other components, we construct the tensors $H_{\mu\nu}$ and $P_{\mu\nu}$ as follows:

$$\begin{pmatrix} 0 & cD_x & cD_y & dD_z \\ -cD_x & 0 & -H_z & H_y \\ -cD_y & H_z & 0 & -H_x \\ -cD_z & -H_y & H_x & 0 \end{pmatrix} = \frac{1}{\mu_0} F_{\mu\nu} + \begin{pmatrix} 0 & cP_x & cP_y & cP_z \\ -cP_x & 0 & M_z & -M_y \\ -cP_y & -M_z & 0 & M_x \\ -cP_z & M_y & -M_x & 0 \end{pmatrix} \qquad [7.124]$$

The Maxwell equations for $\vec{\mathbf{D}}$ and $\vec{\mathbf{H}}$ in a material medium can now be written in terms of $H^{\mu\nu}$ as follows:

$$\partial_\mu H^{\mu\nu} = J^\nu \qquad [7.125]$$

The transformation laws for magnetization and polarization of a moving medium can also be obtained directly (by analogy with [7.86]):

$$\begin{aligned} \vec{\mathbf{P}}'_\parallel &= \vec{\mathbf{P}}_\parallel \\ \vec{\mathbf{M}}'_\parallel &= \vec{\mathbf{M}}_\parallel \\ \vec{\mathbf{P}}'_\perp &= \Gamma\left(\vec{\mathbf{P}}_\perp - \frac{\vec{\mathbf{V}}}{c^2} \times \vec{\mathbf{M}}_\perp\right) \\ \vec{\mathbf{M}}'_\perp &= \Gamma\left(\vec{\mathbf{M}}_\perp + \frac{\vec{\mathbf{V}}}{c^2} \times \vec{\mathbf{P}}_\perp\right) \end{aligned} \qquad [7.126]$$

7.4. Energy–momentum tensor

Consider the case of a free field (i.e. without particles). Since the Lagrangian density $\mathcal{L}_{\text{free field}} = -\frac{1}{4\mu_0} F\mu\nu F^{\mu\nu}$ does not explicitly depend on x, it follows from the definition [7.39] that the canonical energy–momentum tensor is:

$$T^{\mu\nu} = \frac{\partial \mathcal{L}_{\text{free field}}}{\partial\left[\partial_\mu A_\sigma\right]} \partial_\mu A_\sigma - g^{\mu\nu} \mathcal{L}_{\text{free field}} = -\frac{1}{\mu_0} F^{\mu\sigma} \partial^\nu A_\sigma + \frac{1}{4\mu_0} g^{\mu\nu} F_{\sigma\rho} F^{\sigma\rho} \qquad [7.127]$$

is conserved. Defining $\Psi^{\mu\nu\sigma} = \frac{1}{\mu_0} F^{\mu\sigma} A^\nu$, we see that $T^{\mu\nu} + \partial_\sigma \Psi^{\mu\nu\sigma}$ (which we denote by $\theta^{\mu\nu}$) has the following expression:

$$\begin{aligned} \theta^{\mu\nu} &= -\frac{1}{\mu_0} F^{\mu\sigma} \partial^\nu A_\sigma + \frac{1}{\mu_0} F^{\mu\sigma} \partial_\sigma A^\nu + \frac{1}{4\mu_0} g^{\mu\nu} F_{\sigma\rho} F^{\sigma\rho} \\ &= -\frac{1}{\mu_0} F^{\mu\sigma} F^\nu{}_\sigma + \frac{1}{4\mu_0} g^{\mu\nu} F_{\sigma\rho} F^{\sigma\rho} \end{aligned} \qquad [7.128]$$

The tensor is clearly symmetric at μ and ν. Contrary to [7.127], it is also gauge invariant. The manipulation that was just carried out is only valid in the absence of any particles, because it uses the fact that $\partial_\sigma F^{\sigma\nu} = 0$ (which makes it possible to write $\partial_\sigma \Psi^{\mu\nu\sigma} = \frac{1}{\mu_0} F^{\mu\sigma}\partial_\sigma A^\nu$). The case of the field in interaction with particles is addressed in section 7.4.1. It is clearly verified that the tensor $\theta^{\mu\nu}$ is also conserved (in the absence of any current). Indeed, we have:

$$\begin{aligned}\partial_\mu \theta^{\mu\nu} &= \partial_\mu \theta^{\mu\nu} + \frac{1}{\mu_0}\partial_\mu \left(F^{\mu\sigma}\partial_\sigma A^\nu\right) = 0 + \frac{1}{\mu_0} F^{\mu\sigma}\partial_\mu\partial_\sigma A^\nu \\ &= \frac{1}{2\mu_0} F^{\mu\sigma}\left(\partial_\mu\partial_\sigma - \partial_\sigma\partial_\mu\right) A^\nu = 0 \end{aligned} \qquad [7.129]$$

Let us express the terms $\theta^{\mu\nu}$ in terms of electric and magnetic fields. The second term is the product of the metric tensor, which is diagonal, by the invariant $2\left(B^2 - \frac{E^2}{c^2}\right)$, the Lagrangian density of the electromagnetic field. Thus, we obtain:

$$\theta^{\mu\nu} = -\frac{1}{\mu_0} F^{\mu\sigma} F^{\nu}{}_{\sigma} + \frac{1}{2\mu_0} g^{\mu\nu}\left(B^2 - \frac{E^2}{c^2}\right) \qquad [7.130]$$

whose components are:

$$\begin{aligned}
\theta^{00} &= -\frac{1}{\mu_0} F^{0\sigma} F^{0}{}_{\sigma} + \frac{1}{2\mu_0}\left(B^2 - \frac{E^2}{c^2}\right) = -\frac{1}{\mu_0}\left(-\frac{E^2}{c^2} - \frac{1}{2}B^2 + \frac{E^2}{c^2}\right) \\
&= \frac{1}{2}\left(\varepsilon_0 E^2 + \frac{B^2}{\mu_0}\right) \\
\theta^{11} &= -\frac{1}{\mu_0}\left(F^{1\sigma} F^{1}{}_{\sigma} + \frac{1}{2}\left(B^2 - \frac{E^2}{c^2}\right)\right) = -\frac{1}{\mu_0}\left(\frac{E_x^2}{c^2} - B_z^2 - B_y^2\right. \\
&\quad \left. + \frac{B^2}{2} - \frac{E^2}{2c^2}\right) = \varepsilon_0\left(\frac{E^2}{2} - E_x^2\right) + \frac{1}{\mu_0}\left(\frac{B^2}{2} - B_x^2\right) \\
\theta^{22} &= \varepsilon_0\left(\frac{E^2}{2} - E_y^2\right) + \frac{1}{\mu_0}\left(\frac{B^2}{2} - B_y^2\right) \\
\theta^{33} &= \varepsilon_0\left(\frac{E^2}{2} - E_z^2\right) + \frac{1}{\mu_0}\left(\frac{B^2}{2} - B_z^2\right) \\
\theta^{01} &= \theta^{10} = -\frac{1}{\mu_0} F^{0\sigma} F^{1}{}_{\sigma} = \frac{1}{\mu_0}\left(\frac{E_y B_z - E_z B_y}{c}\right) \\
&= \frac{1}{\mu_0}\left(\frac{\overrightarrow{\mathbf{E}} \times \overrightarrow{\mathbf{B}}}{c}\right)_x \\
\theta^{02} &= \theta^{20} = \frac{1}{\mu_0}\left(\frac{\overrightarrow{\mathbf{E}} \times \overrightarrow{\mathbf{B}}}{c}\right)_y
\end{aligned}$$

$$\begin{aligned}
\theta^{03} = \theta^{30} &= \frac{1}{\mu_0}\left(\frac{\vec{\mathbf{E}} \times \vec{\mathbf{B}}}{c}\right)_z \\
\theta^{12} = \theta^{21} &= -\frac{1}{\mu_0} F^{1\sigma} F^{2}{}_{\sigma} = -\frac{1}{\mu_0}\left(\frac{E_x E_y}{c^2} + B_x B_y\right) \\
\theta^{13} = \theta^{31} &= -\frac{1}{\mu_0}\left(\frac{E_x E_z}{c^2} + B_x B_z\right) \\
\theta^{23} = \theta^{32} &= -\frac{1}{\mu_0}\left(\frac{E_y E_z}{c^2} + B_y B_z\right)
\end{aligned} \qquad [7.131]$$

These elements can be grouped in matrix form:

$$\theta^{\mu\nu} = \begin{pmatrix} w_{\text{em}} & \Pi_x/c \;\; \Pi_y/c \;\; \Pi_z/c \\ \Pi_x/c & \\ \Pi_y/c & (\mathcal{T}_{ij}) \\ \Pi_z/c & \end{pmatrix} \qquad [7.132]$$

where:

$$\begin{aligned}
w_{\text{em}} &= \frac{1}{2}\left(\varepsilon_0 E^2 + \frac{B^2}{\mu_0}\right) \\
\vec{\mathbf{\Pi}} &= \frac{\vec{\mathbf{E}} \times \vec{\mathbf{B}}}{\mu_0}
\end{aligned} \qquad [7.133]$$

We clearly recognize here the electromagnetic energy density and the Poynting vector. $(\mathcal{T}_{ij})$ is a purely spatial rank-2 tensor, called the *Maxwell stress tensor*, whose components are given by:

$$\mathcal{T}_{ij} = \mathcal{T}^{ij} = \varepsilon_0\left[\frac{E^2}{2}\delta^{ij} - E^i E^j\right] + \mu_0\left[\frac{B^2}{2}\delta^{ij} - B^i B^j\right] \qquad [7.134]$$

where δ^{ij} is the Kronecker delta.

7.4.1. *Conservation law*

Let us compute the four-divergence of the tensor $\theta^{\mu\nu}$:

$$\partial_\mu \theta^{\mu\nu} = \frac{1}{\mu_0}\left[\partial^\mu\left(F_{\mu\sigma}F^{\sigma\nu}\right) + \frac{1}{4}\partial^\nu\left(F_{\sigma\rho}F^{\sigma\rho}\right)\right] \qquad [7.135]$$

The last term can be written as $\partial^\nu\left(F_{\sigma\rho}F^{\sigma\rho}\right) = \left(\partial^\nu F_{\sigma\rho}\right)F^{\sigma\rho} + F_{\sigma\rho}\left(\partial^\nu F^{\sigma\rho}\right)$. These two terms are, of course, equal (raising and lowering the indices). By expanding the first term, we thus obtain:

$$
\begin{aligned}
\partial_\mu \theta^{\mu\nu} &= \frac{1}{\mu_0}\left[\left(\partial^\mu F_{\mu\sigma}\right)F^{\sigma\nu} + F_{\mu\sigma}\left(\partial^\mu F^{\sigma\nu}\right) + \frac{1}{2}F_{\sigma\rho}\left(\partial^\nu F^{\sigma\rho}\right)\right] \\
&= \frac{1}{\mu_0}\left[\mu_0 J_\sigma F^{\sigma\nu} + F_{\mu\sigma}\left(\partial^\mu F^{\sigma\nu}\right) + \frac{1}{2}F_{\sigma\rho}\left(\partial^\nu F^{\sigma\rho}\right)\right]
\end{aligned} \qquad [7.136]
$$

Using the trivial transformation $F_{\mu\sigma}\left(\partial^\mu F^{\sigma\nu}\right) = \frac{1}{2}F_{\mu\sigma}\left(\partial^\mu F^{\sigma\nu}\right) + \frac{1}{2}F_{\sigma\mu}\left(\partial^\sigma F^{\mu\nu}\right)$ (by swapping the indices), we deduce:

$$
\begin{aligned}
\partial_\mu \theta^{\mu\nu} - J_\sigma F^{\sigma\nu} &= \frac{1}{2\mu_0}\left[F_{\mu\sigma}\left(\partial^\mu F^{\sigma\nu}\right) + F_{\mu\sigma}\left(\partial^\sigma F^{\nu\mu}\right) + F_{\sigma\rho}\left(\partial^\nu F^{\sigma\rho}\right)\right] \\
&= \frac{1}{2\mu_0}F_{\mu\sigma}\left[\partial^\mu F^{\sigma\nu} + \partial^\sigma F^{\nu\mu} + \partial^\sigma F^{\nu\mu}\right]
\end{aligned} \qquad [7.137]
$$

The term on the right is obviously zero (see [7.101]). Finally, we arrive at the following:

$$
\partial_\mu \theta^{\mu\nu} = J_\sigma F^{\sigma\nu} = -F^{\nu\sigma} J_\sigma \qquad [7.138]
$$

This is the local conservation equation for the tensor $\theta^{\mu\nu}$ in the presence of a current. To give a physical interpretation of equation [7.138], we write separately the time component and spatial components in terms of energy density, Poynting vector and Maxwell tensor.

7.4.2. *Time component: conservation of energy*

The time component $\nu = 0$ is simply written as $\partial_\mu \theta^{\mu 0} = -F^{0\sigma} J_\sigma$. Since $F^{0\sigma} J_\sigma = F^{01} J_1 + F^{02} J_2 + F^{03} J_3 = \frac{\vec{\mathbf{E}}\cdot\vec{\mathbf{j}}}{c}$ and $\partial_\mu \theta^{\mu 0} = \partial_0 \theta^{00} + \partial_1 \theta^{10} + \partial_2 \theta^{20} + \partial_3 \theta^{30} = \frac{1}{c}\frac{\partial w_{\text{em}}}{\partial t} + \frac{1}{c}\vec{\nabla}\cdot\vec{\mathbf{\Pi}}$, we obtain:

$$
\vec{\mathbf{j}}\cdot\vec{\mathbf{E}} + \frac{\partial w_{\text{em}}}{\partial t} + \vec{\nabla}\cdot\vec{\mathbf{\Pi}} = 0 \qquad [7.139]
$$

We recognize here the electromagnetic energy conservation equation that was already established in previous chapters. Recall that $\vec{\mathbf{j}}\cdot\vec{\mathbf{E}} = \vec{\mathbf{f}}\cdot\vec{\mathbf{v}}$ (where $\vec{\mathbf{f}}$ is the Lorentz force density) represents the power density transferred from the field to matter. A global balance equation can be written by integrating over a volume τ bounded by a surface S:

$$
\int_\tau \vec{\mathbf{j}}\cdot\vec{\mathbf{E}}\,d\tau + \frac{dW_{\text{em}}}{dt} + \oint_S \vec{\mathbf{\Pi}}\cdot d\vec{\mathbf{S}} = 0 \qquad [7.140]
$$

where $W_{\text{em}} = \int_\tau w_{\text{em}} d\tau$ is the total electromagnetic energy within the volume τ. We therefore conclude that the time variation of the electromagnetic energy is equal to the opposite of the total power transferred to matter plus the incoming flux of the Poynting vector through the surface S. The Poynting vector indeed describes the transport of electromagnetic energy.

7.4.3. *Spatial components: conservation of momentum*

The spatial components are written as $\partial_\mu \theta^{\mu i} = -F^{i\sigma} J_\sigma$, or equivalently:

$$\begin{aligned} \partial_0 \left(\frac{\Pi_x}{c}\right) + \left(\vec{\nabla} \cdot \overleftrightarrow{\mathcal{T}}\right)_x &= -\left(\rho E_x + J_y B_z - J_z B_y\right) \equiv -f_x \\ \partial_0 \left(\frac{\Pi_y}{c}\right) + \left(\vec{\nabla} \cdot \overleftrightarrow{\mathcal{T}}\right)_y &= -\left(\rho E_y + J_z B_x - J_x B_z\right) \equiv -f_y \\ \partial_0 \left(\frac{\Pi_z}{c}\right) + \left(\vec{\nabla} \cdot \overleftrightarrow{\mathcal{T}}\right)_z &= -\left(\rho E_z + J_x B_y - J_y B_x\right) \equiv -f_z \end{aligned} \qquad [7.141]$$

which can also be combined in the form:

$$\frac{\partial \vec{\mathbf{p}}_{\text{em}}}{\partial t} + \vec{\mathbf{f}} + \vec{\nabla} \cdot \overleftrightarrow{\mathcal{T}} = 0 \qquad [7.142]$$

where $\vec{\nabla} \cdot \overleftrightarrow{\mathcal{T}}$ denotes the vector field $\partial_i \mathcal{T}_{ij}$. We have defined here:

$$\vec{\mathbf{p}}_{\text{em}} = \frac{\vec{\mathbf{\Pi}}}{c^2} = \frac{\vec{\mathbf{E}} \times \vec{\mathbf{B}}}{\mu_0 c^2} \qquad [7.143]$$

To understand the physical meaning of equation [7.142], let us integrate, as before, on a volume τ bounded by the surface S. Thus, we obtain:

$$\frac{d\vec{\mathbf{P}}_{\text{em}}}{dt} + \frac{d\vec{\mathbf{P}}_{\text{mec.}}}{dt} + \oint_S \overleftrightarrow{\mathcal{T}} \cdot d\vec{\mathbf{S}} = 0 \qquad [7.144]$$

where $\overleftrightarrow{\mathcal{T}} \cdot d\vec{\mathbf{S}}$ denotes the contraction of the tensor $\overleftrightarrow{\mathcal{T}}$ with the surface element. We have also defined:

$$\frac{d\vec{\mathbf{P}}_{\text{mec.}}}{dt} = \int_\tau \vec{\mathbf{f}}\, d\tau \qquad [7.145]$$

This term therefore represents the variation over time of the momentum of the matter contained within the volume τ. The term:

$$\vec{\mathbf{P}}_{\text{em}} = \int_\tau \vec{\mathbf{p}}_{\text{em}} d\tau \qquad [7.146]$$

can be seen as the time variation of the total momentum of electromagnetic radiation. $\vec{p}_{em}$ therefore simply represents the local momentum density of the field. It is, up to a constant factor, the Poynting vector that describes the transport of electromagnetic energy. The last term in [7.144] can be interpreted as the momentum flowing out per unit time from the volume τ. This vector quantity appears as the flux of a tensorial quantity.

EXAMPLE 7.8. (Electromagnetic momentum of a non-relativistic charge).– Let us calculate the momentum of the field of a non-relativistic point charge moving at a velocity $v \ll c$ along the Oz axis. In the rest frame of the charge, the energy–momentum tensor is written as:

$$\theta'^{\mu\nu} = \begin{pmatrix} w'_e & 0 \\ 0 & \overleftrightarrow{T}' \end{pmatrix} \qquad [7.147]$$

where w'_e is the energy density (generated in the reference frame attached to the particle) of the electrostatic field, and $\overleftrightarrow{T}'$ is the Maxwell stress tensor. It is clear that $\theta'^{0i} = \theta'^{i0} = 0$ for $i = 1, 2, 3$. In the fixed frame $Oxyz$, the components $\theta^{\mu\nu}$ transform as follows:

$$\theta^{\mu\nu} = \Lambda^{\mu}{}_{\sigma}\Lambda^{\nu}{}_{\rho}\theta'^{\sigma\rho} \qquad [7.148]$$

where $\left(\Lambda^{\alpha}{}_{\beta}\right)$ are the components of the Lorentz transformation. Since the particle is slow, these components are given by $\Lambda^0{}_0 = \Lambda^3{}_3 = \gamma \simeq 1$, $\Lambda^1{}_1 = \Lambda^3{}_3 = 1$ and $\Lambda^0{}_3 = \Lambda^3{}_0 = \beta\gamma \simeq \beta$. The other components are zero. In first order in β, we thus obtain:

$$\begin{aligned}
\theta^{00} &= w_{em} = \Lambda^0{}_{\sigma}\Lambda^0{}_{\rho}\theta'^{\sigma\rho} = \Lambda^0{}_{\sigma}\left(\Lambda^0{}_0\theta'^{\sigma 0} + \Lambda^0{}_3\theta'^{\sigma 3}\right) \simeq \Lambda^0{}_{\sigma}\left(\theta'^{\sigma 0} + \beta\theta'^{\sigma 3}\right) \\
&= \Lambda^0{}_0\left(\theta'^{00} + \beta\theta'^{03}\right) + \Lambda^0{}_3\left(\theta'^{30} + \beta\theta'^{33}\right) = \Lambda^0{}_0\theta'^{00} + \Lambda^0{}_3\beta\theta'^{33} \simeq \theta'^{00} = w'_e \\
\theta^{01} &= \Lambda^0{}_{\sigma}\Lambda^1{}_{\rho}\theta'^{\sigma\rho} = \Lambda^0{}_{\sigma}\Lambda^1{}_1\theta'^{\sigma 1} = \Lambda^0{}_0\theta'^{01} + \Lambda^0{}_3\theta'^{31} \simeq \beta\theta'^{31} \\
\theta^{02} &= \Lambda^0{}_{\sigma}\Lambda^2{}_{\rho}\theta'^{\sigma\rho} = \Lambda^0{}_{\sigma}\Lambda^2{}_2\theta'^{\sigma 2} = \Lambda^0{}_0\theta'^{02} + \Lambda^0{}_3\theta'^{32} \simeq \beta\theta'^{32} \\
\theta^{03} &= \Lambda^0{}_{\sigma}\Lambda^3{}_{\rho}\theta'^{\sigma\rho} = \Lambda^0{}_{\sigma}\left(\Lambda^3{}_0\theta'^{\sigma 0} + \Lambda^3{}_3\theta'^{\sigma 3}\right) \simeq \Lambda^0{}_{\sigma}\left(\beta\theta'^{\sigma 0} + \theta'^{\sigma 3}\right) \\
&= \Lambda^0{}_0\left(\beta\theta'^{00} + \theta'^{03}\right) + \Lambda^0{}_3\left(\beta\theta'^{30} + \theta'^{33}\right) \simeq \beta\left(\theta'^{00} + \theta'^{33}\right) = \beta\left(w'_e + \theta'^{33}\right)
\end{aligned} \qquad [7.149]$$

The energy of the field of the moving electron thus remains unchanged (at first order in $\frac{v}{c}$):

$$W_{em} = \int w_{em} d\tau = \int w'_e d\tau = W'_{em} \qquad [7.150]$$

The components P_{em}^1 and P_{em}^2 of the electromagnetic momentum are written as:

$$P_{\text{em}}^1 = \frac{1}{c}\int \theta^{01} d\tau = \frac{v}{c^2}\int \theta^{31} d\tau = -\frac{v}{c^2}\int E'_x E'_z d\tau$$
$$P_{\text{em}}^2 = \frac{1}{c}\int \theta^{02} d\tau = \frac{v}{c^2}\int \theta^{32} d\tau = -\frac{v}{c^2}\int E'_y E'_z d\tau \qquad [7.151]$$

These integrals are obviously zero, which shows that $P_{\text{em}}^1 = P_{\text{em}}^2 = 0$. The component P_{em}^3 leads to a more interesting result:

$$P_{\text{em}}^3 = \frac{1}{c}\int \theta^{03} d\tau = \frac{v}{c^2}\int \left(w'_{\text{e}} + \theta^{33}\right) d\tau = \frac{v}{c^2}\int \left[w'_{\text{e}} + \varepsilon_0\left(\frac{E'^2}{2} - E'^2_z\right)\right] d\tau$$
$$= \frac{v}{c^2}\int \left[w'_{\text{e}} + \varepsilon_0\left(\frac{E'^2}{2} - \frac{E'^2}{3}\right)\right] d\tau = \frac{v}{c^2}\left(W_{\text{ém}} + \frac{1}{3}W_{\text{em}}\right) = \frac{4}{3}\frac{W_{\text{em}}}{c^2}v$$
$$[7.152]$$

If $\vec{\mathbf{v}}$ has any arbitrary orientation in space, the momentum associated with the field of a charged particle is equal to:

$$\vec{\mathbf{P}}_{\text{em}} = \frac{4}{3}\frac{W_{\text{em}}}{c^2}\vec{\mathbf{v}} \qquad [7.153]$$

The inertial mass of the field is therefore equal to $\frac{4}{3}\frac{W_{\text{em}}}{c^2}$. This seems to contradict the result of special relativity, according to which the inertial mass should be equal to the energy divided by c^2. We will return to this result in the next section.

7.5. Electromagnetic mass

In the previous sections, we provided a coherent, simple formulation of electromagnetic theory that naturally satisfies relativistic covariance. As we see in what follows, when dealing with the interaction of charged particles with their own fields within the framework of this theory, certain inconsistencies arise that defy common sense. In particular, it can be shown that this theory violates causality, that is, the principle that causes precede their effects. Until now, it has been agreed to exclude the effect on the particle of its own field on the basis that a particle cannot move by itself. We have also already seen that the radiation field of an accelerated particle is generally not symmetric with respect to the particle and appears capable of existing on its own: even after the extinction of the particle, the electromagnetic fields can therefore have an existence independent of their sources. It would be inconsistent to think that radiation, after escaping from the atom, suddenly ceases to exist at the same time as the atom. The disappearance of these electromagnetic fields would result in a violation of energy conservation. To be convinced of the self-interaction of a particle with its own field, we can consider the example of an atom emitting an electromagnetic wave that is reflected by a mirror. Indeed, it would be incoherent to say that the atom would not interact with the reflected wave because it would surely be impossible to distinguish it from a wave emitted by another atom. In other words, if the field has an existence independent of the particle, then there is no reason to exclude the interaction of that field with the particle. We are thus compelled to reexamine the idea that a particle cannot feel its own field. We can therefore make the following observations:

– When a charged particle is in uniform motion, its field contributes to its momentum since, for a small change in velocity, the momentum of the field and that of the particle change together.

– If a charge radiates after being accelerated by an external force, the latter must supply the energy and momentum of both fields (the proper field and the radiation field), as well as the momentum of the particle. If the particle is to offer resistance to acceleration corresponding to its own momentum and that of the fields, then the fields must, in fact, exert a reaction force on the particle, reducing the acceleration produced by the external forces.

Let us first consider the interaction of a charged particle with its own field. A change in velocity corresponds to a change in the vector potential (and consequently, in the magnetic field), which induces an electric field at the position of the particle. The electric field is calculated from the formula $\overrightarrow{\mathbf{E}} = -\frac{\partial \overrightarrow{\mathbf{A}}}{\partial t}$. A variation $\delta \overrightarrow{\mathbf{v}}$ (in velocity) in the motion of the particle produces a variation in the vector potential:

$$\delta \overrightarrow{\mathbf{A}} = \frac{\mu_0}{4\pi} \frac{q}{r} \delta \overrightarrow{\mathbf{v}} \quad [7.154]$$

at a distance r from the particle. This expression yields an infinite result in the vicinity of a particle. To avoid this problem, we assume that the particle has a radius r_0 (i.e. the vector potential collapses inside r_0), and that the interaction of the particle with its own field occurs at this distance r_0. The (reaction) force experienced by the particle is thus given by:

$$\overrightarrow{\mathbf{F}} = q\overrightarrow{\mathbf{E}} = -q\frac{\partial \overrightarrow{\mathbf{A}}}{\partial t} = -\frac{\mu_0 q^2}{4\pi r_0} \frac{d\overrightarrow{\mathbf{v}}}{dt} = -\frac{q^2}{4\pi\varepsilon_0 r_0 c^2} \frac{d\overrightarrow{\mathbf{v}}}{dt} \quad [7.155]$$

This force resembles the electromotive force in an electric circuit. The form of the force is that of an reaction of inertia:

$$\overrightarrow{\mathbf{F}} = -m^*_{\text{em}} \frac{d\overrightarrow{\mathbf{v}}}{dt} \quad [7.156]$$

with:

$$m_{\text{em}}^{*} = \frac{q^2}{4\pi\varepsilon_0 r_0 c^2} \quad [7.157]$$

A similar estimate of the inertial mass of the field of the particle is obtained by modeling the particle as a sphere charged uniformly on its surface, with radius r_0 and carrying charge q. Comparing the electrostatic energy of this distribution, $W_{\text{e}} = \frac{q^2}{8\pi\varepsilon_0 r_0}$, with the mass energy $m_{\text{em}}c^2$ of the particle, we find that:

$$m_{\text{em}} = \frac{W_{\text{e}}}{c^2} = \frac{q^2}{8\pi\varepsilon_0 r_0 c^2} \quad [7.158]$$

These arguments provide only estimates of the inertial mass of the field of the particle (electromagnetic mass), since r_0 depends on the model used.

It is also possible to establish a relation between W_{em} and m_{em} without involving r_0. This result was discussed in example 7.8. Let us review the essential ideas:

1) The momentum carried by arbitrary fields in vacuum is given by:

$$\vec{\mathbf{p}}_{\text{em}} = \int \frac{\vec{\mathbf{E}} \times \vec{\mathbf{B}}}{\mu_0 c^2} d\tau \quad [7.159]$$

2) In the low-velocity approximation, the magnetic field $\vec{\mathbf{B}}$ can be found from [7.87]:

$$\vec{\mathbf{B}} = \frac{\vec{\beta} \times \vec{\mathbf{E}}}{c} = \varepsilon_0\mu_0 \vec{\mathbf{v}} \times \vec{\mathbf{E}} \quad [7.160]$$

3) The momentum of the field moving with the charged particle is then:

$$\vec{\mathbf{p}}_{\text{em}} = \int \varepsilon_0 \frac{\vec{\mathbf{E}} \times \left(\vec{\mathbf{v}} \times \vec{\mathbf{E}}\right)}{c^2} d\tau = \int \varepsilon_0 \frac{E^2\vec{\mathbf{v}} - \left(\vec{\mathbf{v}} \cdot \vec{\mathbf{E}}\right)\vec{\mathbf{E}}}{c^2} d\tau \quad [7.161]$$

The electric field $\vec{\mathbf{E}}$ must be axially symmetric with respect to the direction of the velocity $\vec{\mathbf{v}}$, chosen here along the Oz axis. Writing $\vec{\mathbf{E}} = E_z\vec{\mathbf{k}} + \vec{\mathbf{E}}_{\perp}$, we observe that $\vec{\mathbf{E}}_{\perp} \cdot \vec{\mathbf{v}} = 0$ and that $\int (E_z v)\, \vec{\mathbf{E}}_{\perp} d\varphi$ also vanishes. Consequently, the second term of the integral [7.161] reduces to:

$$\int \left(\vec{\mathbf{v}} \cdot \vec{\mathbf{E}}\right) \vec{\mathbf{E}} d\tau = \int E_z^2 \vec{\mathbf{v}} d\tau \simeq \frac{1}{3} \int E^2 \vec{\mathbf{v}} d\tau \quad [7.162]$$

It therefore follows that:

$$\vec{p}_{\text{em}} = \frac{4\vec{v}}{3c^2}\int \frac{\varepsilon_0 E^2}{2} d\tau = \frac{4}{3}\frac{W_e}{c^2}\vec{v} \equiv \frac{4}{3}m_{\text{em}}\vec{v} \qquad [7.163]$$

This is a very curious result. Since it is impossible to have a charged particle without its field, the empirical mass is the sum of a bare mass and the inertia of its electromagnetic field. The factor $\frac{4}{3}$ in equation [7.163] conflicts with the principle of mass–energy equivalence of special relativity. In electromagnetism, it seems that $\mathcal{E} = \frac{3}{4}mc^2$ instead of $\mathcal{E} = mc^2$. We will return to the origin of this problem in section 7.7.

7.6. Radiation reaction force from energy balance

It is quite clear that the power radiated by an accelerated particle (taken from its mechanical energy) will cause a disturbance in its motion. Here, we will attempt to describe this disturbed motion caused by radiation. In non-relativistic motion, the total power radiated by an (electrically) charged accelerated particle is given by the Larmor formula:

$$P(t) = \frac{q^2 a^2}{6\pi\varepsilon_0 c^3} \qquad [7.164]$$

where a is the acceleration. Now, we try to qualitatively examine the effect of this power on the motion of the charge. To do this, consider first a charge initially at rest, uniformly accelerated over a period T. The total energy radiated during acceleration is given by PT. On the other hand, the kinetic energy gained (the final mechanical energy of the charge) during the time T is equal to:

$$\Delta\mathcal{E} = \frac{1}{2}m(aT)^2 \qquad [7.165]$$

Radiation will have a negligible influence on motion if $PT \ll \frac{1}{2}m(aT)^2$. This condition can also be written as:

$$T \gg \tau \qquad [7.166]$$

with:

$$\tau = \frac{1}{6\pi\varepsilon_0}\frac{q^2}{mc^3} \qquad [7.167]$$

(the radiated power then being simply $P = m\tau a^2$). τ is called the *characteristic time* and is maximal for the electron (the lightest of charged particles). Note that $c\tau$ is on the order of $\frac{4}{3}r_0$, where r_0 is the classical radius of the charged particle. The extremely small value of τ indicates that the effect of radiation is entirely negligible, except when considering infinitely short acceleration periods.

This qualitative approach can also be extended to a charged particle undergoing sinusoidal motion with frequency ω. The acceleration is then on the order of $\omega^2 d$, where d is the amplitude of the motion. The mechanical energy is of the order of $m\omega^2 d^2$. The radiated power over one period, $\frac{P}{\omega}$, is small compared to the mechanical energy, and therefore the radiation constitutes only a small perturbation of the motion, when $\frac{q^2}{6\pi\varepsilon_0 c^3 \omega}\left(\omega^2 d\right)^2 \ll m\omega^2 d^2$, that is, if:

$$\omega\tau \ll 1 \qquad [7.168]$$

This condition is very well verified for any movement at ordinary frequency. We conclude that radiation has a very weak influence on motion and it is therefore justified to consider, at least over a short enough interval, that the motion of the particle is imposed. However, over long periods of time, the cumulative effect of the radiated power will eventually dampen the motion, whose mechanical energy will slowly decrease.

The question now arises as to how to include these radiation effects in the equations of motion of a charged particle. We will begin with a simple argument based on energy conservation for a non-relativistic charged particle. If radiation emission is neglected, a charged particle of mass m and charge q, subjected to an external force $\vec{\mathbf{F}}_{\text{ext}}$ moves according to Newton's equation of motion:

$$m\dot{\vec{\mathbf{v}}} = \vec{\mathbf{F}}_{\text{ext}} \qquad [7.169]$$

To account for the energy loss (due to radiation) and its effect on the motion of the particle, we will modify Newton's equation [7.169] by adding a radiation reaction force $\vec{\mathbf{F}}_R$:

$$m\dot{\vec{\mathbf{v}}} = \vec{\mathbf{F}}_{\text{ext}} + \vec{\mathbf{F}}_R \qquad [7.170]$$

Although the force $\vec{\mathbf{F}}_R$ has not yet been determined at this stage, we can already identify certain requirements it must satisfy:

1) $\vec{\mathbf{F}}_R$ must vanish if $\dot{\vec{\mathbf{v}}} = 0$, since there is no radiation for uniform motion.

2) $\vec{\mathbf{F}}_R$ must be proportional to q^2, since (a) the radiated power is proportional to q^2 and (b) the sign of the charge plays no role in radiation effects.

3) $\vec{\mathbf{F}}_R$ must in fact involve the characteristic time τ, since it is apparently the only relevant available parameter.

Consider a particle with prescribed motion between instants t_1 and t_2. We can determine the form of $\vec{\mathbf{F}}_R$ by requiring that the work performed by this force in the time interval $t_1 < t < t_2$ be equal to the change in the energy of the particle due to radiation during this period. Using the Larmor formula [7.164], we therefore write:

$$\int_{t_1}^{t_2} \vec{\mathbf{F}}_R \cdot \vec{\mathbf{v}}\, dt = -\frac{1}{6\pi\varepsilon_0}\frac{q^2}{c^3}\int_{t_1}^{t_2} \dot{\vec{\mathbf{v}}} \cdot \dot{\vec{\mathbf{v}}}\, dt \qquad [7.171]$$

The second integral can easily be integrated by parts to give:

$$\int_{t_1}^{t_2} \vec{\mathbf{F}}_R \cdot \vec{\mathbf{v}}\, dt = \frac{q^2}{6\pi\varepsilon_0 c^3}\int_{t_1}^{t_2} \ddot{\vec{\mathbf{v}}} \cdot \vec{\mathbf{v}}\, dt - \frac{q^2}{6\pi\varepsilon_0 c^3}\, \dot{\vec{\mathbf{v}}} \cdot \vec{\mathbf{v}}\Big|_{t_1}^{t_2} \qquad [7.172]$$

If the motion is periodic or such that $\dot{\vec{\mathbf{v}}} \cdot \vec{\mathbf{v}} = 0$ at $t = t_1$ and $t = t_2$, we finally obtain:

$$\int_{t_1}^{t_2} \left(\vec{\mathbf{F}}_R - \frac{q^2}{6\pi\varepsilon_0 c^3}\ddot{\vec{\mathbf{v}}}\right) \cdot \vec{\mathbf{v}}\, dt = 0 \qquad [7.173]$$

It is then possible to identify the radiation reaction force as being:

$$\vec{\mathbf{F}}_R = \frac{q^2}{6\pi\varepsilon_0 c^3}\ddot{\vec{\mathbf{v}}} = m\tau\,\ddot{\vec{\mathbf{v}}} \qquad [7.174]$$

The modified equation of motion then becomes:

$$m\left(\dot{\vec{\mathbf{v}}} - \tau\,\ddot{\vec{\mathbf{v}}}\right) = \vec{\mathbf{F}}_{\text{ext}} \qquad [7.175]$$

Equation [7.175] is sometimes called the *Abraham–Lorentz equation of motion*. It can be regarded as an equation that approximately includes the effects of radiation on motion. Here are a few remarks:

– The radiation reaction force, which is proportional to the time derivative of the acceleration, is quite unusual and must be used with great caution. First of all, it obviously leads to third-order equations of motion in time t. To fully characterize the motion, we must specify not only the position as well as the velocity, but also the initial acceleration of the particle. *Such behavior has no counterpart in Newtonian mechanics.*

– Another problem with the radiation reaction force is the presence, among the solutions of the modified equation of motion, of divergent solutions, even for a free particle. Indeed, consider a particle initially at rest. The modified equation of motion reads:

$$m\dot{\vec{\mathbf{v}}} = m\tau\,\ddot{\vec{\mathbf{v}}} \qquad [7.176]$$

This equation admits the solution:

$$\vec{\dot{\mathbf{v}}} = \vec{\mathbf{a}}_0 e^{t/\tau} \qquad [7.177]$$

If it is not specified that the acceleration of the particle is initially zero, both acceleration and velocity will increase exponentially and extremely rapidly over time. These solutions are called *runaway* solutions and are completely non-physical. It should be noted that the derivation method used in [7.172] shows that the solution [7.177] is unacceptable, since $\vec{\dot{\mathbf{v}}} \cdot \vec{\mathbf{v}} \neq 0$ in t_1 and t_2.

– There is another difficulty with the radiation reaction force. It predicts a zero force for uniformly accelerated motion, and therefore a priori a radiated energy equal to zero. However, for this type of motion, the power predicted by the Larmor formula is non-zero. Of course, we can observe that the integration by parts used to obtain the expression for $\vec{\mathbf{F}}_R$ is not valid in this case.

We thus conclude that the radiation reaction force is a convenient computational tool for estimating the influence of radiation on motion, but that this method must be used with caution. In the next section, we will present a method of calculation, introduced by Abraham and Lorentz, which justifies to some extent the form of the radiation reaction force. In this approach, the charge is considered as a distributed charge density, and the force it exerts on itself is evaluated. After a tedious calculation, we find a contribution proportional to the acceleration that can be interpreted as the mass of the particle and the contribution of the radiation reaction force.

EXAMPLE 7.9. (Lifetime of the hydrogen atom in the Rutherford model).– We will now use the radiation reaction force to estimate the lifetime of atoms in the Rutherford planetary model. For simplicity, we consider a hydrogen atom and assume that the orbits are circular to simplify the calculations. The radiation reaction force should cause the electron in the hydrogen atom to spiral into the nucleus by reducing its mechanical energy. We assume that the mechanical energy $\mathcal{E}$ evolves slowly compared to the orbital motion. Let $U(r) = -\frac{\alpha}{r}$ be the spherically symmetric Coulomb potential, with $\alpha = \frac{e^2}{4\pi\varepsilon_0}$. The time derivative of the mechanical energy is given by the work of the radiation reaction force (note that, for periodic motion, the term discarded in the reasoning leading to $\vec{\mathbf{F}}_R$ is identically zero, provided the integration is performed over a full period: the Larmor formula is therefore strictly equivalent to the work of $\vec{\mathbf{F}}_R$). We then write:

$$\frac{d\mathcal{E}}{dt} = -m\tau \left\langle \vec{\mathbf{a}}^2 \right\rangle \qquad [7.178]$$

where $\langle \cdot \rangle$ denotes the time average over one period. Noting that $\vec{a} = -\frac{1}{m}\frac{dU}{dr}\vec{e}_r$, we finally obtain:

$$\frac{d\mathcal{E}}{dt} = -\frac{\tau}{m}\left\langle \left(\frac{dU}{dr}\right)^2 \right\rangle = -\frac{\tau}{m}\frac{e^4}{(4\pi\varepsilon_0)^2 r^4} \qquad [7.179]$$

We assume that the orbit of the electron remains circular during its fall toward the nucleus. Using the virial theorem, we also write:

$$\mathcal{E} = \frac{1}{2}\langle U \rangle = -\frac{e^2}{8\pi\varepsilon_0}\frac{1}{r} \qquad [7.180]$$

which yields:

$$\frac{d\mathcal{E}}{dt} = \frac{e^2}{8\pi\varepsilon_0 r^2}\frac{dr}{dt} \qquad [7.181]$$

By comparing equations [7.179] and [7.181], it finally shows that:

$$\frac{dr}{dt} = -\frac{\tau}{m}\frac{e^2}{2\pi\varepsilon_0 r^2} \qquad [7.182]$$

The latter equation admits the solution:

$$r^3(t) - r_0^3 = -9(c\tau)^3\frac{t}{\tau} \qquad [7.183]$$

where r_0 is the initial radius. We see that r^3 decreases linearly with time, which means that the radius reaches zero in a finite amount of time. In other words, the fall of the electron into the nucleus occurs in a finite time. Thus, we can estimate the order of magnitude of the lifetime of the hydrogen atom in its ground state by taking the initial radius as the Bohr radius r_0. This yields a lifetime:

$$T = \frac{1}{9}\tau\left(\frac{r_0}{c\tau}\right)^3 \qquad [7.184]$$

7.7. Abraham–Lorentz model

In section 7.6, we introduced the radiation reaction force using a simple energy balance argument. However, this method does not provide much insight into how this reaction force can actually manifest. This force has been interpreted as the action of a particle's own fields upon itself. Therefore, any rigorous model must take into account the structure of the particle and its self-fields. In what follows, we will present a model of a charged particle that correctly predicts the expression of this force and shows that it is, in fact, the result of the forces the particle exerts on itself: non-zero if the particle's acceleration varies. This model is known as the *Abraham–Lorentz model.*

We therefore consider the charged particle as a spherical ball with a volume charge distribution. If the particle is at rest, the volume charge density is simply written as $\rho\left(\overrightarrow{\mathbf{r}}\right)=\rho\left(r\right)$. We assume that the particle is undergoing non-relativistic translational motion. The sphere thus moves as a whole with a velocity $\overrightarrow{\mathbf{v}}\left(t\right)$. The resulting volume charge density is written as $\rho\left(\overrightarrow{\mathbf{r}},t\right)$, and the associated volume current density is $\overrightarrow{\mathbf{j}}\left(\overrightarrow{\mathbf{r}},t\right)=\rho\overrightarrow{\mathbf{v}}$. We choose the origin at the center of the particle at a given instant. The particle is placed in an external electromagnetic field $\overrightarrow{\mathbf{E}}\text{ext}$, $\overrightarrow{\mathbf{B}}\text{ext}$. We denote by $\overrightarrow{\mathbf{E}}_s$ and $\overrightarrow{\mathbf{B}}_s$ the electric and magnetic fields produced by the particle itself (in motion), respectively. The resulting fields are therefore the sum of these fields and the external fields:

$$\begin{aligned}\overrightarrow{\mathbf{E}}&=\overrightarrow{\mathbf{E}}_{\text{ext}}+\overrightarrow{\mathbf{E}}_s\\ \overrightarrow{\mathbf{B}}&=\overrightarrow{\mathbf{B}}_{\text{ext}}+\overrightarrow{\mathbf{B}}_s\end{aligned}\qquad [7.185]$$

We assume that all the energetic or inertial properties of the particle are associated solely with its electromagnetic field. The particle's *mechanical* mass is therefore zero. The equation of motion then requires that the sum of the Lorentz forces due to the external field $\overrightarrow{\mathbf{F}}_{\text{ext}}$ and the self-field $\overrightarrow{\mathbf{F}}_s$ cancel out. We must therefore have:

$$\overrightarrow{\mathbf{F}}_{\text{ext}}=-\int\left(\rho\overrightarrow{\mathbf{E}}_s+\overrightarrow{\mathbf{j}}\times\overrightarrow{\mathbf{B}}_s\right)d\tau\qquad [7.186]$$

where the integral extends over the entire volume of the particle. We will attempt to express this integral in terms of successive derivatives of the velocity of the particle. This makes it possible to obtain an equation that describes the evolution of the velocity of the particle in an external field.

Let us compute the integral of the Lorentz forces produced by the particle itself. It is quite evident, first of all, that for non-relativistic motion, the magnetic forces are negligible. It therefore follows that:

$$\overrightarrow{\mathbf{F}}_{\text{ext}}\simeq-\int\rho\left(\overrightarrow{\mathbf{r}},t\right)\overrightarrow{\mathbf{E}}_s\left(\overrightarrow{\mathbf{r}},t\right)d\tau\qquad [7.187]$$

where $\overrightarrow{\mathbf{E}}_s$ is derived from the retarded potentials V_s and $\overrightarrow{\mathbf{A}}_s$:

$$\begin{aligned}V_s\left(\overrightarrow{\mathbf{r}},t\right)&=\frac{1}{4\pi\varepsilon_0}\int\frac{\rho\left(\overrightarrow{\mathbf{r}}',t-\frac{R}{c}\right)}{R}d\tau'\\ \overrightarrow{\mathbf{A}}_s\left(\overrightarrow{\mathbf{r}},t\right)&=\frac{1}{4\pi\varepsilon_0c^2}\int\frac{\overrightarrow{\mathbf{j}}\left(\overrightarrow{\mathbf{r}}',t-\frac{R}{c}\right)}{R}d\tau'\end{aligned}\qquad [7.188]$$

where $\vec{\mathbf{R}} = \vec{\mathbf{r}} - \vec{\mathbf{r}}'$ and $R = \left|\vec{\mathbf{R}}\right|$. Since the dimensions of the particle are small, a Taylor expansion of the potentials can be performed around time t. For the vector potential, we obtain:

$$\vec{\mathbf{A}}_s\left(\vec{\mathbf{r}},t\right) = \frac{1}{4\pi\varepsilon_0 c^2}\sum_{n=0}^{\infty}\frac{(-1)^n}{n!c^n}\int d\tau' R^{n-1}\frac{\partial^n}{\partial t^n}\vec{\mathbf{j}}\left(\vec{\mathbf{r}}',t\right) \qquad [7.189]$$

We can then easily express $-\frac{\partial \vec{\mathbf{A}}_s}{\partial t}$ and determine the contribution of the electromotive field to the integral of the Lorentz force:

$$\vec{\mathbf{F}}_{\text{ext},\vec{A}} = \frac{1}{4\pi\varepsilon_0}\sum_{n=0}^{\infty}\frac{(-1)^n}{n!c^{n+2}}\int\int d\tau d\tau'\rho\left(\vec{\mathbf{r}},t\right)R^{n-1}\frac{\partial^{n+1}}{\partial t^{n+1}}\vec{\mathbf{j}}\left(\vec{\mathbf{r}}',t\right) \qquad [7.190]$$

Doing the same for the scalar potential V_s, it follows that:

$$V_s\left(\vec{\mathbf{r}},t\right) = \frac{1}{4\pi\varepsilon_0}\sum_{n=0}^{\infty}\frac{(-1)^n}{n!c^n}\int d\tau' R^{n-1}\frac{\partial^n}{\partial t^n}\rho\left(\vec{\mathbf{r}}',t\right) \qquad [7.191]$$

We see that the scalar potential contributes to the Lorentz force through:

$$\vec{\mathbf{F}}_{\text{ext},V} = \frac{1}{4\pi\varepsilon_0}\sum_{n=0}^{\infty}\frac{(-1)^n}{n!c^n}\int\int d\tau d\tau'\rho\left(\vec{\mathbf{r}},t\right)\vec{\nabla}R^{n-1}\frac{\partial^n}{\partial t^n}\rho\left(\vec{\mathbf{r}}',t\right) \qquad [7.192]$$

($\vec{\nabla}$ acts here on $\vec{\mathbf{r}}$). Consider the first two terms in the expression of $\vec{\mathbf{F}}_{\text{ext},V}$. For $n = 0$, the term is proportional to:

$$\int\int d\tau d\tau'\rho\left(\vec{\mathbf{r}},t\right)\rho\left(\vec{\mathbf{r}}',t\right)\vec{\nabla}\left(\frac{1}{R}\right) \qquad [7.193]$$

This term represents the integral over the charge distribution of the Coulomb force exerted by the other parts of the distribution. The principle of action and reaction (which holds for electrostatic forces) then shows that this integral is zero. Indeed, it is readily verified that:

$$\begin{aligned}&\int\int d\tau d\tau'\rho\left(\vec{\mathbf{r}},t\right)\rho\left(\vec{\mathbf{r}}',t\right)\vec{\nabla}\left(\frac{1}{R}\right) = \int\int d\tau' d\tau\rho\left(\vec{\mathbf{r}}',t\right)\rho\left(\vec{\mathbf{r}},t\right)\vec{\nabla}'\left(\frac{1}{R}\right)\\ &= -\int\int d\tau' d\tau\rho\left(\vec{\mathbf{r}}',t\right)\rho\left(\vec{\mathbf{r}},t\right)\vec{\nabla}\left(\frac{1}{R}\right)\end{aligned} \qquad [7.194]$$

because $\vec{\nabla}'\left(\frac{1}{R}\right) = -\vec{\nabla}\left(\frac{1}{R}\right)$, which implies that $\int\int d\tau d\tau' \rho\left(\vec{\mathbf{r}}, t\right)\rho\left(\vec{\mathbf{r}}', t\right)\vec{\nabla}\left(\frac{1}{R}\right) = 0$. The term of order $n = 1$ is also zero, since it involves the gradient of a constant. Eliminating these first two terms and shifting the summation index in [7.192], we get:

$$\vec{\mathbf{F}}_{\text{ext},V} = \frac{1}{4\pi\varepsilon_0 c^2}\sum_{n=0}^{\infty}\frac{(-1)^n}{n!c^n}\frac{1}{(n+1)(n+2)} \times \int\int d\tau d\tau' \rho\left(\vec{\mathbf{r}}, t\right)\vec{\nabla}R^{n+1}\frac{\partial^{n+2}}{\partial t^{n+2}}\rho\left(\vec{\mathbf{r}}', t\right) \quad [7.195]$$

By combining [7.190] and [7.195], we find:

$$\vec{\mathbf{F}}_{\text{ext},V} = \frac{1}{4\pi\varepsilon_0 c^2}\sum_{n=0}^{\infty}\frac{(-1)^n}{n!c^n}\int d\tau \rho\left(\vec{\mathbf{r}}, t\right)\frac{\partial^{n+1}}{\partial t^{n+1}}G_n\left(\vec{\mathbf{r}}, t\right) \quad [7.196]$$

with:

$$G_n\left(\vec{\mathbf{r}}, t\right) = \int d\tau' R^{n-1}\left[\vec{\mathbf{j}}\left(\vec{\mathbf{r}}', t\right) + \frac{\partial\rho\left(\vec{\mathbf{r}}', t\right)}{\partial t}\frac{\vec{\nabla}R^{n+1}}{(n+1)(n+2)R^{n-1}}\right] \quad [7.197]$$

The second term in parentheses in $G_n\left(\vec{\mathbf{r}}, t\right)$ can be transformed, using the charge conservation equation $\frac{\partial\rho}{\partial t} = -\vec{\nabla}'\cdot\vec{\mathbf{j}}$ and noting that $\vec{\nabla}R^{n+1} = (n+1)R^{n-1}\vec{\mathbf{R}}$, into the form:

$$\int d\tau'\frac{\partial\rho\left(\vec{\mathbf{r}}', t\right)}{\partial t}\frac{\vec{\nabla}R^{n+1}}{(n+1)(n+2)} = -\int d\tau'\frac{R^{n-1}\vec{\mathbf{R}}}{n+2}\left[\vec{\nabla}'\cdot\vec{\mathbf{j}}\left(\vec{\mathbf{r}}', t\right)\right] \quad [7.198]$$

Using the integration by parts formula, $\int\vec{\mathbf{A}}\left(\vec{\nabla}\cdot\vec{\mathbf{B}}\right)d\tau = -\int\left(\vec{\mathbf{B}}\cdot\vec{\nabla}\right)\vec{\mathbf{A}}d\tau$, which holds if the fields vanish at infinity (as is clearly the case here), this second term can be rewritten as:

$$\int d\tau'\frac{\partial\rho\left(\vec{\mathbf{r}}', t\right)}{\partial t}\frac{\vec{\nabla}R^{n+1}}{(n+1)(n+2)} = \frac{1}{n+2}\int d\tau'\left(\vec{\mathbf{j}}\cdot\vec{\nabla}'\right)\left(R^{n-1}\vec{\mathbf{R}}\right) \quad [7.199]$$

$$= \frac{1}{n+2}\int d\tau'\left[R^{n-1}\left(\vec{\mathbf{j}}\cdot\vec{\nabla}'\right)\vec{\mathbf{R}} + \left[\left(\vec{\mathbf{j}}\cdot\vec{\nabla}'\right)R^{n-1}\right]\vec{\mathbf{R}}\right]$$

Now, we have $\left(\vec{\mathbf{j}}\cdot\vec{\nabla}'\right)R^{n-1} = -(n-1)R^{n-1}\frac{\vec{\mathbf{j}}\cdot\vec{\mathbf{R}}}{R^2}$ and $\left(\vec{\mathbf{j}}\cdot\vec{\nabla}'\right)\vec{\mathbf{R}} = -\vec{\mathbf{j}}$, which gives:

$$\int d\tau' \frac{\partial \rho\left(\overrightarrow{\mathbf{r}}', t\right)}{\partial t} \frac{\overrightarrow{\nabla} R^{n+1}}{(n+1)(n+2)} = \frac{1}{n+2} \int d\tau' \times \left[-\overrightarrow{\mathbf{j}} R^{n-1} - (n-1) R^{n-1} \frac{\overrightarrow{\mathbf{j}} \cdot \overrightarrow{\mathbf{R}}}{R^2} \overrightarrow{\mathbf{R}} \right] \quad [7.200]$$

Substituting [7.200] into [7.197], we obtain:

$$G_n\left(\overrightarrow{\mathbf{r}}, t\right) = \int d\tau' R^{n-1} \rho\left(\overrightarrow{\mathbf{r}}', t\right) \left[\frac{n+1}{n+2} \overrightarrow{\mathbf{v}}(t) - \frac{n-1}{n+2} \frac{\overrightarrow{\mathbf{v}} \cdot \overrightarrow{\mathbf{R}}}{R^2} \overrightarrow{\mathbf{R}} \right] \quad [7.201]$$

where we used $\overrightarrow{\mathbf{j}} = \rho \overrightarrow{\mathbf{v}}$. For symmetry reasons, this integral must have the direction of velocity $\overrightarrow{\mathbf{v}}$. In other words, only the component parallel to $\overrightarrow{\mathbf{v}}$ can survive to integration. Thus, by writing $\frac{\overrightarrow{\mathbf{v}} \cdot \overrightarrow{\mathbf{R}}}{R^2} \overrightarrow{\mathbf{R}} = \frac{\left(\overrightarrow{\mathbf{v}} \cdot \overrightarrow{\mathbf{R}}\right)^2}{R^2 v^2} \overrightarrow{\mathbf{v}} + \frac{\overrightarrow{\mathbf{v}} \cdot \overrightarrow{\mathbf{R}}}{R^2} \overrightarrow{\mathbf{R}}_{\perp}$, the integral [7.201] reduces to:

$$G_n\left(\overrightarrow{\mathbf{r}}, t\right) = \int d\tau' R^{n-1} \rho\left(\overrightarrow{\mathbf{r}}', t\right) \overrightarrow{\mathbf{v}}(t) \left[\frac{n+1}{n+2} - \frac{n-1}{n+2} \frac{\left(\overrightarrow{\mathbf{v}} \cdot \overrightarrow{\mathbf{R}}\right)^2}{R^2 v^2} \right] \quad [7.202]$$

Moreover, all directions of $\overrightarrow{\mathbf{R}}$ are equally possible. Therefore, it is possible, without changing the final result, to replace the second term in parentheses with the average value of its projection onto $\overrightarrow{\mathbf{v}}$ for all possible orientations of $\overrightarrow{\mathbf{R}}$. This value is proportional to the average cosine squared of the angle between $\overrightarrow{\mathbf{v}}$ and $\overrightarrow{\mathbf{R}}$, which is $\frac{1}{3}$. It therefore follows that:

$$\begin{aligned} G_n\left(\overrightarrow{\mathbf{r}}, t\right) &= \int d\tau' R^{n-1} \rho\left(\overrightarrow{\mathbf{r}}', t\right) \overrightarrow{\mathbf{v}}(t) \left[\frac{n+1}{n+2} - \frac{n-1}{n+2} \times \frac{1}{3} \right] \\ &= \frac{2}{3} \overrightarrow{\mathbf{v}}(t) \int d\tau' R^{n-1} \rho\left(\overrightarrow{\mathbf{r}}', t\right) \end{aligned} \quad [7.203]$$

Let us now substitute this expression for $G_n\left(\overrightarrow{\mathbf{r}}, t\right)$ into that of the Lorentz force. We must take the $(n+1)$th time derivative of G_n. This yields terms with successive derivatives of the velocity, as well as terms with time derivatives of ρ. Since $\frac{\partial \rho}{\partial t} \simeq \overrightarrow{\nabla} \rho \cdot \overrightarrow{\mathbf{v}}$ (with $\overrightarrow{\mathbf{v}}$ being the bulk velocity of the particle), the terms involving derivatives of ρ are thus quadratic in the velocity and its derivatives. For small velocities, these quadratic terms can be neglected, and we can restrict ourselves to the expression:

$$\overrightarrow{\mathbf{F}}_{\text{ext}} = \frac{1}{4\pi\varepsilon_0 c^2} \sum_{n=0}^{\infty} \frac{(-1)^n}{n! c^n} \frac{\partial^{n+1} \overrightarrow{\mathbf{v}}(t)}{\partial t^{n+1}} \frac{2}{3} \int d\tau d\tau' \rho\left(\overrightarrow{\mathbf{r}}, t\right) \rho\left(\overrightarrow{\mathbf{r}}', t\right) R^{n-1} \quad [7.204]$$

To understand the meaning of [7.204], consider the first terms of the expansion:

$$\begin{aligned}
\vec{\mathbf{F}}^{(0)}_{\text{ext}} &= \frac{1}{6\pi\varepsilon_0 c^2}\dot{\vec{\mathbf{v}}}\int d\tau d\tau' \frac{\rho(\vec{\mathbf{r}},t)\,\rho(\vec{\mathbf{r}}',t)}{R} \\
\vec{\mathbf{F}}^{(1)}_{\text{ext}} &= -\frac{1}{6\pi\varepsilon_0 c^3}\ddot{\vec{\mathbf{v}}}\int d\tau d\tau' \rho(\vec{\mathbf{r}},t)\,\rho(\vec{\mathbf{r}}',t) = -\frac{q^2}{6\pi\varepsilon_0 c^3}\ddot{\vec{\mathbf{v}}} \\
\vec{\mathbf{F}}^{(n)}_{\text{ext}} &= \frac{(-1)^n}{6\pi\varepsilon_0 n! c^{n+2}}\vec{\mathbf{v}}^{(n+1)}\int d\tau d\tau' \rho(\vec{\mathbf{r}},t)\,\rho(\vec{\mathbf{r}}',t)\,R^{n-1} \\
&\sim \frac{(-1)^n q^2 b^{n-1}}{6\pi\varepsilon_0 n! c^{n+2}}\vec{\mathbf{v}}^{(n+1)}
\end{aligned} \quad [7.205]$$

where $\vec{\mathbf{v}}^{(n)}$ denotes the nth time derivative of $\vec{\mathbf{v}}$. In the third expression, b is a characteristic length scale corresponding to the extent of the particle's charge distribution (which can be identified with the radius of the particle). Indeed, we approximately write:

$$\int d\tau d\tau' \rho(\vec{\mathbf{r}},t)\,\rho(\vec{\mathbf{r}}',t)\,R^{n-1} \sim b^{n-1}\int d\tau d\tau' \rho(\vec{\mathbf{r}},t)\,\rho(\vec{\mathbf{r}}',t) = b^{n-1}q^2 \quad [7.206]$$

We note that for $n \geq 2$, the terms in the expansion vanish in the limit of a point particle ($b \longrightarrow 0$). Thus, for highly localized charge distributions, only contributions from $n = 0$ and $n = 1$ are considered. The term corresponding to $n = 1$ is simply the radiation reaction force already obtained in [7.174]. This force is independent of the structure of the particle and depends only on its total charge. For $b \neq 0$, the successive terms in the expansion [7.204] are in the ratio:

$$\frac{F^{(n)}_{\text{ext}}}{F^{(n-1)}_{\text{ext}}} = \frac{b v^{(n+1)}}{n c v^{(n)}} = \frac{b\Delta\left(\ln v^{(n)}\right)}{n c \Delta t} \quad [7.207]$$

which indicates that they are only significant when substantial changes in the motion (i.e. in the nth derivative of velocity) occur over time intervals of the order $\Delta t = \frac{b}{c} \sim \tau$ (which is not the case here). Therefore, these terms can also be neglected when $b \neq 0$.

The $n = 0$ term deserves special attention. The double integral is proportional to the electrostatic energy of the charge distribution:

$$W_e = \int d\tau \int d\tau' \frac{\rho(\vec{\mathbf{r}},t)\,\rho(\vec{\mathbf{r}}',t)}{R} \quad [7.208]$$

Consequently, this term can be expressed as follows:

$$\vec{\mathbf{F}}^{(0)}_{\text{ext}} = \frac{4}{3}\frac{W_e}{c^2}\dot{\vec{\mathbf{v}}} = \frac{4}{3}m_{\text{ém}}\dot{\vec{\mathbf{v}}} \quad [7.209]$$

The quantity $\frac{4}{3}m_{\text{em}}$ represents the electromagnetic mass of the particle, accounting for the inertia of its own field. If the Abraham–Lorentz model successfully reproduces the radiation reaction force, it is far from satisfactory for the electromagnetic mass. In fact, this mass is equal to $\frac{4}{3}\frac{W_e}{c^2}$, where W_e is the energy. This result contradicts the principle of mass–energy equivalence in special relativity. This is the *famous* $\frac{4}{3}$ *problem.* Note that the electromagnetic mass diverges as $b \longrightarrow 0$. Its inclusion in the empirical mass of the particle (alongside the bare mass) should eliminate this divergence (renormalization). Newton's equation of motion in the Abraham–Lorentz model then takes the following form:

$$\left(\frac{4}{3}m_{\text{em}}\right)\dot{\vec{\mathbf{v}}} - \frac{q^2}{6\pi\varepsilon_0 c^3}\ddot{\vec{\mathbf{v}}} = \vec{\mathbf{F}}_{\text{ext}} \qquad [7.210]$$

provided that the higher order terms in the expansion [7.204] are negligible. This is the same as [7.175], except for the factor $\frac{4}{3}$ which multiplies the electromagnetic mass.

The troubling factor $\frac{4}{3}$ in the electromagnetic mass was first discovered by Thomson. Let us now take a closer look at the origin of this problem. In section 7.4, we have studied the conservation laws for energy and momentum. We have interpreted the first column (or row) of the energy–momentum tensor $\theta^{\mu\nu}$ as representing the energy density (first component) and the momentum density (the other three components) of the electromagnetic field. It is therefore natural to identify the total electromagnetic energy and the total electromagnetic momentum as the volume integrals over the region where the field is active. Accordingly, the electromagnetic energy and momentum are given by:

$$\begin{aligned} E_{\text{em}} &= \int \theta^{00} d\tau = cP^0_{\text{ém}} \\ P^i_{\text{em}} &= \frac{1}{c}\int \theta^{0i} d\tau \end{aligned} \qquad [7.211]$$

where the integral is taken over all space. However, these variables are clearly non-covariant. We therefore conclude that the Abraham–Lorentz model, or any other model using these variables, must violate the requirements of Lorentz covariance. To construct a covariant theory, let us introduce an energy–momentum four-vector representing the electromagnetic field of the particle, which reduces to the electrostatic energy in the particle's rest frame. To do this, we take the scalar product of $\theta^{\mu\nu}$ with the four-velocity $\frac{U_\nu}{c}$, and then integrate over the volume element $\gamma, d\tau$ (a Lorentz invariant):

$$P^\mu_{\text{em}} = \frac{\gamma}{c^2}\int \theta^{\mu\nu} U_\nu d\tau \qquad [7.212]$$

In terms of the components of $\theta^{\mu\nu}$, the electromagnetic energy and momentum are written as:

$$E_{\text{em}} = \gamma^2 \int \left(w_{\text{em}} - \frac{\vec{\Pi} \cdot \vec{v}}{c^2} \right) d\tau$$
$$\vec{P}_{\text{ém}} = \frac{\gamma^2}{c^2} \int \left(\vec{\Pi} - \vec{v} \cdot \overleftrightarrow{T} \right) d\tau \qquad [7.213]$$

We see that these expressions differ from [7.211] by the terms $\frac{\vec{\Pi} \cdot \vec{v}}{c^2}$, $\vec{v} \cdot \overleftrightarrow{T}$ and γ. It is interesting to note that the factor $\frac{4}{3}$ now disappears from the momentum, as we shall see. In the non-relativistic limit, the self-magnetic field is given by $\vec{B}_s = \varepsilon_0 \mu_0 \vec{v} \times \vec{E}_s$. The first term of the momentum is therefore:

$$\vec{P}_{\text{em},(1)} \simeq \frac{1}{c^2} \int \vec{\Pi} d\tau = \frac{\varepsilon_0}{c^2} \int \vec{E}_s \times \left(\vec{v} \times \vec{E}_s \right) d\tau$$
$$= \frac{\varepsilon_0}{c^2} \int \left[E_s^2 \vec{v} - \left(\vec{E}_s \cdot \vec{v} \right) \vec{E}_s \right] d\tau \qquad [7.214]$$

In fact, we recover the result [7.161]. If the field is assumed to be spherical, the second term can be averaged over the angles to yield $\frac{1}{3}$ of the first term, which leads to $\frac{4}{3} \frac{W_{\text{em}}}{c^2}$, $\vec{v}$, as already seen above. Taking into account the expression of the Maxwell tensor [7.134], the second term of the momentum reduces to:

$$\vec{P}_{\text{em},(2)} \simeq -\frac{1}{c^2} \int \vec{v} \cdot \overleftrightarrow{T} d\tau = \frac{\varepsilon_0}{c^2} \int \left[\left(\vec{E}_s \cdot \vec{v} \right) \vec{E}_s - \frac{1}{2} E_s^2 \vec{v} \right] d\tau = -\frac{1}{3} \frac{W_{\text{em}}}{c^2} \vec{v} \qquad [7.215]$$

The sum of [7.214] and [7.215] immediately gives:

$$\vec{P}_{\text{em}} = +\vec{P}_{\text{em},(1)} + \vec{P}_{\text{em},(2)} = \left(\frac{\varepsilon_0}{2c^2} \int E_s^2 d\tau \right) \vec{v} = \frac{W_{\text{em}}}{c^2} \vec{v} \qquad [7.216]$$

as predicted by Lorentz covariance.

7.8. Abraham–Lorentz equation of motion

Consider a charged particle subjected, in addition to the electromagnetic forces discussed in section 2.7, to an external force F_{ext}. The equation of motion of the particle and its associated field is as follows:

$$\vec{F}_{\text{ext}} = \left(m_0 + \frac{4}{3} m_{\text{ém}} \right) \dot{\vec{v}} - \frac{q^2}{6\pi\varepsilon_0 c^3} \ddot{\vec{v}} \qquad [7.217]$$

where m_0 is the bare mass of the particle. Since the particle cannot be separated from its electromagnetic field, an empirical mass (the actual mass of the particle) is introduced:

$$m = m_0 + \frac{4}{3} m_{\text{em}} \qquad [7.218]$$

The equation of motion then becomes:

$$m\dot{\vec{\mathbf{v}}} - \frac{q^2}{6\pi\varepsilon_0 c^3}\ddot{\vec{\mathbf{v}}} = \vec{\mathbf{F}}_{\text{ext}} \qquad [7.219]$$

This is the *Abraham–Lorentz equation of motion.*

The homogeneous equation has solutions $\dot{\vec{\mathbf{v}}} = 0$ and $\dot{\vec{\mathbf{v}}} = \dot{\vec{\mathbf{v}}}_0 e^{t/\tau}$, with $\tau = \frac{q^2}{6\pi\varepsilon_0 mc^3}$. The so-called *runaway solutions* (the second set of solutions to the homogeneous equation) are not physical and originate from the fluctuation terms neglected in [7.204], which become significant at time scales shorter than τ. Due to these fluctuations, we will study the motion over time intervals Δt much larger than τ, that is, $\Delta t \geq \tau$.

The runaway behavior of the solution can be avoided by taking a time average of [7.219]. To do this, we multiply the equation of motion [7.219] by the factor $e^{-t'/\tau}$ and then integrate over a time interval from t_0 to t:

$$\int_{t_0}^{t} \frac{\vec{\mathbf{F}}_{\text{ext}}(t')\, e^{-t'/\tau}}{m} dt' = \int_{t_0}^{t} \left(\dot{\vec{\mathbf{v}}} - \tau\ddot{\vec{\mathbf{v}}}\right) e^{-t'/\tau} dt' = \int_{t_0}^{t} \frac{d}{dt'}\left(-\tau\dot{\vec{\mathbf{v}}}\, e^{-t'/\tau}\right) dt'$$
$$= -\tau\dot{\vec{\mathbf{v}}}(t)\, e^{-t/\tau} + \tau\dot{\vec{\mathbf{v}}}(t_0)\, e^{-t_0/\tau} \qquad [7.220]$$

which also gives:

$$\dot{\vec{\mathbf{v}}}(t) = \dot{\vec{\mathbf{v}}}(t_0)\, e^{(t-t_0)/\tau} - \frac{1}{\tau}\int_{t_0}^{t} \frac{\vec{\mathbf{F}}_{\text{ext}}(t')\, e^{-(t'-t)/\tau}}{m} dt' \qquad [7.221]$$

Choosing $t_0 = +\infty$ prevents the runaway solution and reduces the time-averaged equation of motion to:

$$\dot{\vec{\mathbf{v}}}(t) = \frac{1}{\tau}\int_{t}^{+\infty} \frac{\vec{\mathbf{F}}_{\text{ext}}(t')\, e^{-(t'-t)/\tau}}{m} dt' \qquad [7.222]$$

This equation states that the acceleration of a particle in time t is determined by the time average of the external force $\vec{\mathbf{F}}_{\text{ext}}$ over a time interval of order τ in the future, which violates causality. Causes precede effects, not the other way around.

7.9. Dirac radiation equation

We now wish to present the covariant version of equation [7.219]. A natural generalization of this expression would take the form:

$$m\frac{dV^\mu}{d\tau} - F_R^\mu = F_{\text{ext}}^\mu \qquad [7.223]$$

where F_R^μ is the covariant generalization of $\overrightarrow{\mathbf{F}}_R = \frac{q^2}{6\pi\varepsilon_0 c^3}\ddot{\overrightarrow{\mathbf{v}}}$. Here, τ denotes the proper time of the particle. Let us generalize F_R^μ as follows:

$$F_R^\mu = \frac{q^2}{6\pi\varepsilon_0 c^3}\left(\frac{d^2U^\mu}{d\tau^2} + \mathsf{S}U^\mu\right) \qquad [7.224]$$

where S is a scalar. The introduction of S is necessary, as we shall see. The scalar S is determined by considering the invariant $F_R^\mu V_\mu$:

– In the rest frame of the particle, we have $F_R^\mu = \left(0, \overrightarrow{\mathbf{F}}_R\right)$ and $U_\mu = (c, 0)$, which gives $F_R^\mu U_\mu = 0$ (in any other frame).

– We evaluate S by writing that $F_R^\mu U_\mu = 0 = \frac{q^2}{6\pi\varepsilon_0 c^3}\left(\frac{d^2U^\mu}{d\tau^2}U_\mu + \mathsf{S}V^\mu U_\mu\right)$, which implies that:

$$\mathsf{S}U^\mu U_\mu = -\frac{d^2U^\mu}{d\tau^2}U_\mu = -\frac{d}{d\tau}\left(\frac{dU^\mu}{d\tau}U_\mu\right) + \frac{dU^\mu}{d\tau}\frac{dU_\mu}{d\tau} \qquad [7.225]$$

Since $U^\mu U_\mu = c^2$ and $U^\mu\frac{dU_\mu}{d\tau} = U_\mu\frac{dU^\mu}{d\tau} = 0$, we deduce that $\mathsf{S} = \frac{1}{c^2}\frac{dU^\mu}{d\tau}\frac{dU_\mu}{d\tau}$.

Thus, the equation of motion [7.223] becomes:

$$m\frac{dU^\mu}{d\tau} - \frac{q^2}{6\pi\varepsilon_0 c^3}\left(\frac{d^2U^\mu}{d\tau^2} + \frac{U^\mu}{c^2}\frac{dU^\nu}{d\tau}\frac{dU_\nu}{d\tau}\right) = F_{\text{ext}}^\mu \qquad [7.226]$$

This is the *Dirac radiation equation*. This equation can also be rewritten in a more symmetric form using the identity $0 = \frac{d^2}{d\tau^2}\left(U^\mu U_\mu\right) = 2U^\mu\frac{d^2U_\mu}{d\tau^2} + 2\frac{dU_\mu}{d\tau}\frac{dU^\mu}{d\tau}$, that is, $\frac{dU_\mu}{d\tau}\frac{dU^\mu}{d\tau} = -U^\mu\frac{d^2U_\mu}{d\tau^2}$. Substituting this result into [7.226], we obtain:

$$m\frac{dU^\mu}{d\tau} - \frac{q^2}{6\pi\varepsilon_0 c^5}U^\nu\left(U_\nu\frac{d^2U^\mu}{d\tau^2} - U^\mu\frac{d^2U_\nu}{d\tau^2}\right) = F_{\text{ext}}^\mu \qquad [7.227]$$

This equation also has the runaway solutions described previously.

7.10. And the story continues...

1) At the beginning of the last century, many researchers, noting the similarity between gravity and electromagnetism, attempted to unify these forces in the same way that electricity and magnetism are unified in Maxwell's theory.

2) Before the formulation of general relativity by Einstein, Nordström proposed a theory by adding a fifth dimension to the flat Minkowski space-time. In this space, he introduced a field with five components that satisfies the Maxwell equations (with a conserved 5D current). The first four components represent the usual electromagnetic four-potential, while the fifth component could be identified with the gravitational potential.

3) Within the framework of general relativity, Kaluza extended Einstein's theory to a 5D space, with the fifth dimension being cyclic. In this theory, the electromagnetic potentials are related to components of the 5D metric tensor. Klein connected the periodicity of the fifth dimension to the Planck constant in order to obtain a primitive quantum theory of 5D relativity. The theory and its extensions to higher dimensions are known as *Kaluza–Klein theories*. *String theories*, which aim to unify all forces, are related to these theories.

4) Quantum mechanics made it possible to develop a quantum theory of the electromagnetic field. The electromagnetic field could therefore be quantized. The equivalence between the field Hamiltonian and that of the harmonic oscillator allowed the introduction of photon creation and annihilation operators.

5) The electrodynamics then made it possible to quantify the entire field. The interaction between particles is described as an exchange of photons between charges – the photons become carriers of the field.

6) Quantum electrodynamics was extended by Weinberg, Salam and Glashow by unifying the weak interactions (standard model).

7) At present, gravity (the weakest of all forces) has not yet been successfully included in the grand unified theory (GUT).

And the story goes on... .

References

Abbel, M.L. and Braselton, J.P. (2005). *Maple by Example*. Elsevier Academic Press, Cambridge.

Abdesselam, B. (2007). *Introduction à la mécanique analytique*. Ellipses, Paris.

Ait-Ameur, K. (2016). *Physique, les rayonnements electromagnétiques : résumé de cours et problèmes corrigés*. Ellipses, Paris.

Alastuey, A., Margo, M., Pujol, P. (2008). *Physique et outils mathématiques : méthodes et exemples, savoirs actuels*. EDP Sciences/CNRS Editions, Paris.

Alfvèn, H. (1950). *Cosmical Electrodynamics*. Oxford University Press, Oxford.

Arfken, G. (1985). *Mathematical Methods for Physicists*, 3rd edition. Elsevier Academic Press, Cambridge.

Arnold, V.I. (1980). *Ordinary Differential Equations*. MIT Press, Cambridge.

Baker, B.B. and Copson, E.E. (2003). *Mathematical Theory of Huygens's Principle*. American Mathematical Society, Providence.

Bérest, P. (1997). *Calcul des variations : application à la mécanique et à la physique*. Ellipses, Paris.

Berthon, A. (2011). *Électromagnétisme-équations intégrales de l'électromagnétisme, applications aux problèmes axisymétriques, antennes et diffusion des ondes*. Ellipses, Paris.

Bitsadze, A.V. (1980). *Equations of Mathematical Physics*. Mir Publications, Moscow.

Bitsadze, A.V. and Kalinichenko D.F. (1980). *A Collection of Problems on the Equations of Mathematical Physics*. Mir Publications, Moscow.

Bok, J. and Hulin-Jung, N. (1975). *Ondes electromagnétiques relativité*. Les Éditions Hermann, Paris.

Boyce, W.E. and DiPrima, R.C. (2009). *Elementary Equations and Boundary Value Problems*. Wiley, New York.

Brewster, H.D. (2010). *Electromagnetism*. Oxford Book Company, Oxford.

Brillouin, L. (1960). *Wave Propagation and Group Velocity*. Elsevier Academic Press, Cambridge.

Butkov, E. (1968). *Mathematical Physics*. Addison-Wesley, Boston.

Cahn, S.B. and Nadgorny, B.E. (1994). *A Guide to Physics Problems: Part 1: Mechanic, Relativity and Electrodynamics*. Wolters Kluwer, Alphen aan den Rijn.

Cardone, F. and Mignani, R. (2004). *Energy and Geometry: An Introduction to Deformed Special Relativity*. World Scientific, Singapore.

Cohen-tannoudji, C. (2004). *Atoms in Electromagnetic Fields*, 2nd edition. World Scientific, Singapore.

Cordier, D. (2007). *Cours de physique – électromagnétisme – volume 2 : phénomènes d'induction et ondes électromagnétiques – cours et exercices corrigés – licence*. Dunod, Paris.

Cornille, P. (2003). *Advanced Electromagnetism and Vacuum Physics*. World Scientific, Singapore.

Coulomb, C.A. (1884). *Collection de mémoires relatifs à la physique*, volume 3. Gauthier-Villars, Paris.

Courant, R. and Hilbert, D. (1989). *Methods of Mathematical Physics*. Wiley, New York.

D'Auria, R. and Trigiante, M. (2016). *From Special Relativity to Feynman Diagrams, a Course of Theoretical Particle Physics for Beginners*. Springer, New York.

Debye, P. (1945). *Polar Molecules*. Dover Publications, New York.

Dirac, P.A.M. (1958). *The Principles of Quantum Mechanics*, 4th edition. Oxford University Press, Oxford.

Durand, E. (1953). *Electrostatique et magnétostatique*. Masson et Cie, Paris.

Edminister, J.A. (1979). *Schaum's Outline of Theory and Problems of Electromagnetics*. McGraw-Hill, New York.

Einstein, A., Lorentz, H.A., Minkowski, H., Weyl, H. (1952). *The Principle of Relativity*. Dover Publications, New York.

Eom, H.J. (2004). *Electromagnetic Wave Theory of Boundary-Value Problems, an Advanced Course on Analytical Methods*. Springer, New York.

Faraday, M. (1965). *Experimental Researches in Electricity*. Dover Publications, New York.

Feynman, R.P., Leighton, R.B., Sands, M. (1979). *Le cours de physique de Feynman – Électromagnétisme 1 et 2*. Dunod, Paris.

Fitzpatrick, R. (2008). *Maxwell's Equations and the Principles of Electromagnetism*. Infinity Science Press LLC, Hingham.

Foot, C.J. (2005). *Atomic Physics*. Oxford University Press, Oxford.

Friedman, B. (2011). *Principles and Techniques of Applied Mathematics*. Dover Publications, New York.

Friedrich, H. (2004). *Theoretical Atomic Physics*. Springer, New York.

Fröhlich, H. (1949). *Theory of Dielectrics*. Oxford University Press, Oxford.

Fujimoto, M. (2007). *Physics of Classical Electromagnetism*. Springer, New York.

Funaro, D. (2008). *Electromagnetism and the Structure of Matter*. World Scientific, Singapore.

Goldstein, H. (1950). *Classical Mechanics*, 2nd edition. Addison-Wesley, Boston.

Grimes, D.M. and Grimes, G.A. (2004). *The Electromagnetic Origin of Quantum Theory of Light*, 2nd edition. World Scientific, Singapore.

Guru, B. and Hiziroglu, H.R. (2009). *Electromagnetic Fields Theory Fundamentals*, 2nd edition. Cambridge University Press, Cambridge.

Hairer, E. and Wanner, G. (2006). *Analyse complexe et séries de fourier*. Geneva University, Geneva.

Halpern, A. and Erlbach, E. (1998). *Theory and Problems of Beginning Physics II: Waves, Electromagnetism, Optics and Modern Physics*. McGraw-Hill, New York.

Heitler, W. (2010). *Quantum Theory of Radiation*, 3rd edition. Dover Publications, New York.

Henley, E.M. and Garcia, A. (2007). *Subatomic Physics*, 3rd edition. World Scientific, Singapore.

Hertz, H. (1962). *Electric Waves: Being Researches on the Propagation of Electric Action with Finite Velocity Through Space*. Dover Publications, New York.

Hoffmann, J.D. (2001). *Numerical Methods for Engineers and Scientists*, 2nd edition. Marcel Dekker, New York.

Iwanenko, D. and Sokolow, A. (1949). *Klassische Fieltheorie*. Akademic-Verlag, Berlin.

Jackson, J.D. (1962). *Classical Electrodynamics*. Wiley, New York.

Jeans, J.H. (2009). *Mathematical Theory of Electricity and Magnetism*, 5th edition. Cambridge University Press, Cambridge.

Jeffereys, H. and Jeffreys, B.S. (2000). *Methods of Mathematical Physics*, 3rd edition. Cambridge University Press, Cambridge.

Johnson, W.R. (2007). *Atomic Structure Theory: Lectures on Atomic Physics*. Springer, New York.

Jordan, E.C. and Balman, K.G. (1968). *Electromagnetic Waves and Radiating Systems*. Prentice Hall, Englewood Cliffs, New Jersey.

Kamal, A.A. (2010). *1000 Solved Problems in Modern Physics*. Springer, New York.

Kay, D.C. (2011). *Theory and Problems of Tensor Calculus*. McGraw-Hill, New York.

Kibble, T.W.B. and Berkshire, F.H. (2004). *Classical Mechanics*, 5th edition. Imperial College Press, London.

King, R.W.P. and Wu, T.T. (1959). *Scattering and Diffracton of Waves*. Harvard University Press, Cambridge.

Landau, L.D. and Lifchitz, E.M. (1962). *Classical Theory of Fields*. Addison-Wesley, Boston.

Landau, L.D., Lifshitz, E.M., Pitaevskii, L.P. (1984). *Electrodynamics of Continuous Media, Volume 8 of Theoretical Physics*. Addison-Wesley, Boston.

Le Bellac, M. (1998). *Des phénomènes critiques aux champs de jauge : une introduction aux méthodes et aux applications de la théorie quantique des champs*. EDP Sciences, Les Ulis.

Lichnerowicz, A. (1955). *Théories relativistes de la gravitation et de l'électromagnétisme, relativité générale et théories unitaires*. Masson et Cie, Paris.

Lorentz, H.A. (1952). *Theory of Electrons*, 2nd edition. Dover Publications, New York.

Mathey, P. (2017). *Electromagnétisme : ondes et propagation guidée, cours, 130 QCM et exercices corrigés, exemples d'applications*. Dunod, Paris.

Maxwell, J.C. (1954). *Treatise on Electricity and Magnetism, volume 2*. Dover Publications, New York.

Moliton, A. (2004). *Applications de l'électromagnétisme dans les milieux matériels*. Hermès-Lavoisier, Paris.

Morin, D. (2008). *Introduction to Classical Mechanics with Problems and Solutions*. Cambridge University Press, Cambridge.

Neff, H.P. (1991). *Introductory Electromagnetics*. Wiley, New York.

Nikiforov, A. and Ouvarov, V. (1983). *Fonctions spéciales de la physique mathématique*. Mir Publications, Moscow.

Nogarède, B. (2005). *Électrodynamique appliquée : bases et principes physiques de l'électrotechnique – cours et exercices corrigés licence 3ème année, master, ecoles*. Dunod, Paris.

Okun, L.B. (2008). *Energy and Mass in Relativity Theory, Illustrated Edition*. World Scientific, Singapore.

Panofsky, W. and Philips, M. (1955). *Classical Electricity and Magnetism*, 2nd edition. Addison-Wesley, Boston.

Pauli, W. (1981). *Theory of Relativity*, New Edition. Dover Publications, New York.

Pérez, J.P., Carles, R, Fleckinger, R. (2020). *Electromagnétisme : fondements et applications, avec 300 exercices et problèmes résolus*, 4th edition. Dunod, Paris.

Provost, J.P. and Vallée, J.P. (2011). *Les maths en physique, la physique à travers le filtre des mathématiques avec éléments d'analyse numérique, cours et applications*, 3rd edition. Dunod, Paris.

Raab, R.E. and De Lange, O.L. (2004). *Multipole Theory in Electromagnetism: Classical, Quantum, and Symmetry Aspects, with Applications*. Oxford Science Publications, Oxford.

Rax, J.M. (2017). *Electromagnétisme : milieux, structures et energie*. De Boeck, Brussels.

Reed, M. and Barry, S. (1980). *Methods of Modern Mathematical Physics*. Elsevier Academic Press, Cambridge.

Reitz, J.R., Milford, F.J., Christy, R.W. (2008). *Foundations of Electromagnetic Theory*, 4th edition. Addison-Wesley, Boston.

Reuse, F.A. (2012). *Electrodynamique : 125 exercices avec solutions – bachelor, licence, classes prépa*. Presses Polytechniques et Universitaires Romandes, Lausanne.

Rudnicki, K. (ed.) (2000). *Gravitation, Electromagnetism and Cosmology: Toward a New Synthesis*. Apeiron, Montreal.

Samarskii, A.A. and Vabishchevich, P.N. (2007). *Numerical Methods for Solving Inverse Problems of Mathematical Physics*. Walter de Gruyter, Berlin.

Schultz, B. (1980). *Geometrical Methods of Mathematical Physics*. Cambridge University Press, Cambridge.

Sénéchal, D. (2007). *Electromagnétisme avancé*. Université de Sherbrooke, Sherbrooke.

Sommerfeld, A. (1952). *Electrodynamics*. Elsevier Academic Press, New York.

Spiegel, M.R. (1968). *Mathematical Handbook of Formulas and Tables*. McGraw-Hill, New York.

Spiegel, M.R. (1974). *Fourier Analysis with Applications to Boundary Value Problems*. McGraw-Hill, New York.

Thirring, W. (1997). *Classical Mathematical Physics, Dynamical Systems and Field Theories*, 3rd edition. Springer, New York.

Van Bladel, J. (2007). *Electromagnetic Fields*, 2nd edition. Wiley, New York.

Vanderlinde, J. (2005). *Classical Electromagnetic Theory*, 2nd edition. Wolters Kluwer, Alphen aan den Rijn.

Verweij, M.D., Van den Berg, P.M., Blok, H. (2006). *Electromagnetic Waves: An Introductory Course*, 2nd edition. VSSD, Delft.

Wrede, R.C. and Spiegel, M.R. (2010). *Advanced Calculus*. McGraw-Hill, New York.

Zhang, M. (1979). *Electromagnetic Field Theory: A Problem Solving Approach*. Wiley, New York.

Index

A

angle
  Cherenkov, 198
  critical, 181–183
  incidence, 14, 31
  solid, 170, 171
antenna, 135, 137, 139
aperture
  circular, 208
  rectangular, 208, 228
approximation
  Born, 194
  electric dipole, 113
  Kirchhoff, 211, 217, 223, 226, 227, 230
  weak-guidance, 54, 55

B, C

Bohr radius, 325
Brehme-Lorentz diagram, 255–259
capacitance, 72, 74, 78
charge
  Noether, 292
  surface, 14, 15, 20, 62, 66
  total, 15, 22, 62, 67
coaxial cable, 1, 76
  core, 77
  impedance, 78
  shield, 77
coefficient
  attenuation, 25, 28, 31, 32, 37, 88
  overvoltage, 95
  reflection, 14, 78, 88, 91
  standing wave ratio (SWR), 92, 93
  transmission, 224–226
complex permittivity, 2
compton wavelength, 279
condition
  Cauchy, 204
  radiation, 210
conductance, 80
conductivity, 24, 32, 64
current
  Noether, 292
  surface, 14, 20, 67
  total, 15, 22, 72

D

density
  electromagnetic energy, 314
  Hamiltonian, 289
  Lagrangian, 286, 287

diffraction
  Fraunhofer, 220
  Fresnel, 220
  pattern, 208, 217
  screen, 208–211, 216–220, 222, 225, 228
dispersion relation, 6, 12, 19, 34, 50, 54, 55, 59

E

effect
  Compton, 278
  Doppler, 253, 255, 269
  edge, 71
  headlight, 165
  Joule, 63, 67, 79, 81
  resonance, 94
  retardation, 101
  skin, 63
Einstein convention, 263
energy radiated, 132, 171, 172, 175, 180
equation
  Abraham–Lorentz (motion), 323, 333
  characteristic, 50, 72
  charge conservation, 328
    covariant form, 295
  conservation
    electromagnetic energy, 315
    tensor, 315
  continuity, 46, 110, 218
  Dirac (radiation), 334
  Euler-Lagrange, 288, 294, 311
  Hamilton, 289
  Hamilton-Jacobi, 276
  Helmholtz, 5, 9, 11, 18, 36, 103, 106, 194, 211
  Klein-Gordon, 290, 293
  Laplace, 7, 73, 211
  Maxwell
    first group, 307
    second group, 311
  Maxwell-
    Ampère, 4
    Faraday, 4
    Gauss, 100, 310
  Thomson, 191, 192
  transmission, 72, 77, 79, 80
  wave, 82, 204
    inhomogeneous, 100–102, 105, 128, 129
experiment
  Bertozzi, 278
  Fizeau, 253

F

factor
  attenuation, 16
  form, 196
  Lorentz, 308
  quality, 64, 65, 68, 97
field
  displacement, 2
  electric, 2, 65
  incident, 195, 218
  line, 7, 15, 20
  magnetic, 2, 65
    excitation, 2, 7
  proper, 319
  radiated, 153
  reflected, 218
  scattered, 195, 236
  transmitted, 218, 219
FitzGerald-Lorentz contraction, 249
force
  Lorentz, 294, 301, 315, 326, 327, 329
    relativistic, 300
  Newtonian, 241
  reaction, 319, 322–325, 330

formula
  Huygens, 210
  Kirchhoff, 210
  Larmor, 154, 167, 189, 321, 323, 324
    relativistic, 167
four-
  acceleration, 267
  current, 268, 295
  force, 281, 298
  gradient, 299
  momentum, 267, 275
  position, 263, 266
  potential, 295, 299, 310
  velocity, 267, 295, 297
four-vector
  component
    contravariant 263, 266
    covariant 263, 266
    wave, 268
Fourier
  series, 175
  transform, 102, 105, 171, 172, 175, 195, 201, 202
    inverse, 105
frequency
  critical, 19, 181
  cutoff, 11, 16, 19, 41, 48, 51
  cyclotron, 301
function
  aperture, 210
  Bessel, 223
    first kind, 33
    second kind, 33, 180, 181
  Green, 103, 145, 201, 204, 219
  Hankel spherical (first kind), 113
  spherical harmonic, 107

G

gauge
  Coulomb, 300
  invariance, 18, 290, 299
  Lorentz, 18, 100, 295, 300, 311
group
  Carroll, 261
  Galilean, 261
  Lorentz, 261
    orthochronous, 262
    proper, 262
  rotation, 261

I

inductance, 72, 74, 78, 80
integral
  action, 274
  Airy, 180
  Dirac delta (representation) function, 204
  Gaussian, 183
  Kirchhoff, 207, 210, 212
  Kirchhoff–Huygens, 214
interval, 245, 246
  space, 246
  time, 246

L

law
  addition of velocities, 250, 251, 262
  Kirchhoff
    first, 79
    second, 79
  Newton, 167, 241, 298, 322
  Ohm, 139
line
  admittance, 82
  current, 71, 74, 77–79, 82, 89
  Heaviside condition, 88
  impedance, 82–84, 88–90
  open-circuited, 84, 92

short-circuited, 83, 94
transmission, 71, 72, 74–76, 82
distortionless, 88
voltage, 71, 74, 77–79, 82, 89

M

mass
apparent, 277
bare, 321, 333
electromagnetic, 320, 331
mechanical, 326
rest, 277
medium
dielectric, 88
HLI, 29
LIH, 2, 37, 71, 77, 100
plasma, 12
method
conformal transformation, 73
images, 73
Kirchhoff, 207
separation of variables, 10, 19, 23, 33, 77
Minkowski space/spacetime, 245, 263, 272, 309
mode
attenuation, 23
azimuthal, 45, 48
degeneracy, 55
dominant, 19, 48
EH, 45, 48, 51
EH_{mn}, 52, 55
HE, 45, 48, 51
HE_{mn}, 53
LP_{mn}, 57, 59
TE_{mn}, 19
TE_{mnp}, 61
TM_{mn}, 11, 15
TM_{mnp}, 62
transverse electric (TE), 6, 9, 44, 46, 51, 60, 127
transverse electric and magnetic (TEM), 6–8
transverse magnetic (TM), 6, 9, 18, 44, 47, 51, 127
model
Abraham-Lorentz, 326, 331
Rutherford, 324
moment
electric
dipole, 113, 114, 118, 154
quadrupole, 120, 122, 125
magnetic dipole, 120, 121, 175
momentum
angular, 132, 134, 135
tensor, 293
electromagnetic, 317
non-relativistic charge, 317

O, P

optical fiber, 1
cladding, 44, 48, 53, 54
core, 44, 53
step index, 44, 50
weakly guiding, 54
potential
Debye, 127, 131
electric, 7, 17, 100, 145
Liénard–Wiechert, 147
retarded, 99, 105, 326
vector, 17, 99, 146
power
diffracted, 224, 231
dissipated, 25, 26, 63, 64, 68, 96
electromagnetic, 25, 81
radiated, 111, 112, 117, 119, 124, 125, 135, 137, 139, 140, 143, 154, 155, 158, 162, 163, 192
scattered, 238, 239
principle
addition of velocities, 243
Babinet, 217, 218, 220, 231

Galileo's of relativity, 242
Hamilton, 287
Huygens, 207, 208
Lavoisier, 243
mass–energy equivalence, 277, 331
relativity, 243
problem
Cauchy, 204, 206
Dirichlet, 211
Neumann, 211

R

radiation
Cherenkov, 197
intensity, 174, 188
resistance, 139
reaction threshold, 278, 279
refractive index, 1
region
illuminated, 232–237, 239
shadowed, 232–234, 238, 239
relativistic
Lagrangian of a free particle, 274
relativity postulate
first, 243
second, 243
third, 246
resonant cavity, 60, 63, 64, 67, 68

S

scattering
Compton, 189, 193
cross-section
differential, 191, 195
Thomson, 191
Rayleigh, 192
Thomson, 189, 193
second Green identity, 205
skin depth, 24
solution
Poisson, 206
runaway, 324, 333, 334
stellar aberration, 252
surface resistance, 27

T

tensor
contravariant, 269
covariant, 270
energy-momentum, 293, 312, 331
Faraday, 298
Levi-Civita, 271, 272
magnetization, 312
Maxwell (electromagnetic), 314
metric, 263, 265, 270
mixed, 270
polarization, 312
theorem
Ampère, 305
Cauchy, 202
divergence, 8, 25, 81, 273
Gauss, 304, 305
Helmholtz–Hodge, 100
Poynting, 80
complex, 25
virial, 325
time
absolute, 243
characteristic, 321
dilation, 249, 255
proper, 250, 296
transformation
Galilean, 241, 242, 244
Lorentz, 245, 247, 248, 256, 259, 262, 264–266, 296, 303, 305

V, W

vector
Poynting, 16, 22, 153, 154, 314, 316
complex, 31, 32, 117
radiation, 109

wave
  amplitude, 3
  transverse electric (TE), 22, 29
  transverse magnetic (TM), 17, 29
  velocity
    group, 12
    phase, 12
  wavefront, 244
  wavelength (cutoff), 35, 37
  wavenumber, 3, 10
waveguide, 1, 3, 8
  cylindrical, 3, 33, 65
  dielectric, 1
  metallic, 1
  rectangular, 1, 3, 9, 23, 60
worldline, 245, 275, 296

Z

zone
  Fraunhofer, 221
  Fresnel, 221
  intermediate, 112
  near, 221
  radiation, 108, 135, 137, 209, 221, 231
  static, 107, 114

Other titles from

iSTE

in

Waves

2025

ABDESSELAM Boucif
Introduction to Classical Electrodynamics 1
Introduction to Classical Electrodynamics 2

2024

BERNARD Jean-Michel L.
Advanced Studies in the Mathematical Theory of Scattering
(Waves and Scattering Set – Volume 3)

JUVÉ Daniel, GALLAND Marie-Annick, CLAIR Vincent
Acoustics of Fluid Media 1: Principles and Applications

RÉVEILLAC Jean-Michel
Synthesizers and Subtractive Synthesis 1: Theory and Overview
Synthesizers and Subtractive Synthesis 2: Application and Practice

2022

GONTRAND Christian
Electromagnetism: Links to Special Relativity

LACAZE Pierre Camille, LACROIX Jean-Christophe
Nanotechnology and Nanomaterials for Energy

RÉVEILLAC Jean-Michel
Recording and Voice Processing 1: History and Generalities
Recording and Voice Processing 2: Working in the Studio

ROYER Daniel, VALIER-BRASIER Tony
Elastic Waves in Solids 1: Propagation

SAKHO Ibrahima
Nuclear Physics 1: Nuclear Deexcitations, Spontaneous Nuclear Reactions

VALIER-BRASIER Tony, ROYER Daniel
Elastic Waves in Solids 2: Radiation, Scattering, Generation

2021

DAHOO Pierre-Richard, LAKHLIFI Azzedine
Infrared Spectroscopy of Symmetric and Spherical Top Molecules for Space Observation 1
(Infrared Spectroscopy Set – Volume 3)
Infrared Spectroscopy of Symmetric and Spherical Top Molecules for Space Observation 2
(Infrared Spectroscopy Set – Volume 4)

2020

DANIELE Vito G., LOMBARDI Guido
Scattering and Diffraction by Wedges 1: The Wiener-Hopf Solution - Advanced Applications
(Waves and Scattering Set – Volume 1)
Scattering and Diffraction by Wedges 2: The Wiener-Hopf Solution - Advanced Applications
(Waves and Scattering Set – Volume 2)

SAKHO Ibrahima
Introduction to Quantum Mechanics 2: Wave-Corpuscle, Quantization & Schrödinger's Equation

2019

BERTRAND Pierre, DEL SARTO Daniele, GHIZZO Alain
The Vlasov Equation 1: History and General Properties

DAHOO Pierre-Richard, LAKHLIFI Azzedine
Infrared Spectroscopy of Triatomics for Space Observation
(Infrared Spectroscopy Set – Volume 2)

RÉVEILLAC Jean-Michel
Electronic Music Machines: The New Musical Instruments

ROMERO-GARCÍA Vicente, HLADKY-HENNION Anne-Christine
Fundamentals and Applications of Acoustic Metamaterials: From Seismic to Radio Frequency
(Metamaterials Applied to Waves Set – Volume 1)

SAKHO Ibrahima
Introduction to Quantum Mechanics 1: Thermal Radiation and Experimental Facts Regarding the Quantization of Matter

2018

SAKHO Ibrahima
Screening Constant by Unit Nuclear Charge Method: Description and Application to the Photoionization of Atomic Systems

2017

DAHOO Pierre-Richard, LAKHLIFI Azzedine
Infrared Spectroscopy of Diatomics for Space Observation
(Infrared Spectroscopy Set – Volume 1)

PARET Dominique, HUON Jean-Paul
Secure Connected Objects

PARET Dominque, SIBONY Serge
Musical Techniques: Frequencies and Harmony

RÉVEILLAC Jean-Michel
Musical Sound Effects: Analog and Digital Sound Processing

STAEBLER Patrick
Human Exposure to Electromagnetic Fields

2016

ANSELMET Fabien, MATTEI Pierre-Olivier
Acoustics, Aeroacoustics and Vibrations

BAUDRAND Henri, TITAOUINE Mohammed, RAVEU Nathalie
The Wave Concept in Electromagnetism and Circuits: Theory and Applications

PARET Dominique
Antennas Designs for NFC Devices

PARET Dominique
Design Constraints for NFC Devices

WIART Joe
Radio-Frequency Human Exposure Assessment

2015

PICART Pascal
New Techniques in Digital Holography

2014

APPRIOU Alain
Uncertainty Theories and Multisensor Data Fusion

JARRY Pierre, BENEAT Jacques N.
RF and Microwave Electromagnetism

LAHEURTE Jean-Marc
UHF RFID Technologies for Identification and Traceability

REMY Jean-Gabriel, LETAMENDIA Charlotte
LTE Standards

REMY Jean-Gabriel, LETAMENDIA Charlotte
LTE Services

SAVAUX Vincent, LOUËT Yves
MMSE-based Algorithm for Joint Signal Detection, Channel and Noise Variance Estimation for OFDM Systems

THOMAS Jean-Hugh, YAAKOUBI Nourdin
New Sensors and Processing Chain

TING Michael
Molecular Imaging in Nano MRI

VALIÈRE Jean-Christophe
Acoustic Particle Velocity Measurements using Laser: Principles, Signal Processing and Applications

VANBÉSIEN Olivier, CENTENO Emmanuel
Dispersion Engineering for Integrated Nanophotonics

2013

BENMAMMAR Badr, AMRAOUI Asma
Radio Resource Allocation and Dynamic Spectrum Access

BOURLIER Christophe, PINEL Nicolas, KUBICKÉ Gildas
Method of Moments for 2D Scattering Problems: Basic Concepts and Applications

GOURE Jean-Pierre
Optics in Instruments: Applications in Biology and Medicine

LAZAROV Andon Dimitrov, KOSTADINOV Todor Pavlov
Bistatic SAR/GISAR/FISAR Geometry, Signal Models and Imaging Algorithms

LHEURETTE Eric
Metamaterials and Wave Control

PINEL Nicolas, BOURLIER Christophe
Electromagnetic Wave Scattering from Random Rough Surfaces: Asymptotic Models

SHINOHARA Naoki
Wireless Power Transfer via Radiowaves

TERRE Michel, PISCHELLA Mylène, VIVIER Emmanuelle
Wireless Telecommunication Systems

2012

BÉCHERRAWY Tamer
Electromagnetism: Maxwell Equations Wave Propagation and Emission

BOUCHET Olivier
Wireless Optical Communications

BOUTIER Alain
Laser Metrology in Fluid Mechanics: Granulometry, Temperature and Concentration Measurements
Laser Velocimetry in Fluid Mechanics

LALAUZE René
Chemical Sensors and Biosensors

LE MENN Marc
Instrumentation and Metrology in Oceanography

PICART Pascal, LI Jun-chang
Digital Holography

SAGUET Pierre
Numerical Analysis in Electromagnetics: The TLM Method

2011

BECHERRAWY Tamer
Mechanical and Electromagnetic Vibrations and Waves

BESNIER Philippe, DÉMOULIN Bernard
Electromagnetic Reverberation Chambers

GOURE Jean-Pierre
Optics in Instruments

GROUS Ammar
Applied Metrology for Manufacturing Engineering

LE CHEVALIER François, LESSELIER Dominique, STARAJ Robert
Non-standard Antennas

2010

BEGAUD Xavier
Ultra Wide Band Antennas

GUNIOT Vincent
Wave Propagation in Fluids: Second Edition – Models and Numerical Techniques

MARAGE Jean-Paul, MORI Yvon
Sonar and Underwater Acoustics

2009

BOUDRIOUA Azzedine
Photonic Waveguides

BRUNEAU Michel, POTEL Catherine
Materials and Acoustics Handbook

DE FORNEL Frédérique, FAVENNEC Pierre-Noël
Measurements using Optic and RF Waves

FRENCH COLLEGE OF METROLOGY
Transverse Disciplines in Metrology

2008

CHARRUAU Stéphane
Electromagnetism and Interconnections: Advanced Mathematical Tools for Computer-aided Simulation

FILIPPI Paul J.T.
Vibrations and Acoustic Radiation of Thin Structures

LALAUZE René
Physical Chemistry of Solid-Gas Interfaces

2007

KUNDU Tribikram
Advanced Ultrasonic Methods for Material and Structure Inspection

PLACKO Dominique
Fundamentals of Instrumentation and Measurement

RIPKA Pavel, TIPEK Alois
Modern Sensors Handbook

2006

BALAGEAS Daniel, FRITZEN Claus-Peter, GÜEMES Alfredo
Structural Health Monitoring

BOUCHET Olivier, SIZUN Hervé, BOISROBERT Christian, DE FORNEL Frédérique, FAVENNEC Pierre-Noël
Free-Space Optics

BRUNEAU Michel, SCELO Thomas
Fundamentals of Acoustics

FRENCH COLLEGE OF METROLOGY
Metrology in Industry

GUILLAUME Philippe
Music and Acoustics

GUYADER Jean-Louis
Vibration in Continuous Media

RIGNEAULT Hervé, LOURTIOZ Jean-Michel, DELALANDE Claude, LEVENSON Juan Ariel
Nanophotonics

Printed and bound by CPI Group (UK) Ltd, Croydon, CR0 4YY

13/09/2026

14961011-0005